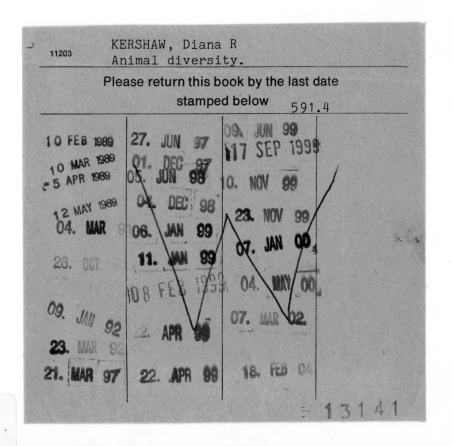

Animal Diversity

Diana R. Kershaw

Lecturer in Anatomy, St. Mary's Hospital Medical School; formerly Lecturer in Zoology, Queen Mary College (University of London)

**With illustrations by
Brian Price Thomas**

University Tutorial Press

019582

Published by University Tutorial Press Limited
842 Yeovil Road, Slough, SL1 4JQ

ISBN 0 7231 0847 1

Published 1983

13141

Photoset, printed and bound in Great Britain by
REDWOOD BURN LIMITED
Trowbridge, Wiltshire

To my parents

Preface

This book has been written with two main purposes in mind, the first being to give a general review of the entire animal kingdom, and the second to give more detailed functional accounts of the anatomy of a representative of each major animal group. It is intended to be used by those who are interested in animals and does not start with the assumption of any great zoological knowledge. It is hoped that it will prove particularly helpful to those studying biology or zoology at 'A' level, or in the early stages of a university course.

Most modern zoological anatomical and physiological research concentrates on 'bits' of animals. This text is designed to help the reader to see the animal as a complete living being within its environment. For this reason both detailed description and illustrations are restricted to a few representative animals and where possible the structure of an animal is illustrated by a series of drawings on a single page. At the same time animals show immense variation and none is truly typical. Some idea of the immense variety of animals is given in the diversity sections, with a synopsis of the classification of each major phylum.

Zoology has a language of its own, which appears highly complicated but in most cases can, in fact, be derived simply from either Latin or Greek. Translations and derivations have been given of a selection of zoological terms; these should be regarded as examples. The interested zoologist may find the use of a Greek and Latin dictionary rewarding.

Finally, zoology is a subject with a long history stretching back to the ancient world. In many modern texts the important contribution of earlier workers is forgotten in favour of concentration on more recent achievement. It is hoped that sufficient historical background has been given to remind the reader of the great value of earlier work.

Diana R. Kershaw
St Mary's Hospital Medical School

The use of this book

Animal Diversity has been planned to be read in three ways, apart from as a unit, with the intention of providing three text books in one.

Firstly, each major phylum includes a detailed description of one representative. These have been chosen with regard to their availability and use in school and university syllabi and can be used together or separately to give a full account of a limited number of animals. Secondly, a review is given of the variation within each phylum. These sections are intended to be used together to give a more overall look at the animal kingdom. Thirdly, major functions and organ systems are discussed within each section under the same headings and in the same order. Used together, these provide a comparative account of the functional anatomy of the animal kingdom.

Acknowledgements

In the preparation of this book I have benefited from suggestions and ideas from many fellow zoologists, colleagues, and students. Among my colleagues special mention must be made of Drs A. D. Hoyes and R. A. Travers, also Professor A. d'A Bellairs. My students at Queen Mary College (University of London) helped me to create the undergraduate courses on which the book was based. Some read parts of the book and convinced me it was worth continuing to write it; in this context I am particularly grateful to N. M. A. Horn, with J. R. Clague and R. A. Matthews.

My husband, Dr P. J. Edwards, read the entire book in manuscript and ensured that the English was comprehensible, also adding some ecological information. Mrs L. A. Wheatley turned my scrawled handwriting into a tidy typescript. Brian Price-Thomas drew the illustrations and went to great pains to find adequate source material to this end, and Mrs A. L. Price Thomas and the Zoology Department of Westfield College provided innumerable specimens and references.

I should like to thank the staff of University Tutorial Press, particularly C. J. Baker, who encouraged me to write the book, and Anne Hollifield and David Blogg for endless help in its editorial and artwork aspects.

It is impossible in a book of this type to credit every piece of information to its original source; indeed in many cases the original source has been lost. I am deeply indebted to many zoologists, without whom the book could not have been written, and particularly to: A. J. Alexander, R. McN Alexander, D. Alkins, R. D. Allen, E. P. Allis, D. T. Anderson, S. B. Barber, E. J. W. Barrington, R. D. Barnes, G. R. de Beer, A. d'A Bellairs, N. J. Berrill, Q. Berry, A. Bidder, W. Bloom, Q. Bone, B. Bracegirdle, A. Brodal, T. H. Bullock, R. M. Cable, A. C. Campbell, J. D. Carthy, G. Causey, G. Chapman, M. Chinery, A. M. Clark, R. B. Clarke, J. A. Clegg, E. H. Colbert, J. L. Corliss, R. P. Dales, C. Darwin, B. Dawes, R. Denison, E. J. Denton, T. Dobzhansky, F. H. Edgeworth, R. Fange, D. Fawcett, W. Fisher, E. A. Fraser, W. H. Freeman, V. Fretter, K. von Frisch, W. G. Fry, G. Fryer, C. Gans, W. Garstang, R. Gibson, T. Gilson, E. S. Goodrich, A. Graham, P. P. Grassé, H. Gray, J. Gray, J. Green, P. H. Greenwood, W. K. Gregory, A. J. Grove, J. Hadzi, H. J. Hansen, E. D. Hanson, R. H. Harrison, B. Hatschek, J. W. Hedgepeth, W. N. Hess, G. Huff, L. H. Hyman, A. D. Imms, H. Isseroff, A. V. Ivanov, J. B. Jennings, C. John, M. Jollie, R. P. S. Jefferies, D. Kennedy, G. A. Kerkut, M. S. Laverack, W. E. LeGros Clark, C. H. Lewis, J. A. Mcleod, K. F. Liem, H. W. Lissman, E. E. Lund, K. H. Mann, S. M. Manton, A. J. Marshall, N. B. Marshall, L. H. Matthews, E. Mayr, P. A. Meglitsch, E. Meyer, R. S. Miles, H. M. Miller, J. E. Morton, J. A. Moy-Thomas, O. Nelsen, T. C. Nelson, G. Newell, P. F. Newell, D. Nichols, G. K. Noble, R. T. Orr, R. Owen, T. S. Parsons, W. P. Pycraft, W. J. Rees, F. S. Russell, W. D. Russell-Hunter, O. W. Richards, A. S. Romer, K. Schmidt Nielsen, A. Sedgewick, G. C. Simpson, M. A. Sleigh, J. E. Smith, J. C. Smyth, R. E. Snodgrass, W. Stephenson, M. F. Sutton, D'A W. Thompson, C. A. Villee, W. F. Walker, R. Warwick, T. H. Waterman, D. M. S. Watson, J. E. Webb, P. S. Welch, T. S. Westoll, P. J. Whitehead, V. B. Wigglesworth, P. H. Williams, D. M. Wilson, H. V. Wilson, C. M. Yonge, J. Z. Young.

I should like to thank Faber for permission to quote from 'Archy and Mehitabel' by Don Marquis, Methuen for permission to quote from 'Now we are six' by A. A. Milne, and Unwin for permission to quote 'The Fly' and 'The Termite' by Ogden Nash.

Contents

Introduction to the subphylum Vertebrata 237

Vertebrate diversity 293

Introduction

The characteristics of living things

At a superficial level it is easy to separate the non-living from the living; no one is, for example, likely to confuse a horse and a rock! However, when all living things are considered from the smallest to the greatest, superficial distinctions are no longer sufficient and more precise definitions have to be made. Briefly, living things can be recognized by the following characteristics: to a greater or lesser extent all are capable of metabolism, growth, reproduction, response to stimuli and movement.

Metabolism is the general name given to a wide variety of chemical processes carried out by all living organisms, and involving transformation of energy. It is useful to distinguish between anabolic processes in which simple chemicals are combined to make more complex substances and energy is stored (for example protein formation) and catabolic processes in which complex substances are broken down with release of energy (for example cellular respiration). Both kinds of metabolism occur continuously in living things, and all other phenomena of life (growth, reproduction, irritability etc.) require an expenditure of energy by the cell.

If the energy entering a system is measured, and compared with that which leaves it, the measurements are found to be the same. Energy is neither created nor destroyed, but is transformed from one form to another. This idea is expressed in the First Law of Thermodynamics, sometimes called the Law of Conservation of Energy. In biological systems the primary energy source is the sun. Radiant energy from the sun is transformed by green plants into chemical energy within carbohydrate molecules, as the energy of the bonds which hold the constituent atoms together. This process is known as photosynthesis (see below). This chemical energy is stored in a biologically useful form in phosphate bonds by cellular respiration and is then used by the cell to do various forms of work. The energy released in this process flows back into the environment as heat.

Growth involves increase in living material, apparent either in the size of an individual, or by the production of new individuals. The ability of an organism to reproduce itself may be regarded as the definitive characteristic of life. A consequence of reproduction is that living things are recognizable by a characteristic form and appearance. A horse is basically similar to all other horses and a pine tree to all other pine trees.

All living things, plants as well as animals, show response to changes in the physical or chemical nature of their surroundings (sometimes termed irritability). One of the commonest responses is to move: immediately and obviously in the case of most animals, rather more slowly and less obviously in the case of most plants. In general the responses of living things are adaptive. They are able either to alter their immediate surroundings by moving from the unfavourable to the more favourable, or may modify themselves to make them better suited to their environment. Obviously this adjustability is not indefinite and no organism can adjust to all conditions of, for example, temperature. However, some organisms are far better at tolerating unfavourable conditions than others.

The study of living things forms the subject of biology (from the Greek *bios* meaning 'life' *logos* meaning 'knowledge'). Modern biology is an extremely diverse subject embracing a vast number of different specialisations ranging from cytology, the study of the detailed structure and function of a cell, to ecology, the study of living things in their natural environment. Zoology is concerned with the biology of animals.

The differences between animals and plants

The differences between the animal and plant kingdoms as a whole are obvious. If it is easy to distinguish between a horse and a rock, it is equally easy to distinguish between a horse and a pine tree. The man in the street would probably make the following distinctions: 'Pine trees are green, horses are not' and 'Horses move, pine trees do not'. However both horses and pine trees are complex representatives of their respective kingdoms, and at the level of the simpler plants and animals the distinction becomes blurred. For example, *Euglena* is green, and also moves: sea anemones usually remain fixed in one place and yet are clearly animals. If there is a real distinction between plants and animals then it must be based on other characteristics.

The most basic difference between animals and plants is their means of nutrition. Green plants, for example, are able to manufacture the organic materials they require from simple inorganic materials using sunlight as a source of energy (autotrophic nutrition). Energy-storing molecules are manufactured from carbon dioxide and water and the process can be summarised by the following equation:

$$6CO_2 + 6H_2O \xrightarrow{\text{energy from sunlight}} C_6H_{12}O_6 + 6O_2$$

carbon dioxide + water → glucose + oxygen

This is a simplification of a process known as photosynthesis which involves a complex series of intermediate changes and uses energy intercepted from sunlight by chlorophyll, in specialised cell organelles, the chloroplasts. Chlorophyll gives the green colouration typical of plants. With this organic starting point the plant can use inorganic salts (nitrates, phosphates and sulphates) to make the wide range of organic substances (amino acids, proteins, carbohydrates, nucleic acids) of which it is composed. Animals, on the other hand, are not capable of photosynthesis although they require essentially the same organic materials as plants. They therefore have to obtain these by feeding on complex organic food substances (heterotrophic nutrition). Certain essential organic substances must be present in the diet and from these the animal can manufacture the precise materials of which it is composed.

This basic difference in feeding behaviour has a direct implication for the structure of the animal. Plants, which do not need to search for their food, remain fixed in one place and obtain the necessary carbon dioxide and light by a large system of leaf-bearing shoots, and water and inorganic substances by a root system. Since plants do not have to move from one location, they are able to show almost unlimited growth, with multiplication of the parts of both root and shoot systems to an indefinite size. Animals have to be able to seek and obtain their food, and the vast majority find it by moving about. They therefore are restricted to a maximum size (which in comparison with many plants is usually small) and a fixed number of parts such as limbs. They must have a means of recognizing their food, and therefore have sense organs and associated nervous systems. To capture and digest food requires many more special structural adaptations. Oxygen is needed in large quantities to provide the energy required for movement and for the metabolism of food, and therefore many animals have evolved elaborate systems to obtain it from the environment. Finally, excretory systems are necessary to eliminate waste products from the body. Thus almost all the structure of an animal can be related to its nutritional needs. In the simplest unicellular animals and plants these differences do not apply. In fact in many cases it is difficult to say to which kingdom an organism belongs, especially as a unicellular form may have the nutritional characteristics of both animals and plants. This problem has led many workers to suggest that unicellular forms should be placed together in a separate kingdom, Protista.

Since the plants (and some bacteria) are the only organisms which can synthesise carbohydrates from inorganic materials, thus converting energy from the sun into a useable form, all animal life ultimately depends on them. The transfer of energy in the form of organic matter from its initial source in a plant through a series of organisms, each of which eats the preceding one and is in turn eaten by the following one, is known as a food chain. Humans are involved in a number of food chains. A simple example is man eats cow eats grass. Rather more complex situations may involve eating fish which eat smaller fish

which eat invertebrates, which eat algae. Obviously, each transfer between organisms involves a dissipation of energy through respiration, and in practice food chains seldom involve more than four or five steps.

In summary, all animals must procure food and oxygen and remove wastes in order to grow and reproduce. All animals are adapted to fulfil these functions in different ways which depend in part upon the kind of environment they inhabit and their mode of life. The size of the animal has a particular modifying effect. The smaller an animal, the less problematical both the process of procuring food and oxygen and the removal of metabolic wastes. Increased size demands increased efficiency, and development of systems of internal transport for food materials, oxygen, and waste products. The response to these demands is the basis for the incredible diversity seen within the animal kingdom. Approximately 1,250,000 species of living animals have been described and undoubtedly very many more remain to be discovered.

The cell

The cell is the basic unit of most living organisms. An amoeba, one of the simplest animals, consists of a single cell, within which all the life activities are performed. In contrast, the human body consists of many millions of cells arranged into organs, such as lungs and heart, each specialized to carry out one or more life activities. However, if a single cell of man is compared with an amoeba many similarities can be found.

The existence of cells was first recognized by Robert Hooke in 1665, when studying a section of cork through a very early compound microscope. Hooke concentrated on the easily visible thick cell wall, characteristic of plants, but also recognized that the 'cells' of many tissues contained 'juices'. Animal cells do not form rigid cellulose cell walls as plants do and their cell boundaries are far less conspicuous. Plant and animal cells therefore appear very different when studied under the light microscope, and were at first regarded as fundamentally different structures. It was not until 1838–39 that Schwann and Schielden postulated the Cell Theory, that the cell is the basic unit of all living organisms.

The term 'protoplasm' for the fluid cell contents, the 'juices', was coined by Purkinje in 1840. It has no clear physical or chemical definition but may be regarded as meaning all the organized constituents of a cell. The structure and functions of protoplasm are enormously complex and have been extensively studied from the standpoints of chemistry and organization. By far the major constituent of protoplasm, and therefore of living things, is water: both a lettuce and the Prime Minister are about 80 per cent water. The chemical composition of protoplasm varies a great deal, especially between plant and animal cells, but the most important elements are hydrogen and oxygen (the constituents of water) carbon and nitrogen; together these make up about 95 per cent of the mass of protoplasm. In addition about 30 other elements are present, sometimes only as minute traces: in animals the most important are sodium, potassium, sulphur, iron, magnesium, iodine and chlorine.

The most important group of structural organic compounds within the cell are the proteins. These are formed from about 30 amino acids. The amino acids differ in the number and arrangement of their constituent atoms but always have an amino group (NH$_2$) and a carboxyl group (COOH). Proteins are formed from sequential chains of amino acids, linked by a peptide bond between the amino group of one and the carboxyl group of the next. These chains may be very long. Cells contain hundreds of different proteins, and each kind of cell has some proteins which are unique to it. Structural proteins include keratin (hair and nails), myosin and actin (muscle), collagen (connective tissue and bone). Many cellular proteins are enzymes, (biological catalysts which control the rate of chemical reactions within the cell).

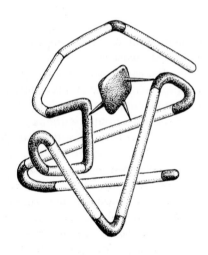

General structure of a protein molecule

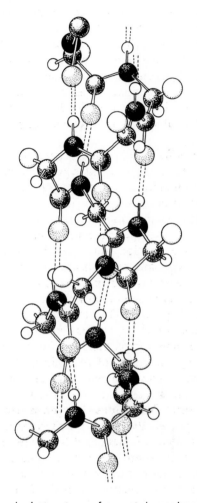

Chemical structure of a protein molecule

The remaining major types of organic substances within the cell are carbohydrates, fats and nucleic acids. Carbohydrates and fats are the chief sources of chemical energy. The carbohydrates are the simplest organic substances, containing carbon, hydrogen and oxygen. The hydrogen and oxygen are always present in a ratio of 2:1. Starches, sugars and celluloses are all carbohydrates. Fats or lipids are also composed of carbon, hydrogen and oxygen, but the proportion of oxygen is smaller. They are important both as a fuel source, and as structural elements of cell components such as membranes and are virtually insoluble in water. Nucleic acids are large molecules, containing carbon, oxygen, hydrogen, nitrogen and phosphorus. They are composed of units or nucleotides. There are two classes of nucleic acid, ribose nucleic acid (RNA) containing ribose, and deoxyribonucleic acid

(DNA) containing deoxyribose, which differ in function and in structural detail. DNA is particularly important in the formation of the hereditary units of the cell, the chromosomes, which carry the organism's genetic code.

From the organizational standpoint, protoplasm is very complex and is composed of a variety of structures. The majority of the cells are too small to be seen by the naked eye and vary between 10 and 50 micrometres in diameter though some of the largest unicellular animals (protozoans) are cells about 1 mm across. (An exception to this small size is found in the eggs of animals such as birds in which a large inert mass of food substance, the yolk, is present). Because of their small size the study of cells relies upon the use of microscopes. The size limit for seeing by the human eye is between 1 mm and 100 nm, about the size of the largest unicellular forms. The light microscope extends vision to between 0.17 and 0.25 μm, allowing the study of cells and larger cell structures. Various chemicals can be applied to the cell to stain its various chemical constituents, making differentiation between structures easier. Electron microscopy permits the study of objects as small as 1 nm in diameter, thus including large molecular structures and the finer points of protoplasmic detail.

The unit membrane is an important feature of many cellular structures. In particular the plasma membrane which surrounds each cell is composed of a single unit membrane. Unit membranes are formed from lipoprotein and appear under the electron microscopes as two dense lines separated by the less dense region, the whole being about 7.5 nm in thickness. This structure was originally interpreted as two regularly arranged layers of phospholipid, each one molecule in thickness and associated with a protein layer on each surface. More recent interpretations postulate a less regular arrangement with proteins distributed both at the surface and within the phospholipid layer. The protein gives strength and elasticity, and the lipid impermeability to the structure. The fine detail of the membranes is probably very complex and they play an important role in controlling the passage of chemical substances into and out of cell structures.

EXTERIOR OF CELL

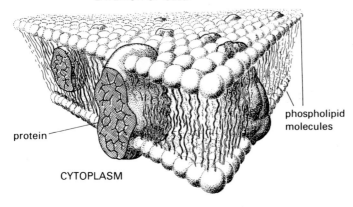

Molecular structure of a plasma membrane

If a cell is examined through the light microscope an immediate distinction can be made between the nucleus as a densely staining, ovoid, spherical or bun-shaped region and the less dense granular cytoplasm which surrounds it. The nucleus is surrounded by a nuclear envelope which separates the nucleus from the cytoplasm and is a double unit membrane, perforated by large pores formed by fusion of the inner and outer membranes. The pores may be filled with a plug of denser material which includes various fibrillar arrangements. The pores provide a route for chemical interaction between the nucleus and the cytoplasm.

The nucleus is of major importance in controlling the cell's activities. It contains the chromosomes (the hereditary units of the cell, composed of DNA, RNA and protein) which carry the organism's genetic code within the DNA molecules. The nucleus also includes the nucleoli, which are small spherical structures made up of RNA and proteins, concerned with protein synthesis and the transfer of genetic messages from the nucleus to the cytoplasm. The chromosomes and nucleoli lie within a chemically complex, amorphous substance, the nucleoplasm.

The cytoplasm may contain droplets of fat and other stored substances such as glycogen. It also includes several kinds of sub-cellular structures known as organelles.

Mitochondria (singular mitochondrion) were amongst the earliest organelles to be recognized. They are present in all cells and are concerned with the major catabolic process of the cell, cellular respiration, whereby energy stored in food or reserve products is made available for use by the cell. They are elongated and have a double membrane. The outer membrane is smooth and the inner one is folded into numerous elongated projections, the cristae. The innermost part of the mitochondrion is the mitochondrial matrix.

An intricate system of membrane-limited cavities (cisternae) extends throughout the cytoplasm. These are continuous with each other and possibly with the nuclear envelope. The system is known as the endoplasmic reticulum. In some regions the endoplasmic reticulum has a rough or granular appearance due to the presence of ribosomes on its external surface. These ribosomes are made up of RNA molecules and about 50 different proteins and are involved in protein synthesis. Smooth endoplasmic reticulum consists of membranes without ribosomes. The functions of the endoplasmic reticulum are poorly understood but are thought to include intracellular transport and the orderly arrangement of enzymes, as well as protein synthesis.

The Golgi apparatus is sometimes thought of as a specialised region of the endoplasmic reticulum. It consists of stacks of flattened cisternae, with usually 5–8 to a stack. It is responsible for the collection, modification and subsequent export of metabolic products from the cell, and is involved in maintaining its water balance. In addition the Golgi apparatus packages substances such as proteins and polysaccharides into membranous vesicles. Some of these form secretory vacuoles and are released from the cell; others remain and function in the cell. Lysosomes, which contain digestive enzymes, are important examples of such structures.

Two small, darkly-staining structures, the centrioles, are found in the cytoplasm near the nucleus of animal cells, but are not found in the cells of some plants. They are cylindrical in shape with a wall and an amorphous translucent centre. In cross-section the wall can be seen to contain nine groups of longitudinally oriented microtubules, each group consisting of three connected tubules. The centrioles are typically orientated to be perpendicular to each other and are involved in cell division.

Microtubules are present in many cells. These are

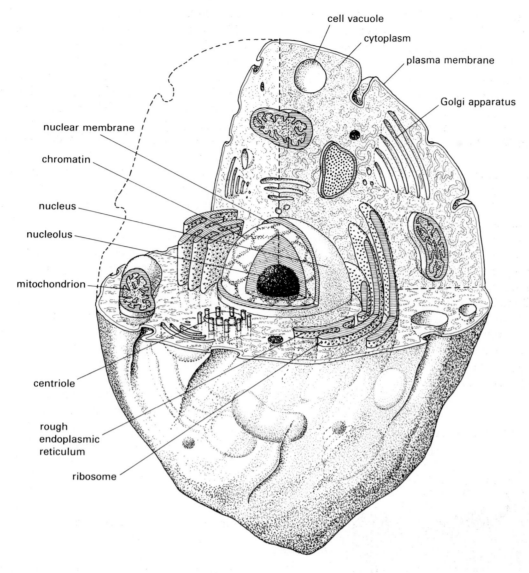

cell vacuole
cytoplasm
plasma membrane
Golgi apparatus
nuclear membrane
chromatin
nucleus
nucleolus
mitochondrion
centriole
rough
endoplasmic
reticulum
ribosome

Generalised animal cell

cylindrical cytoplasmic structures composed of several types of protein. Tubulin is the major protein, but others including dyrein and moxin have been isolated and named. Microtubules are important in maintaining and controlling the shape of the cell, as well as occurring in locomotory organelles (flagella, cilia).

In addition to the cell organelles, cytoplasm may contain droplets of fat, protein granules, and crystalline substances all of which are simply stored for future use.

The maximum size of a cell is limited by its metabolism. The metabolic activities of a cell are proportional to its volume, and necessary nutrients and metabolic by-products must enter and leave through the cell wall. As a sphere increases in size, its surface increases as the square of its radius, and its volume as the cube of its radius. At a certain size the surface area can no longer provide sufficient area for exchange. At this point the cell must either stop growing or divide.

Cell division

Since the size of individual cells is limited by the necessity for a high surface to volume ratio, growth of tissue or a whole organism is generally brought about by an increase in cell numbers. This occurs through cell division. First, the nucleus divides to give two daughter nuclei, then this is followed by a division of the cytoplasm to give two daughter cells. There are two different kinds of nuclear division, known as mitosis and meiosis, but mitosis is the process that occurs in normal growth. The cytoplasmic division that follows is called cytokinesis.

Mitosis

If a cell about to undergo mitosis is stained and examined under the microscope, the chromosomes are visible as paired, darkly staining threads (chromatids) within the nucleus. Each chromatid consists of bead-like swellings, the chromomeres, along a central strand, the chromonema. A chromosome is formed from a pair of chromatids constricted by a small clear zone, the centromere.

Mitosis may be regarded as consisting of four stages: prophase, metaphase, anaphase and telophase, each of which merges imperceptibly into the next. (In early prophase the chromosomes are elongated and the individual chromomeres are visible. Later they contract to become shorter and thicker and the chromomeres lie so close together that they can no longer be distinguished). Early in prophase the centrioles migrate to lie on opposite sides of the cell and a structure known from its appearance as a spindle forms between them. The spindle is composed of protein threads and is broad at the equator of the cell, narrowing to a point at either pole. By the end of prophase, the spindle is formed and the chromosomes have become short and thick. This process usually takes between 30 and 60 minutes in most animal cells. Metaphase follows: the nuclear membrane disappears and the chromosomes line up along the equatorial plane of the spindle. The centromere of each chromosome divides and the two chromatids become separate daughter chromosomes. Metaphase is much shorter than prophase, lasting between 2 and 6 minutes.

In anaphase the daughter chromosomes move apart, one of each pair going to each pole, moving along the spindle threads. As they move they adopt a V shape, with the centromere at the angle of the V pointing towards the pole. The spindle threads act as guide rails, ensuring that all the daughter chromosomes are incorporated in the daughter nuclei. They are also thought to be contractile, pulling the centromeres towards the poles. This stage takes between 3 and 15 minutes.

The final phase of mitotic division, telophase, takes between 30 and 60 minutes. Several processes take place. A nuclear membrane forms around the group of chromosomes at each pole, and the chromosomes resume their original extended form. Finally the cytoplasm divides, beginning at a furrow which encircles the cell at the level of the equatorial plate. The furrow deepens and the two new daughter cells separate.

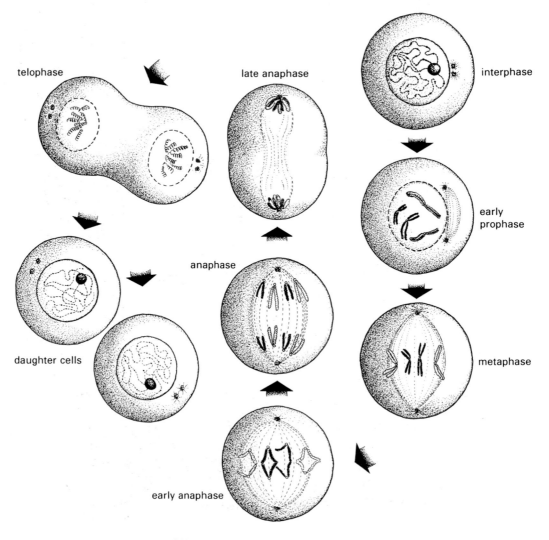

Mitosis in a hypothetical animal cell

Mitosis results in the formation of two cells from a single cell, each with exactly the same number and kinds of chromosomes as the parent cell. In multi-cellular organisms cell division occurs continuously though with variable speed and frequency according to the tissue type, and the stage of development. The process of cell growth, mitotic division, and the subsequent growth of daughter cells is the cell cycle. Typically the whole process, including a period of growth between divisions when the nucleus is said to be resting (interphase), takes 20 hours. During interphase the daughter chromosomes undergo replication, simple separation of the chromatids produced occurring in mitosis.

The most frequent cell divisions are seen in the early stages of embryonic development, when they may be as frequent as once every 30 minutes. In the adult the rate varies between tissues. For example vertebrate bone marrow cells are thought to divide frequently, but nervous tissue cell division is extremely rare.

Meiosis

The development of a new individual involves a different type of cell division. During the development of eggs and sperm two successive cell divisions take place, the whole process being known as meiosis. Fertilized eggs and all normal body cells have two sets of each kind of chromosome, a condition known as diploid. For example the diploid number for humans is 46, the 46 chromosomes totalling over 2 m in length! In eggs and sperm only one of each kind of chromosome is present, a condition known as haploid, meiotic division bringing about the necessary reduction in chromosome number. The fusion of eggs and sperm at fertilization restores the original diploid condition.

Each of the two divisions involved in meiosis includes prophase, metaphase, anaphase and telophase as in mitosis. However, there are important differences, particularly in prophase, of the first meiotic division. At this stage the two

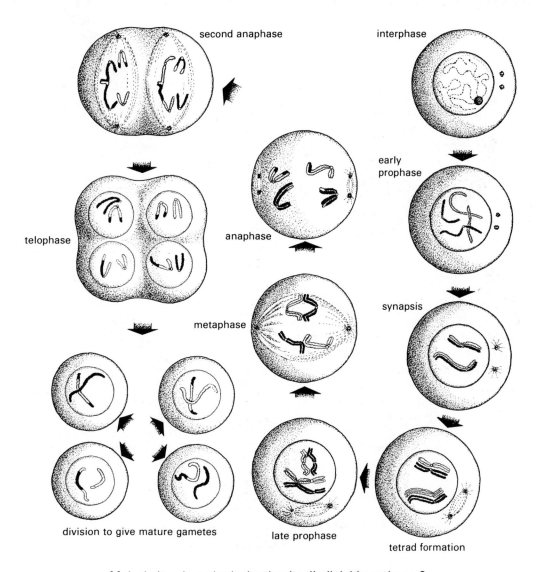

second anaphase

interphase

telophase

anaphase

early prophase

metaphase

synapsis

division to give mature gametes

late prophase

tetrad formation

Meiosis in a hypothetical animal cell; diploid number = 2

chromosomes of each kind (homologous pairs) come to lie close together along their length and twist round each other, a process known as synapsis. They then shorten and thicken, and each doubles to give rise to a bundle of four chromatids (a tetrad). The centromeres do not double or divide, so each tetrad has only two centromeres. As in mitosis the centrioles pass to opposite poles, a spindle forms, and the nuclear membrane dissolves.

When the tetrads line up at the equator of the spindle the cell is in metaphase. In anaphase the tetrad separates and the daughter chromatids, still joined by the centromeres, move to opposite poles of the spindle. Thus the homologous chromosomes, but not the daughter chromatids, are separated. A short telophase follows, with the arrival of the haploid number of double chromosomes at each pole.

Prophase of the second meiotic division is short. The centrioles divide and a spindle forms in each cell at right angles to the position of that formed in the first prophase (prophase 1). The chromosomes come to lie along the equatorial plate, the centromeres divide, and the chromatids separate. The cell enters a second anaphase and the chromatids move to opposite poles. The cytoplasm divides and the nuclear membranes are formed.

The two divisions which make up the process of meiosis result in the formation of four nuclei, each with a haploid set of only one individual of each type of chromosome. Each cell thus formed is a mature gamete, able to take part in the formation of a new individual by sexual reproduction.

Animal classification

The species

Every living organism is regarded as a member of a species; the individuals of any one species are more closely related to one another than to members of any other species. A working definition of a species could be 'a series of individuals sharing sufficient genetic homology to be able to produce fertile offspring by sexual reproduction'. This definition, whilst clear, is limited in application. It can only be applied to organisms which reproduce sexually, and in practice, to those whose breeding behaviour is known. More usually the determination of a species has to be made on a basis of comparison with any other species, and considering particularly its anatomical structure. This has lead some biologists to distinguish between the biospecies, to which the definition above strictly applies, and the morphospecies, which is defined on the basis of its form.

Another important concept is genotype (from the Greek gen- which means 'to be born, to become', and typos, meaning 'type') which is the genetic complement of an individual. This can be distinguished from the phenotype (Greek phaino: 'to show') which is the visible expression of this hereditary constitution, subject to environmental influences. Any organism we look at is a phenotype, where development has been determined partly by its genetic constitution, and partly by its environment. The distinction between genotype and phenotype is important because environmental effects may have a considerable influence on the appearance of an individual. If species are distinguished solely by their anatomy there is always the possibility of splitting up a single species into many, simply because of phenotypic differences. For example the brown trout found in the British Isles, *Salmo trutta*, has been referred to as the Eastern sea trout *Salmo albus*, the Western sea trout, *S. cambricus*, the gillaroo of Ireland, *S. stomachius*, and the brook trout, *S. fario*. All of these are now regarded as being members of the single species, *Salmo trutta*.

Taxonomy

A classification is any man-made system for the organization of objects and information. For the satisfactory study of the many hundreds of thousands of biological species each must have an internationally recognized name and a place within a classification. The science of classification is taxonomy.

Naming a species
The binomial system of naming a species dates from the eighteenth century, and was established by Carl von Linné (latinized as Linnaeus in line with the practice of his day) in two works 'Species Plantarum' (1753) and 'Systemae Naturae' (1758). Both these works were enormously important and best sellers, 'Systemae Naturae' running into twelve editions. The binomial system gives every species a name consisting of two parts, e.g. modern man is *Homo sapiens*. The first part is the name of the genus, written with a capital letter in the form of a Latin singular noun, irrespective of the origin of the word. This name may apply to several closely related species and may be used by itself to describe the group. The second, the specific or trivial name, is also latinized as an adjective in apposition with the generic name. In effect the same binomial arrangements occur in popular names, for example polar bear and brown bear as subdivisions of the genus bear, although the order is reversed. The Latin form was used by Linnaeus because Latin was the international language of the educated classes of that time. Indeed Bacon used Latin when writing his 'Opus Magnus' in the seventeenth century because he did not believe that English would survive as a language.

The bewildering assortment of names facing any biologist probably arouses more resentment than any other aspect of the subject; a feeling well expressed by the Gnat in Lewis Carroll's "Through the Looking-Glass and what Alice found there": 'What's the use of their having names if they won't answer to them?' To which Alice replied as a taxonomist might, 'No use to them, but it's useful to the people that name them, I suppose'.

Many of the generic names in translation refer to some easily recognizable anatomical feature of the animal or plant; for example *Acanthocephalus* means 'spiny headed', *Ostracion* means 'a little box'; or to some fanciful resemblance between the animal and a character from Greek or Roman mythology (*Nereis, Hydra, Aphrodite*). Specific names may also be descriptive, or refer to the geographical region in which the species was first discovered (*hispanicus, japonicus*) or to its habitat (*lacustris* meaning 'of the lake', *fluviatilis* meaning 'of the river'). Unfortunately not all scientific names are as helpful. Many animals have been called after distinguished scientists. There is one group named after some of the characters of J R R Tolkien's 'Lord of the Rings', and one biologist has even used anagrams of his wife's name, Caroline (*conilera, nerocila*). With so many different species the problem of inventing names is not always easy.

Although the same generic name cannot be used for two different groups of animals it sometimes happens that an animal and a plant may have the same name, for example *Bougainvillea* has been applied to both a colonial hydrozoan animal and to a tropical shrub. The same specific name may be applied to different genera. As we have seen many are descriptive and a word like *lacustris* is used for many plants and animals, for example the two different lake fish *Craterocephalus lacustris* and *Melanotaenia lacustris*.

The full scientific name of an organism may have two further pieces of information besides the binomial; the name of the author who first described it, and the year in which the description was made. Linnaeus described his species by reference to one individual specimen known as a 'type'. The type system is still used by present day taxonomists, the type specimen being deposited in a central place such as one of the large museums, for reference by later workers. The system has the advantage of eliminating many ambiguities; however these can still occur, often through the same name being applied to two different animals, or the same species being given two different names. It was to solve such problems that the international rules of zoological nomenclature were established in 1931. The use of scientific names, although they may appear confusing at first, is necessary for a variety of reasons. Many organisms, including the majority of fossils and the less well known living forms, do not have a popular name at all. Furthermore, popular names differ not only between languages, but also in different regions where the language is fundamentally the same. English people would call the North American hedgehog a porcupine, and whilst the robin is recognized in both North America and England the name is not applied to the same species of bird.

Animal groups

Linnaeus arranged species into groups, on the basis of anatomical similarities, believing that he was studying the pattern of creation. He recognized seven major groups, empire, kingdom, class, order, genus, species and variety. Empire and variety are rarely used by zoologists, though plants are often classified as far as variety. Modern classifications rest on the following hierarchy:

    Kingdom
      Phylum
        Class
          Order
            Family
              Genus
                Species

For many animal groups, these categories are insufficient. They may be increased by the use of prefixes such as 'super-', sub-' or 'infra-' or by the addition of further categories such as cohort or tribe.

The full classification of the domestic cat is as follows. The scientific name of the domestic cat, *Felis catus* links the domestic cat with other closely related forms, for example the wild cat, *Felis silvestris*, within the genus *Felis*. Cats are related to the lions and tigers, which are included with them in the family Felidae. The Felidae are related to other flesh eaters, for example dogs, foxes, wolves, seals and weasels, and included with them in the order Carnivora. Like many other animals, carnivores suckle their young. All these animals form the class Mammalia. Mammals are related to the other animals with backbones and are included in the subphylum Vertebrata within one of the major divisions of the Animal Kingdom, the phylum Chordata.

This arrangement of animals has several advantages. Originally it was regarded merely as a convenient filing system, bracketing animals into larger and larger assemblages, and which could readily be turned into a system for identifying animals. In addition a classification of this sort, if founded on basic characters rather than superficial resemblances, can express evolutionary relationships. Classification therefore is a useful basis for the study of evolution. However we must always be aware of the dangers of circular argument. If we base further classification on assumed evolutionary development we cannot at the same time use this as evidence for evolution.

The origins and interrelationships of animals

Organic evolution is the name given to the theory which regards present day organisms as having arisen from earlier forms. To a large extent modern classifications (such as that given for *Felis catus* in the previous section) are based upon our ideas of the evolutionary relationships of organisms. The idea of change is an ancient one which can be traced back in a modified form to the ancient Greek philosophers. For example Euripedecles, a pre-Socratean philosopher, envisaged various parts such as arms, legs, shoulders, and eyes, wandering around in a kind of limbo in which they came together by chance, the best adapted living and reproducing. Thales (640–548 BC) regarded all forms of life as having developed from aquatic animals. Aristotle, (384–322 BC) who was one of the most influential Greek philosophers and a great naturalist, studied and described the collections made by Alexander the Great in his campaigns and may also have thought along evolutionary lines: statements such as 'The embryo is an animal before it is a particular animal – its general characteristics appear before the special' certainly suggest that he did. However even the most brilliant of the Greeks were limited by their lack of faunal knowledge. Aristotle recognized only 18 species of fish, out of over 30,000 described today; Pliny (AD 23–79), whilst thinking that 'In the sea and the ocean, vast as it is, there exists, by Hercules! nothing that is not known to us', recognized only 176 species in his 'Natural History'.

An equally ancient and opposing view is that of Special Creation which holds that the world and present day fauna and flora were created by an outside influence or god and have remained more or less unchanged ever since. This idea is found in many mythologies, the story of the Garden of Eden in Genesis (circa 4,000 BC) being only one example. One reason for the dominance of this idea is that, until very recently, man had no conception of the age of the Earth or of how long life has existed on the planet. For example, in 1600, Shakespeare reflected the thinking of his time when he wrote in 'As You Like It', 'the poor old world is almost six thousand years old'. Learned men spent a great deal of time in calculating the exact moment of creation from clues given in the Bible; in 1644 the then Vice-chancellor of Cambridge concluded that creation occurred in 3928 BC, on the 17th of September, at 9.00 a.m! Although the Greek philosophers speculated about evolution, since they knew relatively little biology their ideas were rather vague and cannot be regarded as foreshadowing the present theory of organic evolution. Aristotle, perhaps, approached it most nearly when he observed that animals and plants could be arranged in a graded series from lower to higher, and inferred that one evolved from another. However, he held the metaphysical belief that this evolution occurred as a result of nature striving for perfection. Aristotle's ideas did not become generally accepted, first because the ancients had not observed species change and later because Judaeo-Christian ideas of special creation became firmly established in Western culture.

In the seventeenth century, increasing knowledge of animals led scientists such as Ray (1627–1705), Hooke (1635–1703), Buffon (1707–88) and Lamarck (1744–1829) to consider evolution more favourably. One of the earliest theories of organic evolution was that of Lamarck, a French zoologist who published 'Philosophie Zoologique' in 1809. He believed that living things were endowed with a vital force which controlled the development and functioning of their various parts and enabled them to respond to the demands of their environment, the size of the organ being proportional to its use or disuse. Traits acquired by an organism in its lifetime could be passed on to its offspring and, over succeeding generations, lead to a considerable change in form. One of Lamarck's examples concerned the evolution of the giraffe. In his view, an ancestral giraffe took to feeding from trees rather than from the ground. In doing so, the continuous stretching elongated its neck. This process, repeated over many generations, resulted in modern giraffes. Lamarckism provides an explanation for the adaptation of animals to their environment; however, there is no evidence to support the concept that acquired characters can be inherited.

Darwin's 'Origin of Species' (1859) probably represented the greatest single event in the history of biology for he not only demonstrated, with a wealth of evidence, the possibility of evolution but proposed the mechanism of natural selection by which it could occur. Based on observations that more organisms are born than could possibly obtain food to survive and that population sizes remain fairly constant, Darwin's conclusion was that those better fitted to particular conditions would survive in greater numbers than the less fit, an idea summarized as 'survival of the fittest'. One of the problems with this idea is a definition of fitness, since all those that survive are fit by definition, nevertheless this work marks a watershed in science and the idea of evolution is a fundamental premise behind much of modern biology. Since 1859 the evidence for organic evolution has become overwhelming and much more is known of the detailed processes by which it takes place.

In the 1930s workers such as Simpson (a vertebrate palaeontologist), Dobzhansky (a population geneticist) and Mayr (a systematist) produced ideas which contributed to a synthesis of evolutionary views. These have led to the contemporary theory of evolution, neo-Darwinism, in which natural selection is regarded as by no means the only ingredient, others being mutation (the source of new variation), genetic drift (random variability leading to the loss of traits), and population dynamics.

Another important scientific development of the past 150 years has been the geological evidence for the long history of the Earth's development, and of the enormous changes which have occurred. It is now clear that the span of life on Earth is extremely long though no-one can say with certainty when it all began. Modern evidence suggests at least 3,000 million years ago, and probably earlier. During this period the major animal groups emerged, adapted to many modes of life, and in many cases became extinct. With the passing of time the evolutionary tree of the animal kingdom branched repeatedly; the modern evolutionary zoologist is left with the problem of reconstructing that tree from its outermost, youngest twigs and branches. It is a daunting task and the evidence available is often slim and fragmented.

The only direct evidence of the course of evolution comes from the fossil record (from the Latin *fossilium* meaning 'a dug up thing'). Fossils have been objects of interest to man since the earliest times. They have been found in palaeolithic burial places and were studied and described by the Ancient Greeks. Anaximander (610–547 BC) described fossil shells and fish in rocks in Sicily and at Samos as proof of the existence of ancient oceans which had been burned out by the sun. Empedocles (484–424 BC) described fossil

Hippopotamus bones as the remains of a race of giants, and Pliny believed fossilized sharks' teeth to be fossilized tongues (*Glossopterae*). Many fossils were accorded magical properties. *Glossopterae* were thought to cure snake bites, and fossil molluscs were used for fortune telling.

Palaeontology (from the Greek *palaeos*, meaning 'ancient' and *on*, meaning 'being' can be regarded as having begun with the work of Leonardo da Vinci (1452–1519), who refuted many old superstitions. The Comte de Buffon (1707–88) was the first to appreciate the immense length of geological time and to confirm that some animal groups were essentially fossil. However, he was subsequently forced by the Church to recant and to accept the Book of Genesis as fact.

Unfortunately, the fossil record is far from complete as a record of past animal life since the most successful fossilization is of hard parts. Soft-bodied animals rarely survive as fossils. Thus of all the invertebrate phyla known today only five groups: Foraminifera (Protozoa), Trilobitomorpha (Arthropoda), Mollusca, Brachiopoda and Echinodermata, have left a fossil record of any size. The vertebrate fossil record is rather better but even so the story is far from complete. A study of animal evolution based only on the fossil record is comparable with a description of the history of a village based on the remains in a churchyard where grave robbers have been at work and the grave diggers frequently on strike! Furthermore, very few animal fossils are known from earlier than 600 million years ago and by that time many of the major phyla were established; thus there is no direct proof of the evolutionary relationships between them. A summary of the fossil record and the order in which the major animal groups appear is shown in the table on page 12.

In the absence of direct evidence, the student of animal relationships has to derive them from a study of living forms. Broadly speaking four lines of evidence are available: a study of comparative anatomy, of embryonic development, of physiological and biochemical similarities, and geographical distribution. All sources have something to offer; all can prove misleading if followed uncritically. Animals with similar anatomical adaptations may be closely related but may equally well be unrelated and show similar adaptations to the same mode of life. This is called convergent evolution. A crude example of convergence which is easily demonstrated by close examinations is the superficial similarity of a bird and a bat. Embryos and larvae are equally likely to show similar adaptive modification. Physiological and biochemical studies are comparative newcomers in the field of evolutionary theory. However some of the results given are strange to say the least and it is increasingly clear that convergence is as possible at the chemical as it is at the anatomical level. Furthermore it is difficult to decide which line of evidence should be given the greatest weight and here all the prejudices of individual workers can come into play.

Strictly any taxonomic group should be monophyletic; that is a natural entity with a single origin. However groups with the most complete fossil records, the reptiles for example, suggest that evolutionary advances arise not once, but several times and that many of the phyla recognized in a modern classification are not true taxonomic groups but levels of organisation reached independently by several organisms. The probable evolutionary inter-relationships of each animal group are discussed briefly within the chapters. Obviously some groupings are more generally accepted than others. The origin of multicellular animals, for example, is particularly controversial, and such discussions will inevitably be superficial. A synopsis of the animals included in each taxonomic group and included in the text will be given at the end of each chapter.

Animals and their environment

The study of the relationship of living organisms to each other and to their natural habitat is the subject matter of ecology, a word derived from the Greek *oikos* meaning 'a house'. A basic ecological concept is the ecosystem, which comprises all aspects of the physical surroundings (habitat) and all the organisms which live in it (biotic community). The significance of the concept is that it emphasizes the close environmental interrelationships of living organisms which exist since all living things depend upon limited resources of chemical materials and energy. For example, within an ecosystem we can study how energy from the sun is accumulated by green plants through photosynthesis, and may pass into the bodies of animals which feed on plants (secondary consumers) and animals (tertiary consumers). At the same time we can trace the pathway of chemical substances which are taken up by plants as they grow, and eventually return to the environment through decomposition. Since the ecosystem is a theoretical concept which provides a framework for studying the ecological relationships of living things, its size and boundaries are largely a matter of convenience, for example a small aquarium or an entire ocean can be regarded as an ecosystem; for some purposes it is even useful to consider the planet Earth as an ecosystem. Ecological niche is the term applied to the role of an individual within the ecosystem and defines not only where it lives but also what it does: how it moves, feeds, survives. In general we can expect all organisms in an ecosystem to occupy slightly different niches since two animals with the same ecological requirements would compete with each other, though niches can overlap.

Five major habitats can be distinguished: the sea, estuaries, and freshwater (sometimes combined as the hydrosphere), land, and air. No animal or plant is adapted to live in all five, although some, for example seabirds, combine two or three. All habitats impose anatomical and physiological constraints on the animal, many of the structural and physiological similarities are adaptations to a

Summary of plant and animal life through geological time

Era	Period	Epoch	Time from beginning of period to present in millions of years	Plants	Animals
Archaeozoic			3600	No recognizable fossils	
Proterozoic	Pre-Cambrian		1600	Algae and fungi.	Various protozoa, worms, molluscs and other invertebrates.
Palaeozoic	Cambrian		600	Algal diversification.	Most modern invertebrate phyla established.
	Ordovician		500	Abundant marine algae.	Abundant corals, trilobites, molluscs; first vertebrates.
	Silurian		425	Radiation of early terrestrial plants.	Invasion of land by arthropods; first jawed vertebrates.
	Devonian		405	Considerable land plant diversity.	First winged insects, abundant sharks, bony fish, lungfish. First Amphibia.
	Carboniferous (lower)		345	Extensive lowland forests.	Abundant primitive echinoderms. Primitive amphibian radiation.
	Carboniferous (upper)		320	Extensive coal swamp forests.	Abundant insects and primitive amphibians, first reptiles.
	Permian		280	Decline of some early plant forms.	Decline of many early animal forms. Modern insect orders established. Reptile radiation (including some mammal-like forms).
Mesozoic	Triassic		230	Gymnosperms, cycads, ginkgos. Dominant conifers.	Extinction of primitive amphibians, dinosaur appearance and radiation, emergence of primitive mammals.
	Jurassic		181	Modern conifers established.	Maximum reptile radiation. Origin of birds.
	Cretaceous		135	Appearance and rapid diversification of flowering plants.	Extinction of many reptiles. First modern birds, abundant primitive mammals.
Cainozoic	Tertiary	Palaeocene	63	Widespread tropical vegetation.	Evolution of modern birds, primitive mammals dominant. Some new marine invertebrate appearance and radiation.
		Eocene	58	Widespread tropical vegetation.	Radiation of hoofed and carnivorous mammals.
		Oligocene	36	Widespread forest.	Extinction of primitive mammals with establishment of most modern mammals. Origin of anthropoid forms.
		Miocene	25	Increased grassland.	Radiation of grassland mammals. First man-apes.
		Pliocene	13	Spread of grassland.	Evolution of man.
		Pleistocene	1	Rapid vegetational fluctuations with tundra at high altitudes.	Most modern species established. Extinction of large mammals. Emergence of human social life.
	Quaternary	Recent	0·011		Modern man.

similar mode of life (convergent evolution). For example the torpedo-like body shape of fish, whales, squid, and shrimps, allows them all to move more easily and quickly through water.

Of the major habitats, the sea is the least demanding and is generally believed to be the place where life first evolved. Some taxonomic groups of animals are entirely marine and every major phylum includes at least some marine representatives.

Various physical factors combine to influence the environment of an organism, though temperature, light and salinity have the most far-reaching effects. Temperature is probably the most important single factor affecting the distribution of animals; it can determine their distribution both directly and also indirectly through its effect on vegetation. Salinity forms the basis for the principal subdivisions of the hydrosphere: sea, fresh and brackish water (particularly common in estuaries). Very many animal groups are restricted by salinity: amphibians and some fishes to freshwater; echinoderms and cephalopods to the sea. However some animals, for example the polychaete worm *Nereis diversicolor*, can tolerate wide salinity changes whilst others, for example the salmon or the eel, undergo migrations for breeding purposes between freshwater and the sea. Other important physical factors include humidity, availability of oxygen, nature and direction of winds or ocean currents, or various chemical peculiarities such as availability of calcium of phosphate or acidity of water or soil. All may act on an animal directly, or affect it indirectly through plants. Consideration of such factors is essential to understanding the animal's mode of life, and hence its structure.

Phylum Protozoa

The study of protozoans did not begin until the end of the seventeenth century when Antonj van Leeuwenhoek (1632–1723), a somewhat testy, self-educated Dutchman became fascinated by a newly invented microscope. Using homemade instruments he carried out a wide range of investigations with varied results. His attempts to observe what puts the heat into pepper or the 'bang' into gunpowder failed but he had great success with other studies including such varied subjects as the scum from pond surfaces, scrapings from his daughter's teeth, and the contents of frog's intestines. He was intrigued by his discovery that millions of actively swimming minute organisms occurred in all sorts of unlikely places; even in a strange infusion which he made by soaking peppercorns in water and called 'pepper water'. Many of van Leeuwenhoek's 'little animals' have subsequently been identified as protozoa and he is therefore credited with the discovery of the phylum.

Protozoa is a diverse group of minute animals found wherever there is moisture. They are cosmopolitan, and often the same species occurs in several continents. There are 50,000 or so known species which have adopted many different modes of life from free-living to parasitic. Protozoans are characterized by a 'body' composed of a single unit rarely larger than 1 mm across. Since primordial life is regarded as having originated as an undivided blob of protoplasm, protozoans are regarded as some of the simplest of living forms, hence their name, which means 'first animals'. In most essential features protozoans resemble the cells of which the bodies of higher organisms are formed. They have a nucleus and a series of cell organelles, many of which are identical with those of the Metazoa. However a protozoan represents more than the equivalent of a single cell in a multicellular animal. Functions, such as feeding and water balance, which in higher animals are performed by separate organs, are carried out by specialized regions of the protoplasm. Thus a protozoan can be a fairly complex structure.

It is a mistake however to regard even the simplest living protozoan as representing the earliest living forms, since all are specialised to a greater or lesser degree to fill present day ecological niches. The genus *Amoeba* is an example of one of the least specialized of modern protozoans.

Amoeba: a protozoan of simple structure

Amoeba is a widely distributed genus with species occurring in the sea, freshwater and soil, and with related forms living as commensals or parasites. *Amoeba proteus*, one of the larger species, lives mainly on the bottom of shallow lakes and ponds and is just visible to the naked eye. A living animal examined under the microscope is seen to be colourless and transparent, with an irregular, constantly changing shape.

The greater part of *Amoeba* is cytoplasm. The innermost cytoplasm, the endoplasm, is fluid with a distinctly granular appearance, due to the presence of cellular organelles such as mitochondria and the Golgi apparatus, and an endoplasmic reticulum of numerous small vesicles with ribosomes on their surface. Minute crystals of an excretory product, carbonyl diurea, are also present. The endoplasm is surrounded by a clear gelatinous layer, the ectoplasm, and

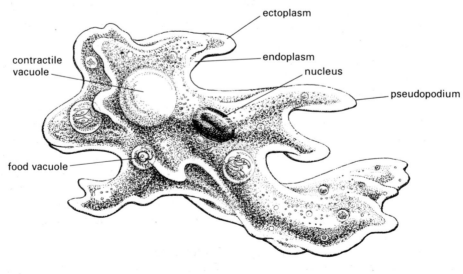

Amoeba

the outermost edge of the animal is formed by a simple unit membrane layer, the plasma membrane or pellicle. The nucleus is visible within the endoplasm, as a regular, bun-shaped object.

Locomotion

Amoeba moves by means of blunt, tubular pseudopodia ('false feet') which are responsible for its constantly changing shape. The characteristic amoeboid movement is slow (about 20 mm per hour) and the animal does not continue in any direction for long. Viewed from above *Amoeba* appears flat, like a blob of oil lying on the surface on which it lives. However, if viewed from the side it can be seen that a large part of its body is above the substrate, the pseudopodia extending downwards like little legs.

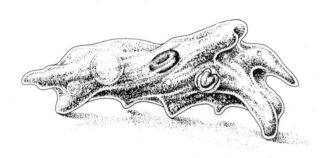

Amoeba in lateral view

Until recently, pseudopodia were thought to form simply as the result of a change in the colloidal nature of the cytoplasm from a plasmasol (or fluid) state to a plasmagel (or solid) state at the point where the pseudopodium appears. The reverse process was assumed to take place at the hind end of the animal. *Amoeba* could therefore be thought of as a tube of plasmagel containing the more liquid plasmasol which was squeezed forwards from the rear, the tube continuously extending at the temporary anterior end and reliquefying at the temporary posterior end. An alternative and more recent theory derives from the observation that sometimes endoplasmic granules near the front end begin moving before those at the hind end, so that progression cannot be due to squeezing from behind forwards. This theory, the 'fountain zone contraction theory' regards movement as resulting from a pull in front rather than a push behind. The evidence for it is based on detailed studies of the internal movement and physical properties of the cytoplasm in different parts of the stream. The 'fountain zone', a region near the tip of the pseudopodium where the everting endoplasm is gelled to form a wall, is a place where tension is developed and this tension is transferred posteriorly along the axial endoplasm. Recent electron microscopical studies have shown that amoeboid cytoplasm contains two types of filament: short, thick fibres of 16 nm diameter, and long, thin fibres of 7 nm diameter. These diameters are similar to those of the

myosin and actin characteristic of vertebrate muscle and there is good reason to interpret them in the same way. Vertebrate muscles work by the filaments sliding along each other; it is possible that a similar method is used in amoeboid movement.

Nutrition

Amoeba feeds on the abundant microscopic organisms, such as algae, bacteria, or other protozoa which are found on the mud on which it lives. When *Amoeba* comes near suitable prey it flows around it, forming a cup-shaped projection with its pseudopodia. The edges of the cup eventually meet and join to surround the food. Since the pseudopodia are not in close contact with the food, a drop of water is enclosed as well. The resulting structure is a food vacuole and several may be present at one time. The digestive process is essentially similar to that of higher animals. The organic contents of the food vacuoles are digested by enzymes secreted into the food vacuoles by lysosomes (bodies 0·2–0·8 μm in diameter with a single unit membrane containing various hydrolytic enzymes with a common optimal pH). The use of indicator dyes has shown that the vacuole fluids are at first acid, and then become alkaline as digestion proceeds. Digested material in solution is absorbed into the surrounding cytoplasm, leading to the shrinkage of the vacuole until it contains nothing but undigestible remains. Eventually it reaches the surface and is left behind as the animal moves.

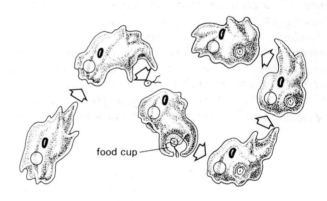

food cup

Amoeba engulfing a flagellate

Minute vacuoles may also be formed on the surface at the tips of the pseudopodia. In this way food material in colloidal form which becomes adsorbed onto the surface of the plasma membrane can be taken into the endoplasm. This process, known as pinocytosis, enables *Amoeba* to thrive in the absence of solid food. A similar process known as micropinocytosis is thought to be involved in the absorption of digested food from the food vacuoles.

Osmoregulation

The osmotic potential of the cytoplasm of a freshwater protozoan such as *Amoeba proteus* is greater than that of the surrounding medium. Consequently water continually passes by osmosis through the pellicle into the organism, in addition to any water taken in with the food or produced internally as a metabolic by-product. To remove this excess water from the cytoplasm *Amoeba* has a contractile vacuole, which functions as a small pump. This is a spherical cavity containing liquid and appears, when viewed through the light microscope, as a clear circular area in the endoplasm. The cavity is surrounded by a membrane with the same ultrastructure as that which forms the outside of the animal. The region immediately surrounding the vacuole contains mitochondria as well as many smaller vesicles which discharge into the main vacuole. The vacuole may be seen to enlarge gradually and then collapse suddenly, expelling the water out of the animal. A contractile vacuole is absent from most marine amoebae because the osmotic potential of the surrounding medium is closer to that of the cytoplasm, and a contractile vacuole would therefore serve no useful purpose.

Gaseous exchange and excretion

As it is a small animal *Amoeba* has a large surface area relative to its bulk, and therefore needs no special feature to ensure the diffusion of dissolved substances in and out of the cytoplasm. Since the whole of the surface is in contact with the surrounding water, which contains oxygen in solution, all the oxygen necessary for respiration can pass through the plasma membrane by simple diffusion. As the oxygen inside is used, its concentration is lower than in the surrounding water, thus a diffusion gradient is maintained and oxygen tends to pass into the animal. In the same way carbon dioxide and other soluble waste products of metabolism (principally ammonia) diffuse out of the animal.

Sensitivity and behaviour

While *Amoeba* has no distinct sense organs, experiments have shown that it is sensitive to a variety of stimuli over the whole of the cytoplasm. Generally reaction to stimuli is such as to keep it within a favourable environment. Its preferred light is dim and diffuse and *Amoeba* therefore moves away from any sudden increases in light intensity. Stimulus with dilute chemical solutions, for example changes in pH, or the addition of various sugars or salts, may initially cause the production of pseudopodia in the direction of the stimulus with stronger solutions resulting in movement in the reverse direction. Some chemicals, generally those of nutritional value, stimulate formation of food cups.

Mechanical stimulation causes varying responses. Gentle movement, such as that produced by possible prey, causes the formation of a food cup which is widely opened for the capture of motile prey, particularly when this movement is combined with a chemical stimulus. Violent jolting or electric shocks cause the withdrawal of pseudopodia. If an *Amoeba* is gently prodded with, for example, the point of a needle, it will move away from the stimulus.

Within limits the speed of movement increases with a rise in temperature so that the rate approximately doubles for a 10° C rise in temperature thus roughly following van't Hoff's formula for the relation between temperature and velocity of a chemical reaction. Laboratory studies have shown that freshwater protozoa from temperate climates are generally unable to survive at temperature above 30° to 35° C, although in nature certain protozoans have been recorded at much higher temperatures.

Reproduction and growth

An actively feeding *Amoeba* assimilates more food than is required for survival and the excess is used for growth: when an optimum size is reached the animal reproduces. In *Amoeba* the reproductive process is very simple; the nucleus divides mitotically and the two newly-formed nuclei move apart. The contractile vacuole is replicated and the cytoplasm separates into two approximately equal parts, one around each nucleus, so that one large *Amoeba* is replaced by two small ones.

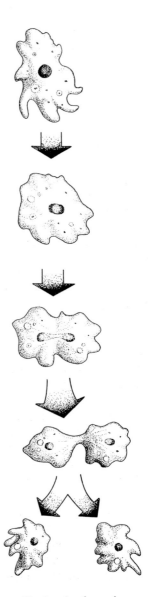

Fission in *Amoeba*

Certain amoebae, especially soil-dwelling forms, will, under adverse conditions, round up and secrete a tough protective wall to form a cyst. These cysts can survive both desiccation and a wide range of temperatures, and also provide a means of dispersal since they may be blown by the wind in dry conditions or transported in the mud on the feet of various animals. Multiple fission (sometimes called sporulation) has been described as occurring within large cysts formed by old amoebae. In this way many new amoebae are produced when conditions for growth become favourable.

Many of the main adaptations of *Amoeba* have been summarised, as follows:

When we were a soft amoeba, in ages past and gone,
'Ere you were Queen of Sheba, or I King Solomon,
Alone and undivided, we lived a life of sloth,
Whatever you did, I did, one dinner served for both.
Anon came separation, by fission and divorce,
A lonely pseudopodium, I wandered on my course.

Allegedly by Sir Arthur Shipley,
published anonymously in 'Life'.

A general consideration of protozoan structure

The main structural features already described in *Amoeba* appear, with some modifications, through the whole phylum Protozoa.

The cytoplasm invariably shows a distinction between ectoplasm and endoplasm, although there may be considerable variation in the endoplasmic inclusions. A cell membrane is always present, although its structure may show considerable elaboration in more advanced members of the group. In addition, non-living outer shells or tests are known in many forms, and some protozoans (the phytoflagellates) have cellulose cell walls.

The nucleus may be densely packed or vesicular, and contains chromosomes and one or more nucleoli within the nucleoplasm. The more primitive protozoans have a single nucleus, but two or several nuclei occur in some forms. These multiple nuclei may be of more than one kind and serve distinct functions.

Contractile vacuoles are characteristic of protozoa which have cytoplasm which is hypertonic to the surrounding medium, but are usually absent in marine and parasitic forms. A protozoan may have one or several. In the many protozoans which have a more rigid body surface than *Amoeba* they occupy a fixed position within the animal's body, opening to the outside through permanent pores and filling through a system of canals. The rate of filling and emptying varies, according to the size of the individual and its habitat, from every few seconds in the smaller freshwater species, to once or twice an hour in some of the larger endosymbiotic forms.

All types of nutrition occur in the Protozoa. In holozoic forms there is variation from the relatively simple and transitory structures found in such forms as *Amoeba*, to elaborate and permanent 'mouths' in more complex species. More confusingly, some protozoans are holophytic and have specialized organelles (chromoplasts) for synthesizing food from inorganic substances using energy from sunlight, a type of nutrition characteristic of plants! A very large number of species live as parasites, with hosts drawn from almost every animal group.

Pseudopodia are characteristic of *Amoeba* and the many protozoans related to it. They may occur in a variety of forms from blunt-ended to fine and needle-like projections. Two other kinds of organelle associated with movement occur in protozoans and the classification of the Protozoa is largely based on these structures. Cilia are exceedingly fine, hair-like structures which have been shown by electron microscopy to consist of a longitudinal bundle of fibrils (the axoneme) embedded in cytoplasm and enclosed within a membrane which is continuous with the surface membrane of the cell. The axoneme may be seen in cross section to consist of two central fibrils and nine peripheral fibrils, the central fibrils being thinner than the outer ones. Each peripheral fibril is a double structure composed of an entire microtubule (A), and a partial microtubule (B). Structures known as arms project laterally from the A microtuble to the B microtubule of the adjacent pair, and in addition the fibrils are joined by a ladder like arrangement of circumferential links. The A microtubule also bears structures known as spokes or radial links which appear to link the peripheral structures to the central fibrils of the axoneme.

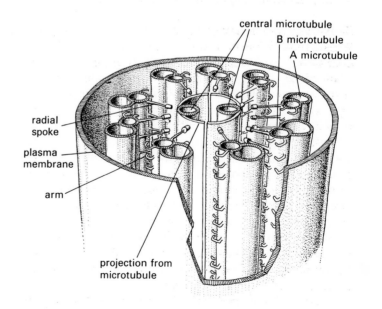

Cilium structure

Flagella are longer than cilia and are usually present in much smaller numbers. They are structurally the same as cilia, except for minor functional differences. Since the flagellate protozoa are regarded as more primitive, it is probably reasonable to regard cilia as short, specialized flagella. Neither cilia nor flagella are restricted to protozoa, but occur throughout the animal kingdom as well as in many plants. However they can only exist in some sort of fluid medium because of the naked plasma membrane. Their function is always the propulsion of this fluid. It is also worth noting that the basic '9 plus 2' fibril structure appears to be characteristic of all eucaryotic organisms and of many motile structures occurring in higher animal groups, for example the fine structure of the tail of a spermatozoan is very similar to that of a flagellum.

Asexual reproduction by fission similar to that described for *Amoeba* is common in the protozoans. However sexual reproduction, characteristic of the animal kingdom as a whole, also occurs in many representatives of the group.

Amoeba was selected as an introduction to the Protozoa because of its simplicity of structure. *Euglena* is an example of a very different mode of life.

Euglena: a 'plant-like' protozoan

Amoeba feeds on organic matter, including other living organisms, especially small unicellular plants, for example *Chlamydomonas*. Although these microscopic plants are also very simple structures, they show clear points of contrast with *Amoeba*. For example, the plasma membrane is usually enclosed within a layer of cellulose, and the cytoplasm contains specialized organelles known as chloroplasts or chromoplasts which contain chlorophyll and other photosynthetic pigments.

Euglena also contains chlorophyll, but in other respects, for example its lack of cellulose, it is rather more animal-like. It may be regarded as being neither plant nor animal, showing some characteristics of both groups. Organisms of this type do not fall easily into a zoological or botanical classification and there is frequent disagreement amongst biologists as to their precise position. Indeed, some workers avoid the problem altogether and include all unicellular forms together as Protista.

Euglena is one of the phytomastigophoreans (or phytoflagellates), a group of protozoa which show plant-like characteristics. *Euglena viridis* is one of the commoner species and occurs in stagnant water with a high content of nitrogenous organic matter such as farmyard ponds and puddles where it may be so numerous as to give a green coloration to the water. *E. viridis* is minute (about 40 μm in length) and spindle-shaped, with a blunt anterior end and a more pointed posterior end. The cytoplasm is surrounded by a pellicle, a thin but firm structure which, whilst giving a definite form to the organism, is sufficiently elastic to allow some changes of shape. The detailed structure of the pellicle is more complex than the simple plasma membrane of *Amoeba*. Electron microscopy of sectioned cells has established that the euglenoid pellicle consists of flexible, flat, interlocking strips arranged helically along the cell.

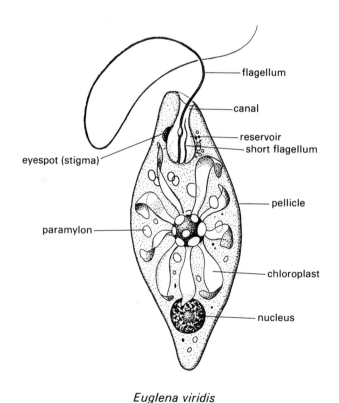

Euglena viridis

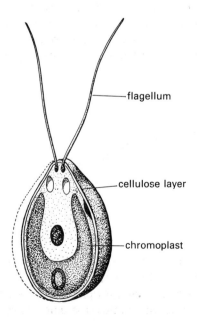

Chlamydomonas

These strips lie immediately below the plasma membrane and are therefore not equivalent to a plant cell wall which always lies outside the plasma membrane. Rows of mucilage producing bodies are present in the pellicle from which narrow canals pass into a groove in the pellicular strip and thence to the exterior. These produce a thin layer of slime which coats the animal and may have a protective function, as well as being involved in cyst formation. In addition a series of microtubules is arranged parallel to the pellicular strips.

At the anterior end of the body there is an infolding or invagination consisting of a narrow tubular portion, the canal, and a pear-shaped chamber, the reservoir. Associated with this structure is a prominent red eyespot (stigma) lying close against the reservoir wall, and the contractile vacuole. The reservoir is the only region of the animal limited by a plasma membrane but no pellicle, and it is here that discharge of the contractile vacuole takes place. The anterior invagination is not associated with ingestion of food particles in most species, and the use of a term implying such a function (cytopharynx, gullet) should be avoided.

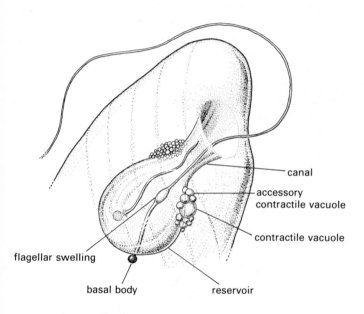

Euglena : details of reservoir

Euglenoid flagella have a relatively wide diameter and are inserted in the anterior invagination on the wall of the reservoir, opposite the stigma. Two are present, a very long flagellum and a second which is so short that it does not emerge from the canal. Basal bodies are present at the points of flagellar insertion into the body wall. The longer flagellum of *Euglena* has long, fine, hair-like threads (mastigonemes) along one side of the emergent length whose function may be to increase its surface area.

The nucleus is large and situated near the pointed end of the organism. It is spherical and vesicular, and contains a large nucleolus. The green colour of *E. viridis* is due to the presence in the cytoplasm of chloroplasts which are rod-shaped and radiate from a central zone giving a star-shaped structure. This arrangement may be lost in cultured forms, the rods dissociating into disc-like structures. At the centre of the chloroplasts and scattered through the cytoplasm generally, are large ellipsoidal granules of paramylon (a carbohydrate similar to glycogen), which is a food storage product.

Locomotion

Euglena has two methods of locomotion. The most usual is a fairly rapid swimming motion (about 3.6 mm/minute) by means of the long flagellum, which essentially acts as a propeller. The long flagellum trails obliquely to the rear. Waves pass along it from base to tip at a rate of about twelve per second, and drive the animal forwards. *Euglena* is unable to reverse its direction of movement but can alter course by flexure of the anterior end.

The other form of movement is the characteristic changes of shape resulting from contraction of one part of the body and dilation of the rest. This is called euglenoid movement whether it occurs in *Euglena* or in other protozoans and is thought to be the result of cytoplasmic flow within the confines of an elastic pellicle of definite shape. There is no evidence that the microtubules associated with the pellicle have a contractile function or are concerned in euglenoid movement. In fact their function is thought to be more probably skeletal or concerned with intracellular transport.

Euglenoid movement

Nutrition

E. viridis is holophytic, synthesising carbohydrates from carbon dioxide and water like any green plant. However, some species of *Euglena*, for example *E. gracilis*, can also live saprozoically, and in this case pinocytosis may occur in the reservoir. There are also a number of colourless euglenoid flagellates which are holozoic, ingesting food through a permanent cytostome lateral to the canal and reservoir, for example *Peranema*.

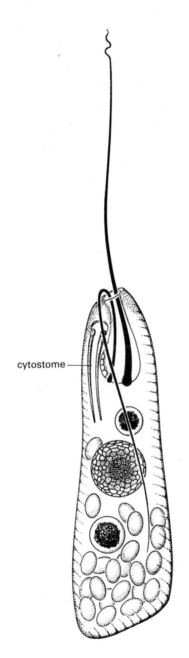

cytostome

Peranema

Gaseous exchange and excretion

As in the other protozoans, gaseous exchange and the removal of excretory products is by simple diffusion.

Osmoregulation

A single spherical contractile vacuole is present which discharges its contents periodically into the reservoir. It is surrounded by accessory vacuoles which coalesce with each other and the main vacuole as they fill with water.

Sensitivity

E. viridis reacts to a variety of stimuli, and usually gives an avoiding reaction to those which are unpleasant. Like most motile photosynthetic organisms, *Euglena* detects and responds to changes in light intensity. A specialised organelle, the eyespot, is thought to function as a primary light receptor and contains orange-red pigments which are β-carotene derivatives. In contrast with *Amoeba*, *Euglena* exhibits a phototactic response and tends to move towards the optimal light intensity for photosynthesis. At the base of the flagellum, opposite the eyespot, there is a flagellar swelling which may also be involved in phototaxis. It has been suggested that the eyespot is not the primary light receptor organelle, but a light-absorbing screen which intermittently shades the swelling which is itself light sensitive. This point awaits clarification.

Reproduction

Euglena reproduces by binary fission. The organism stops moving, withdraws its flagellum and encloses itself in a mucilaginous coat. After division of the nucleus and chloroplasts a longitudinal division of the cytoplasm, beginning at the reservoir, produces mirror image daughter cells. Euglenoids are not known to reproduce sexually.

Under adverse conditions encystment may occur and division take place within the cyst. The cysts are dispersed in similar ways to those described for *Amoeba*.

Paramecium: a complex protozoan

A good example of a protozoan showing an elaborate organization within the confines of the basic unicellular form is provided by *Paramecium*. Members of the genus *Paramecium* are very common in freshwater ponds, where there is abundant decaying organic matter. Of the several species known, *Paramecium caudatum* is the one most usually studied. Like *Amoeba proteus* it is just visible to the naked eye and its body consists of a single cell. However, *Paramecium* differs from *Amoeba* and *Euglena* in having a definite and constant shape, in outline resembling the sole of a slipper (hence the common name, 'slipper animalcule'). It swims through the water, rotating as it does so, with apparent great rapidity.

Paramecium is shaped like a short, partially flattened cigar with one end more pointed than the other. During movement the rounded end is usually directed forwards and may therefore be called anterior, although the animal can reverse its direction. Despite its definite shape, *Paramecium* is sufficiently flexible and elastic to squeeze itself through narrow openings.

The pellicle is a far more elaborate structure than that of *Amoeba* or *Euglena* and consists of a plasma membrane with several associated organelles. The pellicle itself has well marked depressions and ridges giving a lattice-like appearance to its surface. A cilium (or in some species of the genus two cilia) protrudes from the centre of each

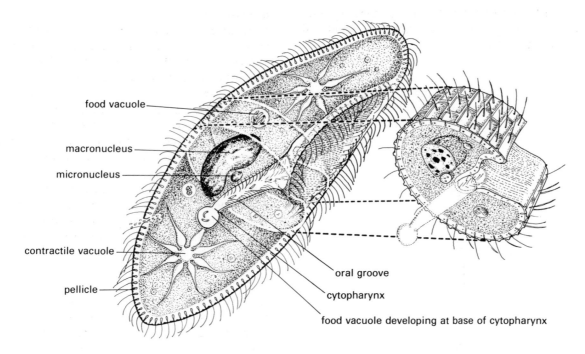

food vacuole

macronucleus

micronucleus

contractile vacuole

pellicle

oral groove

cytopharynx

food vacuole developing at base of cytopharynx

Structure of *Paramecium*

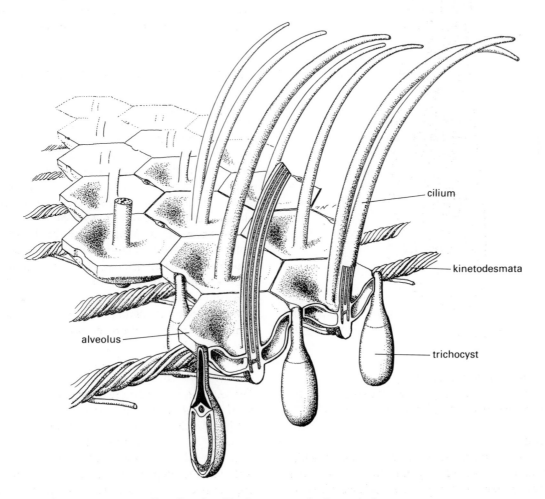

cilium

kinetodesmata

trichocyst

alveolus

Details of pellicular system in *Paramecium*

depression. The detailed arrangement of the pellicle in *Paramecium* is very complex and is best understood by reference to the drawing. The body cilia are arranged in longitudinal rows called kineties. Beneath the outer plasma membrane there are closely-packed, inflated vesicles (alveoli) with outer and inner membranes which, in effect, form the middle and inner membranes of the pellicle, and not only contribute to its stability but may also limit its permeability. The cilia emerge between the alveoli and are connected together at their bases by an elaborate infraciliary system lying beneath the alveoli. This consists of a small granule at the base of each cilium (the kinetosome) which is continuous with the axoneme, and a series of composite fibres (kinetodesmata) which run parallel to the kineties receiving contributory fibres from each kinetosome. Each kinetodesmos ends when it has passed four or five kinetosomes. A system of microtubular fibrils may underline the kinetodesmata. The function of this complex system is to provide a firm anchorage for the cilia.

Embedded in the ectoplasm, at right angles to the surface and beneath the alveoli, are a series of small, bottle-shaped structures called trichocysts. These can be stimulated to eject long, fine, striated threads surmounted by barbs. The thread is not evident in the undischarged trichocyst, and is probably formed during discharge. The function of these threads remains uncertain. It has been suggested that they serve to anchor *Paramecium* during feeding to prevent the ciliary beat from moving it from a favourable feeding area. In some ciliates the trichocysts are thought to have a defensive function, and in predaceous species, for example *Dileptus*, they may serve to attach and paralyse other small protozoans.

The cytoplasm of *Paramecium* is differentiated into ectoplasm and a very fluid endoplasm. Two nuclei are present which differ in size, appearance and function. The larger macronucleus is kidney-shaped with the smaller micronucleus lodged in a depression at one side. The macronucleus controls the ordinary activities of metabolism and growth while the micronucleus is concerned with reproduction. Thus if the micronucleus is removed the animal can survive, but not reproduce.

Locomotion

The animal moves through the water by beating its cilia, and may also use them to glide over a solid substrate. In slow motion a cilium may be seen to stiffen and bend over rapidly to lie nearly parallel with the body surface. It then becomes relatively limp and returns more slowly to its original position in an anti-clockwise motion. A cilium therefore moves in three different planes during a complete cycle. These positions have been recorded in scanning electron micrographs of freeze-dried *Paramecium*. The down stroke of the cilium, acting against the resistance of the water, provides the impetus. The recovery stroke offers less water resistance and may be compared to feathering an oar. The ciliary beat is synchronized with waves of movement progressing down the length of the body from anterior to posterior. This is known as a metachronal rhythm and gives a rippling effect rather like that of successive gusts of wind on a field of standing corn. The significance of the metachronal rhythm is that it prevents interference between cilia and increases their general effectiveness.

Paramecium swims through water at a speed of about 60 mm per hour. In absolute terms this may seem rather slow, but expressed as the number of times the length of its body is traversed in unit time (four times its body length/sec) it compares favourably with that of subsonic aircraft. The direction of the waves of movement is slightly oblique, taking a spiral course along the body and this causes *Paramecium* to move forwards in a spiral path about a straight line, at the same time continually rotating about its own axis. The direction of the ciliary beat can be reversed, and the animal then moves backwards.

The precise mechanism by which cilia beat is not yet fully understood. However the ultrastructure of both cilia and flagella must be closely related to their function in the same way as the design of the limbs of the higher animals is to theirs. Chemical analysis of both cilia and flagella has shown that protein predominates and it is reasonable to assume that the nine peripheral fibrils (page 17) are protein macromolecules. Movement was thought to result from contractile changes in these. However a more recent hypothesis assumes that the microtubules remain constant in length, and that they slide relative to each other. This sliding action is caused by ATP–powered changes in shape of the molecules forming the links between the microtubule pairs (cf. muscle contraction). The central fibrils with the radial strands restrict the effects of sliding and limit the region and amount of bending. The co-ordination of adjacent cilia results from the pulse of water that the ciliary action produces bringing the movement of adjacent cilia into phase (hydrodynamic linkage). In the protozoan

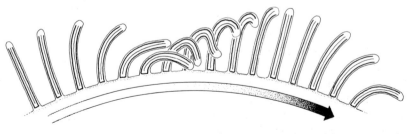

Action of a cilium

Opalina, the cilia have been observed to begin their beat with random timing though constant direction, adjacent cilia becoming synchronous within a few beats.

Nutrition

The free living ciliates are almost entirely holozoic. In *Paramecium*, unlike *Amoeba*, the 'mouth' or cytostome is a permanent structure, lying at the base of the oral groove and opening into a short canal, the cytopharynx, which extends into the endoplasm. The walls of the cytopharynx are strengthened by bundles of microtubules which surround it like the staves of a barrel. Food vacuoles are formed at the terminal end of the cytopharynx. The beat of the body cilia causes a current of water to move towards the animal, so that a cone-shaped volume of water with particles in suspension is drawn towards the oral groove. The cilia on the oral groove walls are arranged in longitudinal rows to form the undulating membranes, and these waft particles into the cytopharynx which is not ciliated. When sufficient particles have collected they are taken into the endoplasm in a drop of water and become enclosed as a food vacuole. This increases to an optimum size and then breaks free from the cytopharynx, a new one forming in its place. The food vacuoles circulate along a fairly definite course through the endoplasm during which the food passes through an acid and then an alkaline digestive process like that of *Amoeba*. This well defined course led early workers to the logical but mistaken belief that a digestive tract was present in *Paramecium*. Eventually undigested remains are discarded at a definite anal opening on one side of the animal near the posterior end. Digested food reserves are stored in the form of glycogen and fat droplets within the endoplasm.

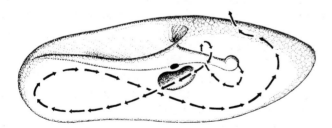

Path traced by food vacuoles in *Paramecium*

Although any suspended particles of a suitable size will be swept into the cytopharynx in the feeding current, *Paramecium* shows some selection of the particles which are actually engulfed. Non-nutritive particles may be rejected or, if ingested, be eliminated from the vacuole before the digestive sequence starts. Furthermore, certain kinds of food materials are selected in preference to others, as can be shown from the greater or lesser number of food vacuoles that are formed.

Gaseous exchange and excretion

As in *Amoeba*, gaseous exchange and the removal of nitrogenous waste takes place by diffusion through the pellicle. Urea and ammonia are the chief waste products.

Osmoregulation

Paramecium has both anterior and posterior contractile vacuoles in the innermost regions of the endoplasm, which empty through distinct pores in the pellicle; thus unlike the contractile vacuole of *Amoeba* they occupy a fixed position in the animal. The vacuoles themselves are complex, with about six radiating canals feeding the main vacuole. These canals are surrounded by a layer about one micron thick of fine, branching tubules which communicate at their ends with tubular components of the endoplasmic reticulum. The posterior vacuole discharges more frequently than the anterior one as the cytopharynx delivers extra water into the posterior region.

Sensitivity

Paramecium responds to similar kinds of stimuli as *Amoeba*, including certain dissolved chemicals, light and touch. Its range of behaviour in general is limited and predictable, but well adapted for its survival within a favourable environment. Like *Amoeba*, *Paramecium* has no special receptor organelles, although the anterior end appears to be more sensitive than other parts. However there seems to be no relation between the direction of movement and the position of the stimulus. A simple avoiding reaction in which the animal stops, moves backwards a short distance, turns slightly and moves forward again is the only response. The effect of this behavioural response is to cause the *Paramecium* to congregate in favourable surroundings. The avoiding response is also produced by collision with solid objects.

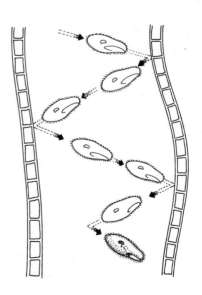

Avoiding reaction in *Paramecium*

Reproduction

The most usual method of reproduction in *Paramecium* is
binary fission. The micronucleus divides mitotically; this is
followed by an amitotic division of the macronucleus,
usually by constriction. Important organelles including the
cytopharynx and the contractile vacuole duplicate
themselves, and the animal becomes divided by constriction
in the transverse plane at right angles to the ciliary rows.
Each half draws apart and then assumes the form of a
normal *Paramecium*.

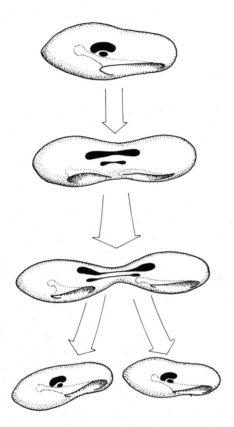

Fission in *Paramecium*

Under optimum conditions *Paramecium* multiplies
asexually very rapidly, dividing two or three times a day to
produce a series of descendants (a clone). After about 350
generations cytoplasmic abnormalities become increasingly
common, and the rate of division slows down and eventually
ceases.

In *Paramecium* one form of sexual reproduction is termed
conjugation. This involves two sexually compatible
members of the same species coming together as conjugants
apparently at random during swimming, and adhering
together in the oral region. Adhesion is initially brought
about by a sticky secretion of the cilia and is later followed
by cytoplasmic fusion. Attachment lasts for several hours,
and during this period reorganization and exchange of
nuclear material takes place. The details of the process vary
between species. In *Paramecium caudatum* the

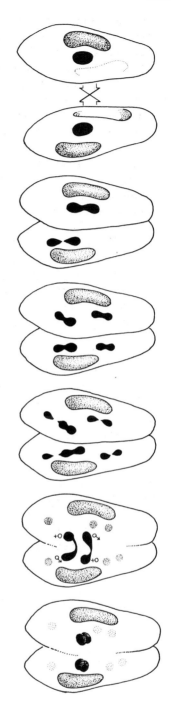

Conjugation in *Paramecium*

micronucleus separates from the macronucleus which then
disintegrates and is absorbed into the endoplasm. The
micronucleus of each individual divides meiotically to form
four haploid nuclei. In each conjugant all but one of the
nuclei degenerate and the survivor divides again to give two
identical micronuclei. Nuclei are then exchanged between
the conjugants, one nucleus from each passing across the
fused cytoplasm. The wandering nucleus then fuses with the
stationary nucleus, forming a nucleus resembling a diploid
zygote and known as a synkaryon. The two animals then

separate, each being called an exconjugant. After separation the synkaryon divides three times to produce eight daughter nuclei of which four become macronuclei and four micronuclei. Three micronuclei degenerate and disappear. The cytoplasm divides twice to give four new individuals, the single micronucleus dividing mitotically at each cytoplasmic division. Each individual receives one of the products of division of the micronucleus, and one macronucleus.

Sexual reproduction, which is absent in many protozoans, introduces a new aspect into their life history as two individuals are involved. Further, the production of a synkaryon foreshadows a reproductive process which is seen throughout the Metazoa and provides for an exchange of genetic characteristics. The wandering and stationary nuclei may be compared with the special reproductive structures (gametes) of the Metazoa, with the synkaryon corresponding to the product of fusion of male and female gametes, the zygote. However it should be emphasized that a zygote does not undergo reduction division after its formation.

A second sexual process is autogamy. The behaviour of the nuclei is essentially similar to that of conjugation, with degeneration of the macronucleus and division of the micronucleus. However the entire process takes place in one individual and there is therefore no exchange of nuclear material between partners.

Dispersal

Paramecium is widely distributed in freshwater ponds including those subject to periodic drying and this suggests that it is capable of forming cysts which are resistant to drought and may be easily dispersed. Further support to this view is given by the frequent appearance of *Paramecium* in hay infusions.

Monocystis and *Plasmodium*: parasitic Protozoa

The protozoans discussed so far are free-living active forms. These compete with each other, and with other organisms, for food and in addition have the problem of avoiding predators. The changing conditions of the habitat are a further problem: temperatures may rise or fall, water supplies dry up, food supplies increase or decrease. There are many protozoans which live an entirely different mode of life: within the body of another organism, where competition is reduced and many of the problems of an independent existence are absent. In some cases the presence of the intruder is not harmful and may even be advantageous, in which case the alliance is commensalism. However the most common relationship between animal and animal is parasitism in which the intruder lives at the expense of the host and is harmful to it to a greater or lesser degree. A parasitic mode of life is reflected in both the structure and the life history of the animals which adopt it

and makes an interesting contrast to that of the free-living forms already described. The majority of parasitic protozoans are found in the subphylum Sporozoa and it is from this group that the two examples given are drawn.

All sporozoans have a complex life cycle involving both sexual and asexual generations, and many of them are responsible for human diseases. *Monocystis* illustrates the sporozoan life cycle in a relatively simple form and is readily accessible for study, occurring commonly in the seminal vesicles (part of the male reproductive system) of the earthworm. *Plasmodium*, the organism responsible for the occurrence of malaria in man, is an example of an animal which illustrates, in a complex way, the parasitic mode of life.

Monocystis

A description of the life history of *Monocystis* (figure overleaf) may begin at the feeding stage when the parasite is a minute organism (the trophozoite) found in the central cytoplasm of developing spermatozoa, the sperm morula. At first the trophozoite lives on the cytoplasm, although it does not ingest its food, having no structure for so doing. It is thought to exude digestive enzymes which act upon the cytoplasm in its immediate vicinity, the soluble products of digestion being absorbed through the pellicle. Thus unlike the free-living *Amoeba* and *Paramecium* it does not form food vacuoles but digests its food outside its body. This method of digestion is common amongst parasitic animals. Eventually the trophozoite grows too large for the sperm morula (the host cell) and then lives in the seminal fluid from which it absorbs nutriment in the same way as do the developing sperms. Reserve food material is stored in the endoplasm in the form of granules of carbohydrate.

The mature trophozoite is spindle-shaped with a thick, smooth pellicle, cytoplasm, and a nucleus. Since it does not need to search for food, elaborate locomotory mechanisms would be no advantage and it is capable only of wriggling, probably by means of the large number of microtubular fibrils which reinforce the pellicle. Gaseous exchange and excretion are carried out as in other protozoa. Oxygen is brought to the seminal fluid via the circulatory system of the earthworm, and carbon dioxide and excretory products are removed in the same way.

Full grown trophozoites come together in pairs, and secrete a double-walled cyst. At this stage the trophozoites are known as gamonts, and the cyst as a gametocyst. Fusion does not occur and instead each gamont undergoes subdivision to form large numbers of nucleolated amoeboid gametes. The gamont nuclei each divide into two, four, and so on until as many as sixty-four small nuclei are produced. These migrate to the edge of the gamont where each becomes surrounded by a little cytoplasm so that the gamont becomes somewhat blackberry-like in appearance. Finally the gametes become separated from the rest of the cytoplasm. The two gamonts in the cyst usually produce the

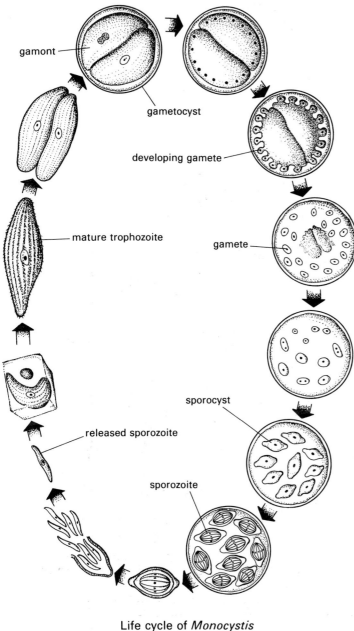

gamont

gametocyst

developing gamete

gamete

mature trophozoite

sporocyst

released sporozoite

sporozoite

Life cycle of *Monocystis*

Two important aspects of parasitism emerge from this description of *Monocystis*. Firstly, the adult parasite has a very simple structure in comparison with a free-living protozoan such as *Paramecium*; and secondly, reproduction yields a vast number of potential individuals. Both characteristics are related to the parasitic mode of life. Food is freely and easily available with competition virtually absent, and surroundings more uniform. This ideal situation does have one drawback: the earthworm does not have an unlimited number of seminal vesicles and the progeny must be transferred to another host worm. This necessitates the parasite entering the outside world where there are dangers of predation. Furthermore, the probability of an individual being ingested by a host worm is extremely small. Large losses are bound to occur in this period but the production of a large number of offspring increases the probability that some will successfully infect new hosts.

Plasmodium

More than fifty species of *Plasmodium* are known, all of which attack vertebrates. Four species are parasitic on man but of these only *Plasmodium vivax*, one of the malarial parasites, need be described in detail. The life history of the malarial parasite is complex and involves two hosts, man and a mosquito of the genus *Anopheles*. There are three main divisions: a sexual phase beginning in man and continuing in the mosquito, an asexual phase (sporogony) and a phase of growth with further asexual multiplication in the red blood cells and liver cells of man. Infection of man occurs following a bite by an infected mosquito with the injection of sickle-shaped sporozoites into the blood with the mosquito's saliva. An incubation period of approximately ten days follows during which no symptoms of malaria are shown. The sporozoites disappear completely from the blood, having entered parenchyma cells in the liver within half an hour of infection. After between five and fifteen days each sporozoite grows rapidly to form a schizont which then divides (schizogony) to produce numerous merozoites. A total of between ten and thirty thousand merozoites may be produced during the course of this first phase of reproduction in the liver, the pre-erythrocytic phase. The merozoites normally pass into the general blood circulation although some may reinfect liver cells and go through one or more stages of multiplication (exo-erythrocytic phase). The persistence of the parasite in the liver means that a reservoir of infection exists that can prolong the disease in a latent form indefinitely, relapses occurring periodically when resistance is low. Since they are within the host's liver cells, the parasites are virtually unharmed by drugs such as quinine which can operate successfully only when the parasite is in the blood cells.

The merozoites released into the general bloodstream attack and enter the red cells within which each grows to form a trophozoite. The trophozoites feed on the cytoplasm of the cell, digestion being external as in *Monocystis*. A

same number of gametes and these fuse together in pairs as zygotes. The fusion is complete, involving both cytoplasm and nucleus, and the zygote secretes a tough resistant covering to become a sporocyst. Within the sporocyst the diploid zygote divides three times producing eight fusiform (cigar-shaped) nucleolated haploid structures, the sporozoites. After the death and decay of an infected earthworm, the sporocysts are released. Their resistant covering enables them to survive in the soil for a considerable time. Further development depends on the sporocysts being eaten by another earthworm. The outer covering is then digested in the anterior part of the gut and sporozoites released. They pass through the intestinal wall and travel via the body cavity to the seminal vesicles, where they enter the sperm morulae.

SALIVARY GLANDS

BLOOD

LIVER

MIDGUT

Life cycle of the malarial parasite, *Plasmodium*

vacuole appears in the middle of the red cell cytoplasm which is pushed to one side, and the red cell assumes a characteristic 'signet ring' appearance. Growth continues and the trophozoite may thrust out pseudopodia into the cytoplasm. After about 36 hours the growth rate slows and the mature trophozoite multiplies by schizogony (fission). In this condition, as a schizont, the nucleus undergoes repeated divisions, portions of the cytoplasm aggregating round the daughter nuclei, to form between 6 and 24 merozoites. After about 48 hours the red cell bursts open and the merozoites are freed into the plasma together with toxins. Here they attack fresh red cells and the cycle is then repeated.

Eventually simple asexual multiplication is replaced by the onset of a sexual phase. Some of the merozoites develop as rounded compact gamonts. Two types of gamont may be distinguished: macrogamonts, each with a small nucleus and dense, food-laden cytoplasm, and microgamonts, each with a large nucleus and a little clear cytoplasm. These remain dormant in the blood corpuscle and ultimately degenerate and die, unless they are taken into a mosquito.

Before ovulating, the female *Anopheles* must feed on blood, obtaining it by piercing the skin, and then sucking, using specially adapted mouth parts. If the source of blood is infected with malaria and contains gamonts the parasite will develop in the mosquito's stomach. Any other stages of the parasite will be digested but the gamonts are resistant to digestive fluid. The macrogamont becomes a spherical macrogamete without division. The microgamont undergoes three nuclear divisions to give eight daughter nuclei and develops eight flagellar projections. One nucleus enters each projection forming a motile microgamete. Fertilization of the macrogamonts takes place in the stomach of the mosquito and a spherical zygote is formed. This becomes elongated and active, and burrows into the stomach wall to lie near the outer surface between the epithelium and subepithelial tissues. It becomes enclosed in a cyst derived both from the zygote and from the stomach wall. Growth of the zygote is followed by sporogony, an asexual process resulting in the production of large numbers of sporozoites. These break out of the cyst and are discharged into the mosquito's body cavity. Most of the sporozoites enter the salivary glands and eventually pass into the salivary duct with the saliva. If the mosquito then bites a human the sporozoites will pass into the blood stream as some saliva is always injected into the wound.

Malaria is an acute intermittent fever, the fever being caused by toxins released into the blood with the trophozoites. Four different forms are known. *Plasmodium vivax* and *Plasmodium ovale* both cause a tertian fever with 48 hours between climaxes, this being the period required for the trophozoite to grow and undergo schizogony. Quartan fever (72-hour intervals) is caused by *P. malariae*. These types are not usually fatal although the recurrent fevers are debilitating. *P. falciparum* is the most pathogenic species, probably causing more human deaths in the tropics than any other organism. The fever is irregularly continuous. No harmful effects have been observed in *Anopheles* which appears to have tolerance to the parasite.

Plasmodium, like *Monocystis*, has many of the features associated with a parasitic mode of life, such as its extremely simple structure, and the very active phases of multiplication. An additional feature in *Plasmodium* is the introduction of a second host, which, because it feeds on man, serves as a means of active transfer from one human being to another. Such an intermediary or vector has proved a successful feature in the life cycle of *Plasmodium* as may be demonstrated from the prevalence of malaria in tropical areas where it has not been controlled.

The classification of the Protozoa

The phylum Protozoa is usually divided into four subphyla: Ciliophora, Sporozoa, Cnidospora, and Sarcomastigophorea. Of these the first three are clearly animal groups, but the last contains organisms with both animal and plant-like characteristics (for example *Euglena*), as well as undoubted animal species such as *Amoeba*. The Sporozoa and Cnidospora (a relatively minor group, formerly regarded as Sporozoans and omitted from this discussion) are entirely parasitic and at some stage in their life cycle form sporozoites. However, the occurrence of parasitism in the Protozoa is not restricted to this group, the other main groups each having some parasitic members.

The subphylum Sarcomastigophorea includes all flagellate and amoeboid forms and is generally regarded as the most primitive group. Within this group the superclass Mastigophora includes all organisms which possess flagellae as adult locomotory organs and divides into the Phytomastigophorea, which are plant-like with chromoplasts, and the Zoomastigophorea which are animal-like, lacking chromoplasts. The Sarcodinids (superclass Sarcodina) all possess pseudopodia of one type or another in the adult form.

The subphylum Ciliophora is the largest and most homogeneous of the protozoan groups and is characterized by cilia at some stage in the life cycle.

The concept of the phylum Protozoa presents a problem in that the unicellular level of organization is the only characteristic by which the phylum can be described. Virtually all motile unicellular organisms have been included in this huge phylum with little regard to their evolutionary relationships. Because of this the members of the Protozoa show much more diversity than is usual between the members of a phylum. Indeed there is a good case for regarding the Protozoa as representing not one, but several phyla, at the same level of organization.

Protozoan diversity

It is impossible in the space available to give more than a limited idea of the diversity achieved in both mode of life and morphology by the phylum Protozoa. About a quarter of the known species are parasitic. All vertebrates and most of the invertebrates are susceptible to infection by Sporozoa. As a group these are characterized by elaborate life cycles of the type described for *Monocystis* and *Plasmodium*. Even protozoans themselves may be parasitized by other protozoa, as well as by bacteria or fungi. Of the free-living protozoans, two-thirds are marine. About 250 species live in the water film which surrounds soil particles.

As we have already seen, within the protozoa there are a variety of types of body structure, which reflect the varying modes of life adopted by members of the group. Whilst the 'typical' protozoan has a body unit with a single nucleus and associated cytoplasm, there are many protozoans with two or more nuclei. An extreme multinucleate form is seen in *Opalina*. It is a commensal in the gut of amphibians and was one of the protozoans discovered by Van Leeuwenhoek in frog faeces. Superficially it looks like a ciliate with many identical nuclei. It is laterally flattened and has cilia arranged in rows extending backwards from its right anterior margin, so that it was originally regarded as a ciliate. However it differs from 'normal' ciliates in other aspects of its morphology as well as in its mode of division and life cycle, and is now regarded as a sarcomastigophorean.

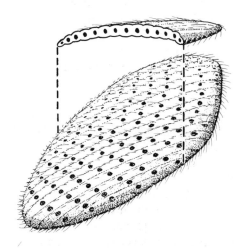

Opalina : a multinucleate sarcomastigophorean

Many protozoans live in groups or colonies. The colonial organization is remarkably sophisticated in forms such as *Volvox*, a hollow spherical colonial flagellate composed of up to 10,000 individuals (zooids) each very similar to *Chlamydomonas* in structure. The zooids are joined together by fine cytoplasmic strands. The flagella of the zooids beat synchronously and the colony always swims with the same end directed forwards. This is therefore called the anterior pole. The zooids at the anterior pole have a large, anteriorly placed eyespot, but the eyespots of the posterior members of the colony are small and variable in position.

Volvox is also interesting in that it shows a certain degree of cellular specialization of the type usual in multicellular animals. A limited number of the posterior zooids are specialized for reproduction: a maximum of 8 for asexual reproduction, 5 or more for the production of male gametes and between 1 and 15 for the production of female gametes. The female gamete is large, immobile and egg-like, and the male is small with two flagella. Less elaborate colonies are common amongst the flagellates, including *Pandorina*, a globular colony of up to 16 individuals, and *Gonium* which has a colony in the form of a flat plate.

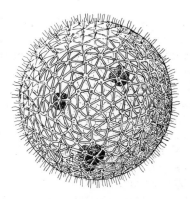

Volvox : a colonial flagellate

Non-living outer coverings are common in the flagellates. *Volvox* is amongst the many plant-like forms which have an outer wall of cellulose. Many of the dinoflagellates have a cellulose outer wall in the form of plates cemented together. This armour, or theca, may be sculptured and in some forms may form long projections. *Ceratium*, an abundant constituent of marine plankton, has a single curved anterior extension and two curved posterior extensions, which serve as flotation devices. Other flotation devices in the protozoa may take the form of drops of oil secreted in the cytoplasm or a highly vacuolated endoplasm.

Whilst the phytoflagellates are typically autotrophic, there are many exceptions. *Peranema*, a colourless freshwater euglenoid, feeds on a wide variety of living organisms including *Euglena* which it ingests through a greatly extensible cytostome. *Chrysamoeba* is a flagellate which also feeds like *Amoeba*. Many dinoflagellates show a tendency towards heterotrophic nutrition. *Ceratium* feeds both autotrophically and heterotrophically forming a net of pseudopodia for food capture, the pseudopodia being extruded through holes in the theca. Some dinoflagellates are ectoparasites, living on the gills of fish, marine annelids or Crustacea. The euglenoid *Phacus* is a gut parasite of tadpoles.

Parasitism is common amongst the zooflagellates. The most notable examples are *Leishmania* and *Trypanosoma*. *Trypanosoma* causes African sleeping sickness (trypanosomiasis) and has an elaborate life cycle involving two hosts, the Tsetse fly (as vector) and a mammal.

Trichonympha is a symbiont in the gut of termites which superficially resembles a ciliate, with hundreds of flagella attached to ribbon-shaped kinetosomes. Some flagellate parasites infect other protozoans.

The choanoflagellates are examples of protozoan filter feeders. They are unusual in that they have a collar made up of a mesh of parallel contractile fibrils surrounding the base of the flagellum. Particles are attracted to the collar by the flagellar beat and are trapped on the outer surface of the mesh and move downwards into a food vacuole at the base of the collar. Like most filter feeders throughout the animal kingdom, the majority of choanoflagellates are sessile. They are generally colonial, and often, as in *Codosiga*, the colonies are attached to the substrate by a stalk. However *Proterospongia* consists of a floating gelatinous mass in which the individual collar cells are embedded.

There are four main types of structure within the Sarcodina, of which *Amoeba* has already been described. Many amoebae are enclosed in non-living shells or tests, which are generally vase-shaped, a large opening at one end providing for the extrusion of the pseudopodia. The shell may either be composed of mineral particles, which the animal collects and cements together, or of silicious or proteinaceous substances. For example, *Arcella*, a common freshwater amoeboid, secretes a flattened dome-shaped test of keratin.

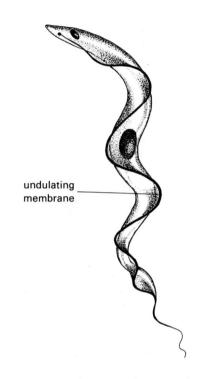

Trypanosoma : the causative organism of African sleeping sickness

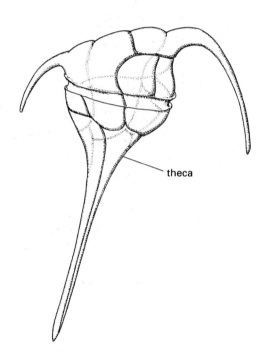

Ceratium : a dinoflagellate

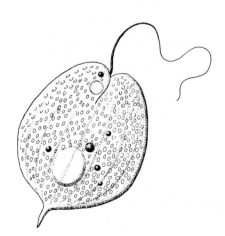

Phacus : a euglenoid gut parasite of tadpoles

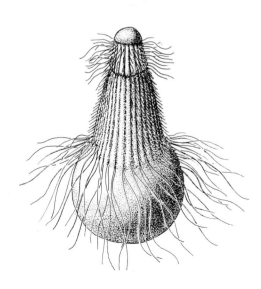

Trichonympha : a symbiotic zooflagellate

The Foraminifera secrete highly elaborate shells made mainly of calcium carbonate, which differ from the amoeboid tests, in that the majority are many chambered. Some of the largest protozoans known are extinct Foraminifera with a shell size of 50 to 60 mm across. Foraminiferan shells are of geological significance in that practically the whole of the ocean floor at 2,500 m to 4,800 m depth is covered by a deposit composed mainly of the shells of dead foraminiferans, especially the genus *Globigerina*, from which the name of the deposit, *Globigerina* ooze is derived.

Two sarcodinid groups, the Heliozoa and the Radiolaria, have elongated radiating pseudopodia. Heliozoa (sun animalcules) such as *Actinosphaerium* have fine needle-like axopodia, each supported by a central axial rod composed of a bundle of protoplasmic fibres.

Like the Heliozoa, the Radiolaria (which are entirely marine) have long radiating pseudopodia. A skeleton, usually silicious, is present, and from this radiolarian ooze, a common ocean bed deposit below 4,500 m is derived. The skeleton is prolonged into radiating spines with elaborate patterns and this makes radiolarians amongst the most beautiful protozoa. The radiating pseudopodia in Foraminifera, Radiolaria, and Heliozoa act as nets for prey capture. Organisms which come into contact with the net stick to it.

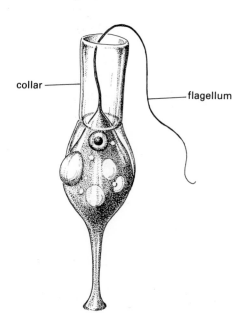

A choanoflagellate : *Codosiga*

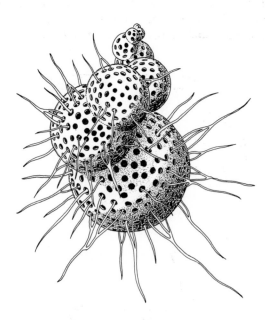

Globigerina : a foraminiferan

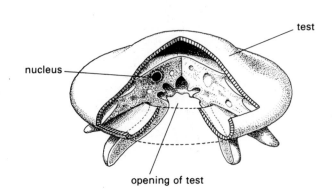

Arcella : a testate amoeboid

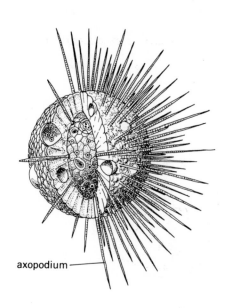

Actinosphaerium : a heliozoan

Parasitic sarcodinids are mainly intestinal inhabitants of man and other vertebrates, and also of some invertebrates. The majority are assigned to the genus *Entamoeba*, represented in man by *E. histolytica*, the causative organism of amoebic dysentery, and *E. coli* in the colon. *E. gingivalis* is a mouth parasite. Amoeboid parasites are transmitted directly by means of cysts which are passed out of the host's intestines with the faeces. Amoeboid parasites in animals other than man include *Entamoeba ranarum* (of frogs), *Hydramoeba hydroxena* (ectoparasite of the cnidarian *Hydra*) and *Entamoeba blattae* (of cockroaches).

Ciliophora is the largest and most homogeneous group of protozoans, with some 6,000 species. They are generally the most active members of the phylum, either swimming or using modified cilia (cirri) for crawling or walking over surfaces. Free-swimming forms tend to be spherical whilst the creeping forms tend to be flattened. Whilst the majority of ciliates are free-living and solitary, representatives of the group are sessile, colonial, and both ecto-commensals and endo-commensals are known. Typically the ciliates are holozoic, feeding on either small particles, or prey. *Paramecium* is a relatively simple particulate feeder. More elaborate adaptations are seen in forms such as *Stentor*, a sessile ciliate with a highly developed region of fused cilia, the buccal membranelles, with which the water current is created.

Didinium is one of the most highly studied of the ciliate predators. It feeds chiefly on *Paramecium*, and is able to engulf prey larger than itself. Its anterior end is prolonged as a proboscis at the end of which is a mouth by which *Didinium* attaches to its prey. Another group of raptorial ciliates, the Suctoria, for example *Acineta*, are unusual in that cilia are only present in the immature stages. The adult

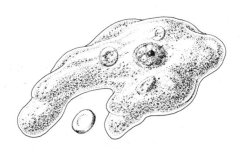

Entamoeba : a sarcodinid intestinal parasite

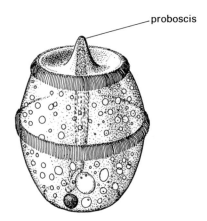

proboscis

Didinium

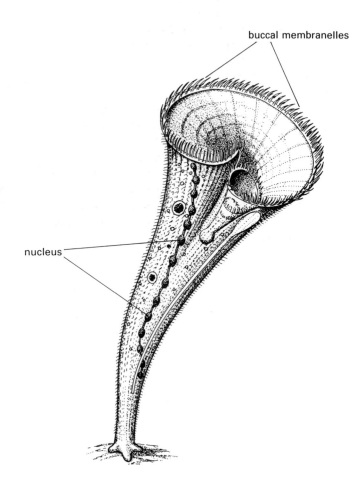

buccal membranelles

nucleus

Stentor : a sessile ciliate

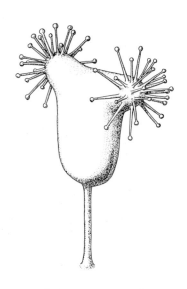

Acineta : a suctorian

has a globular body form, bearing tentacles, and consisting of a rigid tube surrounded by a contractile sheath. The prey sticks to the tentacles and is sucked into the body through a central tube.

There are very few parasitic ciliates. One example, *Foettingeria*, lives in the gastromuscular cavity of sea anemones. Young *Foettingeria* encyst on small crustaceans and are transferred to new sea anemones when the crustaceans are eaten. Another example is seen in the suctorian *Sphaerophyra*, which lives inside the endoplasm of *Stentor*. Ecto- and endo- commensals are more common. *Kerona* is a crawling ciliate which lives on the surface of *Hydra*. *Balantidium coli* is endocommensal in pig intestines and transferred by means of cysts in the pig's faeces. *Diplodinium ecaudatum* is a highly elaborate ciliate which lives as a commensal in the rumen of cattle. In many ways it may be regarded as showing the greatest complexity achieved by a single cell.

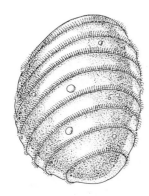

Foettengeria : a parasitic ciliate

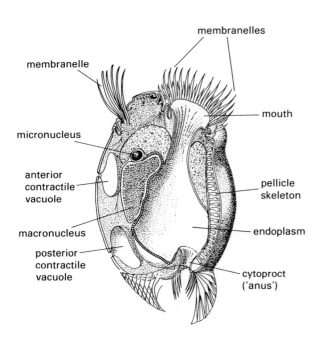

membranelles

membranelle

micronucleus

anterior
contractile
vacuole

macronucleus

posterior
contractile
vacuole

mouth

pellicle
skeleton

endoplasm

cytoproct
('anus')

Diplodinium ecaudatum : a commensal ciliate

Synopsis of the Protozoa

This includes only those minor groups included in the discussion of protozoan diversity.

Subphylum Sarcomastigophorea
    Superclass Mastigophora
        Class Phytomastigophorea: various orders
          including Dinoflagellida (*Ceratium*), Euglenida
          (*Euglena, Peranema, Phacus*), Volvocida
          (*Volvox, Chlamydomonas*)
        Class Zoomastigophorea: orders
          Choanoflagellida (*Codosiga, Proterospongia*),
          Kinetoplastida (*Trypanosoma, Leishmania*)
    Superclass Opalinata
        Opalina
    Superclass Sarcodina
        Class Rhizopodea
        Subclasses include Lobosa: orders Amoebida
        (*Amoeba, Entamoeba*, Arcellinida, *Arcella*);
        Granuloreticulosia: order Foraminiferida
        (*Globigerina*) Class Actinopodea.
        Subclasses include Radiolaria (*Sphaerozoum*),
        Acantharia (*Acanthometra*), Heliozoa
        (*Actinosphaerium*), Proteomyxidia (*Vampyrella*).
Subphylum Sporozoa
    Class Telosporea
        Subclass Gregarinia (*Monocystis*).
        Subclass Coccidia (*Plasmodium*).
Subphylum Cnidospora.
Subphylum Ciliophora
    Class Ciliatea
        Subclass Holotrichia includes orders
        Gymnostomatida (*Dileptus, Didinium*)
        Trichostomatida (*Balantidium*),
        Hymenostomatida (*Paramecium*).
        Subclass Suctoria, Order Suctorida (*Acineta*)
        Subclass Spirotrichia includes orders
        Heterotrichida (*Stentor, Spirostomum*),
        Entodiniomorphida (*Diplodinium, Entodinium*),
        Hypotrichida (*Stylonychia*).

Phylum Porifera

The protozoan ciliates demonstrate the greatest complexity that has evolved in a unicellular organism. Further elaboration demands an increase in size and for this a multicellular structure has proved necessary. Subdividing the body mass into a number of smaller parts permits an increase in size for two main reasons. Firstly, the mere presence of cell walls provides some measure of support to the protoplasm, which when unsupported has a fluid consistency similar to that of egg white. Secondly, the size of an individual unit of protoplasm is limited by the effective diffusion path of oxygen, carbon dioxide, and nitrogenous waste. Above a certain size, about equal to that of the largest protozoans, the diffusion distance becomes too great and the cell would be unable to support life. The majority of animals are multicellular and are known collectively as the Metazoa.

The simplest multicellular animals are the sponges, the phylum Porifera. The traditional bath sponge (which should not be confused with modern synthetic sponges!) comes from a member of this phylum, though by the time it reaches the bathroom the organism bears little resemblance to its natural state. In fact the bath sponge represents the supporting structure of the living organism. In life most sponges look rather unstructured, and appear in a variety of colours: green, yellow, red, purple, and orange being common. Because of their apparent lack of movement and their 'un-animal-like' coloration, sponges were regarded as plants by ancient naturalists such as Aristotle and Pliny. The animal nature of the Porifera was not established until 1867 and even then they were regarded merely as aggregations of protozoan flagellates. Not until 1875 did the work of T. H. Huxley establish them as truly multicellular forms. However, because of certain unique features of their organisation they are usually regarded as a separate sidebranch from the Metazoa, the Parazoa.

Sponges have neither the true tissues nor organs characteristic of the metazoans and their cells retain a considerable degree of independence. However, unlike the Protozoa the various kinds of cells do not act entirely as individuals but show a definite social organisation with individual cells specialised for the functions of feeding, support or reproduction. This specialisation of function increases the efficiency of the organism as a whole, though at the expense of the ability of individual cells to lead an independent life.

Sponges lead a sedentary existence submerged in freshwater, shallow seas, or the ocean depths. A fully grown sponge varies according to the species from about 1 cm in height to a large spreading mass, 3–4 m tall. Many sponges are colonial and show plant-like indefinite growth.

The best introduction to sponge morphology is provided by a primitive member of the group such as *Leucosolenia*. *Leucosolenia complicata* occurs in groups of tubular individuals attached to objects such as seaweeds and shells, in shallow water in areas like the English Channel and the North Atlantic coast. Each individual is vase-shaped with a central cavity (the paragaster or spongocoele) and a large opening (the osculum) at the opposite end of the attachment (stolon).

The sponge wall consists of three layers: an outer pinacoderm; a supporting gelatinous protein matrix, the mesohyal; and an inner feeding layer of choanocytes, which are closely similar to the choanoflagellates. The pinacocytes are highly contractile, and thus the animal is able to increase or decrease in size. Further contractability is provided in some species by the collenocytes, which have long cytoplasmic strands extending across the spaces through

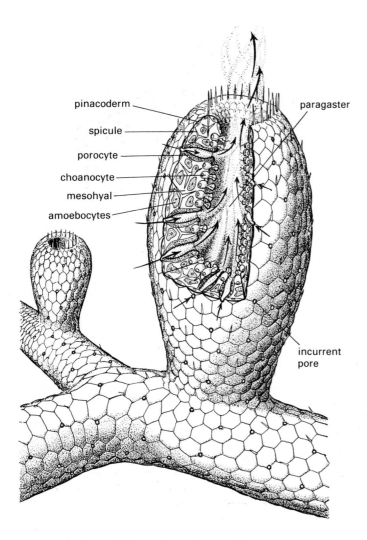

pinacoderm
spicule
porocyte
choanocyte
mesohyal
amoebocytes
paragaster
incurrent pore

A partially sectioned asconoid sponge

which water passes and thus may be used to reduce the size of those spaces.

(The mesohyal contains skeletal elements and wandering amoeba-like cells (amoebocytes). In *Leucosolenia* the skeletal elements are needle-like or three-branched spicules of calcium carbonate, but sponge skeletons may also be made of silica, or organic fibres of spongin, a fibrous protein related to keratin and collagen. The skeleton is developed by specialized amoebocytes. The sponge wall is pierced by microscopic pores (ostia), each surrounded by a single tubular cell, the porocyte. Porocytes are modified pinacocytes. The name of the phylum, Porifera, is a reference to these pores which provide for an inflow of water to the paragaster. This water leaves through the osculum. The osculum is so positioned that the excurrent is as widely separated from the incurrent as possible.)

All sponges are based on this general asconoid plan as outlined above for *Leucosolenia*. However in its simple form this design imposes an upper size limit and more advanced syconoid and leuconoid sponges show various elaborations of the water current system, achieved through folding of the body wall. The folds in the syconoid sponges

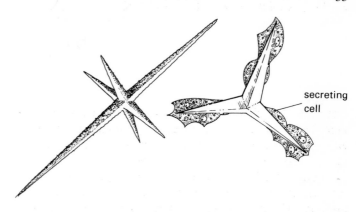

secreting cell

Examples of variation in sponge spicules

(for example *Grantia compressa*, a small intertidal species commonly found under rocky overhangs on the Atlantic coast) form a number of parallel alternating canals. In the leuconoid sponges extra folds of the pinacoderm create spaces (incurrent canals) which divide to form multiple, flagellated chambers lined with choanocytes. Short canals from the chambers join to form excurrent canals which

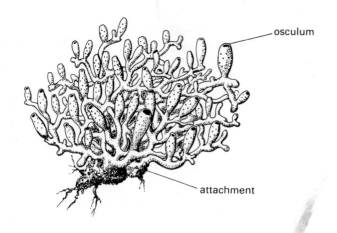

osculum

attachment

Leucosolenia complicata colony

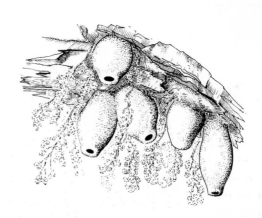

Grantia compressa

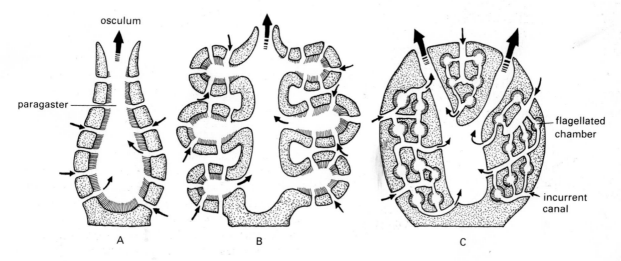

osculum

paragaster

flagellated chamber

incurrent canal

A   B   C

Sponge morphology : 'A' asconoid type, 'B' syconoid type, 'C' leuconoid type

ultimately unite at the osculum. In this case the folding of the body, giving rise to the excurrent canals, has virtually obliterated the paragaster. The majority of sponges have the more complex leuconoid structure, thus demonstrating its value as an adaptation. Porocytes are generally absent in the most complex leuconoid forms.

Nutrition

Feeding in the sponges is holozoic, the chief food source being very fine particulate matter and planktonic organisms. They are sedentary animals which obtain their food by filtering the waters which surround them. The feeding current is created by the continuously beating flagella of the choanocytes, and water enters the ostia or incurrent canals, leaving through the osculum. While very little pressure is created, the total volume of water which flows through a sponge may be considerable. For example, it has been estimated that a sponge with a volume of about 8 cm^3 can filter as much as 23–28 litres of water each day.

The flagellum beats spirally from base to tip, causing a stream of water to flow away from the tip, bringing further water towards the base of the choanocyte. The base of the flagellum is surrounded by a collar which the electron microscope reveals to be a fibrillar mesh, and which filters out particles as they arrive at the cell. Food particles are ingested by the choanocytes and digested intracellularly, as in free-living protozoans. The amoebocytes act as storage centres for food reserves in the form of lipoprotein or glycogen, and wander through the matrix distributing food to other regions of the sponge. Undigested food material is discharged into the water current.

Gaseous exchange

The water flow produced by the choanocytes also serves as a respiratory current. Gaseous exchange occurs by simple diffusion between the water and the sponge cells. Although sponges generally need a high supply of oxygen, it appears that some species can survive periods of oxygen deprivation by depressing their metabolic rate.

Excretion and osmoregulation

The chief form of nitrogenous waste in sponges is ammonia, a common excretory product in aquatic animals. It leaves the body in the water currents. Contractile vacuoles have been recorded in the choanocytes and amoebocytes of fresh water sponges but have not been recorded in any marine sponge. These function in the same way as protozoan contractile vacuoles.

Sensitivity

The cells of sponges are sluggish in their reactions, and even severe injury produces little more than concentration or withdrawal of the protoplasm in a region extending a few millimetres from the damage. The cells which surround the osculum are more sensitive than others and will contract when stimulated, thus closing the opening. Because, however, the sponges lack any structure which could be regarded as a nervous system, reactions are generally local and unco-ordinated. The amoebocytes respond to stimuli rather like individual amoebae.

Since sponges have no obvious protective devices and no ability to escape they might be regarded as particularly vulnerable to predators. In fact they are avoided by most potential predators such as fish, since their sharp spicules will penetrate any soft tissue exposed to them and they apparently taste and smell unpleasant. Indeed many small animals, for example crabs, take advantage of this to protect themselves. For example, some small marine sponges are always found in association with hermit crabs. This is a commensal relationship which benefits both, the hermit crab gaining protection and the sponge gaining mobility.

Reproduction

The Porifera show both sexual and asexual reproduction. In many sponges the capacity for asexual reproduction is considerable, new individuals being produced from buds which may develop in any region of the body or from fragments of the parent sponge. Another common method of asexual reproduction involves the formation of gemmules. These structures develop as aggregations of amoebocytes with large food reserves (archaeocytes) surrounded by other amoebocytes which differentiate into a layer of columnar cells. In freshwater forms they secrete a hard, protective membrane, supported by spicules formed by amoebocytes from the parent sponge. Gemmules may be very resistant and able to survive through periods of adverse conditions. When conditions improve the archaeocytes emerge through a pore (micropyle) in the gemmule wall, and new sponges are formed. Such gemmules are normally formed in the autumn. In marine sponges gemmules begin in the same way (as a mass of archaeocytes) but the outer layer is of spongin reinforced by spicules, or may be absent.

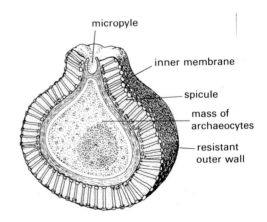

Section through a gemmule of a freshwater sponge

Sponges have remarkable powers of regeneration. If a sponge is separated into single cells by squeezing it through a piece of silk, the cells will migrate together to form a cell mass which will then grow into a new sponge by repeated cell divisions. The amoebocytes, which are the least specialised of the cells and truly 'Jacks of all trades' are chiefly responsible for this. Even when the cells of two different species are separated and mixed together they will eventually reform as two sponges. Thus new individuals can redevelop from small fragments of the body; for example fragments of *Leucosolenia complicata* of approximately 1 mm in diameter can develop into whole sponges. This ability to regenerate is exploited in the commercial propagation of bath sponges.

Sexual reproduction occurs in all sponges. Egg cells are formed from the amoebocytes which accumulate food materials in their cytoplasm and enlarge. The origin of sperm is less clear and the process may vary between species. Both amoebocytes and choanocytes have been recorded as involved in sperm production. The sperm are formed in spermatocysts consisting of spermatogonia surrounded by a capsule of flattened cells. They are discharged into the canals and transferred by the water current to other sponges. Although most sponges are hermaphrodite, sperm and ova mature at different times and this ensures that self-fertilization does not occur. On arriving in the second sponge, each sperm is taken into a choanocyte or an amoebocyte and loses its tail. The engulfing cell, known as a nurse cell, migrates to an egg cell nearby in the mesohyal, and makes contact with it. The sperm is transferred and fertilization takes place. Development begins immediately after fertilization and after a short period the young sponge emerges from the osculum as a flagellated free-swimming larva (the amphiblastula) which soon settles and develops to an adult sponge. Clearly this motile stage is particularly valuable for dispersal in an animal in which the adult is sedentary.

Classification

The classification of sponges is based on the structure of the skeleton. Four classes are recognized of which one is limited to a very few species.

Members of the class Calcarea are generally small in size and have spicules of calcium carbonate. *Leucosolenia* and *Grantia* are both examples of calcareous sponges.

Hexactinellida have skeletons based on six-pointed spicules of silica and hence are commonly called the glass sponges. The spicules may be fused to form elaborate latticed skeletons which are generally symmetrical flask- or vase-shaped structures. These remain intact even if the soft parts rot away and are very beautiful. One hexactinellid, *Euplectella* (Venus's flower basket) sometimes shows an interesting relationship with a species of shrimp. Shrimps enter the sponge as juveniles and after growth are unable to

leave through the osculum which is covered by a sieve plate. They thus spend their entire adult life imprisoned within the sponge feeding on plankton in its feeding current. Such sponges were used in Japan as a symbol of marital fidelity.

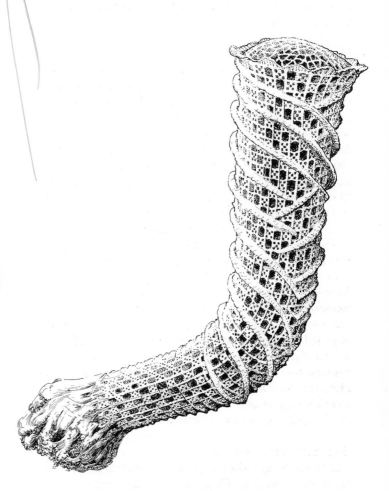

Skeleton of *Euplectella*

The class Demospongiae has a skeleton composed of spongin fibres or simple silicious spicules, or both. The bath sponge, *Spongia*, with a skeleton composed only of spongin fibres, is a member of this group.

Finally the class Sclerospongiae has been proposed for a few species of sponge which, whilst having an internal skeleton similar to that of the Demospongiae, have an outer casing of calcium carbonate.

Synopsis of the Porifera

Class Calcarea (*Leucosolenia, Grantia*)
Class Hexactinellida (*Euplectella*)
Class Demospongiae (*Spongia*)
Class Sclerospongiae

Radiata

Two phyla, Cnidaria and Ctenophora, are often considered together as the radiate phyla or Radiata. This classification recognizes a closer relationship between them than with other invertebrate groups and is based on their shared radial symmetry. The term 'Coelenterata', commonly used as an alternative to 'Cnidaria', is actually synonymous with the term Radiata. Whereas Porifera are multicellular forms which function at a cellular level of organization, the radiate phyla are more advanced. These phyla function at the cell tissue level, with similar cells aggregated into definite tissues.

Phylum Cnidaria

The phylum Cnidaria is a group of relatively simple aquatic metazoans, including such well known forms as jellyfish, sea anemones and corals. Their combination of radial symmetry and brilliant coloration makes them some of the most beautiful of the marine animals.

The group shares two basic metazoan features and in these respects shows an advance on the sponges. They have an internal space for digestion, the gastrovascular cavity, which has a single extensible opening, the mouth. They can therefore feed on a much greater range of food sizes than is possible for either the Protozoa or the sponges. The mouth is surrounded by a circle of tentacles which are extensions of the body wall and serve for capture and selection of food. Feeding is generally a more active and selective process than in the Porifera. The body wall itself is composed of tissue layers, an outer ectoderm or epidermis and an inner endoderm or gastrodermis, with a third layer, the mesogloea (middle layer) between them. The mesogloea varies in thickness from a thin non-cellular layer to a thick, fibrous, jelly-like structure forming the major part of the bulk of the animal. The presence of stinging threads in cnidarians was recognized by Aristotle and the name of the phylum (Cnidaria) is based on this feature, *cnide* being the Greek for 'stinging nettle'.

Like the sponges many cnidarians have a superficially plant-like appearance and as a result their true animal nature was recognized only comparatively recently. For many years they were called 'Zoophyta' (animal-plants) and thought to be intermediate between the plant and animal kingdoms. Part of the confusion was due to the presence of two structural types within the phylum. One, the polyp, is sessile. The other, the medusa, is free-swimming. The typical polyp is cylindrical in shape, attached at one end and with the mouth and tentacles placed upwards. The medusoid form is umbrella shaped, with the tentacles hanging down at the lateral margin and the mouth in the centre of the concave under surface. While the polypoid mesogloea is thin, the medusoid mesogloea is extremely thick and forms the main bulk of the animal, hence the common name for the medusoid form, 'jellyfish'.

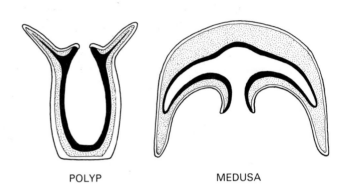

POLYP   MEDUSA

Comparative polyp and medusoid form

Some cnidarians are polypoid, others medusoid, and many pass through both stages during their life cycle. This is an example of polymorphism, a single species being represented by two or more clearly distinguishable forms during its life cycle.

While cnidarians are predominantly marine and attain their greatest diversity and importance in the sea, *Hydra*, a freshwater form belonging to the class Hydrozoa, is one of the most common laboratory animals and a description of *Hydra* is a useful point at which to begin a study of the group.

Hydra: an example of a polyp

Hydra was amongst the 'animalcules' described by van Leeuwenhoek in 1702. He observed *Hydra* growing on water plants and described it in a letter to the editor of the Journal of the Royal Society. In 1704 further observations and drawings were sent to the same journal by 'An anonymous gentleman from the country'. However after this *Hydra* was virtually forgotten. In 1740 Trembley, the tutor to the sons of a Dutch magnate, found *Hydra* growing on water plants. He was unable to decide whether it was an animal or a plant. Since in his day animals were regarded as being virtually incapable of regeneration, Trembley cut a *Hydra* in two, horizontally, believing that if it were a plant it would regenerate above the cut. Greatly to his amazement he grew not one *Hydra*, but two! Trembley was fascinated by his discovery and published minutely detailed

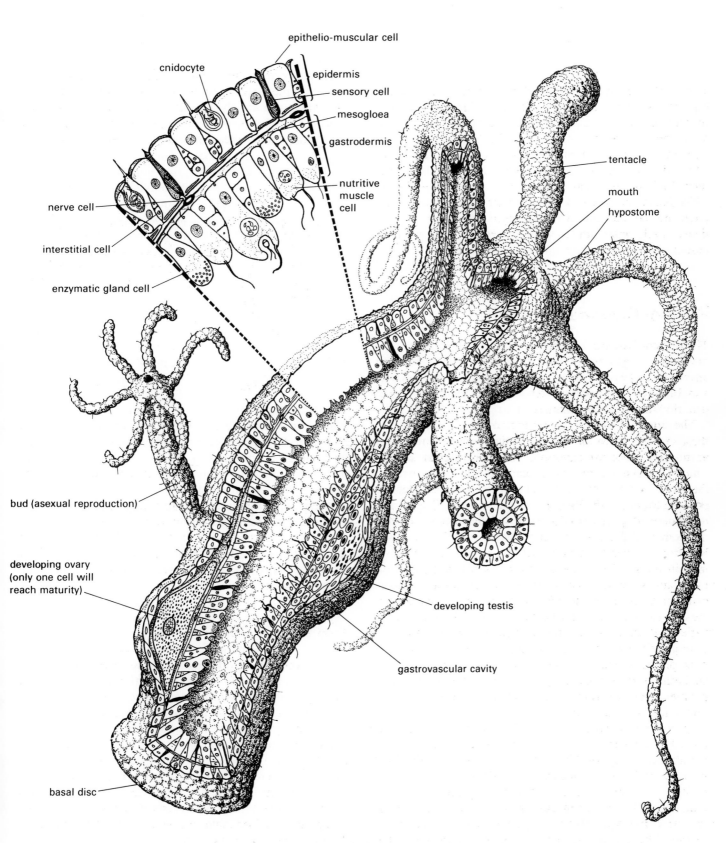

epithelio-muscular cell

cnidocyte

epidermis

sensory cell

mesogloea

gastrodermis

nutritive muscle cell

nerve cell

interstitial cell

enzymatic gland cell

tentacle

mouth

hypostome

bud (asexual reproduction)

developing ovary (only one cell will reach maturity)

developing testis

gastrovascular cavity

basal disc

Hydra

descriptions of *Hydra*, many of which still stand to this day. He used the term *Hydra* (the many-headed monster of Greek mythology) in his description, and the term was taken by Linnaeus in 1758 as the animal's generic name.

Hydras are amongst the few cnidarians that have invaded fresh water. They are small polyps (about 10 mm in length when fully extended), and are found in ponds, lakes, or streams attached to rocks or water plants. They are generally transparent and therefore difficult to spot, but *Chlorohydra viridissima (Hydra viridis)* has a bright green colour due to the presence of a unicellular symbiotic alga (*Chlorella*) in very large numbers (about 150,000 per individual) in the gastrodermis.

The basic structure of the hydras is close to that of the simplest type of polyp. The body is cylindrical with a basal disc for attachment. At the opposite (oral) end, a circle of up to ten long hollow tentacles surrounds a slightly raised area, the hypostome, in the centre of which is the mouth. The mouth opens into a large gastrovascular cavity which occupies a high proportion of the total volume of the animal. When the mouth is closed the fluid within the gastrovascular cavity provides support, functioning as a simple hydrostatic skeleton.

The body wall is built on the typical cnidarian plan, with epidermis, gastrodermis and between them mesogloea or 'middle jelly'. There is considerable differentiation between the cells of the epidermis, and to a lesser extent, the gastrodemis. Five principal types of cell make up the epidermis of which the epithelio-muscular cell is the most important in terms of body covering. These are wedge-shaped cells and are highly characteristic of cnidarians, although they also occur in some of the primitive Platyhelminthes. Their bases rest against the mesogloea and the outer margins fuse with adjacent cells forming most of the tough outer surface of the epidermis. Each cell has two, three or more basal extensions, each containing a contractile fibril or myoneme with the same proteins, actin and myosin, that are found in vertebrate muscle (page 245). The extensions are orientated lengthwise parallel to the

longitudinal axis of the body and tentacles and thus form a cylinder of contractile fibres. While this is not composed of true muscle cells it is effectively a layer of longitudinal muscle. Similar contractile extensions are found at the bases of the principal gastrodermal cells, the nutritive muscle cells, but in this case the extensions are orientated at right-angles to the long axis of the body and thus in effect form a circular muscle layer. They are generally more delicate than those of the epidermis and are most highly developed in the hypostome and tentacles. The remaining four types of epidermal cell (interstitial, cnidocyte, sensory and nerve) are interspersed amongst the epithelio-muscular cells. Interstitial cells are small and rounded with relatively large nuclei. They are versatile and may give rise to sperm and eggs, or develop to replace any other type of cell; they are therefore comparable with the amoebocytes of sponges.

The cnidocytes are highly specialized cells which are unique to the phylum (with one exception, see page 55). They usually occur in groups or batteries embedded in the superficial layers of the epidermis and are present in especially large numbers on the tentacles. They are rounded or ovoid cells with a basal nucleus and each contains a stinging structure known as a nematocyst. This is a capsule covered by a lid and containing a coiled tube. The nematocysts are discharged from the cnidocytes and can be used for prey capture, dispersal or anchorage. They vary in structure and from a functional standpoint can be divided into three major types: volvent, penetrant and glutinant. Volvents have a simple closed tube which wraps round and entangles the prey. Penetrants have an open tube, which may in addition be armed with barbs or spines, and on discharge the thread penetrates the prey tissues and injects a paralysing toxin. Glutinants are open-ended and sticky and are used for anchorage.

At the surface of each cnidocyte is a cnidocil, a small bristle-like process which acts as a trigger and has an ultra-structure very like that of a flagellum. There are supporting rods along the length of the cnidocyte and its base is anchored to the extensions of one or more epithelio-muscular cells. During discharge the nematocyst is expelled from the cnidocyte and the thread turned inside out in the process. The discharge mechanism is not fully understood but is thought to involve sudden changes in the permeability of the cell membrane. As a result, fluid rushes in, the lid opens and the tube is everted by hydrostatic pressure. After discharge a nematocyst consists of a bulb representing the original capsule and a thread-like tube, generally attached at the base to the epidermis.

A nematocyst can be used only once and the cnidocyte must then be replaced. Discharged nematocysts move into the gastrodermis and are digested. Large numbers of nematocysts may be used at any one time, for example *Hydra littoralis* has been shown to discharge 25 per cent of the nematocysts of its tentacles during the capture and ingestion of a brine shrimp. Cnidocyte replacement is one of the main functions of the interstitial cells, taking about 28 hours. The tentacular cnidocytes are usually formed

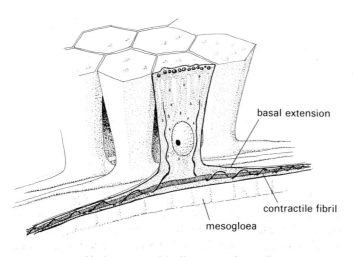

Hydra : an epithelio-muscular cell

basal extension

contractile fibril

mesogloea

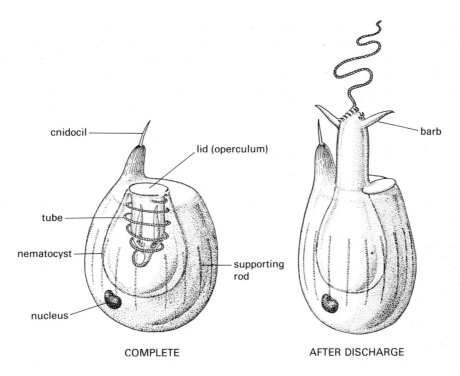

Structure of a cnidocyte

elsewhere in the *Hydra* and migrate by amoeboid movement, usually through the epidermis, to their final position.

Sensory and nerve cells are both found in *Hydra*. The sensory cells are elongated, and placed at right angles to the surface, where they have a sensory sphere or bristle. The base of each cell gives rise to a number of neuron processes. The nerve cells lie at the base of the epidermis and are elongated with two or more processes orientated parallel to the mesogloea.

The basal disc of *Hydra* is composed of mucus-secreting cells, with myonemes similar to those of the epithelio-muscular cells.

The histology of the gastrodermis is similar to that of the epidermis: interstitial cells, nerve and sensory cells (although in far smaller numbers) and mucus-secreting cells being present. Mucus-secreting cells are particularly abundant round the mouth where they aid in swallowing. The nutritive muscle cells correspond to the epithelio-muscular cells of the epidermis but are usually flagellated and can also form pseudopodia. Digestive enzymes are secreted by the enzymatic gland cells which do not have basal contractile processes.

Locomotion

Hydra is usually attached by the basal disc and from this position can extend or contract or bend itself from one side to the other. The gastro-vascular cavity functions as a hydrostatic skeleton and movement is brought about by the action of the myonemes. The mesogloea has visco-elastic properties. It helps maintain the animal's body form and also facilitates the myoneme action.

Hydra can move from place to place. The most rapid means of progression is a series of somersaults in which the body extends and bends over, attaching its tentacles to the substrate by means of the glutinant nematocysts. The basal disc is then freed and swung over to a new attachment on the substrate near the attached tentacles. The tentacles are then freed and the process repeated. In some species of *Hydra* the animal can move by extending its tentacles to catch hold of an object, then loosening the base and contracting the tentacles, thus drawing up the body. Another common method is by floating. The basal disc detaches and the animal floats to the surface where it may float about upside down for some time.

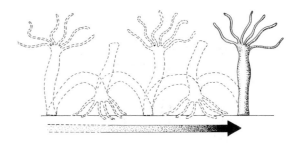

Hydra : somersaulting movement

Nutrition

Like most cnidarians, hydras are carnivores, feeding mainly on small crustaceans which they capture by means of their nematocysts and tentacles. When a suitable prey comes into contact with the tentacles, nematocysts are discharged and the prey is entangled and paralysed. The prey is then pulled towards the mouth which opens widely and engulfs it. This process is helped by the secretion of the mucus-producing cells round the mouth. The enzymatic gland cells in the gastrovascular cavity produce trypsin-like enzymes and the prey is gradually digested to a thick soup-like broth. Mixing is brought about by the beating of the flagella of the nutritive muscle cells. The extracellular digestive phase is very effective so that a small crustacean, like *Daphnia*, may be broken down 4 hours after ingestion. Digestion is completed intracellularly, the nutritive muscle cells producing pseudopodia which engulf the partly digested particles by forming food cups and food vacuoles in the same way as that described for *Amoeba*. The food vacuoles undergo the acid and alkaline phases already described in protozoans. Digestive products are circulated by cellular diffusion and undigested materials are passed back into the gastrovascular cavity and ejected through the mouth.

Gaseous exchange and excretion
Gaseous exchange takes place through the plasma membranes of individual cells as in the protozoans and sponges. This is effective as the body walls are thin and the arrangement and size of the gastrovascular cavity ensure an adequate water supply to the internal cells. Water is constantly circulated in the gastrovascular cavity by both body movements and the flagellar beat. Similarly, no special structures are required for excretion. Nitrogenous waste is largely in the form of ammonia and this diffuses through the general body surface. The fluid contents of the gastrovascular cavity are hypo-osmotic to the body cells. Periodically the contents of the gastrovascular cavity are discharged, and this enables *Hydra* to osmoregulate, the whole cavity performing the function of a contractile vacuole.

Nervous system
The nervous system of *Hydra* shows considerable advances over the forms already described. The nerve cells are arranged in an irregular net, and are particularly concentrated around the mouth. Synaptic junctions exist between neurons, and between neurons and muscle fibres and cnidocytes. The sensory cells receive stimuli and relay impulses to the nerve net though the transmission of impulses is slow and diffuse. In general the most definite movements in *Hydra* are seen in response to food stimuli. The presence of nerve cells and myonemes enables *Hydra* to achieve a level of co-ordinated movement impossible for sponges.

Growth and reproduction
The commonest method of reproduction in *Hydra* is asexually by budding. The bud develops as a simple evagination of the body wall enclosing an extension of the gastrovascular cavity. Mouth and tentacles are formed and eventually the bud detaches from the parent and becomes an independent hydra. Budding normally takes place during the warmer months of the year when food is plentiful.

Hydra also shows sexual reproduction which generally takes place in the autumn since the fertilized eggs are the means by which *Hydra* overwinters. The interstitial cells of the epidermis in particular regions specialize to form eggs and sperm. Ovaries are formed towards the basal disc, testes nearer the oral end, and both may be present in the same animal. Many sperm are present in the testis, but only a single cell out of the many potential ova in the ovary reaches maturity. As the egg enlarges, the overlying epidermis splits and the egg is exposed. Sperm are released into the water from the testis and the egg is fertilized within the ovary. The egg then undergoes cleavage and is covered by a chitinous shell. The embryo then drops off the adult and overwinters in the casing, a young *Hydra* emerging at the beginning of the spring.

Regeneration
Hydra exhibits considerable powers of regeneration with even quite small parts being capable of reforming complete animals providing both epidermal and gastrodermal tissue are present. A classic experiment involved turning a *Hydra* inside out by means of a thread drawn through the basal disc. The gastrodermal and epidermal cells reorientated themselves in a surprisingly short space of time.

Obelia: a representative hydrozoan

While *Hydra* forms a useful introduction to a study of the cnidarians, in many respects it is unusual. More typical is *Obelia*, one of many small marine hydrozoans which form branching colonies attached to rocks, shells or seaweeds. *Obelia* shows two important points of contrast with *Hydra*; it has both polyp and medusoid phases in its life cycle, and the polyp is colonial rather than solitary. The lack of a medusoid form in *Hydra* is itself an unusual feature and is thought to be an adaptation to life in fresh water where a dispersal stage is not such an advantage.

The polyp phase
Obelia begins its existence as a small polyp, superficially similar to *Hydra*. The polyp reproduces by budding, the buds remaining attached to form a colony which develops as a central hollow stem, the hydrocaulus, attached to the substrate by a branched rooting structure, the hydrorhiza. Individuals bud off in a branching fashion so that the whole structure is tree-like in appearance. The polyps are of two types, feeding polyps (gastrozooids) and reproductive polyps (gonozooids) connected by tubular extensions of the gastrovascular cavity. *Obelia* shows specialization of parts and division of labour of the kind found in many colonial organisms.

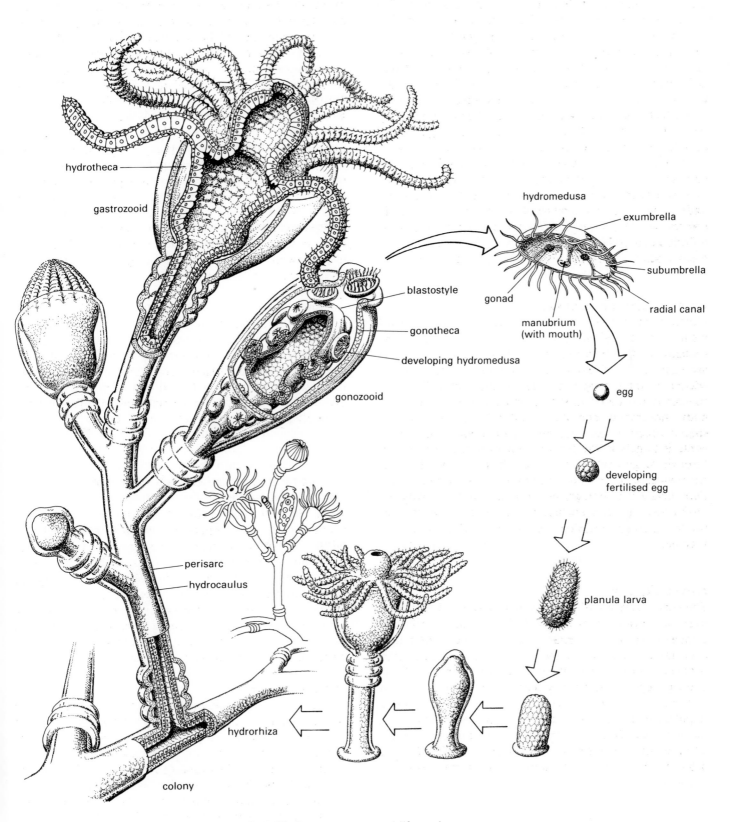

hydrotheca

gastrozooid

blastostyle

gonotheca

developing hydromedusa

gonozooid

hydromedusa

exumbrella

subumbrella

gonad

radial canal

manubrium
(with mouth)

egg

developing
fertilised egg

perisarc

hydrocaulus

planula larva

hydrorhiza

colony

Obelia : structure and life cycle

The gastrozooids resemble *Hydra* in the possession of a mouth, hypostome and tentacles, but differ in that the tentacles form a flexible supporting structure against which the longitudinal muscle fibres are able to act. The tentacles capture prey to feed the entire colony and also have a defensive function.

As the colony develops, the epidermis secretes a flexible protective covering, the perisarc. This expands to form cups (hydrothecae) round the gastrozooids, into which the tentacles can be withdrawn. Flexibility is provided by rings in the perisarc at the base of each polyp and these allow the animal to adjust to water currents.

The gonozooids have a reproductive function. They have a central stalk, the blastostyle, enclosed within a flask-shaped gonotheca and produce medusae by budding. Mature medusae are freed from the blastostyle and pass to the outside through the opening in the gonotheca.

The medusoid phase

The hydrozoan medusa (hydromedusa) is disc-shaped, with an outer convex exumbrella surface and a concave subumbrella surface. In most hydromedusae an inward projection of the margin forms a velum but this structure is rudimentary in *Obelia*. A short projection, the manubrium, hangs from the centre of the subumbrella and the mouth is at its apex. The mouth opens into the gastrovascular cavity which is considerably more complex than that of the polyp. Four radial canals extend from a small central cavity (the stomach) and join a ring canal running round the margin of the umbrella. The mouth, manubrium, stomach and canals are all lined with gastrodermis. The exumbrella and subumbrella are covered by an epidermis in which the cells are flattened.

Numerous solid tentacles, with swollen bases (the tentacular bulbs) due to accumulation of interstitial cells, hang at the margin of the medusa. These form a reservoir for the renewal of the nematocysts with which the tentacles are liberally supplied. Nematocysts are also found on the manubrium.

As in all medusoid forms the mesogloea is thick and gelatinous and makes up the main bulk of the animal, bringing it near to neutral buoyancy. The mesogloea is separated from both epidermis and gastrodermis by a thin non-cellular membrane and contains fibres that are thought to be secreted by both gastrodermis and epidermis.

Locomotion

The medusa shows a rather more specialized muscular system than the polyp in response to the demands of a free-swimming mode of life. There are no contractile extensions in the gastrodermis, and those of the epidermis are so orientated as to form definite muscle tracks. The muscular system is best developed around the bell margin and at the subumbrella surface, the fibres forming a radial and a circular system. Contractions of the muscles reduce the volume of the subumbrella cavity, driving water out from beneath the subumbrella. The shape of the animal is then restored to its original form by the elastic properties of the mesogloea. The velum reduces the aperture and thus increases the force of the jet. Swimming is largely vertical, a period of pulsation being followed by a period of sinking, with horizontal movement depending chiefly on water currents.

Nutrition

The medusa is carnivorous and feeds on all sorts of animals which come into contact with the tentacles. The digestive process is essentially the same as that of *Hydra*, with most of the intracellular digestion taking place in the gastrodermis of the manubrium, stomach and tentacular bulbs, the radial and ring canals being of lesser importance.

Gaseous exchange and excretion

Like *Hydra*, the medusa has no special organs for respiration or excretion.

Nervous system

As would be expected in a free-swimming form, the nervous system shows greater specializations than in *Hydra*. The epidermal nerve cells are concentrated into two nerve rings above and below the attachment to the velum, which are connected by nerve fibres. The upper ring controls the rhythmic pulsation of the bell and the lower ring is associated with eight specialized sensory organs, the statocysts, which are equally spaced at the margin and associated with eight tentacle bases. Each statocyst consists of a small gastrodermal sac, lined with sensory cells with bristles projecting into the lumen containing a particle of calcium carbonate, the statolith. Statocysts are receptor organs which enable the medusa to detect changes in its orientation: when the bell tilts, the statoliths respond to gravity and stimulate the sensory bristles. The animal then brings itself back to a horizontal position by muscular contractions.

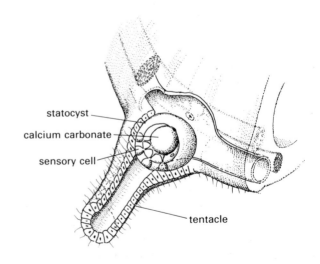

statocyst

calcium carbonate

sensory cell

tentacle

Statocyst of a hydromedusa

Reproduction

Reproduction in the medusa is sexual. Four reproductive organs are present, one in the middle of the course of each radial canal and formed as a ventral diverticulum from it. As in *Hydra*, these gonads are simple clusters of developing gametes, but the sexes are separate in *Obelia*. The gametes originate in the blastostyle epidermis. Sperm and eggs are shed into the sea where fertilization occurs. The fertilized ovum undergoes cleavage and forms a hollow blastula which eventually becomes a ciliated planula larva. The planula settles and develops into the colonial form by budding. The life history of *Obelia* thus consists of an alternation of the colonial polypoid phase and the solitary medusoid phase.

Characteristics of the Cnidaria

Hydra and *Obelia* serve to demonstrate between them the basic cnidarian features in a simple form. However complex a cnidarian may be, and some of them are very complex indeed, it never deviates from a basically two-layered structure, with a mesogloea secreted by the epidermis, the whole arranged concentrically round an oral-aboral axis to form a sac with a single opening. This fundamental radial symmetry may however undergo some bilateral or biradial modification. The circulatory and digestive systems are always combined in a single gastrovascular system, and respiratory and excretory systems are always absent. Sexual and asexual reproduction are both common, with individuals being typically monoecious (both sexes in one individual).

The structural diversity of cnidarians depends mainly on varying cellular and tissue differentiation and the degree of development of the mesogloea. The widespread polymorphism of the group allows the cnidarians to adjust to a greater variety of habitats than is possible for the Porifera, the possession of both sessile and floating phases making both surface- and bottom-dwelling existences possible. The development of different phases within the life cycle of a species also compensates in some degree for the lack of organs (and consequential division of labour) present in higher metazoa.

While retaining the basic structures outlined above, the class Scyphozoa, the animals most commonly referred to as 'jellyfish', show many points of contrast with the Hydrozoa.

Class Scyphozoa

In the Scyphozoa the medusa is always the dominant conspicious individual in the life cycle, the polyp being either reduced to a strictly larval phase or absent altogether. All scyphozoans are marine and they are generally larger and more elaborate than the hydromedusae already described. *Aurelia*, the common purple jellyfish, a scyphozoan of extensive distribution, whilst unusual in some features, may be used as a representative of the group.

Aurelia: a representative scyphozoan

Aurelia is a transparent animal with a relatively shallow saucer-shaped bell ranging in size from 70 to 600 mm across. While showing basic similarities to the hydromedusa, *Aurelia* has a generally more complex structure. For example, the arrangement of the epidermis and gastrodermis is similar to that of the hydromedusa but the mesogloea is more elaborate. It contains true muscle cells as well as wandering amoebocytes and may be regarded as a true cellular layer. In addition, buoyancy is provided by the inclusion of fluid-filled spaces within the mesogloea.

Locomotion

The muscular system and swimming movements are basically like those of the hydromedusa, with locomotion being brought about by a band of circular muscle, the coronal muscle, at the subumbrella margin. The radial fibres, which are more marginally placed than in the hydromedusa, pull in the edge of the bell before the circular muscles contract and hence narrow the opening and make the outflow more jet-like. Further narrowing is provided by the pseudovelum. Longitudinal muscle fibres are restricted to the tentacles and manubrium. Horizontal movement in *Aurelia* is brought about by the currents or waves in the sea. Swimming is generally vertically upwards, downward movement occurring by sinking.

Nutrition

Unlike the vast majority of cnidarians, *Aurelia* is a suspension feeder, feeding on small copepods and other planktonic organisms which become entangled in mucus on the ciliated subepidermal surface as the animal sinks, and on the exumbrella surface as it swims (although much of the plankton collected in this way may be lost). Plankton and mucus are swept to the bell margin by the action of the cilia and collected in a marginal groove. The corners of the mouth are drawn out into four trailing oral arms, which are thin-walled troughs containing a ciliated groove. The oral arms may collect plankton directly, but in addition they scrape plankton and mucus from the marginal groove and transfer it to the mouth via their ciliated grooves. The feeding mechanism is very effective, a medusa 100 mm in diameter being capable of removing the larger planktonic organisms from 700 ml of water in less than an hour.

Like the hydromedusa, *Aurelia* has a small central stomach cavity with radial canals extending to the margin of the bell. However, the arrangement of the canals is more elaborate than in a hydromedusa, with eight branched and eight unbranched canals arranged radially to join a circular canal at the umbrella margin. The current flows to the circular canal via the unbranched canals, returning through the branched canals. The typical scyphozoan stomach is subdivided by septae into four gastric pouches, but that of *Aurelia* is undivided. The current flows to the circular canal via the unbranched canals, returning through the branched canals. Numerous gastric filaments project into the stomach

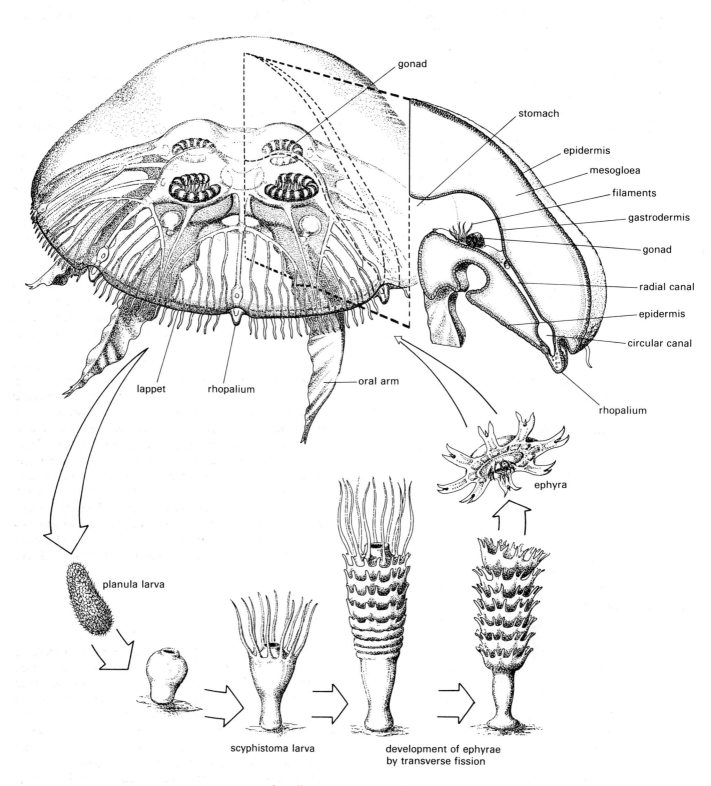

gonad

stomach

epidermis

mesogloea

filaments

gastrodermis

gonad

radial canal

epidermis

circular canal

rhopalium

lappet   rhopalium   oral arm

ephyra

planula larva

scyphistoma larva   development of ephyrae
                               by transverse fission

Aurelia : structure and life cycle

lumen from the stomach periphery and these are the source of extracellular enzymes. Digestion is essentially the same as that already described for the hydromedusa. Partially digested food is circulated through the canal system together with water drawn in through the mouth. The current serves not only for food transport but for respiration and removal of nitrogenous waste. It therefore combines the functions of the digestive and circulatory systems of more advanced animals.

The majority of scyphozoans feed on prey such as small fish which they entangle in their oral arms which are covered with nematocysts. The manubrium functions as a flexible food collecting stalk. The gastric filaments have nematocysts and are therefore able to paralyse any prey which is alive when it reaches the stomach.

Nervous system

Schyphozoans have two nerve nets present which are histologically distinct. One is diffuse and consists of multipolar nerve cells spread through the epidermis of the bell, tentacles, oral arms and manubrium. This net controls feeding, and other irregularly occurring activities. The other is a synaptic or giant fibre nerve net composed of bipolar cells, responsible for pulsation control during swimming, and narrowly associated with the swimming muscles, and with the rhopalia. These are characteristic scyphozoan sense organs, derived from modified tentacles. *Aurelia* has eight rhopalia, equally spaced round the margin in indentations; each lies between a pair of small specialized rhopalial lappets and is covered by a hood. (See drawing) The rhopalium bears two sensory pits with concentrations of general sensory cells, a statocyst, and a simple ocellus (eyespot) containing pigment and photoreceptor cells. In

the typical scyphozoan, impulses are transmitted rapidly from the rhopalia to the swimming muscles all around the bell. In addition the frequency of discharge of the synaptic nerve net is governed by the diffuse nerve net, thus relating swimming to activities such as feeding.

Reproduction and life cycle

Four pink, horseshoe-shaped gonads lie in the floor of the stomach at its periphery. Unlike those of *Hydra*, the gonads of *Aurelia* are gastrodermal in origin. In the majority of scyphozoans (which have divided stomachs) one gonad lies within each gastric pouch. The sexes are separate, and in most scyphozoans eggs and sperm are released into the stomach, and leave through the mouth. However, in *Aurelia* the eggs pass into pits in the oral arms where fertilization takes place, and the zygote develops into a planula larva. This is released and after a brief, free-swimming existence settles and develops into a small polypoid larva, the scyphistoma, which may survive for several years. The scyphistoma larva has a superficial resemblance to *Hydra*: it is trumpet-shaped, attached by a basal disc, and has an expanded oral end with a mouth and a short manubrium, surrounded by a ring of tentacles. This larva feeds and may produce further scyphistomae by budding, throughout autumn and winter.

In winter and early spring the scyphistoma forms immature medusae called ephyrae by transverse fission (strobilisation) at the oral end. The ephyrae become stacked up like a pile of saucers and are released from the scyphistoma one by one. This type of development is characteristic of the Scyphozoa. The ephyrae are minute (one to several millimetres in diameter), with incomplete adult structures. They have deeply incised margins, and therefore look like eight-pointed stars. They feed on small crustaceans which they catch by means of their eight arms. One or several arms catch food and bend inwards towards the manubrium which moves towards them. Development to a mature medusa takes between six months and two years.

Class Anthozoa

The anthozoans are the sea anemones and the reef-building corals, and their allies: colonial or solitary polyps with no medusoid phase. They are generally both larger and more heavily built than the hydroid polyp, and may have chalky skeletons. Whilst they show modifications of the polyp structure as shown by a form such as *Hydra*, the basic structural relationships remain easily recognizable. Like scyphozoans, anthozoans are exclusively marine.

Actinia occurs commonly on rocks and in crevices and is one of the chief sea anemones of temperate regions. It is solitary, like other sea anemones. *Actinia equina*, the beadlet anemone, may occur in several colours; brown, red, orange, green, or may be red with green-yellow spots (the strawberry variety).

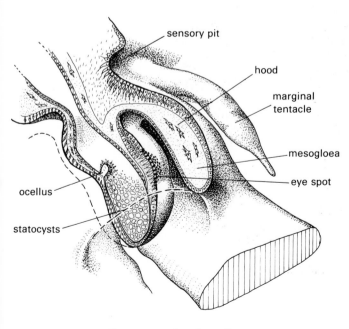

sensory pit

hood

marginal tentacle

mesogloea

eye spot

ocellus

statocysts

Structure of a rhopalium

Actinia: a representative anthozoan

The major part of the sea anemone is a heavy cylindrical column, attached to the substrate by a flattened pedal disc. The column is differentiated into an upper, short, thin-walled region, the capitulum, and a lower thick-walled column, separated by a fold, the collar. At the upper or oral end the column is slightly flared to form the oral disc, and at the junction between the oral disc and the column are about 200 solid tentacles, in five or six circlets and bearing large numbers of nematocysts. The tentacles can be folded

inwards and covered by the collar when the animal is not covered by water or if it is disturbed.

The mouth lies in the centre of the oral disc, separated from the tentacles by a smooth region, the peristome, and is slit-shaped. Ciliated grooves called siphonoglyphs lie at each end of the mouth. These cilia beat downwards, forcing a water current into the gastrovascular cavity which maintains the internal pressure of the gastrovascular cavity and may also provide for gaseous exchange from the gastrodermis. The mouth leads into the pharynx, a long stout tube extending through approximately one-third of

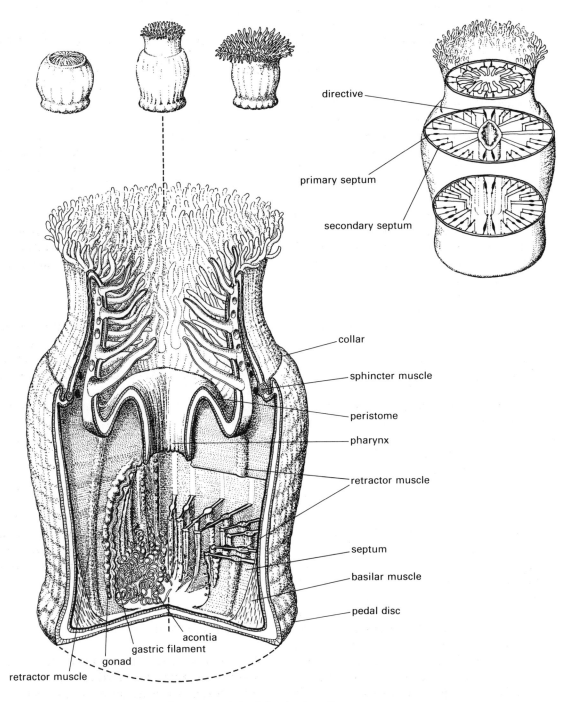

directive

primary septum

secondary septum

collar

sphincter muscle

peristome

pharynx

retractor muscle

septum

basilar muscle

pedal disc

acontia

gastric filament

gonad

retractor muscle

Structure of *Actinia,* the beadlet anemone

the length of the column. Like the body wall, the pharyngeal walls are made up of epidermis and gastrodermis with a thin layer of mesogloea between them. The pharynx is kept closed and flat by the water pressure within the gastrovascular cavity when the animal is not feeding. The siphonoglyphs have extra-thick mesogloea and heavy epidermal cells and therefore remain open.

The gastrovascular cavity is divided by longitudinal radiating partitions, the septae or mesenteries, formed of a double gastrodermal layer separated by mesogloea. The number of mesenteries present increases with age; six pairs are present initially to which others are added. A pair of mesenteries at each end of the pharynx are known as the directives. The primary mesenteries extend from the body wall to the pharynx. Further subdivisions of the gastrovascular cavity are formed by shorter secondary or tertiary mesenteries, which are connected only to the body wall. The arrangement of mouth, siphonoglyphs and mesenteries means that *Actinia* is bilaterally symmetrical. Openings in the directives and primary mesenteries in the upper part of the pharyngeal region allow water circulation. The free edges of the mesenteries are folded, thickened, and extended to form gastric filaments with a trilobed structure. The lateral lobes are ciliated, but the middle lobe consists of nematocysts and enzymatic gland cells. At the bottom of the mesentery the middle lobes of the gastric filaments are prolonged to form thread-like acontia.

The muscle system shows advances in comparison with the simpler cnidarian arrangement as seen in *Hydra* and is largely gastrodermal. Longitudinal muscle bands (retractors) are present in the mesenteries and contract to shorten the column and pull the tentacles and oral disc inside the polyp. The mesogloea contains fibres and wandering amoebocytes and thus approaches the complexity of the connective tissue of higher metazoans. The basic cnidarian distribution of muscle fibrils as illustrated by *Hydra* is retained in the oral disc and the tentacles. The epidermal cells are orientated to form a series of radial fibres in the oral disc and longitudinal fibres in the tentacles. This more superficial arrangement permits the more precise movement of the tentacles needed for feeding. Radial muscles in the mesenteries open the mouth and pharynx for swallowing.

Nutrition

The pharynx enables *Actinia* to swallow relatively large prey without damage to itself. Its soft, distensible body enables it to engulf prey which is nearly as large as itself. Other, larger species are capable of swallowing fish. Prey is paralysed by nematocysts, caught in the tentacles and carried to the mouth. Extracellular digestion proceeds further than in other cnidaria, both proteins and fats being digested by digestive enzymes produced by the middle lobe of the gastric filaments. The food broth is ingested by the gastrodermis of the gastric filaments, and digestion is completed intracellularly. The presence of numerous cilia on the general gastrodermal surface allows adequate

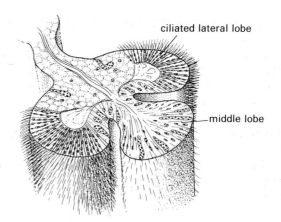

Sea anemone : detail of gastric filament

circulation and mixing. Excess food is stored as fat, chiefly in the gastrodermis.

Locomotion

Actinia is basically sessile, but is capable of limited locomotion by creeping on the pedal disc or walking on its tentacles. During these processes the gastrovascular cavity acts as a hydrostatic skeleton.

Nervous system

Actinia has few specialized sense organs. The nervous system follows the basic cnidarian plan, with a generalized sub-epidermal nerve net, and a gastrodermal nerve net, at least in the mesenteries.

Reproduction

The gonads are found on the faces of the mesenteries. The egg is normally fertilized in the body cavity and a tiny, oval, ciliated planula larva develops. The larva has a mouth, pharynx, siphonoglyphs and a small number of mesenteries. After a brief free-swimming period it attaches at the aboral pole and develops into an adult polyp. *Actinia* may also reproduce asexually, either by leaving parts of the pedal disc behind as the animal moves, or by longitudinal fission.

Cnidarian classification and diversity

There are between 9,000 and 10,000 known species of cnidarians and of these, the majority (about 6,000) are members of the class Anthozoa. Hydrozoa numbers about 2,000 species, including all the freshwater cnidarians, and Scyphozoa about 250 species.

Class Hydrozoa

The simplest level of cnidarian organization is found amongst the Hydrozoa. Hydrozoan polyps always have a thin non-cellular mesogloea, simple nerve nets, poorly developed muscle and gonads of epidermal origin. The hydromedusae are also relatively simple. One result of the thin-walled structure and undivided gastrovascular cavity of the hydrozoan polyp is that they are generally small in size. However some fairly large forms do occur, an extreme example being *Branchiocerianthus*, a polyp between one and two metres in length which is found in still oceanic water at a depth of several thousand metres. In many of the larger hydroids some measure of extra support is provided by a mass of parenchymous (packing) tissue within the gastrovascular cavity.

In spite of their basic simplicity the hydrozoa show remarkable structural and ecological diversity. Throughout the group there is a tendency to suppress the medusa; for example *Campanularia* has a medusa which remains attached to the parent polyp, and *Tubularia* and *Hydractinia* have medusae which are present only in a reduced form. Certain genera such as *Hydra* do not produce medusae at all. However an exception to this common trend is found in the order Trachylina which occur as small medusae in both marine and freshwater, and in which the polypoid phase is usually entirely suppressed.

In the many colonial hydrozoans, polymorphism is not restricted to alternating polypoid and medusoid phases. Since all the members of the colony are connected by hollow extensions of the individual polyp, food caught by the gastrozooids is available to the whole colony and therefore individual polyps can be specialized to fulfil a variety of functions. Specialization of this kind occurs in *Obelia* but more complex examples are found in the order Siphonophora, a group of pelagic colonial hydrozoans. In *Physalia*, the Portuguese man-of-war, three types of polyps are suspended beneath a float formed by a highly specialized polyp with a gas secreting gland. The gastrozooids, or feeding polyps, have a mouth and a single very long tentacle which arises (unusually) at their base, In cnidarians generally there is a trend for increased tentacle length to be correlated to reduced tentacle number, and *Physalia* with tentacles several metres in length is an extreme example of this. Dactylozooids, specialized for defence, have tentacles with batteries of nematocysts but no mouths. These can cause extremely painful stings so that *Physalia* is notorious amongst swimmers. The gonozooids produce medusae which remain attached to the colony.

Physalia feeds on epipelagic fish which become trapped in its tentacles. It is remarkable that one species of fish, the *Nomeus gronovii*, may be found living safely amongst the tentacles and appears not to cause the nematocysts to be discharged. Also, some octopodes can acquire and use *Physalia* tentacles. Siphonophores have no specializations for locomotion and are entirely dependent on wind and wave action, although some species can regulate the amount of gas in the float and so can migrate vertically. In other species modified medusae called nectophores move the colony by pulsating.

The Hydrozoa are normally divided into five orders. The majority are members of the order Hydroida, and it is this group that includes both *Obelia* and *Hydra*. Within the

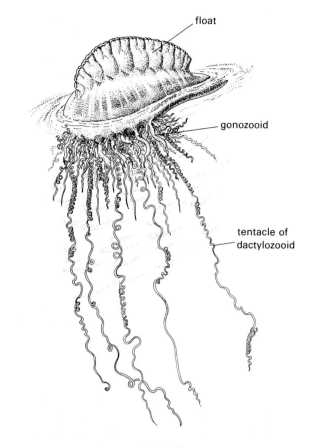

Physalia

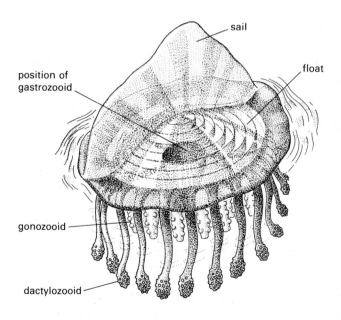

Velella

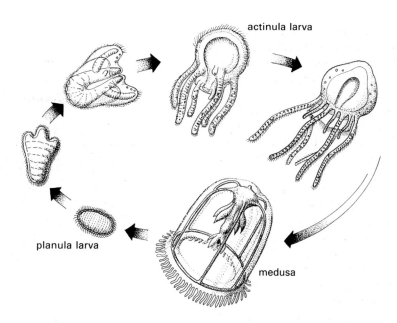

actinula larva

planula larva

medusa

Hydrozoan life cycle. *Aglaura,* order Trachylina, in which the polypoid
phase is entirely suppressed

group a distinction is made between the athecate forms (suborder Anthomedusae) such as *Hydra*, in which the polyps are not surrounded by a horny perisarc, and the thecate forms (suborder Leptomedusae) such as *Obelia*, in which a perisarc is present. The Limnomedusae is another suborder of the Hydroidea in which the polypoid form is minute, and the medusa dominant. Two common examples are *Gonionemus*, a marine form with hollow tentacles with adhesive pads with which it clings to sea weed, and '*Craspedacusta*' a freshwater form. Many members of this group were originally thought to have no polypoid phase; for example, the freshwater '*Craspedacusta*' has only recently been shown to be the medusoid phase of a tiny polyp, *Microhydra*.

Finally hydroideans include one small group of pelagic colonial forms, the suborder Chondrophora, a well known example being *Velella*, the by-the-wind sailor. *Velella* is similar to *Physalia* in the possession of a gas-filled float and was previously regarded as a siphonophoran. However the float is flat and discoid, with a horny skeleton and a vertical extension supporting a sail. A single gastrozooid in the centre of the disc is surrounded by a circle of gonozooids which are unusual in the possession of mouths. The colony is completed by an outer ring of dactylozooids. Like most siphonophores, the chondrophores appear to have no specialized structures for swimming.

The most primitive hydrozoan order is thought to be the Trachylina. The adult is a medusa but the order has an actinula larva which is polyp-like in form. Since the polypoid phase is entirely absent, the zygote develops to give a planula larva. This becomes an actinula larva by the development of a mouth and tentacles, from which the adult form develops directly.

The members of order Hydrocorallia, (suborders Milleporina and Stylasterina) are colonial forms which secrete massive external calcareous skeletons and are common components of coral reefs in shallow tropical seas. They are often brightly coloured and very beautiful. The basic structure is similar to the hydroids, with the addition of a calcareous skeleton, which in the stylasterinids is covered by a thick layer of tissue.

Finally, the order Actinulida is a minor group of tiny, interstitial (living within the sand) hydrozoan polyps which lack a medusoid phase and derive their name from a resemblance to the actinula larva.

Class Scyphozoa

The scyphozoans include the largest known cnidarian, the gigantic *Cyanea*, with a bell size of up to 2 or 3 m in diameter, and 800 or more tentacles each up to 60 m in length.

Whilst the group is primarily composed of free-swimming organisms, the Stauromedusae, (for example *Haliclystus*, the stalked jellyfish!) are bottom dwellers and basically sessile. The medusa resembles a polyp in appearance and is attached to the substrate by a short stalk with its mouth facing upwards, the whole forming a trumpet-shaped structure.

Generally the scyphozoans show a greater degree of elaboration, but less variation than the hydrozoans. The most familiar scyphozoans, such as *Aurelia* and *Cyanea* are members of the order Semaeostomae, a group generally coastal in distribution. In addition there are three minor orders, Coronatae, Rhizostomae, and Cubomedusae.

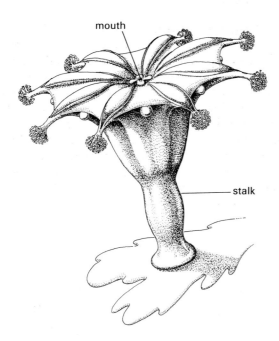

Haliclystus : the stalked jellyfish

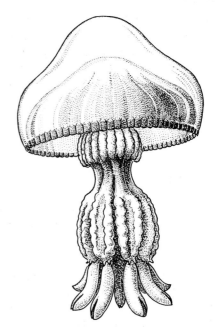

Rhizostoma

Coronatae are mainly deep sea forms, although some occur in surface waters. The bell is divided into two parts by a coronal groove running around the exumbrella. They also have pedalia (sing. pedalium), which are gelatinous leaf-shaped structures, bearing a single tentacle, at the bell margin.

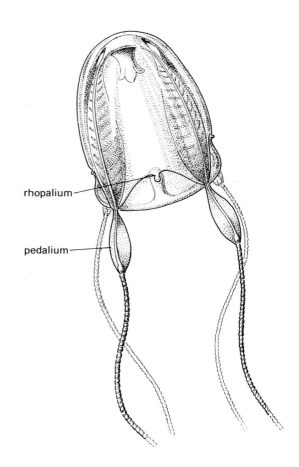

rhopalium

pedalium

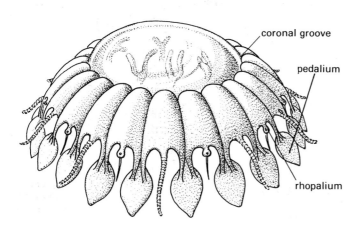

A coronatean scyphozoan, *Nausithoe*

A representative cubomedusan, *Carybdea*

The Rhizostomae have fused oral arms and no marginal tentacles. Instead of a single central mouth they have suctorial 'secondary mouths' leading into the oral arms and hence to the stomach. *Rhizostoma* is widely distributed in shallow waters, and the group as a whole are fairly common shallow water scyphozoans in the tropics and subtropics.

The Cubomedusae are most unusual scyphozoans, and some authorities have suggested that they should be regarded as a separate class, the Cubozoa. They have a bell with four flattened sides and four exumbrella ridges and are therefore basically cuboidal in form. At the base of each exumbrella ridge is a pedalium bearing one or several tentacles armed with nematocysts, the group being noted for the violence of its stings. The rhopalia are located in the centre of each side immediately above the margin, and each has a statocyst and six ocelli. Cubomedusae are atypical amongst the scyphozoans in their possession of a velum.

Class Anthozoa

The Anthozoa include many different types of sea anemone with a basic plan similar to that already described, but the majority of species are corals. The most conspicuous and important are the stony corals which resemble small sea anemones but live in colonies connected together by lateral folds of the body wall. The epidermis secretes a calcium carbonate skeleton, and this forms a series of cups into which the polyps can withdraw. The colony may be either low and spreading or tall and branching.

Some corals occur as isolated colonies, but in some areas vast assemblages of corals are responsible for the formation of coral reefs. The reef-building corals are very important in both tropical and subtropical waters. The reefs are classified according to their location, the most common being fringing reefs which develop directly from the shore. Barrier reefs, which are rather more rare, grow parallel to the coastline, but are separated from it by a lagoon. Atolls are coral reefs in the form of a ring surrounding a lagoon and are formed by the gradual submergence of volcanic islands on which the coral then grows.

An interesting example of commensalism is found amongst the Anthozoa, involving a sea anemone, *Calliactis parasitica*, and a hermit crab. Whilst both partners can live apart, they are commonly found together. A similar association is seen in *Adamsia palliata* which is almost always found associated with the crab *Eupagurus prideauxi*. In this case the crab has been observed feeding its anemone partner by putting pieces of food into its mouth. Fish may also be found in association with sea anemones: for example the gigantic *Stoichactis*, which may be up to a metre across and lives on the Great Barrier reef, is frequently found to have fish amongst its tentacles; each fish defending a sea anemone as its territory.

The anthozoans are divided into two subclasses: Octocorallia (or Alcyonaria) which are almost entirely colonial and characterized by the possession of eight pinnate (branched) tentacles, and Zooantharia, a subclass including both solitary and colonial forms with more than eight tentacles. The zoantharians include the main sea anemone order (Actinaria) with two other sea anemone-like groups, the small zooanthideans which encrust the surfaces of rocks and shells, and the burrowing ceriantharians with greatly elongated bodies. These live in soft bottoms within heavy, secreted, mucous tubes, from which they project their tentacles and oral discs for feeding.

The stony corals or true corals (order Madreporaria or Scleractinia) are closely related to the sea anemones. Some, for example *Fungia* (so called due to the superficial resemblance of its skeleton to the under-surface of a mushroom) are solitary and relatively large, reaching 25 cm in diameter. However the majority are small and colonial, the whole colony becoming very large indeed. The corallimorphs resemble true corals in structure but lack skeletons. The antipatharians (black corals) form upright plant-like colonies. Their polyps are arranged round an axial skeleton, which is made up of a black, horny material and bears thorns.

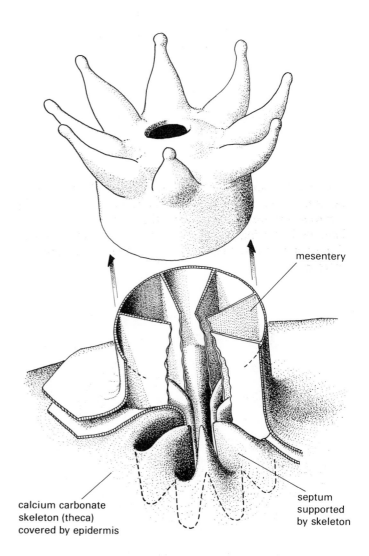

mesentery

calcium carbonate skeleton (theca) covered by epidermis

septum supported by skeleton

Structure of a stony coral

The octocorallians include such common marine forms as the sea fans, sea pens, whip corals and pipe corals. The polyps are rather small, and connected by a thick mass of mesogloea called the coenenchyme which is perforated by gastrovascular tubes.

The gastrovascular tubes of the coenenchyme are continuous with the gastrovascular cavities of the polyps, and the coenenchyme is covered by a layer of epidermis, continuous with the polyp epidermis. The octocorallians include the order Gorgonacea, for example the sea feathers, sea fans and the red coral, *Corallium*, used in jewellery making. Generally gorgonaceans have an upright plant-like form and are supported by a central rod composed of gorgonin, a horny substance, but the supporting structure in *Corallium* is of fused calcareous spicules. The sea pens and sea pansies are members of the order Pennatulacea which have a fleshy, flattened or elongate body, the rachis, with a skeleton of calcareous spicules. A skeleton of calcareous spicules is also found in the Alcyonacea, the soft corals, and in a minor order, the Telestacea. Stolonifera do not have a coenenchyme, and the polyps arise from a stolon instead. A calcareous tube skeleton may be present, as in *Tubipora*, the organ pipe coral. Finally the order Coenothecalia is represented only by *Helipora*, an Indo-Pacific form with a massive calcareous skeleton.

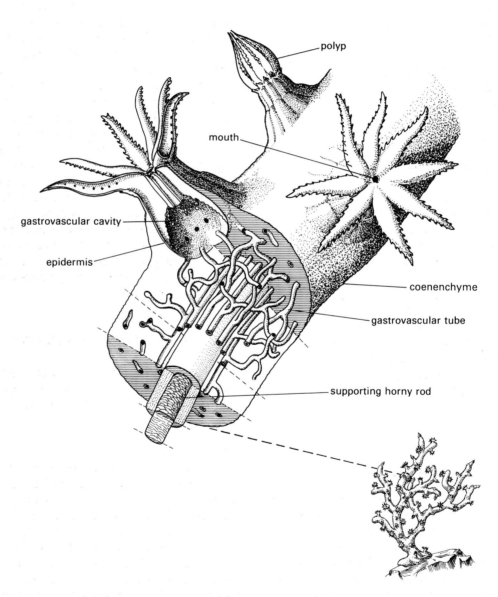

Structure of an octocorallian

Synopsis of phylum Cnidaria

Class Hydrozoa
    Order Trachylina (*Aglaura*)
    Order Hydroida
        Suborder Limnomedusae (*Gonionemus, Craspedacusta*)
        Suborder Anthomedusae (*Hydra, Branchiocerianthus, Tubularia*)
        Suborder Leptomedusae (*Obelia, Campanularia*)
        Suborder Chondrophora (*Velella*)
    Order Actinulida
    Order Siphonophora (*Physalia*)
    Order Hydrocorallia
        Suborder Milleporina (*Millepora*)
        Suborder Stylasterina (*Stylaster*)
Class Scyphozoa
    Order Stauromedusae (*Haliclystus*)
    Order Cubomedusae (*Carybdea*)
    Order Coronatae (*Nausithoë*)
    Order Semaeostomae (*Aurelia, Cyanea*)
    Order Rhizostomae (*Rhizostoma*)
Class Anthozoa
    Subclass Octocorallia or Alcyonaria
        Order Stolonifera (*Clavularia*)
        Order Telastacea (*Telesto*)
        Order Alcyonacea (*Alcyonium*)
        Order Coenothecalia (*Heliopora*)
        Order Gorgonacea (*Gorgonia, Corallium*)
        Order Pennatulacea (*Pennatulacea*)
    Subclass Zoantharia
        Order Zoanthidea (*Epizoanthus*)
        Order Actiniaria (*Actinia, Calliactis*)
        Order Madreporaria (*Fungia, Caryophyllia*)
        Order Rugosa (fossil forms)
        Order Corallimorpharia (*Corynactis*)
        Order Ceriantheria (*Cerianthus*)
        Order Antipatharia (*Antipathes*)

PHYLUM CTENOPHORA

The Ctenophora is one of the minor invertebrate phyla with only about fifty known species. Whilst at least one genus, *Beroë*, is likely to have been known to the ancient Greek biologists, the first clear account did not come until 1870 when they were recorded off Spitzbergen by a ship's doctor called Martens. Like many of the lower invertebrates their early classification was confused: Linnaeus regarded them as 'Zoophytes' and included them with a mixture of lower invertebrates in the genus *Volvox*. Only in 1829-33 was the distinct nature of the group established by the work of Eschscholtz who, after an extensive period of collecting, described three orders: Discophorae (which included all known medusae), Siphonophorae, and Ctenophorae.

Ctenophores are small, transparent, gelatinous animals, commonly known as the comb jellies. Modern biologists regard them as an evolutionary offshoot of the Cnidaria, and at about the same level of organisation. The group is exclusively marine, and whilst some ctenophores have adopted a bottom-dwelling, creeping mode of life, the majority are pelagic carnivores, paralleling the cnidarian medusae.

Like the anthozoans the ctenophores show a superficially radial symmetry, which can be shown on closer study to be a biradial symmetry. Cnidarians and ctenophores are therefore frequently grouped together as the 'Radiata' or 'Coelenterata'.

In many respects ctenophores can be compared to the cnidarian medusa. They have a body composed of two, thin cellular layers, the epidermis and the gastrodermis, with a well developed mesogloea (which may be regarded as an adaptation to pelagic life) between them. Like the medusa they have a canalised gastrovascular system although the details of its arrangement are peculiar to the group. The majority of ctenophores have a single pair of tentacles, which are also very long, and can be withdrawn into epidermal pouches. Nematocysts are present in one representative *Euchlora rubra*. In other species there are specialised thread-like adhesive cells called colloblasts present in large numbers on the tentacles these cells may be regarded as paralleling the nematocysts in many respects.

Ctenophores do not show the extensive polymorphism characteristic of the cnidarians, and there is no equivalent to the cnidarian polyp within the group. Many ctenophores show luminescence and this may be regarded as one of the features of the group.

Pleurobrachia: a representative ctenophore

One of the most generalized and commonly occurring ctenophores is the roughly gooseberry-shaped *Pleurobrachia* or 'sea gooseberry' which lives in coastal waters.

General structure
Pleurobrachia is about 20 mm in length and completely transparent. The body wall consists of an epidermis of cuboidal or columnar cells, with sensory and mucus gland cells. Beneath this is a thick jelly-like mesogloea with fibres, a network of muscle cells, and amoebocytes. A thin gastrodermis is also present. The mouth is situated at the end of the body that is directed forwards when the animal swims (the oral surface) and opens into a complex series of digestive canals embedded in the mesogloea.

The body is divided into eight equal sections by eight ciliated bands, the comb rows or costae. Each costa consists of a succession of transverse plates of long, fused cilia, the ctenes or combs. They are unique to the phylum, and the phyletic name is derived from them. The costae extend about four-fifths of the distance from the aboral surface to the oral surface, ending some distance from the mouth. At each side of the aboral surface a long tentacle extends from the base of a deep, ciliated epidermal pouch, running at 45° to the longitudinal axis of the body.

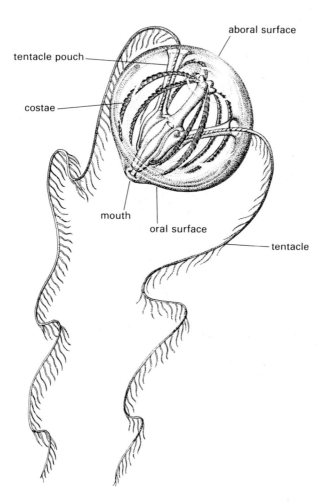

Pleurobrachia : the sea gooseberry

Locomotion
Ctenophores normally swim in straight lines vertically upwards or downwards. Propulsion is provided by the cilia, which beat in waves along the costae, beginning at the aboral end. The cilia are lifted rapidly towards the oral pole and then slowly lowered so that the effective sweep of the cilia is towards the aboral pole and the animal swims with its mouth directed forwards. If a ctenophore encounters an obstacle the beat is temporarily reversed. Since *Pleurobrachia* uses cilia for propulsion the muscular system is less well developed than that of the cnidarian medusa which relies on muscular contraction for movement. Thin layers of fibres arranged longitudinally

and latitudinally lie beneath the epidermis, and radially-placed fibres link the epidermis and gastrodermis.

Nutrition
Like most cnidarians, ctenophores are carnivorous. *Pleurobrachia* feeds on small planktonic organisms such as crab and oyster larvae, arrow worms, and copepods. During feeding *Pleurobrachia* hovers motionless in the water and captures its prey by means of the highly contractile tentacles. These consist of a core of mesenchyme with bundles of muscle cells, which is covered by epidermis. The tentacles can be extended up to 0.5 m and the area for contact with the prey is further extended by a series of lateral filaments up to 40 mm in length. Special adhesive cells, the colloblasts, are present on both the main tentacles and the tentacular filaments. Colloblasts are approximately pear shaped, and positioned with their narrow end anchored to the mesenchyme and their hemispherical heads projecting above the epithelial surface. Within the cell, a helical thread is wound in a coil parallel to the long axis. At its distal end this gives rise to many radiating fibres, each terminating in a granula filled with an adhesive mucus which is freed when the colloblast comes into contact with prey. Colloblasts are manufactured continuously throughout the life of the animal.

Prey is caught on the tentacles which are then hauled in and wiped across the mouth. The prey is then swallowed. The mouth is connected to the stomach by an elongated pharynx with folded walls lined with enzyme-secreting cells. The initial phase of digestion is extracellular as in the cnidarians; a partly digested broth is passed into the stomach and the canal system where digestion is completed intracellularly. The stomach is the centre of the canal system, with horizontal, vertical and meridional canals being present. A series of horizontal canals connects the stomach with eight meridional canals lying beneath the costae. Two canals originate from the stomach and run in the tentacular plane where each branches to give two interradial canals which in turn give rise to two adradial canals, making eight in all. Each ends in a meridional costal canal, lying beneath each costa. Apart from the various radial canals, a number of additional vertical canals arise from the stomach. Two pharyngeal canals extend towards the mouth, on either side of the pharynx, and two tentacular canals run along the base of the tentacular pouches. In addition a single aboral canal extends towards the aboral pole, where it divides into four short branches, the ampullae. Two end blindly, and the other two open by small pores. Indigestible wastes leave the animal by these two anal pores, and also through the mouth. This elaborate canal system serves for both digestion and circulation, with special canals running to areas with high energy requirements such as the pharynx (where enzymes are produced), the tentacles (an area of active muscular action) and the gonads which lie along the meridional costal canals. The meridional canals are also the site of light production in many bioluminescent ctenophores.

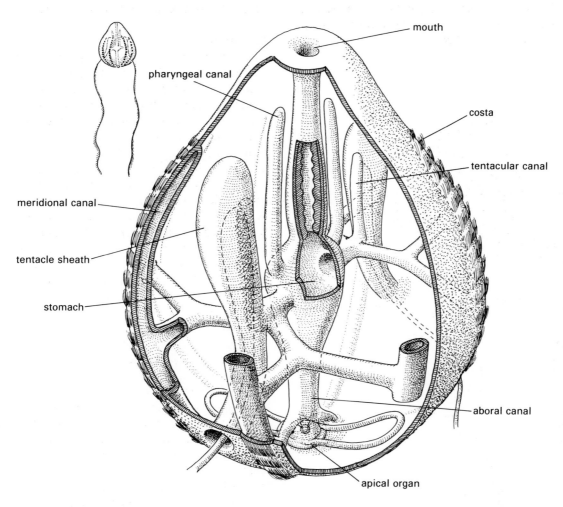

mouth

pharyngeal canal

costa

tentacular canal

meridional canal

tentacle sheath

stomach

aboral canal

apical organ

The digestive system of a ctenophore

Nervous system

Pleurobrachia has a sub-epidermal nerve net which is particularly well developed beneath the costae, and extends into the tentacles. The only sense organ present is the apical statocyst. This contains a statolith composed of several hundred grains of a calcareous substance and resting on four tufts of cilia. The frequency of impulses along the costae is controlled by four groups of balancer cilia connected to the ciliary tufts of the apical organ. If *Pleurobrachia* comes to lie at an angle to the vertical the statolith comes to rest unevenly on the balancer cilia, which causes an increased rate of impulses at the point where it rests most heavily. *Pleurobrachia* responds to these impulses by resuming an upright position. In many respects the ctenophoran nervous system is comparable with the double nerve net of the cnidarian medusa. The sub-epidermal nerve net is mainly concerned with feeding, stimulating muscle contraction and inhibiting the movement of the ctenes. Locomotion is controlled by the ciliated conducting cells associated with the apical organ and the ctenes.

Reproduction

Pleurobrachia, like all ctenophores, is hermaphrodite and has an ovary and a testis in the outer thickened wall of each meridional canal. The eggs and sperm are shed through the mouth, with fertilization taking place outside the animal. A free-swimming cydippid larva develops, closely similar in form to the ovoid ctenophore adult.

Ctenophore diversity

Whilst the basic characteristics described for *Pleurobrachia* are present in the majority of ctenophores, the group as a whole shows a certain degree of diversity. *Pleurobrachia* is thought to show the more primitive features of the group, and the spherical shape has been lost in many ctenophores. Several show some lateral flattening, and in forms such as *Cestum*, the Venus's Girdle, the body form has become so flattened and expanded that the animal swims not only by the cilia of the costae but also by muscular undulation.

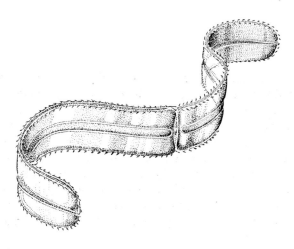

Cestum

Cestum includes some of the largest species of ctenophores, some of which reach a length of over 1 m. The order Platyctenea includes some highly unusual forms which are flattened in the oral-aboral plane and have taken up a bottom-living, creeping mode of life. The comb rows are either reduced or absent in the adult.

The phylum is divided into two classes: Tentaculata, which includes the majority of ctenophores with tentacles; and Nuda, ctenophores without tentacles.

There are four tentaculate orders: Cydippida, the spherical forms including *Pleurobrachia*; Lobata, in which the body form shows a moderate degree of compression; Cestida, the ribbon-like ctenophores and Platyctenea, which are bottom-dwelling. The class Nuda includes a single order, Beroida. *Beroë* is mitre-shaped with a very large mouth and feeds on other ctenophores.

Synopsis of the Ctenophora

Class Tentaculata
    Order Cydippida (*Pleurobrachia*)
    Order Lobata
    Order Cestida (*Cestum*)
    Order Platyctenea
Class Nuda
    Order Beroida (*Beroë*)

The acoelomate bilateral phyla

Cnidarians and ctenophores share a basically radial symmetry appropriate to a relatively simple mode of life, in which feeding is more a process of sitting and waiting rather than active seeking, and locomotion is consequently less important. If an animal has a mouth at one end, and attaches itself by the other, there is no apparent reason why it should not be radially symmetrical since no direction is more important than any other. An essential modification for a more active and wide ranging existence where food is actively sought is the development of a more elongated body form with a specialised sensitive region, the head, which is always directed forwards when the animal moves. This results in a new type of organisation, with bilateral symmetry, in which the animal can be divided along only one plane of symmetry to give two halves which are mirror images of each other. Furthermore one side of the body is always kept upwards (the dorsal surface) and one downwards (the ventral surface), the latter frequently being specialised for locomotion.

The animals commonly known as 'worms' are amongst the simplest animals showing this type of organisation. The term 'worm' does not now have any significance in a study of animal relationships, being the common term for an elongated, bilaterally symmetrical, cylindrical invertebrate. This was not always so. In the eighteenth century Linnaeus and Cuvier both established the worms as important animal groups, Linnaeus including all known invertebrates except the arthropods in the 'Vermes', and Cuvier, some 40 years later, applying the term 'Vers' to all except the arthropods, coelenterates, molluscs, echinoderms and tunicates, which by then had been established as separate phyla. The worms are now known to include several distinct phyla, representing three different levels of organisation.

Phylum Platyhelminthes

The Platyhelminthes may be regarded as being at a simple level of organisation and in fact occupy a unique position in the animal kingdom in that they express morphological features which appear, with modification, throughout the more complex animal groups. Apart from the development of a specialised head region (cephalisation) and bilateral symmetry, the Platyhelminthes show two major differences from the radiate phyla. Their tissues are aggregated into organs with specialised functions, several organs combining as an organ system to perform a single life activity such as feeding or reproduction. Digestive, reproductive, excretory and nervous systems are all developed to some degree in at least some representatives of the phylum. Secondly, they lack the mesogloea characteristic of the coelenterates and have three well-developed cellular layers: the epidermis, gastrodermis and mesoderm. The mesoderm is represented by a loosely-packed mass of relatively undifferentiated cells, the parenchyma or mesenchyme, which fills the space between the body wall and the internal organs. It gives the Platyhelminthes a generally more robust structure than many coelenterates and is regarded by some authorities as an evolutionary development of cnidarian amoeboid cells. Mesenchyme is sufficiently plastic to allow some change of body shape.

Platyhelminthes translated means 'flatworms' (*platys* means 'flat', *helminthos* means 'worm'). The group is easily recognisable in that all members are extensively flattened in the horizontal plane (dorsoventrally flattened) and the common names applied to the animals, such as fluke (an old name for a flatfish), tapeworm, planarian, all emphasise this characteristic. The Platyhelminthes comprise between 12,000 and 15,000 species and are an interesting and economically important phylum because they include a large number of common parasites. Naturally this has meant that the group has attracted attention from the earliest times with parasitic worms (probably tapeworms) being recorded in an Egyptian papyrus dated 1550 BC. Of the three platyhelminth classes, Turbellaria, Trematoda (the flukes) and Cestoda (the tapeworms), only the first is not parasitic. Both Trematodes and Cestodes are thought to have evolved independently from the turbellarians at an early stage and are extensively modified in association with the parasitic habit. A consideration of the Turbellaria is therefore a suitable place to begin to look at the phylum as a whole.

Class Turbellaria

Turbellarians are chiefly marine, with some species living in freshwater. There are a few terrestrial forms which have no protection against desiccation and are therefore restricted in distribution to humid areas such as tropical rain forests and to suitable habitats such as hummocks of moss. They are nocturnal, living under logs or leaf-mould during the day and feeding at night. Although no turbellarian is parasitic the group does include commensal forms, one such being *Bdelloura* which lives in the gills of horse-shoe crabs and takes some of their food.

The majority of turbellarians are small in size, being less than 10 mm in length. *Dugesia*, the form commonly studied in laboratories, is somewhat larger than most, growing to about 20 mm in length, and some terrestrial turbellarians may be as much as 600 mm in length.

In spite of a certain superficial uniformity the turbellarians are extremely diverse, and the class includes a large number of different orders. Three are of particular interest: the primitive Acoela, the marine Polycladida and the Tricladida. The last includes relatively large turbellarians of marine, freshwater and terrestrial habitats. Both *Bdelloura* and *Dugesia* are members of this group. The freshwater triclads, known as planarians and including *Dugesia*, will be taken as examples of the Turbellaria as a whole.

Class Turbellaria

The Planarians: examples of free-living platyhelminths

Planarians generally have an elongated form with a well developed head with lateral projections, the auricles, which gives them a characteristic arrow shape in dorsal view. *Dugesia* is commonly found crawling over the bottom of slow streams, lakes or ponds.

The outer surface of the body is formed by the epidermis which is ciliated. In *Dugesia* the cilia are restricted to the ventral surface, but some other forms are ciliated on both surfaces. Mucus and other gland cells are present, both within the epidermis and beneath it in the mesenchyme. In this case an elongated neck extends from the body of each gland cell to the surface of the epidermis. Rod-like structures, the rhabdites, are present in most of the epidermal cells, arranged at right angles to the surface. These structures have an extremely complex ultrastructure

and their function is still uncertain though they appear to have a role in defence. They are formed with specialised gland cells and discharged if the worm is attacked. They then swell up to form a slimy coating. Within the basement membrane of the epidermis, fibres believed to be collagen form an elaborate trellis and this is of considerable importance in maintaining the shape of the animal.

Several layers of muscle cells complete the body wall, with fibres orientated to run in a circular direction, diagonally, or in a longitudinal direction. In addition blocks of fibres extend between the dorsal and ventral surfaces within the parenchyma to form the dorsoventral muscles. In contrast with forms such as *Hydra* the muscle fibres are not extensions of the epidermal cells, but are independent cells specialised for contraction. Furthermore they are not derived from the epidermis but develop independently from the mesenchyme.

Locomotion

Dugesia glides over the substrate by means of the epidermal cilia, over a thin film of mucus produced by glands in the epidermis and mesenchyme.

By this means it can crawl slowly, at about 1.5 mm per second. Slightly faster movement is produced by muscular action, in which waves of contraction pass from the head down the length of the animal along the ventral longitudinal muscle, alternately raising and lowering the body. This is essentially the same mode of movement seen in crawling snails (see page 176) though these belong to an entirely different phylum. The flattened shape of *Dugesia* is advantageous for both types of locomotion, with increased size producing a greatly increased surface area for both movement and support.

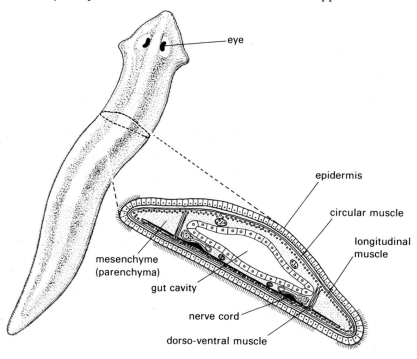

Dugesia : general body organisation

Nutrition

Planarians are carnivores, either predatory or scavenging. *Dugesia* is an effective predator, attacking and capturing small invertebrates such as insect larvae, small earthworms, water fleas, and snails. Potential prey may get caught in the trail of mucus left behind the animal as it crawls. *Dugesia* is able to wrap the anterior part of its body tightly round its prey, thus immobilising it, or can entangle it in slime and pin it to the substrate. In addition *Dugesia* feeds on the dead bodies of larger animals.

In contrast with that of the coelenterates, the digestive system is elaborate and well developed, although the gut must still be regarded as a simple structure in that only a single opening, the mouth, is present. As in the cnidarians, waste matter is egested through the mouth. This is situated in the mid-line of the animal's ventral surface and opens into a tubular pharynx with a thick muscular wall, housed in a cavity in the body. The pharynx is protruded during feeding and is used to engulf smaller prey. It is inserted into larger food matter, the insertion being facilitated by the action of proteolytic enzymes secreted by the pharyngeal glands. Food is then pumped into the intestine. *Dugesia* has a basically three-branched intestinal sac, one branch extending anteriorly and two posteriorly; all three have

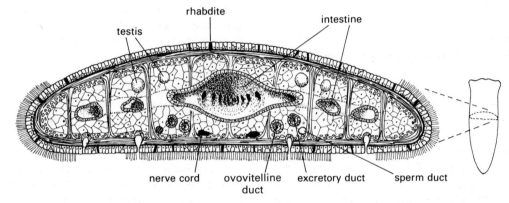

T.S. *Dugesia*

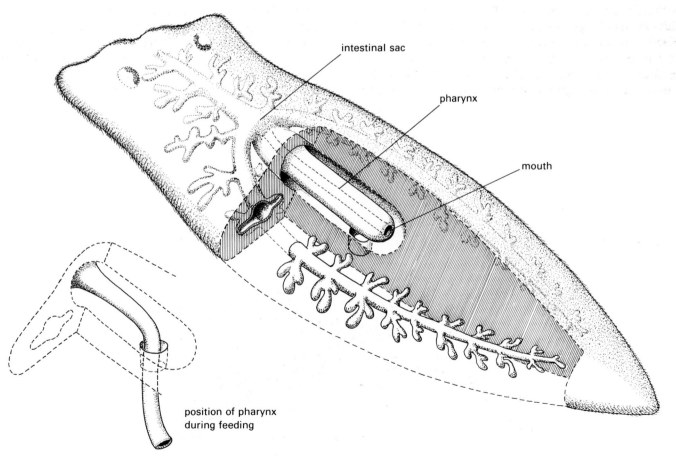

Dugesia : digestive system

numerous smaller lateral extensions. The gut is lined with an epithelium of gland and phagocytic cells. The digestive process is comparable with that of the coelenterates, an initial period of extracellular digestion being followed by intracellular digestion in the gastrodermal cells. The gland cells secrete a peptidase similar to that secreted by the pharyngeal glands and extracellular digestion takes place in an acid medium (optimum pH about 5). The partly digested food is taken up in food vacuoles by the phagocytic cells where protein digestion continues in an acid medium. After this phase is completed the pH rises to about pH 7, and fats and carbohydrates are broken down.

The products of digestion are transferred through the body by diffusion. This is facilitated by the flattened shape of the animal and by the numerous intestinal branches resulting in short diffusion pathways between any cell and an area of intestine.

Gaseous exchange

The flattened shape gives planarians a large surface area in relation to volume; simple diffusion is therefore adequate for gaseous exchange. Therefore in this respect they show no advance upon the coelenterates, in spite of their increased metabolic requirements due to the presence of the parenchyma. The oxygen consumption of an average planarian is ten times that of an average coelenterate of the same weight.

Excretion and osmoregulation

Nitrogenous waste is removed by a specialised excretory system, consisting of a number of branching blind-ending tubules, the protonephridia. The ends of the tubules are expanded into hollow bulbs, each of which contains a close-packed bundle of 35–90 flagella which beat in unison in

waves from the base to the tip. These structures are visible under the light microscope in living animals, and are called the flame bulbs, as they appear to flicker. Several flame bulbs may be derived from a single cell; thus their cavities are intracellular, as are those of all except the largest of the tubular structures in the system. The protonephridia unite to form a series of ducts which are presumed to open to the exterior by nephridiopores on the surface of the body, although electron microscopical study has failed to reveal these openings.

The flame bulb has groups of slits in the wall, connecting its lumen to the intercellular spaces of the surrounding parenchyma. The slits appear to be crossed by closely spaced filaments. It is thought that fluid is withdrawn from the intercellular spaces by the current caused by the flame beat, and that ultra-filtration occurs, with the passage of larger molecules being prevented by the filaments.

Bundles of flagella are found at intervals along the protonephridial ducts, and cause a current to flow through them. The distal parts of the ducts are lined with cells with deeply infolded outer cell membranes and many mitochondria. This type of cell structure is frequently associated with active concentration of salts against a concentration gradient across the membrane, and the cells may perform this function in *Dugesia*.

The protonephridial system is better developed in freshwater planarians than in marine forms, and it is therefore likely that it has an osmoregulatory function. In addition it probably controls the elimination of metabolic waste products such as nitrogenous compounds.

Nervous system

The nervous system shows considerable advances on that of the cnidarians, with that of the majority of turbellarians

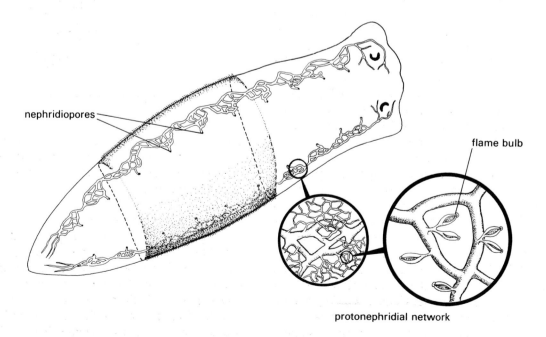

nephridiopores

flame bulb

protonephridial network

Dugesia : excretory system

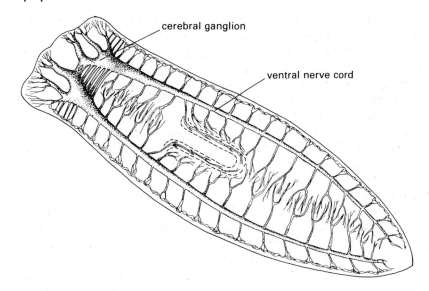

Dugesia : nervous system

being strongly bilateral in arrangement. In addition to a delicate sub-epidermal nerve plexus, the planarians have a pair of longitudinal ventral nerve cords extending from an anterior bilobed concentration of nervous tissue, the cerebral ganglion or 'brain'. The cords are linked to the sub-epidermal plexus by peripheral connections, and to each other by a ladder-like arrangement of medial commissures. The nerve cords and branches lie immediately below the muscles of the body wall. There is also a pharyngeal nerve net which coordinates pharyngeal feeding movements.

Several types of sense organs are present. Ciliary tactile receptors are distributed over the entire body with particular concentrations on the margins and auricles. In addition ciliated slits on either side of the auricles are thought to function as chemoreceptors. Two 'eyes' are present on the dorsal surface of the head on either side of the mid-line. They consist of a cup of pigment cells surrounding many photosensitive cells. These are bipolar nerve cells with a rounded distal end marked with longitudinal striations due to evaginations which probably increase the surface area for photochemical reactions. The eyes are very simple structures which function only to distinguish light from dark and certainly do not produce an image like the eyes of more complex animals.

The behaviour of planarians is comparatively simple although showing advances on forms such as *Hydra*. Their chief responses are to move themselves away from light and towards food. They also orientate themselves so that their head is directed towards a water current. The cerebral ganglion has an important role in behaviour: if it is removed the animal is still capable of movement, but appears to be unable to recognise food although food may be swallowed if touched by the pharynx. Planarians have been shown to be capable of learning: *Dugesia* for instance can be taught to make avoiding movements under special circumstances although it cannot be taught to go into conditions it

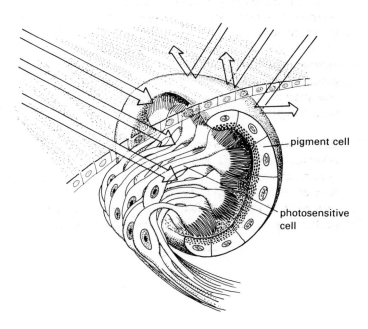

Dugesia : pigment cup ocellus

normally avoids, for example towards the light. Learning requires long conditioning and acquired responses are not retained for long. The learning response is prevented by removing the brain.

Reproduction and growth

The reproductive system is complex, and since it can vary even between closely related species a general description follows. The Turbellaria are generally hermaphrodite which is advantageous in that if any two animals meet during the breeding season both can be fertilized instead of just one or neither. This is an adaptation frequently found in animals with a reduced chance of meeting due to being slow moving or parasitic.

The male organs consist of paired testes with their associated ducts, the vasa deferentia, joining and opening into the penis. The penis is a protrusible ejaculatory organ, opening into a pocket, the genital atrium, which opens on the ventral surface, posterior to the pharynx, at the genital pore. Platyhelminthes are unusual in having biflagellate sperm, the flagella themselves having a $9 + 1$ arrangement of microtubules instead of the more usual $9 + 2$ arrangement. The sperm are stored in dilated regions known as spermiducal vesicles.

The female system comprises an egg-producing structure, the ovary or germinarium, and a yolk-producing structure, the vitelline gland or vitellarium. These may be united or occur as separate structures. In either case the egg is surrounded by a number of yolk cells after release. The oviduct opens into the genital atrium. Dilated areas may be present in the female ducts for sperm storage after fertilization. The fertilized eggs develop to hatch either as a free swimming larva or as the adult form.

Dugesia has a line of testes on each side of the body. A single ovary is present on each side near the anterior end and the oviducts are lined with yolk glands, forming ovovitelline ducts. Copulation takes place, each partner inserting its penis into the genital pore of the other. The process lasts a few minutes, sperm being transferred to a sperm storage area, the copulatory bursa. On entering the copulatory bursa, the sperm become active and travel up the ovovitelline ducts where fertilization takes place. Several ova and hundreds of yolk cells are enclosed in a capsule formed from droplets in the yolk cells. This hardens and is given a sticky coating produced by a special gland, the cement gland. As the eggs are laid the coating becomes drawn out to form a stalk and this is used to attach them to the underside of stones. Young worms, which are minute but have the adult form, normally hatch after two to three weeks.

Some planarians, including *Dugesia*, show a form of asexual reproduction. While the rear end adheres firmly to the substrate the front continues to move forward and the animal pulls apart at a region behind the pharynx. Regeneration follows to give two complete worms. Obviously for this process to occur planarians must possess high powers of regeneration and *Dugesia* has been used in many experiments on this subject. For example if an animal is cut transversely into three equal parts the head will regenerate a new tail, the tail a new head and the middle region a new head and tail, the polarity of the original animal being retained. If a planarian is split along the antero-posterior axis from the head to half way along the body length and the two pieces kept apart for about 24 hours, each part will regenerate the missing half and a two-headed animal will result. The experiment can even be

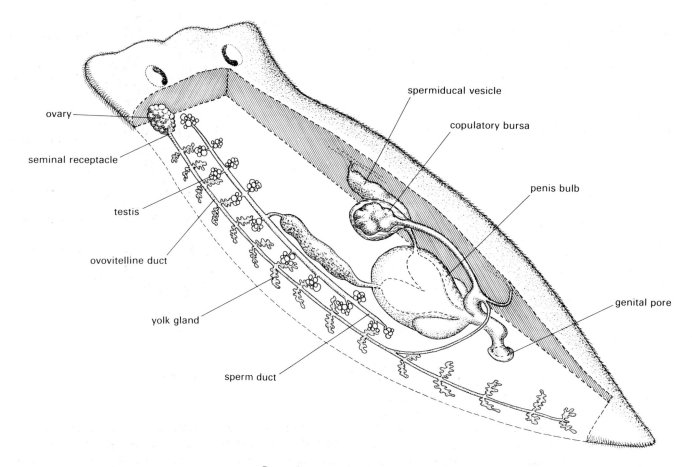

Dugesia : reproductive system

repeated with the same animal with subsequent splitting to give up to sixteen heads on one individual.

Characteristics of the Platyhelminthes

The basic structure described for the planarians is found throughout the Platyhelminthes. All the members of the phylum are bilaterally symmetrical and dorsoventrally flattened; all have a solid (acoelomate) form of construction with parenchyma filling the space between the body wall and the internal organs. In all Platyhelminthes there is a protonephridial excretory system and a hermaphrodite reproductive system. The gut, when present, has only a single opening.

However, considerable variety is seen in the details of organisation throughout the group and particularly in the remaining platyhelminth classes, the Trematoda (flukes) and Cestoda (tapeworms). The creeping habit and basically vermiform (worm-like) body shape have to a considerable extent determined the evolutionary trend of the group as both are valuable pre-adaptations for endoparasitism. Some of the demands of the parasitic mode of life have already been discussed in the context of the parasitic protozoa, and similar constraints and problems apply to the parasitic Platyhelminthes. In addition, metazoan parasites have special problems arising from their increased size. They must avoid being dislodged from the host either by currents or by the movements of the gut. They are also liable to attack by the host, either by the digestive enzymes (in the case of gut parasites) or by immunity reactions. Both the Trematoda and the Cestoda are highly successful and well adapted parasites; their economic and social importance is enormous since they parasitise man and his domestic animals.

Class Trematoda

Platyhelminths belonging to the class Trematoda, the flukes, comprise one of the major groups of metazoan parasites. They show some modification of their structure, and extensive modification of life style in contrast with free-living forms. Structural adaptations include the development of suckers as a means of attachment to the host, and the reduction of the sensory system. Elaborate life cycles including both sexual and asexual reproduction in two hosts have already been described in the context of protozoa as useful adaptations for successful parasitism. There are similar complexities among the trematodes, though with important differences between the three orders which comprise the group: Monogenea, Apisthrobothrea, and Digenea. The monogenetic trematodes live only in one host, and have a free-swimming ciliated larva in their life cycle. They normally live in accessible sites on the host, for example on the skin or in the gills of fishes, or in regions such as the buccal cavity, cloaca or bladder of frogs. The Apisthrobothrea is a small order of flukes, which are also usually restricted to a single host (which may be a reptile, fish or snail) in their life cycle. The majority of flukes are members of the order Digenea, and are all endoparasites with a life history involving two or three hosts, of which one is a vertebrate and another, generally, a snail. Between hosts there is always a free-living stage.

Structure of a generalized trematode

Trematodes show a general similarity of structure. They are often leaf-shaped in outline and the majority are only a few centimetres in length. They are characterized by the possession of suckers (acetabula) for attachment with typically two being present, an oral sucker surrounding the mouth and a ventral sucker on the ventral surface. A few mesenchymal gland cells are found in association with the suckers.

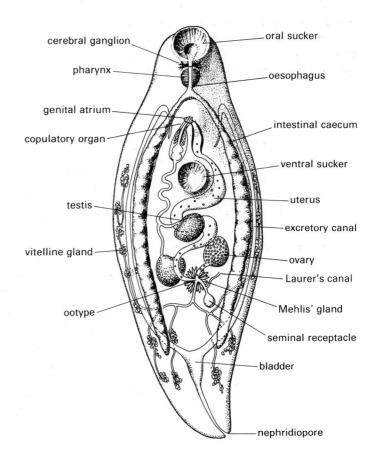

Structure of a generalized trematode

In contrast to the ciliated epidermis characteristic of turbellarians, flukes have a tegument, with a highly elaborate syncytial structure, arranged in several distinct layers. The outermost layer, the matrix, is of cytoplasm,

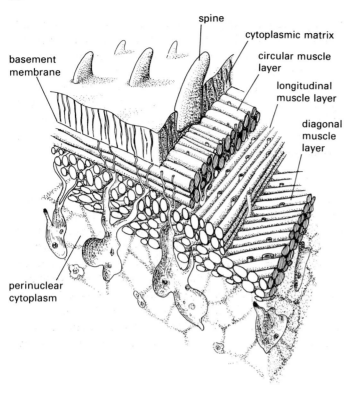

spine
cytoplasmic matrix
circular muscle layer
longitudinal muscle layer
diagonal muscle layer
basement membrane
perinuclear cytoplasm

Tegument of *Fasciola hepatica*

rich in mitochondria and pinocytotic vesicles. Beneath this is a thin basement membrane, strengthened by fibrils which give shape to the animal and to which the circular, diagonal and longitudinal muscles of the body wall attach. Fine strands of cytoplasm extend through the basement membrane and muscle layers to an inner nucleolated region, the perinuclear cytoplasm. The tegument is covered by a secretion of mucopolysaccharide, and the whole structure serves to protect the parasite from the host, particularly from the host's digestive enzymes in gut-inhabiting species.

Digestive system
A well developed gut is present in trematodes. The mouth is either terminal or slightly ventral in position and leads into a funnel-like mouth cavity and then into a sucking pharynx with strong radial musculature which pumps food into a short oesophagus. In blood flukes the pharynx may be reduced or absent. Two intestinal caeca fork from the oesophagus. These normally take the form of simple tubes, but are branched in some species, for example in liver flukes. The intestinal wall is lined by an epithelium and frequently longitudinal and circular muscles are present.

Flukes feed on cells, blood, mucus, or tissue fluids of the host and digestion is predominantly extracellular.

Excretory system
Trematodes have a protonephridial excretory system similar to that of turbellarians. Two longitudinal and commonly recurring excretory canals are present and open either through two dorsolateral nephridiopores (Monogenea) or unite at a median bladder and open via a single posterior nephridiopore (Digenea).

Nervous and sensory systems
The nervous system is essentially similar to that of the turbellarians and is unusually well developed for a parasitic animal. Paired cerebral ganglia lie dorsally to the pharynx and from these extend up to three pairs of longitudinal nerve cords. Both suckers and pharynx are well supplied with nerves.

Trematode sense organs are generally poorly developed but numerous tactile receptors are present, especially in association with the suckers. In addition eyes are present in some ectoparasitic flukes.

Reproductive system
The reproductive system is similar in basic design to that of turbellaria and like that is subject to considerable variation. The majority of trematodes are hermaphrodite; many are capable of self-fertilization although cross-fertilization appears to be more common.

The ovary and vitelline glands are separate, the latter being scattered throughout the parenchyma. The vitelline glands produce both yolk and a part of the materials from which the egg capsule is formed; further material involved in capsule formation is produced by Mehlis's gland. The ovovitelline duct is modified as a uterus for the storage of fertilized eggs and is greatly elongated and coiled. The testes are scattered. The male ducts unite to connect with a copulatory organ, and male ducts and uterus open together at the genital atrium placed in the anterior region of the ventral surface. A vagina opens separately and leads to a receptacle for sperm storage; in digenetic trematodes the vaginal opening is some distance from the genital atrium and the structure is known as Laurer's canal.

A central point in the female system, a chamber called the oötype, receives eggs, via the oviduct, yolk and material for making the egg capsule from the vitelline duct, and secretions from Mehlis's gland. Here the eggs are fertilized if this has not already occurred. The fertilized egg is encapsulated in the oötype and passes to the uterus for storage. The capsule is formed of a tanned protein, sclerotin, and has a small operculum or lid, which opens at hatching to allow the larva to escape.

Trematode life cycles

Order Monogenea
Adult monogenes are distinguished by the possession of a large, muscular, posterior, adhesive organ, the opisthaptor, which bears well-developed suckers as well as a number of sclerotized hooks. An oral sucker, the prohaptor, surrounds the mouth. Apart from this their structure is very similar to that described for the generalized trematode.

Monogenes have a relatively simple life cycle, with a single host and a free-swimming larva, the onchomiracidium, which is the agent of reinfection. There is no larval multiplicative phase. The possession of a free-swimming larva restricts monogenes to aquatic hosts such as fish and amphibians.

Polystoma integerrinum is a common parasite in the urinary bladder and cloaca of frogs and toads and has a life cycle which shows remarkable synchronization with that of its host. The parasite sheds its eggs when the amphibian returns to the water to breed. Each egg capsule contains a single fertilized egg and a number of yolk cells. After three to four weeks the egg hatches and the onchomiracidium is released. This has four eyes, a pair of protonephridia, a hooked opisthaptor, and swims by means of bands of cilia. Onchomiracidia can swim actively for up to 48 hours during which time they must attach to a new host, a tadpole rather than an adult frog. The onchomiracidium attaches to the gills of the tadpole by means of its opisthaptor and remains there, feeding on blood and mucus until the tadpole undergoes metamorphosis. The parasite then leaves the gills and migrates along the ventral surface to the cloaca and thence to the bladder where it develops its adult form by losing its eyes and cilia and by developing the posterior opisthaptor of the adult form. Onchomiracidia which hatch early in the year may become sexually mature while retaining their juvenile form (a process known as neoteny) and reproduce whilst attached to the tadpole gills. Larvae produced in this way settle on tadpoles later in the year, and develop to the adult form in the same way.

Order Digenea

The life history of digenetic trematodes is more complex than that of the monogenes with the introduction of at least a second host, with some digenes having as many as four host species in their life cycle. The presence of a second host duplicates the hazards involved in reinfection and in compensation for the possible losses incurred the digene life cycle includes a series of larval reproductive stages. Primary hosts (the host of the adult stage) are found in any of the vertebrate groups, with intermediate hosts being usually invertebrate and frequently snails. Eggs produced by the

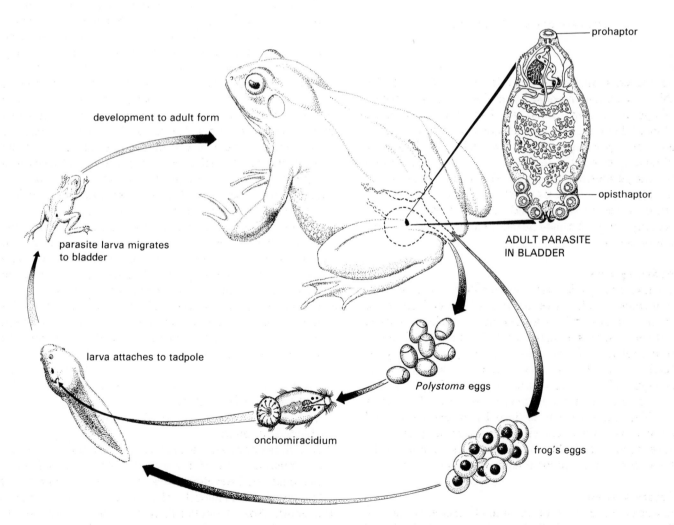

Life cycle of a representative monogene, *Polystoma*

adult parasite hatch to give a free-swimming ciliated larva, the miracidium, which infects the intermediate host. Within this host the miracidium gives rise to a second larval stage, the sporocyst. The germinal cells of the sporocyst develop into embryonic masses each of which forms a redia (a third larval stage). Each redia gives rise to a number of cercarian larvae (a fourth larval stage). The cercaria has a digestive tract, suckers and a tail. It may enter a second intermediate host and in any case encysts as a metacercaria. If the metacercaria is eaten by the primary host it is released from the cyst and will develop to the adult form. It has been estimated that one miracidium can give rise to as many as 10,000 cercariae as a result of the successive multiplications at each larval stage.

The life history of digenetic trematodes is both complex and highly variable, as may be demonstrated by three examples. *Fasciola hepatica*, the liver fluke, has a life history which is particularly well studied because of the great economic importance of this parasite. It infects cattle and sheep worldwide, causing 'liver rot' to an extent which may reach epidemic proportions amongst flocks of sheep. The adult *Fasciola* live in the liver and its ducts, feeding mainly on blood and also on liver cells. The structure of the adult is very similar to that described for the generalized trematode. It grows to a length of about 50 mm and is leaf-shaped with a conical projection at the anterior end at the apex of which is the oral sucker which surrounds the mouth. The ventral sucker is a short distance behind the oral sucker in the mid-line of the ventral surface. The tegument has backwardly-directed spines, which are thought to aid the fluke in moving through the bile ducts. The gut has paired intestinal caeca with many branches. The excretory tubules unite at a single nephridiopore near the posterior end.

Fasciola hepatica is hermaphrodite, and the reproductive system and process of fertilization are closely similar to that described for a generalized trematode. A considerable amount of body space is occupied by the paired, many branched, testes. A single ovary lies at one side of the mid-line at about a third of the total body length from the anterior end with a pair of diffuse vitelline glands. Fertilized eggs are stored in the uterus until they are shed, when they pass down the bile ducts of the host into the gut and leave in the faeces. While they remain in the faeces little development occurs, but if they are washed clean and remain moist they will hatch within a few weeks to give a miracidium. The miracidium is capable of swimming but can only survive for a few hours unless it finds and enters an intermediate host. In the case of *F. hepatica*, the intermediate host is a snail of the genus *Lymnaea*, the species varying between countries. In Britain *Lymnaea trunculata* occurs commonly in damp pastures, particularly in the puddles in muddy areas surrounding gateways and is the usual intermediate host. The miracidium bores its way into the tissues of the snail, digesting the epidermis by means of enzymes produced at its anterior tip by specialized penetration glands. This takes about 30 minutes. After penetration, the miracidium loses its cilia and enters the next stage in its life cycle, the sporocyst. The sporocyst has neither mouth nor gut, and absorbs food directly through the body wall. As it grows, several rediae (the next larval stage) develop inside it. The redia is an elongated larva with a suctorial pharynx and a simple gut. In due course rediae burst out of the sporocyst and travel to the snail's digestive gland where they remain, feeding on the digestive gland tissue and growing to a length of about 2 mm. Throughout the summer they reproduce asexually, producing either further rediae or the fourth larval stage, cercaria. In this way a very heavy infestation is built up within the snail host. Cercariae are heart-shaped, tailed structures, with oral and ventral suckers, a two-branched gut, an excretory system, and a rudimentary reproductive system. They leave the redia through an aperture and collect in a boil-like region near the snail's anus which they leave when the snail is in water. They then swim to vegetation to which they attach themselves. They lose their tails, become metacercariae, and encyst; in this condition they can survive for up to twelve months. If the cysts are swallowed by a mammal, normally a sheep or cow, the cyst walls are digested and metacercariae are released. They bore through the gut wall into the body cavity and after about two days reach the liver and tunnel through it to the ducts. By this time they are adult flukes.

The Chinese liver fluke, *Clonorchis sinesis*, is an example of a digenetic trematode with two intermediate hosts. It is found in the liver of man in Eastern Asia, especially amongst populations which include raw fish in their diet. The eggs are passed with the faeces, but do not hatch unless they are eaten by an aquatic snail, usually a member of the genus *Parafossarulus*. The miracidium bores through the gut wall and enters the digestive gland where it transforms into a sporocyst. Redian and cercarian stages follow, and the cercariae leave the snail. They swim in the water and find the second intermediate host, a freshwater fish which is generally a member of the carp family. The cercariae burrow through the fish's skin into the body wall muscles where they lose their tails and encyst. If the fish is eaten by man the young flukes escape, and make their way up the bile duct and so to the liver, where they feed on blood.

Blood flukes live in the blood of their hosts. The genus *Schistosoma* includes several species which are parasitic in man and cause the disease known as schistosomasis or Bilharzia. This can be extremely widespread in the tropics and sub-tropics, with up to 90% of the population being infected in some areas. *Schistosoma* is also known to infect rodents and monkeys. *Schistosoma mansoni* and *S. japonicum* live in the veins of the hepatic portal system which drain the intestine, and *S. haemotobium* in the veins draining the bladder.

Schistosoma mansoni is the most common and best known form. Unlike most platyhelminths the sexes are separate, although males and females are generally found together, the smaller, more slender female being carried within a ventral groove on the larger male, which is about 6–20 mm in length. At the time when eggs are laid, the

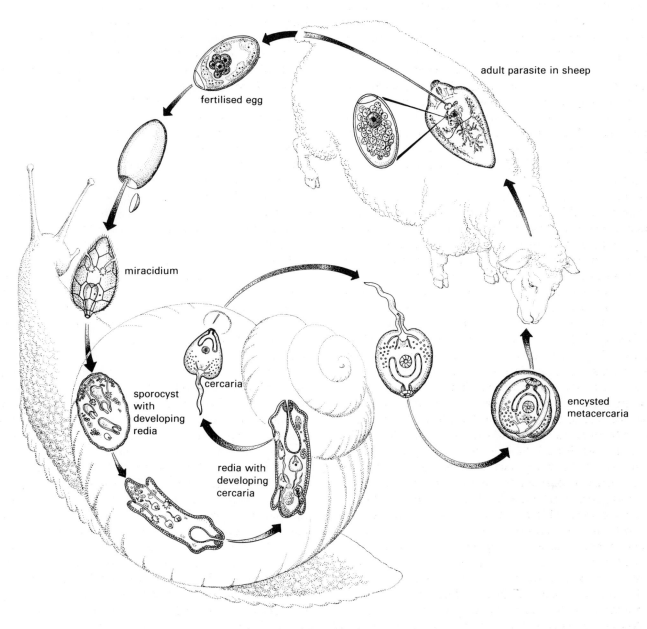

adult parasite in sheep

fertilised egg

miracidium

sporocyst
with
developing
redia

cercaria

redia with
developing
cercaria

encysted
metacercaria

Life cycle of *Fasciola hepatica*

flukes move towards the gut through smaller and smaller vessels until prevented by the decreased diameter and release their eggs into the host's blood. Each egg bears a spine and breaks through the walls of the veins and through the tissues causing considerable damage. Some reach the intestinal cavity and leave with the faeces. If these are diluted with water, miracidia hatch, infect freshwater snails and develop as sporocysts. Two generations of sporocysts follow each other in the snail but the redia stage is absent. The cercariae escape through the body wall. They do not encyst, but if they come into contact with human skin, they penetrate by means of enzymes and muscular action, a process which takes about 7 minutes. The cercaria then sheds its tail and is carried by the blood stream via the lungs and liver to the hepatic portal veins, during which time it develops to the adult form. The life cycles of *S. japonicum* and *S. haematobium* are both similar to that of *S. mansoni*, but the eggs of *S. haematobium* leave through the bladder with the urine.

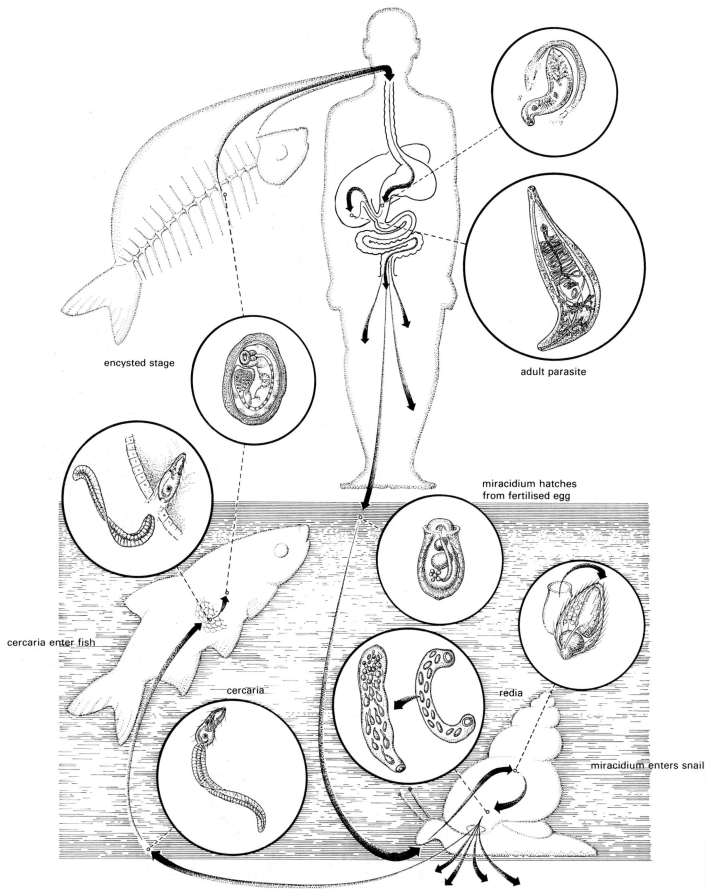

adult parasite

encysted stage

miracidium hatches
from fertilised egg

cercaria enter fish

cercaria

redia

miracidium enters snail

Life cycle of Chinese liver fluke, *Clonorchis*

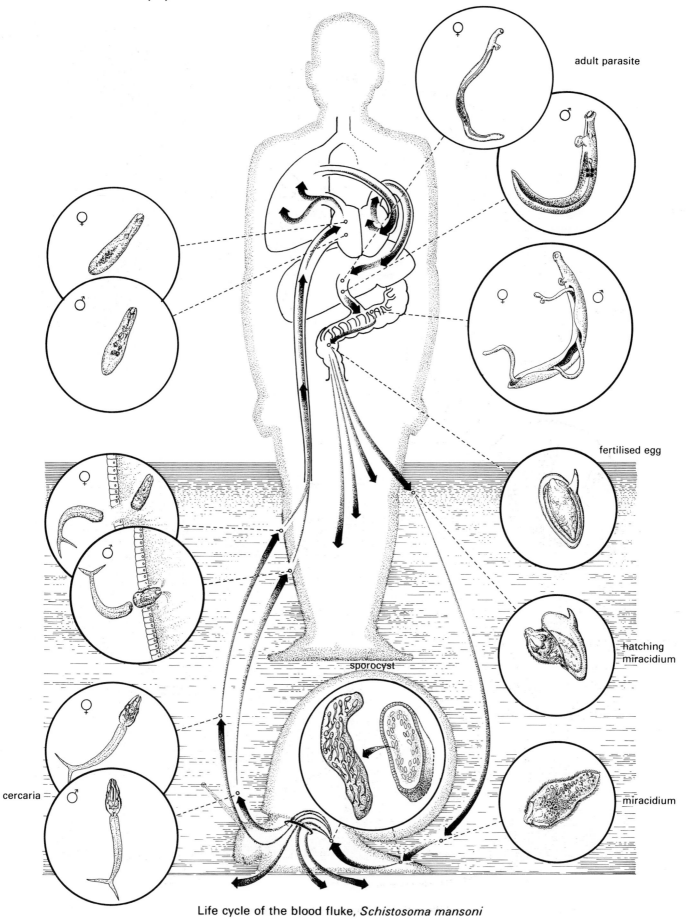

adult parasite

fertilised egg

hatching
miracidium

sporocyst

miracidium

cercaria

Life cycle of the blood fluke, *Schistosoma mansoni*

Class Cestoda

The Cestoda (the name is derived from the Latin *cestus*, meaning 'girdle') are endoparasites which live in the small intestine of vertebrates and differ from the other two platyhelminth classes in their total lack of a gut. The majority (members of the subclass Eucestoda) are long and ribbon-like in form and are mostly intestinal parasites of carnivores. A few (subclass Cestodaria) have a superficial resemblance to the flukes, with a leaf-like form, and are parasitic on cartilaginous and primitive bony fish.

Subclass Eucestoda

Eucestodes, for example *Taenia*, are very elongated, with body lengths of up to 10 metres having been recorded in some species, and are commonly known as tapeworms. The body is divided into three regions. The most anterior region is the scolex, a small knob-like structure adapted for attachment to the host's intestinal wall, with well developed suckers and hooks. A narrow neck gives rise to a third region, the strobila, which makes up the greater part of the animal's length. The strobila consists of many linearly arranged sections, the proglottids. These are formed by a series of transverse constrictions at the neck (a process known as strobilization) so that the youngest proglottids are at the anterior end. A proglottid may be regarded in some ways as being equivalent to an individual trematode with a complete set of male and female reproductive organs and part of the nervous and excretory systems. As they reach maturity they are shed from the posterior end. Eucestodes are more obviously anatomically adapted for parasitism than trematodes; apart from the total absence of a gut, both nervous and excretory systems show some reduction. The tegument is well developed and similar in structure to that of

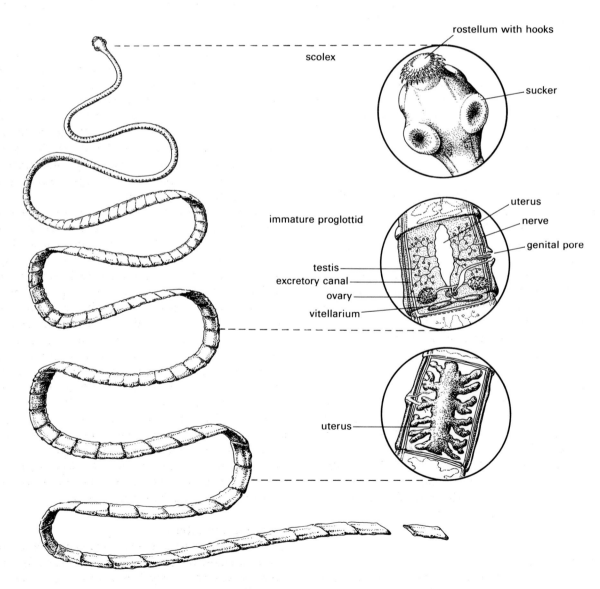

rostellum with hooks

scolex

sucker

immature proglottid

uterus

nerve

genital pore

testis

excretory canal

ovary

vitellarium

uterus

Structure of *Taenia solium*

the trematodes. However, the surface membrane is folded in a series of microvilli which greatly increase its surface area. This is related to the increasing importance of the tegument in food absorption in cestodes, in the absence of a gut. In addition to the circular and longitudinal muscles which lie beneath the basement membrane, tapeworms have an additional parenchymal musculature of longitudinal, dorsoventral and transverse fibres.

Tapeworms have a life cycle which includes two or more intermediate hosts, the animals involved being vertebrates and arthropods. The life cycle includes two basic developmental stages, the oncosphere larva which hatches from the egg and the cysticercus which gives rise to the adults. Like that of the flukes, the tapeworm life cycle is subject to variation between different species. Some of the best known tapeworms are species of the genus *Taenia*, which is a parasite of man. *Taenia solium*, the pork tapeworm, has the pig as an intermediate host and was until fairly recently a common (human) intestinal parasite in Europe. *Taeniarhynchus saginatus* is also a parasite of man and has a similar life cycle, but with the cow as an intermediate host.

Taenia solium

The adults of *T. solium* live in the small intestine of man, and can reach lengths of up to 7 metres. It attaches to the intestinal wall by means of a small muscular knob-like scolex, which bears four circular suckers, and a double row of curved, chitinous hooks on a terminal rostellum.

Excretory system

The excretory system consists of protonephridia with flame bulbs, scattered throughout the parenchyma. These drain via a system of collecting tubules into four longitudinal canals, two dorsolateral and two ventrolateral. In young *Taenia* the canals end at a median bladder which opens via a median nephridiopore. The dorsolateral canals are normally lost in the mature proglottids, and the broken ends of the ventral canals serve as nephridiopores. The ventrolateral canals are linked by a transverse canal near the posterior margin of each proglottid, and also join in the mid-line of the scolex.

Nervous system

The nervous system is reduced and there are no special sense organs although sensory nerve endings are present over the body surface and are especially concentrated on the suckers. Two lateral longitudinal nerve cords, a pair of accessory lateral nerves and paired dorsal and ventral longitudinal nerves extend from an anterior nerve mass in the scolex and are connected by ring commissures in each proglottid. Nerves extend from the nerve mass to each of the suckers.

Reproductive system

The reproductive system has the same basic plan as that of other platyhelminths with a single bilobed ovary, scattered rounded testes, and scattered vitellaria. The uterus is a blind ending sac with many branches. The system develops progressively from anterior to posterior ends of the strobila, with the male organs maturing before the female, so that functional male organs are present in the anterior proglottids, male and female organs in the centre of the strobila, and a greatly enlarged uterus containing egg capsules and embryos at the posterior end. The genital atrium opens irregularly on either side of the body and fertilization can occur between proglottids, or within the same proglottid, in the lower part of the oviduct, though cross-fertilization is the general rule. The fertilized eggs are stored in the uterus, and development begins immediately to give a six-hooked hexacanth embryo within the egg capsule, the whole structure being known as the oncosphere. The proglottids containing these embryos are shed by the strobila and pass out with the host's faeces. The eggs do not develop further unless eaten by a suitable secondary host, which in *T. solium* is normally a pig but may be another animal or even man. On hatching, the larvae bore into the intestinal wall by means of their hooks and enter the host's blood stream. After being transported to striated muscle the larva leaves the blood and develops to the cysticercus stage, or bladderworm. This is oval, about 10 mm in length and consists of a fluid-filled bladder surrounding an inverted scolex. If infected pork which is insufficiently cooked is eaten by man, the bladder is digested, the scolex everts itself, attaches to the intestinal wall, and begins to bud off proglottids. During the course of its adult life a tapeworm produces large numbers of eggs and is an excellent example of one strategy used by a parasite to ensure its successful reproduction.

The beef tapeworm, *Taeniarhynchus saginatus*, is a closely-related genus with a similar structure and life cycle. However the scolex lacks both rostellum and hooks, and the intermediate host is bovine. The fish tapeworm, *Diphyllobothrium latum*, has a wide distribution occurring in the gut of many carnivores including man, bears, seals and porpoises. Unlike *Taenia*, it has a free-swimming oncosphere and two secondary hosts. If fertilized eggs are deposited in water the free-swimming oncosphore hatches after approximately ten days. On ingestion by a copepod crustacean, for example *Cyclops strenuus*, the oncosphere burrows through the intestinal wall to develop to a six-hooked procercoid within the body cavity. If the procercoid is ingested by a freshwater fish, which may be a pike, salmon, or trout amongst many others, it develops to a plerocercoid, which looks like an unsegmented tapeworm, in the striated muscle. This will develop to an adult form (which may attain lengths of 800 mm and have over 3,000 segments) if ingested by a primary host.

Subclass Cestodaria

The cestodarians are a small group of gut parasites of fishes, which resemble the trematodes in their general body form. Like the cestodes, they lack a digestive system and are normally regarded as a subclass of that group.

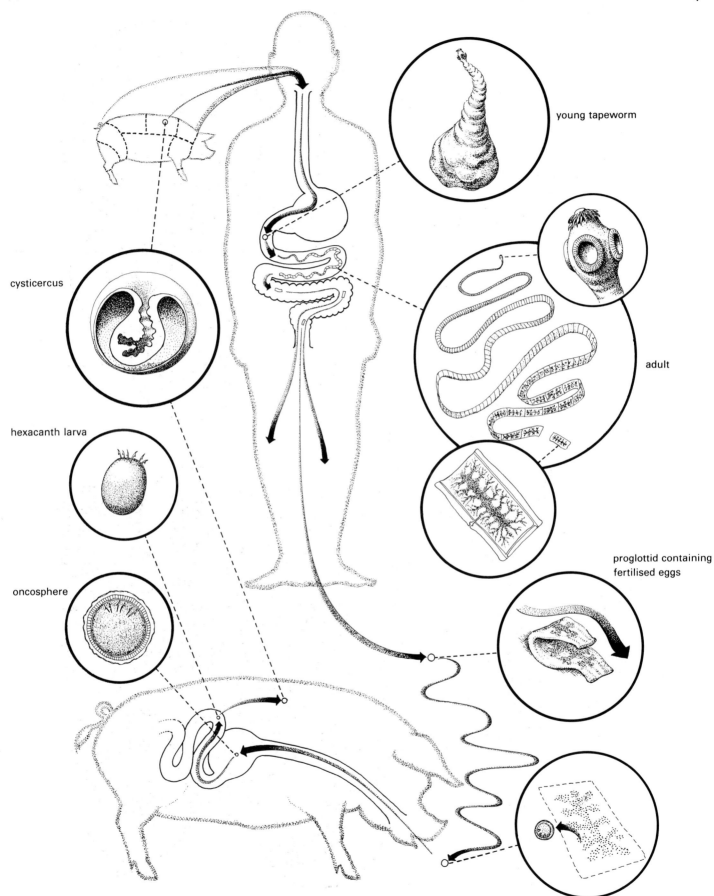

young tapeworm

cysticercus

hexacanth larva

oncosphere

adult

proglottid containing
fertilised eggs

Life cycle of the pork tapeworm, *Taenia solium*

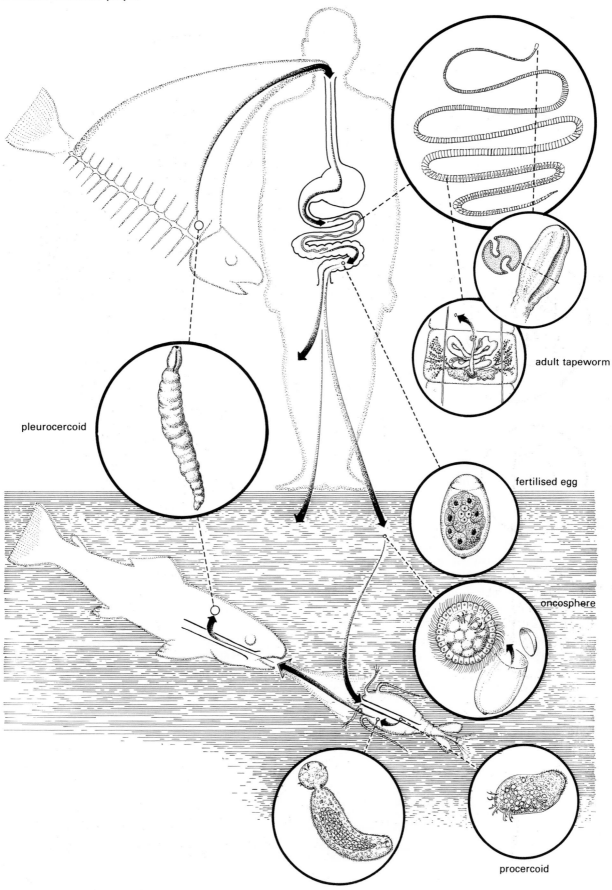

pleurocercoid

adult tapeworm

fertilised egg

oncosphere

procercoid

Life cycle of the fish tapeworm, *Diphyllobothrium latum*

Synopsis of the Platyhelminthes

Class Turbellaria
    Many orders of which the most important are:
    Order Acoela
    Order Polycladida
    Order Tricladida (*Dugesia*)
Class Trematoda
    Order Monogenea (*Polystoma*)
    Order Digenea (*Fasciola, Clonorchis, Schistosoma*)
    Order Aspidobothrea
Class Cestoda
    Subclass Eucestoda (*Taenia, Taeniarhyncus, Diphyllobothrium*)
    Subclass Cestodaria

PHYLUM NEMERTINEA

The phylum Nemertinea is a small group of animals of about 600 species occurring commonly in the intertidal zone of temperate oceans. They are comparable with the platyhelminthes in their general organization, being vermiform with a ciliated epidermis (with rhabdites in some species), parenchyma, and a protonephridial flame bulb excretory system. Like platyhelminths, they have a creeping, burrowing mode of life. However, nemerteans show important differences in organization compared to the platyhelminths. The efficiency of the digestive system is increased by the possession of a posterior opening, the anus. This has several advantages: it allows ingestion and egestion to take place simultaneously; it avoids the mixing of indigested and digested food inevitable in the gastrovascular cavity characteristic of platyhelminths and coelenterates; and it permits specialization of the gut into regions. In a gastrovascular cavity the entire lining epithelium must be capable of digestion and absorption. A one-way system can have areas specialized for the separate functions of ingestion, digestion, absorption and egestion. The gastrovascular cavity, as its name implies, is also responsible for internal distribution and transport of metabolites. In nemerteans these tasks are performed by a circulatory system. The increased efficiency of the digestive and circulatory systems makes nemerteans capable of a more active mode of life and this requires better developed and more highly organized muscle systems. In association with their increased activity, nemerteans show an increased level of cephalization.

Nemerteans are active and effective predators, catching their prey by means of an elongated proboscis. Indeed the name of the phylum, Nemertinea, is based on their effective use of this structure. Cuvier, who first introduced the term 'Nemertes' named them after a Greek sea nymph, Nemertes, 'the Unerring One'. Some authorities prefer the term Rhynchocoela which was coined by Schultze, the first zoologist to describe the anatomy of the group in 1850–51. The name refers to the possession of a 'rhynchocoele' (described below) in which the proboscis lies when not extended. Common names include 'proboscis worms' and 'ribbon worms'.

Nemerteans are predominantly marine, living along the shore or in shallow water, in cracks, under stones or amongst seaweeds. Some, for example *Tubulanus*, secrete mucous tubes for protection. The group as a whole shows little variation. The majority are a few centimetres in length, although some are restricted in size to a few millimetres. *Lineus longissimus*, the boot-lace worm, is remarkable for growing up to 5m although specimens reaching lengths of between 20 and 30m have been recorded. The majority of nemerteans are dull or pallid in colour, but the group does include a few brightly-coloured species.

The nemerteans are divided into two classes: Anopla, with an unarmed proboscis; and Enopla, in which the proboscis is armed with sharply-pointed stylets. The order Heteronemertini, within the Anopla, includes several common genera including *Lineus* and *Cerebratulus*, and may be taken as representative of the phylum. *Lineus ruber* is a common species living under stones or in muddy gravel, both on the shore and in deeper water. It is of moderate size (up to 160 mm in length) and of a reddish colour. *Cerebratulus fuscus* is approximately 100 mm long, and is grey or brown, living amongst seaweeds, shells or pebbles.

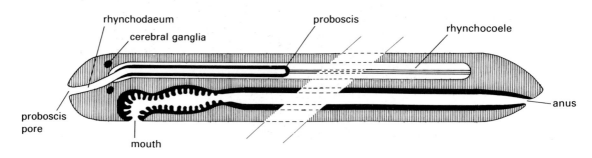

Longitudinal section of a nemertean

General body organization and histology

The majority of nemerteans are vermiform and their bodies are highly extensible; *Lineus* has been shown to extend itself up to ten times its contracted length. The anterior region may have side lobes, making it marginally wider than the rest of the body. However this structure cannot be regarded as a head, as it frequently does not incorporate the 'brain'. Deep lateral cephalic grooves are present on the anterior

ends of many nemerteans and are particularly well-developed in *Lineus*.

Apart from the additional organ systems already described the general body plan is similar to that of the Turbellaria. Gut, excretory and circulatory systems and the reproductive organs are embedded in a mass of parenchyma. In comparison with the Turbellaria, the parenchyma shows a reduction in the number of cells present and an increase in the interstitial fluid.

The body wall is composed of three distinct regions: epidermis, dermis and musculature. The epidermis is formed by an epithelium of wedge-shaped closely ciliated cells, with microvilli between the cilia. The spaces between the narrowed cell bases are filled by a mass of small, branched, interstitial cells which sometimes form a syncytium. Rhabdite-forming cells, similar to those of platyhelminths, are present. In addition numerous mucus glands are present which are either unicellular flask-shaped or elongated structures within the epidermis, or subepidermal pocket glands with ducts to the exterior. Like the turbellarians, nemerteans leave extensive slime-trails as they move.

A dermis of connective tissue lies beneath the epidermis; this is differentiated into an outer gelatinous layer in which nuclei, fibres, and cells are embedded, and an inner fibrous layer. The fibres (probably collagen) are arranged in an elaborate lattice of right- and left-handed geodesic helices. They are attached to each other where they cross, and help to maintain the shape of the animal. The body wall musculature is thick and powerful, and the precise arrangement of the muscle layers is a diagnostic feature in nemertean classification. In heteronemerteans two longitudinal muscle layers lie on either side of a bed of circular muscle. Dorsoventral muscles connect the dorsal and ventral body walls and surround the digestive tube.

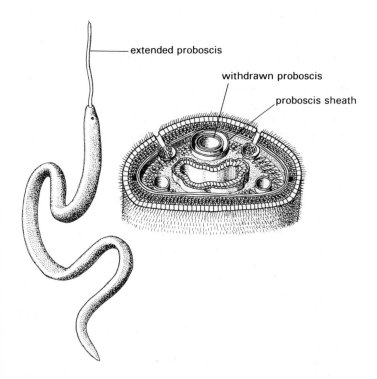

General body organisation of a heteronemertean

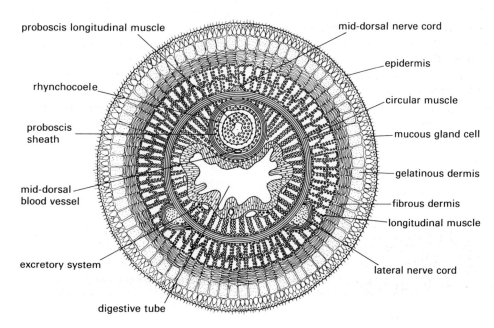

Transverse section of a heteronemertean

The structure of the proboscis

The proboscis, with its associated structures, is characteristic of the phylum and is used both to trap prey and for defence. Three components are recognizable: the proboscis itself, the proboscis sheath, and the rhynchodaeum.

The heteronemertean proboscis is an eversible tubular structure, the same length as the body or longer when fully extended. When not in use it lies dorsal to the gut in a completely closed, fluid-filled cavity, the rhynchocoele. This is surrounded by muscular walls, and the whole structure forms the proboscis sheath. The base of the proboscis is attached to a short canal, the rhynchodaeum, which appears to the exterior at the anterior tip of the worm through the proboscis pore. Usually a sphincter muscle marks the junction of the proboscis and the rhynchodaeum. The blind end of the proboscis (which forms the anterior tip when the structure is everted) is attached to the posterior end of the proboscis sheath by a retractor muscle. The histology of the proboscis wall is similar to that of the body wall, consisting of an outer epithelium of columnar cells, a thin layer of connective tissue, and muscle layers. The circular muscle layer is generally poorly developed and in some species, for example *Lineus rubescens*, the innermost longitudinal muscle layer is absent.

Many of the epithelial cells of the proboscis contain rod-shaped secretions, similar in appearance to the rhabdites found in the epidermis. They may be aggregated in groups and can be projected from the surface when the proboscis is everted. They are thought to have a role in wounding potential prey. Structures similar in appearance to nematocysts are also present in longitudinal wells in the anterior proboscis of some species of both *Cerebratulus* and *Lineus*. They produce stinging filaments and assist the proboscis in gripping its prey. Once they have been discharged they are replaced.

The proboscis is protruded by contraction of the circular muscle of the proboscis sheath (and possibly of the body wall) which raises the pressure within the rhynchocoele and causes the proboscis to shoot out through the proboscis pore, everting as it does so. Whilst the proboscis is in use the pressure in the rhynchocoele is maintained by blood vessels, the rhynchocoele villi, which lie on either side of the ventral midline of the rhynchocoele and become distended with blood. The proboscis is drawn back into the sheath by the retractor muscle at the same time as the hydrostatic pressure is reduced.

Locomotion

Like turbellarians, many small nemerteans move by ciliary action over a mucus film produced by the numerous mucus glands. This method of progression is extremely slow (less than 10 mm per minute) and larger forms move by muscular contraction. The muscles have undergone considerable development in comparison with those of platyhelminths. *Lineus* has been recorded as moving at 100 mm per minute by peristaltic contraction, its body being alternately flattened and elongated by means of alternate waves of contraction of the longitudinal and circular muscle. This is a very effective means of travel over a substrate or within a burrow. Bulges produced by longitudinal muscle contraction act as fixed points against which other parts of the body can push. Contraction of the circular muscles elongate part of the body which is therefore pushed forwards. The incompressible gelatinous parenchyma functions as a hydrostatic skeleton, enabling muscular contraction to bring about a change in the worm's shape rather than an alteration in its size. In addition the fibrous lattice system of the dermis assists in maintaining a constant volume.

Digestive system

Nemerteans are predators and scavengers, feeding on many kinds of small invertebrates which they attack by means of their proboscis and swallow whole. Generally they are not selective in their diet; *Lineus corrugatus* has been shown,

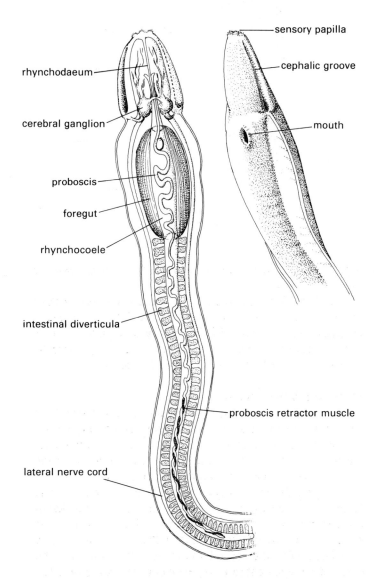

rhynchodaeum

cerebral ganglion

proboscis

foregut

rhynchocoele

intestinal diverticula

lateral nerve cord

sensory papilla

cephalic groove

mouth

proboscis retractor muscle

General anatomy of *Cerebratulus*

under laboratory conditions, to take diatoms, sponges, sea anemones, annelids, amphipods, isopods, gastropods, fish, seal blubber and skin, and sardines in tomato sauce! However this diversity of diet is probably exceptional. Food is detected by either sight or scent; for example *Lineus ruber* can detect living prey by sight at a distance of 20 to 30 mm, and damaged prey by scent at a distance of 80 mm.

The alimentary tract extends the full length of the body and can be divided into fore-gut (with a buccal cavity, oesophagus and stomach), mid-gut (intestine) and a short hind-gut leading to the anus. In *Lineus*, as in the majority of nemerteans, the regions of the fore-gut cannot be distinguished externally. The mouth is a highly elastic structure which lies ventrally near the anterior tip and has thick glandular lips. The epithelium of the lips and the buccal cavity resembles the epidermis in structure with mucus- and acid-secreting glands. The oesophagus has a thin epithelium without gland cells; the stomach is highly glandular, the glands producing similar secretions to those of the buccal cavity. *Lineus* kills its prey by acid secretions in the fore-gut and digestion takes place in the intestine.

The intestinal wall is composed of elongated cells with cilia which move the food along the gut, and microvilli which increase the absorption surface. There are numerous intestinal gland cells towards the anterior end of the intestine, and fewer posteriorly. The surface area of the intestine is increased by lateral caecae of the same histological structure as the main part of the intestine.

The digestive process is very rapid with extracellular digestion playing an important part. The food is broken down by enzymes, secreted by the intestinal gland cells, which digest protein, carbohydrate and fat. The digestive process is completed intracellularly.

Circulatory system

Unlike platyhelminths, nemerteans have a distinct circulatory system with two types of vessels: true tubular vessels, and larger lacunae, which are embedded in the parenchyma. Nemertean blood is usually colourless (though green, yellow, red and orange blood occurs) and consists of a clear fluid containing nucleolated blood corpuscles and amoebocytes. *Lineus* has four different types of amoebocytes.

The walls of the blood vessels vary in structure. The lacunae have very thin walls formed by a thin endothelium of flattened cells. The larger tubular vessels are contractile and have a four-layered structure: endothelium, connective tissue, circular muscle and a nucleolated surface layer. Non-contractile tubular vessels have an endothelium and a surface layer separated by a thin membrane.

In *Lineus* the circulatory system consists basically of three longitudinal vessels, two placed laterally and one mid-dorsally. The longitudinal vessels are connected at the anterior end by the cephalic lacuna, and at the posterior end by the anal lacuna. Paired rhyncocoele blood vessels extend on either side of the mid-dorsal vessel from the cephalic lacuna to form the rhynchocoele villi. In the intestinal

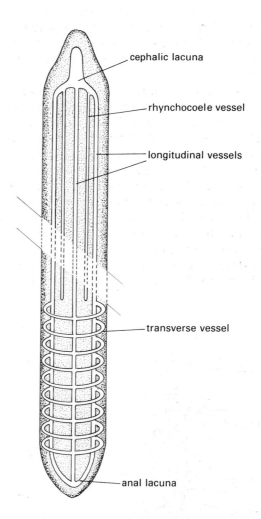

Lineus : circulatory system

region, transverse vessels connect the mid-dorsal vessel to the lateral vessels.

The blood flow is not regular but usually flows forwards in the mid-dorsal vessel and back in the lateral vessels, a common arrangement in invertebrates. Circulation is brought about by local contractions in the larger vessels, assisted by the action of the body wall muscles.

Osmoregulation and excretion

Like the platyhelminths, nemerteans have a flame bulb system. However, as a circulatory system is present, it is not necessary for the flame bulbs to be scattered throughout the parenchyma, and they are normally concentrated near the lateral blood vessels in the anterior region of the worm.

A single pair of protonephridial tubules extend anteriorly from a pair of nephridiopores which lie one on either side of the foregut. Branches from the inner side of each tubule each give rise to terminal flame bulbs, which are closely associated with the lateral blood vessels. However, the detailed arrangement can vary even between closely-related species.

The primary function of the flame bulb system is probably osmoregulation, but in addition waste metabolites are transported to the flame bulbs by the blood.

Gaseous exchange

Gaseous exchange takes place all over the body surface, by direct diffusion through the mucus produced by epidermal and sub-epidermal glands. The gases are transported via the circulatory system in physical solution in the blood.

Some of the bulkier nemerteans, for example *Cerebratulus lacteus*, have a simple type of 'breathing system'. Water is pumped in and out of the fore-gut, gaseous exchange taking place through the gut wall which is richly supplied with blood.

Nervous system

While the basic plan of the nervous system is similar to that of the turbellarians it shows some advances, with greater cephalization and extensive nerves associated with the well developed body musculature. There is a higher degree of centralization (concentration of nervous tissue) with the development of larger cerebral ganglia and a single pair of ganglionated nerve cords.

The cerebral ganglia are four-lobed, with dorsal and ventral lobes surrounding the rhynchodaeum on either side of the mid-line. They are joined to form a ring by the dorsal and ventral commissures. The ventral lobes are extended posteriorly over the full body length as lateral nerves and unite at the anal commissure above the intestine. They are also connected by nerve plexi at intervals.

In addition to the main system described there are several minor nerves, including well developed paired oesophageal, cephalic, and proboscis nerves, and a median dorsal nerve along the body length. Peripheral nerves extend to the epidermis and epidermal sense organs, eyes, and body wall musculature.

The sense organs are similar to those of the turbellarians. Tactile sense organs, each with a single, projecting cilium, are widely scattered in the epidermis with chemoreceptors concentrated in the cephalic grooves. Pigment cup ocelli are common, *Lineus* having four or six at its anterior end.

Cerebral organs are peculiar structures of unknown function, characteristic of nemerteans. They consist of a pair of invaginated epidermal canals opening to the exterior via the cephalic grooves and with their inner ends embedded in a mass of glandular and nervous tissue. They are closely associated with the blood system and the cerebral ganglia. Water currents are maintained through the canals. The cerebral organs may be very large indeed; those of *Lineus* are particularly well developed and are thought to be chemoreceptors or endocrine organs.

Reproductive system

The reproductive system of nemerteans is extremely simple, without any of the complicated glands and ducts present in platyhelminths. The sexes are generally separate, fertilization is external and the eggs and sperm are passed directly to the exterior. Reproduction normally takes place in the summer.

The gonads develop from parenchyma cells as small, thin-walled, flask-shaped sacs lying between the branches of the intestine. When the animal is sexually mature a short duct grows from each gonad to open at the side of the body. In *Lineus* up to fifty eggs are produced by each ovary and these are laid in strings of mucus produced by the epidermal glands. They are squeezed out of the ovaries by muscular contraction of the body wall. The eggs hatch to give a small gelatinous pilidium, which is helmet shaped and ciliated along its margins. It has a simple sac-like gut with a ventral mouth and no anus, and an apical sense organ with a tuft of stiff cilia comparable in some with the sense organ of ctenophores. Metamorphosis to the adult is a remarkable process. A series of invaginations develop from the surface of the pilidium to surround the gut. A larva is thus formed around the pilidium gut and this escapes and develops as an adult worm.

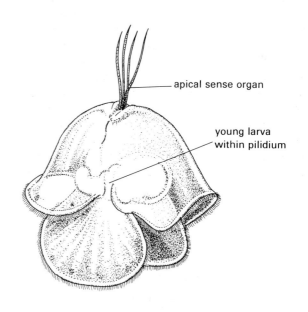

apical sense organ

young larva within pilidium

A pilidium

Some species of *Lineus* are ovoviviparous; that is to say the eggs develop within the female to give rise to a miniature adult.

Regeneration and asexual reproduction

Nemerteans are capable of regeneration to a greater or lesser degree. *Lineus ruber* can only develop a new posterior end; on the other hand *Lineus socialis* can undergo successive regeneration of fragments to produce over a hundred individuals. *Lineus socialis* is capable of asexual reproduction by fragmentation of the adult followed by regeneration to give around twenty individuals. The larger fragments regenerate without any protection but smaller pieces are protected during the process by a mucous cyst.

Nemertean diversity and classification

As a phylum, the nemerteans show far less diversity than platyhelminths or cnidarians; their anatomy is relatively constant and most modifications are only minor variations in organ arrangement. The majority are marine bottom-dwellers in littoral or coastal waters, although some pelagic and some abyssal forms are known. *Lineus ruber* is commonly found on the British coast in estuarine sand and mud, as well as in the intertidal zone, and appears to have a wide range of salinity tolerance. A few have invaded estuarine, freshwater and even terrestrial habitats.

The majority of freshwater species are members of the genus *Prostoma*, which is of very wide distribution. The terrestrial species belong to a single genus, *Geonemertes*, which is largely restricted to the tropics and sub-tropics, although *Geomertes dendyi* is found in parts of the British Isles as an introduction from Australia. Nemerteans are poorly adapted to a terrestrial mode of life and their dependence on copious quantities of mucus for survival and movement restricts them to humid areas. Their sole water-saving device is the ability to produce a hardened mucus cocoon in times of drought. *Malacobdella* leads a specialized life within the mantle cavity of various bivalve molluscs; it has developed a posterior sucker for attachment to its host and lacks special sense organs. *Carcinonemertes* has a close association with certain crabs, and is regarded by some authorities as being truly parasitic, since the adult appear to feed extensively on the eggs of the host.

Some of the greatest nemertean variation is seen in their reproductive behaviour and life cycle, though the reproductive system and life cycle described is fairly typical. *Prostoma* is unusual in being both hermaphrodite and viviparous. Internal fertilization is characteristic of abyssal and terrestrial nemerteans where encounters between free gametes would be few.

The pilidium is normally regarded as a means of dispersal though its importance is variable and three common species of shore-dwelling nemerteans have different life histories. *Amphiporus lactifloreus* has no larval stage and the young leaves the egg at the crawling stage; the pilidium of *Lineus ruber* is only short lived, and *Cephalothrix* larvae survive in the plankton for up to three weeks.

The phylum comprises two classes, each with two orders. The members of the class Anopla have a simple proboscis. They include the Palaeonemertini, which are the most primitive members of the phylum, and the Heteronemertini, with many of the most common and extensively studied nemerteans (including *Cerebratulus* and *Lineus*). Within the class Enopla the members of the order Hoplonemertini all have a more specialized proboscis armed with a heavy barb or stylet. Many of the ecologically interesting nemerteans belong to this group, as well as several from the more usual marine habitats. The order includes both *Prostoma* and *Geonemertes*, as well as *Carcinonemertes* and the pelagic nemerteans. *Malacobdella* is the only genus in the order Bdellonemertini. Although the proboscis is not armed, it is believed to be derived from the armed form.

Synopsis of phylum Nemertinea

Class Anopla
    Order Palaeonemertini (*Tubulanus*).
    Order Heteronemertini (*Cerebratulus, Lineus*).
Class Enopla
    Order Hoplonemertini (*Geonemertes, Prostoma, Carcinonemertes, Amphiporus*).
    Order Bdellonemertini (*Malacobdella*).

Phylum Mesozoa

The Mesozoa are a peculiar group of minute, worm-like animals, including about fifty known species. They are all endoparasites of a variety of invertebrate groups including platyhelminths, nemerteans, annelids, molluscs and echinoderms, and are characterized by a highly unusual, simple, solid two-layered structure and complex life cycles reminiscent of those of parasitic protozoa. There are two mesozoan orders, with very different structures and life cycles. The simplest is the order Orthonectida. Adult orthonectids have an outer ciliated layer and an inner mass of reproductive cells, the sexes being separate. Sexually mature adults leave the host and the eggs are fertilized within the female by sperm produced by the male. The

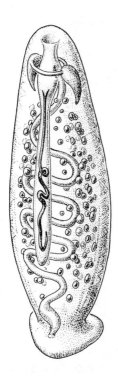

Malacobdella

fertilized egg develops to give a ciliated larva and this infects a new host. After penetrating the host's tissues, the larva loses its cilia and becomes a syncytial plasmodium and gives rise to both male and female gametes which develop to the adult form.

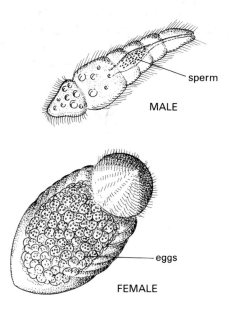

Phylum Mesozoa : adult male and female orthonectids

The mesozoans of the order Dicyemida are restricted in distribution to the kidneys of cuttlefish, squids and octopodes. Within the juvenile host the dicyemid exists as a nematogen. This has an outer layer of ciliated cells surrounding a single elongated central axial cell. The anterior end is used for attachment and consists of an eight- or nine-celled polar cap, with two parapolar cells. In mature nematogens the axial cell divides to form germ-cell nuclei. Agametes form round these, and produce daughter nematogens, the vermiform larvae, which have the same form as the adult and reproduce in a similar way.

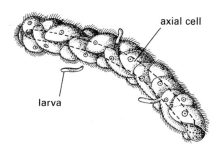

Adult dicyemid

In mature hosts, the parasite produces a different type of larva through a sexually reproductive stage. This develops an infusigorium, which may be interpreted as a single gonad producing both eggs and sperm, within the axial cell. Fertilization occurs within the axial cell, and a minute infusorium larva develops. This is a short, oval structure, generally ciliated but with two unciliated apical cells and with a central cavity, the urn cavity. The larvae leave with the host's urine, and, though the rest of the life story is unknown, it is probable that they infect a new host.

In terms of this classification mesozoans are a problematic group, and two interpretations have been given of their evolutionary position. Simplicity of structure may arise for two different reasons. The animal may be primitive or, like many parasites, having once been more complicated it may be degenerate. Some authorities regard the mesozoans as degenerate flatworms, in which case they should be regarded as a class of the Platyhelminthes, others regard them as an early metazoan offshoot from the Protozoa, with a long independent history of parasitism.

Phylum Gnathostomulida

Gnathostomulida is a small phylum (about eighty species) of recently discovered acoelomate marine worms, which live in the spaces between sand grains. They are microscopic (0.5 to 1 mm) and have an elongated thread-like body form. They have a ciliated epidermis but a highly unusual feature

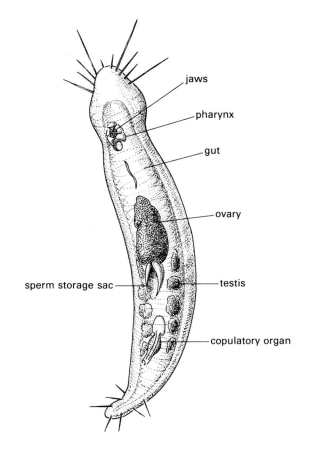

Gnathostomula

is that each epithelial cell has only one cilium. Gnathostomulids move by ciliary action, but can also swim by contractions of their simple bands of longitudinal muscles. Like flatworms they have parenchyma, though it is poorly developed, and a simple gut with a mouth but no anus. They feed on bacteria and fungi which they seize by a pair of toothed jaw-like structures. Gnathostomulids lack excretory or respiratory organs; they are thought to be capable of anaerobic respiration since most of them occur in sands where no oxygen is present.

Gnathostomulids are hermaphrodite with simple reproductive systems. The female system consists of an ovary and a sac for sperm storage; the male system has a pair of testes and a copulatory organ. The precise method of copulation is not known and probably varies somewhat within the group; some species have a vagina, but others may be impregnated directly through the body wall. After fertilization, the body wall is ruptured and the eggs laid. The worms have considerable powers of regeneration.

The relationships of the phylum are obscure; in this book they are placed with other aceolomate groups for convenience.

Animal body cavities

The phyla discussed so far have been characterised by a 'solid' type of body construction with either mesogloea or mesenchyme filling the space between the external body wall and the gastrodermis. All the remaining animal phyla have a body cavity separating the body wall from the gut.

The possession of a body cavity has two major advantages. Whereas organs that lie in solid tissue are squeezed or compressed every time the animal moves, a body cavity permits them freedom of movement, for example providing for more efficient transport of gut contents. Secondly, a body cavity provides space for the gonads to develop, and for sperm and eggs to be stored in a constant environment. The animal is therefore able to restrict its breeding period to the most favourable times of the year without any loss of offspring in comparison with animals which breed all the year round.

There are additional advantages if the body cavity is filled with fluid. The fluid can function as a transporting medium, improving exchange of metabolites and excretory products between the internal organs and the body wall. Structures such as excretory organs can therefore be reduced in number, as there is no longer the need to produce a network of these structures throughout the animal. Furthermore a fluid-filled body cavity can function as a highly efficient hydrostatic skeleton. The digestive cavity serves this function in forms like *Hydra*, but this is far from ideal, and possible only because the body walls are very thin and muscular contraction is able to compress the fluid contents directly. With the development of a thicker body wall and of internal organs, most of the force produced by contraction of the body wall muscles would be wasted in compressing these tissues. The parenchyma of the Platyhelminthes behaves more or less like a liquid, although its cellular nature means that pressure changes are damped down to a certain extent and the alterations in body shape are mechanically inefficient and therefore slow. With the development of a fluid-filled body cavity the hydrostatic skeleton is able to lie in contact with the muscles that relate to it, with considerable increase in mechanical efficiency and rapidity of movement.

Two types of body cavity are found in animals. The commonest is a coelom, occurring in several major phyla: annelids, arthropods, molluscs, echinoderms and chordates. A coelom is a space within the mesoderm, which forms both a lining to the body wall and a cover to the gut tube. The gut tube therefore has well developed muscle layers and the internal organs are held in place by mesodermal mesenteries. The other type of body cavity, the pseudocoelom, occurs in one major phylum, Nematoda, and several minor phyla. It occupies a space between the mesoderm, which forms the muscle of the body wall, and

the endoderm which forms the gut wall. The gut wall therefore has no muscle layers and the internal organs are not supported by mesenteries.

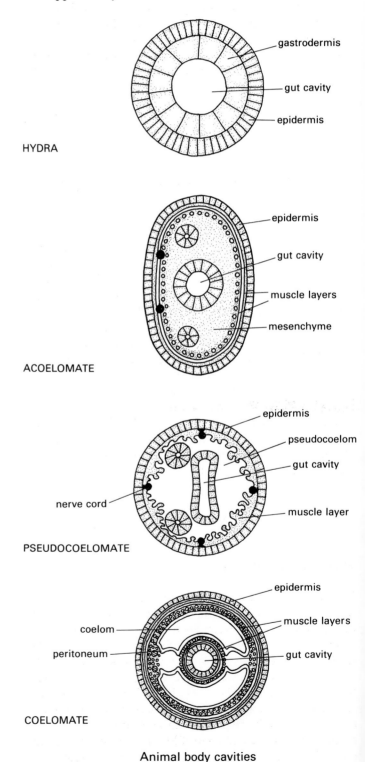

Animal body cavities

The pseudocoelomate phyla

Six invertebrate phyla possess pseudocoeloms and may be conveniently discussed together as the pseudocoelomates. However this is a grouping of convenience and of no evolutionary significance. Five phyla (Rotifera, Gastrotricha, Kinorhyncha, Nematoda and Nematomorpha) whilst not sharing a common plan do show many similarities and may be grouped together as the superphylum Aschelminthes (the cavity worms). The remaining phylum, the Acanthocephala (the spiny-headed worms) although possessing a pseudocoelom, differs from the Aschelminthes in many important aspects.

SUPERPHYLUM ASCHELMINTHES

Aschelminths are worm-like animals lacking definite heads and with some degree of radial symmetry at their anterior ends. They have a complete digestive system, a simple reproductive system and (generally) a protonephridial excretory system. Both respiratory and circulatory systems are absent. An unusual feature is a strong tendency towards a limited number of cells in the animal, with the number of cells in the individual organs being consistent within a species. Another characteristic of the group is the presence of a highly developed scleroprotein cuticle lying on an epidermis which is frequently syncytial.

Aschelminths have a very wide distribution, and occur almost everywhere where moisture is present. Five of the phyla are small: Rotifera has 1500 species, Gastrotricha 150 species, Kinorhyncha 100 species and Nematormorpha 250 species. One aschelminth phylum, Nematoda, is a large and successful group including many important parasites.

Phylum Nematoda

The Nematoda are a vast group of animals. While 10,000 species have been recorded this is unlikely to represent their full numbers, which have been estimated at 500,000. In general they are inconspicuous animals which are difficult to identify to species and have not been very thoroughly studied. In addition to their very large species numbers, there are vast numbers of individuals. It has been calculated, for example, that thousands of millions of nematodes may be present in an acre of soil, 90,000 in a single rotting apple and up to 40,000 in the roots of a potato plant. They may well be the most abundant metazoans.

Whilst the better known nematodes are parasitic, and a parasitic nematode, *Ascaris*, will be described as an example of the group, the vast majority of nematodes are free-living. Their distribution parallels that of the protozoans, extending geographically from north to south and occurring in an enormous variety of habitats from hot springs to icy tundra, ocean depths to mountain tops. One nematode, *Turbatrix aceti*, is found in old vinegar, and a similar form has only been found in German beer mats. Nematodes are dependent on water even when living in the soil, and are common in places where organic matter is found. Perhaps these requirements have lead to many species adopting a parasitic mode of life; they are highly successful parasites of both animals and plants, to the extent that hardly a single species is free of nematode parasites.

Nematodes are highly conservative in structure. They are precisely circular in cross-section and it is from this feature that their common names of roundworms or eelworms are derived. Whether they are free-living or parasitic, microscopic or large, their detailed body plan remains the same, such specialization as there is being restricted to their means of attachment and to their life cycle.

In one respect the phylum is unique. This is its total lack of either flagella or cilia; even the sperm are amoeboid rather than flagellate. Characteristically they show a great economy in cell number, having fewer than 500 cells making up their body wall, and only 2 cells in their excretory system. The group as a whole forms an isolated assemblage characterised by features which are so unusual that it is difficult to establish either the inter-relationships of the phylum or its evolutionary background.

Like parasitic platyhelminths, the parasitic nematodes have been studied for a very long time. Nematodes are mentioned in an Egyptian papyrus dated 1500 BC and appear in ancient Hebrew writings. An outbreak of disease described in the Book of Numbers is regarded by many authorities as a description of a nematode infection among the Israelites. The ancient Greeks (Hippocrates 400 BC, Aristotle 350 BC) knew of parasitic nematodes, as did the Roman physicians who described various attempts to treat nematode infections. The first detailed anatomical study of a nematode was performed by Tyson in 1685 on *Ascaris*, which has been chosen as a representative of the group.

Ascaris lumbricoides: a parasitic nematode

Ascaris lumbricoides is found in the small intestine of various mammals including man and pigs. The sexes are separate, the male being usually about 120 mm and the female about 180 mm long, although specimens of up to 400 mm have been recorded. Male *Ascaris* may also be distinguished by their posterior ends which curve towards the ventral surface.

Body plan

The body wall is composed of a tough, semi-transparent cuticle, a hypodermis and blocks of longitudinal muscle fibres.

The cuticle is a multilayered, proteinaceous structure. It is subdivided into four main regions of which three, the cortex, matrix layer and fibre layer, may be distinguished under the light microscope. The outermost region, only detectable under the electron microscope, consists of a thin

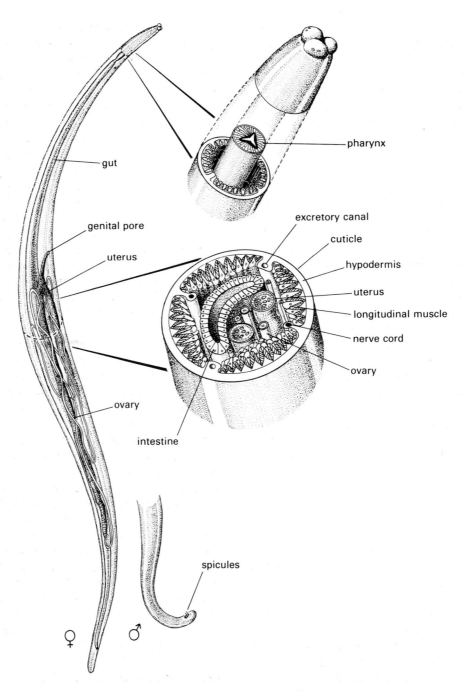

Ascaris : anatomy of female and posterior end of male

(almost 100 mm) layer of lipid and is thought to be protective in function. The cortex is composed of the protein keratin, and the matrix of elastic tissue. The fibre region consists of three layers of collagen-like fibres, arranged to form a geodesic lattice and comparable with the collagen fibrils of nemerteans and turbellarians. The fibres of each layer lie parallel to each other and the layers are arranged so that they cross each other diagonally at 75° to the longitudinal axis of the body. Beneath the fibre region is a basal lamella.

Functionally the nematode cuticle is a remarkable structure. It is elastic, supporting and tough yet flexible. It is permeable both to respiratory gases and to water, though water is not discharged through it.

The hypodermis is a thin syncytial layer of cells which secretes the cuticle and is joined to the underlying muscle by fibres. It projects inwards as four thickenings, the dorsal, ventral and lateral chords, which are visible as faint lines through the cuticle and which divide the body wall musculature into four blocks. The dorsal and ventral chords enclose the nerve chords, and the lateral chords enclose the excretory canals.

The body wall musculature is unusual, both in the large size of individual cells and their arrangement. Only longitudinal muscles are present and each consists of a single cell with an outer contractile part adjacent to the hypodermis and a non-contractile granular cytoplasmic part which contains the nucleus and is prolonged as a long, thin strand. The strands converge towards the dorsal and ventral mid-lines, separating the muscle cells into four columns. The pseudocoelom is large and fluid-filled and is formed from a few highly vacuolated cells, the vacuoles running together. The body cavity of *Ascaris* may be regarded as a series of irregular connecting cavities within the mesenchyme. The internal pressure of the fluid is considerable, ranging from 16 to 125 mm of mercury with a mean value of 70 mm. The combination of a high internal pressure with tension produced by the elastic cuticle accounts for the constant elongated shape and circular cross section found in the nematodes.

Locomotion

Nematodes move by undulatory propulsion in which sinusoidal waves pass backwards along the body length. This type of locomotion is of wide importance in the animal kingdom, being found in such diverse groups as fish, snakes and leeches, as well as nematodes. It is particularly suitable for movement in a fluid medium and is therefore a valuable pre-adaptation for parasitism. Contraction of the longitudinal muscles would normally cause the worm to become shorter and fatter. However, in *Ascaris* the longitudinal muscles act against a force exerted by the cuticle which is stretched to its limit by the high pressure of the pseudocoelomic fluid. The pseudocoelom therefore functions as a hydrostatic skeleton. Local contraction and shortening within the muscle results in a displacement of body fluid. This exerts pressure against the elastic force of

the body wall, resulting in lengthening opposite the region of contraction, and the body is therefore thrown into a curve. Alternate contraction of the dorsal and ventral muscles produce sinusoidal waves down the body, sufficient to allow *Ascaris* to insinuate itself through the host's gut contents.

Digestive system

The gut consists of a simple tube with some regional specialisation into buccal cavity, pharynx, intestine and rectum.

The mouth is at the extreme tip of the anterior end and is guarded by three lips, each bearing pairs of papillae. It leads into a long narrow buccal capsule, formed from a single epithelium with a thin sclerotised cuticle which is thickened to form 'teeth'. The pharynx (sometimes referred to as the oesophagus) has a triangular lumen in cross section. Numerous radial muscle fibres extend from its outer membrane to the cuticle and contraction of these expands the pharyngeal lumen which is closed by the hydrostatic pressure of the pseudocoelom. Because of the high pressure within the body cavity *Ascaris* pumps food forcibly into its intestine. The pharynx functions both as a highly efficient sucker and a pump, well adapted for the exploitation of liquid food. The front half is filled with the intestinal contents of the host by opening the mouth. The posterior half is then dilated and filled with food by the contraction of the anterior half. The posterior half then contracts and forces the food through the open pharyngo-intestinal valve into the intestine. The radial muscles of *Ascaris* contract twenty times a second.

The intestine is a simple epithelial tube with an inner surface thickly coated with microvilli. Digestion is mainly extracellular and extremely rapid, and it is probable that a lot of the ingested food is not absorbed. The digested food is absorbed through the intestinal wall and diffuses directly into the pseudocoelomic fluid. Food reserves, which are generally considerable, occur in the gut wall in the form of glycogen or fat as well as in the non-contractile region of the muscle and the epidermal cords. The gut contents are transferred along the intestine to a narrow flattened rectum with a slit-like anus. Transfer is by simultaneous contraction of the dorsal and ventral musculature. Cilia, the normal means of transporting gut contents in invertebrate animals, are absent, perhaps another consequence of the high internal pressure which would render them ineffective. An intestinal-rectal valve leads into the anus. The anus is opened by a special muscle and the faeces are expelled from the intestine by internal pressure of the body cavity. This is so great that a defaecating *Ascaris* in air may shoot its faeces a distance of two feet!

Excretion and osmoregulation

The economy of cells characteristic of nematodes is taken to its most extreme in the excretory system which is apparently derived from only two cells. Two intracellular longitudinal canals lie with the lateral chords. Transverse branches extend from the canals to unite at a short common excretory canal which opens at a mid-ventral pore a little posterior to the mouth, so the entire system is therefore H-shaped. There are no flame bulbs, which have probably been lost in response to the high hydrostatic pressure. The method of working of the excretory system is not understood though it is thought that the high pressure within the pseudocoelom provides for some form of ultra-filtration. Another suggestion is that the primary function of the so-called excretory system is for osmoregulation, and that excretory products leave the animal via the gut.

Respiration

Ascaris lives in an environment where the supply of oxygen is limited. It therefore respires anaerobically and breaks down carbohydrate into carbon dioxide and a variety of organic acids which are excreted through the body wall. At higher oxygen concentrations *Ascaris* is also capable of aerobic respiration for at least a limited period. In this case oxygen is transferred through the body wall into the pseudocoelomic fluid. Haemoglobin is present to enable *Ascaris* to capture low amounts of oxygen.

Nervous system

Like many of its other organ systems, the nervous system of *Ascaris* is derived from a small number of cells. It consists of four ganglia (dorsal, ventral and paired lateral) which form a ring round the pharynx. From this ring six sensory papillary nerves extend forwards to the anterior papillae and six longitudinal nerves extend backwards along the length of the worm. Paired lateral nerves extend above and below the lateral epidermal chords, a dorsal nerve lies within the dorsal chord, and a ventral nerve within the ventral chord. The ventral nerve is the most important and gives off branches to the gut and copulatory organs, terminating at the anal ganglion.

Development of sense organs necessitates modification of the cuticle in sensory regions: the cuticle takes the form of a sensory pit or pouch; or may be inflated to form sensory papillae or extended into spines.

As an endoparasite *Ascaris* has lost most of its sense organs. Specialised sense organs are found mainly at the anterior end, for example on the lips which surround the mouth, each of which bears pairs of sensory papillae. Tactoreceptors, photoreceptors and chemoreceptors are all present. A single unicellular chemoreceptor, the phasmid, opens on either side of the tail.

Reproduction and growth

The reproductive system lies within the pseudocoelom. The female has paired ovaries and the male a single testis. Both the ovary and the testis consist of long coiled tubes, and the germ cells are derived from a single cell or germinal area at their innermost end. The ovaries widen to paired oviducts, and then to uteri with an epithelial lining and a circular muscular layer which is most developed near the female genital pore. A portion of each uterus acts as a seminal receptacle for sperm storage and egg fertilization. The uteri unite at a short median vagina which opens at the genital pore. The testis enlarges into a sperm duct which widens to form a seminal vesicle. This opens via an ejaculatory duct at the cloaca. Both the seminal vesicle and the ejaculatory duct have layers of circular muscle in their walls.

In the male a pair of stiff horny spicules project from dorsal pockets with the cloaca. These have protractory and retractory muscles, and are pushed into the female vagina during copulation and used to transfer the sperm.

Egg production is exceedingly high as might be expected in a parasite. A female *Ascaris lumbricoides* may produce eggs for several months at a rate of 200,000 a day. The eggs are fertilised in the oviducts and then enter the uteri. Here they are surrounded by protein secreted by glands in the uterine wall and start to develop. Two further layers are added, the outer one being of chitin, and the resulting 'shell' makes the eggs highly resistant and capable of surviving for up to a year. The eggs leave the host with the faeces.

If the eggs are swallowed juvenile worms hatch in the intestine of the new host. They then undergo a complex migration before they return to the intestine as adults. They burrow through the host's intestinal wall into the intestinal blood vessels and are transported by the host's circulatory system to the lungs. There they burrow into the bronchial passages, wriggle up the trachea into the mouth and are swallowed, thus reaching the intestine. Much of the damage inflicted by the parasite on its host occurs during this migration.

Juvenile *Ascaris* moult four times during their development, the fifth stage being the adult which increases in length but does not moult. In most animals, growth normally requires continuous multiplication of cells. However in nematodes cell multiplication is confined to the early stages, and when each body system reaches the cell number characteristic of the species further increase in size is brought about only by increase in cell volume. After the last moult the young *Ascaris* is 20 mm in length, and attains 400 mm in a matter of weeks by cell expansion.

Nematode diversity

The cosmopolitan nature of nematodes has already been emphasised. The free-living species probably owe their wide distribution to the fact that they are extremely easily dispersed. They are sufficiently small and light to be blown by the wind as well as being carried by floating debris or attached to other animals, for example in the mud on the feet of wading birds. Parasitic nematodes are of considerable medical and economic importance. About fifty species occur in man, although the majority of these are not of pathogenic importance. It has been calculated that probably only two per cent of the human race have never acted as host to a nematode!

Nematodes show little anatomical variation but an immense variety of life cycles. The phylum is divided into two classes on the basis of the presence or absence of phasmids, a pair of unicellular chemoreceptive glands, one opening on either side of the tail. The class Aphasmida includes the majority of freshwater and marine nematodes. The Phasmida (in which phasmids are present) include the majority of the parasitic nematodes and the soil-dwelling forms.

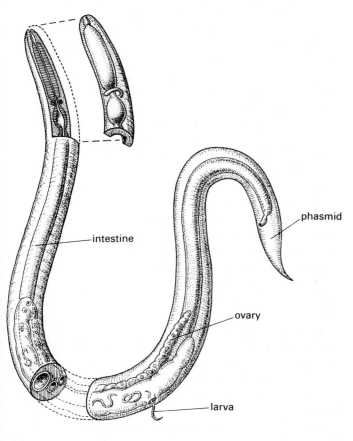

Rhabditis, a phasmid nematode

Free-living nematodes have a simple life cycle in which the egg hatches to a first stage juvenile, which grows and moults to become a second stage juvenile, and so on. Whilst the same basic four-moult pattern is normally retained in the parasitic nematodes, in other respects their life cycles show immense variability. Some may be parasitic on plants, or on animals, or include both plant and animal hosts in their life cycle.

Plant ectoparasites, for example *Longidorus*, feed on the external cells of the roots which they puncture with long, slender stylets. They then suck out the cell contents. Quite apart from the damage they inflict on the plant, forms such as *Longidorus* may infect the plant with viruses; for example the tomato black ring virus is carried by *Longidorus*.

Plant endoparasites enter the plant as juvenile worms and develop within it, feeding on the cells. The plant may respond by developing a gall around the parasite, or the plant tissue may die. Reproduction takes place inside the plant and the juvenile stage then leaves to infect another plant. There are vast numbers of nematodes within this category. One example is the potato root eelworm, *Heterodera rostochiensis*, which causes immense damage and loss to the potato crop each year, invading potato rootlets and feeding on the cortical cells. *Aphelenchoides ritzema-bosi* is a common greenhouse pest living in the tissue of leaves and buds of flowers such as African violets and chrysanthemums.

The ascaroid nematodes are gut parasites of many vertebrates including man, pigs, dogs, cats, horses and cattle, and also of some invertebrates. They feed on the gut contents and have life histories similar to that described for *Ascaris lumbricoides*. The strongyloid nematodes are unusually harmful. Like the ascaroids they inhabit the gut, but they feed on blood from the gut wall, causing continual haemorrhage and anaemia. *Anclyostoma duodenale*, an Old World hookworm, has a wide distribution in the tropics and subtropics, particularly in Asia where 59 million Asians are estimated to be infected. The life history involves a similar migration through the host's tissues as that described for *Ascaris*. The fertilized eggs leave with the host's faeces and hatch in warm, damp places. The newly-hatched juveniles bore through the skin of the feet or hands and travel in the blood system via the lungs to the intestine where the adults develop. *Necator americanus*, the New World hookworm, infected the 'poor white trash' of the American South for generations; indeed the name means 'American killer'. It has a similar mode of life and life history to *Anclyostoma*.

Many nematodes show an alternation between parasitic and free-living forms, or include one or two intermediate hosts in their life cycle. *Heterotylenchus aberrens* is parasitic on both onion flies and onion plants. An essential feature of a life cycle of this type is that both the adult and larval insect stages feed on the same plant. The juveniles enter the plant when the adult insect feeds. Maturation of the nematode and copulation both take place in the plant, and the parasite then enters the larval onion fly. After the insect metamorphoses, the parasites produce eggs which hatch to give juveniles and reinfection occurs. Alternatively, in other forms the juvenile is zooparasitic and the adult a plant parasite. Many of the medically important parasites have an insect vector. *Wucheria bancrofti*, a filarial nematode, is a human parasite of the tropics and subtropics, living in the lymph nodes. Severe filarial infestation blocks the lymph vessels causing elephantiasis, commonly in the legs, breasts and scrotum, in which these parts are grotesquely enlarged. The female is 80 to 90 mm in length and the male 30 to 40 mm. First stage larvae (microfilariae) are produced by the female but development does not progress unless these are sucked up by a night-flying mosquito, *Culex fatigans*. At the normal biting time of this mosquito, the microfilariae migrate to the peripheral blood system whilst at other times they retreat to deeper tissues, especially the lungs. This synchronisation of parasite and host behaviour is common amongst successful parasites. The parasite enters the mosquito with the host's blood and migrates from the gut to the thoracic muscle and finally back to the proboscis. During this period it undergoes two further moults and develops to the infective stage. When the mosquito bites, the worms creep onto the skin and are able to enter the wound; they are not injected into the host via the proboscis.

Loa loa is another filaroid nematode. Its common name, the 'African eye-worm', comes from the fact that it may be seen crossing the eyeball as it migrates about the host's tissues. It is transmitted by the daytime-biting fly, *Crysops* and migrates into the superficial blood system during the day. Heavy infections of these worms can also cause elephantiasis.

The dracunculoids, for example the guinea worm, *Dracunculus medinensis*, believed to be the biblical 'fiery serpent', are highly elongated and thread-like worms living in the connective tissue and body cavities of vertebrates in Africa and Asia, often in the limbs. The female can attain up to 1.2 m in length. The larvae are discharged into the water through skin ulcers formed by toxic substances secreted by the female, often on the feet and ankles. They swim freely until they are eaten by a freshwater copepod, *Cyclops*, in which development continues. Reinfection occurs if the infected *Cyclops* is swallowed by the primary host. *Dracunculus* infects man and has to be removed surgically. An ancient method of removal, still practised, involves winding the parasite out on a matchstick; it is a lengthy and dangerous process which can result in severe inflammation, limb loss and death from secondary infections if the worm breaks.

Some parasitic nematodes include two animal hosts in their life cycle, although this is less common. *Dioctophyma renale* lives in the kidney and coelom of carnivorous mammals, its eggs leave with the host urine and hatch. The secondary hosts belong to a group of highly aberrant annelids, the branchiobdellids, which live attached to crayfish gills. If the young nematodes are ingested by one of these they penetrate through its gut into the coelom where development continues. For further development it is necessary for the annelid to be eaten (with the crayfish) by a fish, whereupon the nematode encysts in the fish coelom. If the fish is eaten by the primary host the nematode migrates from the gut to the kidney or the coelom. Zooparasitic nematodes can include both free-living and parasitic stages in their life cycle and may even have both free-living and parasitic adults. A simple version of such life cycles is found in the many saprophagic nematodes in which the young worms enter an invertebrate host at a late juvenile stage and live within the host, feeding on its tissues when it dies. Alternatively the female may enter an invertebrate host after copulation to produce the next generation. Members of the family Mermithidae have free-living adults, but are parasitic on invertebrates, commonly insects, during their juvenile stages. They may attain so great a length that they almost completely fill the body of their insect host.

The adult *Rhabdias bufonis* lives in frog lungs. The eggs are coughed out of the lung and swallowed, hatching in the intestine. The larvae leave with the faeces and develop into a free-living adult which reproduces, the eggs hatching in the soil. The larvae re-enter the frog through the skin, and are transported via the lymphatic and blood systems to the lungs where the adult develops.

Synopsis of phylum Nematoda

The classification of the nematodes is extremely complex, and there is still no complete agreement about it. The following synopsis is restricted to the examples used in the above discussion on nematode diversity and should not be regarded as comprehensive. Some authorities do not recognise the distinction between Aphasmida and Phasmida, which is included here.

Class Phasmida
   Order Rhabditida (*Rhabdias, Rhabditis*).
   Order Tylenchida (*Heterodera, Aphelenchoides, Heterotylenchus*).
   Order Strongylida (*Anclyostoma, Necator*)
   Order Ascarida (*Ascaris*).
   Order Spirurida (*Dracunculus*).
   Order Filarida (*Wucheria, Loa loa*).
Class Aphasmida
   Order Dioctophymatida (*Dioctophyma*)
   Order Mermithida: family Mermithidae.

MINOR PSEUDOCOELOMATE PHYLA

Phylum Rotifera

Rotifers are pseudocoelomate metazoans but are comparable in size with the ciliates with which they live and compete. Like the ciliates, they were popular objects for study by the early microscopists and were included amongst van Leeuwenhoek's 'little animals'. Rotifera means 'wheel bearer' and the common name for the group is 'wheel animalcules'. Both names refer to an anterior ciliated structure, the corona, or crown, which, when the cilia are beating, looks like a spinning wheel.

The phylum includes about 1500 species. The majority live in fresh water although a few marine forms are known and some live in damp soil or moss. Whilst in comparison with any phyla the rotifers are a small group, they are numerous in terms of individuals. Almost any sample of natural water studied under the microscope will reveal some rotifers. They are truly cosmopolitan in distribution, occurring on oceanic islands, on the tops of mountains, and in polar regions. They may be pelagic or bottom dwelling, and a few are parasitic.

Epiphanes is one of the most common rotifers. It is a bottom dweller and lives in ponds contaminated with manure where it feeds on flagellates such as *Euglena* which are common in such ponds.

Epiphanes: a common rotifer

Body form

Like the majority of rotifers *Epiphanes* is elongated and roughly cylindrical in shape, tapering at the posterior end. The body is divided into three main regions: the head which bears the corona, the trunk which surrounds the viscera, and a small segmented and highly retractile posterior foot with a pair of 'toes'. A slight constriction, the neck, separates the head and trunk.

The body is covered by an elastic scleroprotein cuticle, secreted by a syncytial epidermis, and comparable with the nematode cuticle. The body wall does not consist of distinct longitudinal and circular layers. Instead, incomplete bands of circular muscle are present, with longitudinal retractor muscles associated with the head and foot. The corona can be pulled into the body and the anterior end closed over it by means of a sphincter muscle. The viscera are held in place by visceral muscles which attach them to the epidermis.

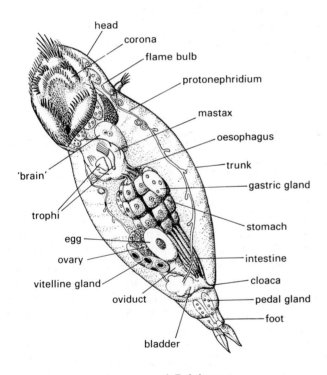

head
corona
flame bulb
protonephridium
mastax
oesophagus
trunk
gastric gland
stomach
intestine
cloaca
pedal gland
foot
'brain'
trophi
egg
ovary
vitelline gland
oviduct
bladder

Anatomy of *Epiphanes*

Internally there are several points of similarity between the rotifers and the nematodes, and the turbellarians. They have a pseudocoelom, filled with pseudocoelomic fluid and containing a network of branched amoeboid cells. Like nematodes, rotifers show a restricted number of cells or nuclei in their various systems which is constant within a species. *Epiphanes* has 280 nuclei in the epidermis, with 246 in the nervous system, 104 in the muscles, 14 in the protonephridia and 157 in the gut.

Locomotion

Like the majority of rotifers (a few are sessile) *Epiphanes* moves either by crawling or by swimming. The corona bears two rings of cilia, separated by a space. The outer ring, the circumapical band, consists of single cilia, but the inner one, the pseudotroch, has ciliary membranelles comparable with those of some ciliates. The coronal cilia beat outwards in a metachronal rhythm around the rings, and this enables the animal to swim. Alternatively the animal can crawl by alternately attaching the head and foot to the substrate and extending the body by contraction of the circular muscles.

Digestive system

Epiphanes uses the corona to create a feeding current. It attaches its foot to solid substrate, probably using the secretion of a pair of pedal glands. The ciliary beat then brings potential food near the mouth which is situated ventrally, between the circumapical band and the pseudotroch. Food is detected by chemical stimuli or by touch. The mouth opens into the mastax, a modified pharynx into which open salivary glands. The mastax is muscular and lined with cuticle. It contains cuticularized biting structures, the trophi, which can be protruded through the mouth to seize protozoans and are also used for grinding.

A tubular oesophagus leads to the stomach where, in an alkaline extracellular medium, digestion and absorption takes place. Associated with the stomach are a pair of bean-shaped gastric glands which produce digestive enzymes. The stomach is enclosed in a network of muscles and its contents are stirred by constriction of these. Digestion is rapid, and food reserves in the form of protein or fat droplets are stored in the stomach wall. A short intestine completes the gut which opens posteriorly at the cloaca.

Excretion and osmoregulation

The excretory system of rotifers is comparable with that of Turbellaria. It consists of a pair of syncytial protonephridia with flame bulbs, which open into a bladder. The bladder has four muscle cells around it which, on contracting, constrict the bladder and expel its contents through the cloaca, discharging about four times a minute.

Gaseous exchange and circulation

Since *Epiphanes* is minute, gaseous exchange takes place through the general body surface and no circulatory system is necessary.

Nervous system and sense organs

The nervous system consists of a bilobed cerebral ganglion, the 'brain', which lies in the head on the dorsal surface of the mastax. From this extend two lateral longitudinal nerves and fibres to the sense organs and muscles. *Epiphanes* has a pair of rudimentary eyes and three short tentacles which bear ciliated sensory cells. In addition there are ciliated sensory cells scattered on the corona.

Reproduction and growth

The process of reproduction in rotifers is somewhat unusual. In the case of the Monogonanta, the group to which *Epiphanes* belongs, both males and females are known but the males are degenerate and are not always present in a population. They are smaller than the females, do not feed, and survive only briefly.

For most of the time an *Epiphanes* population consists entirely of females which reproduce parthenogenetically, producing diploid ameiotic eggs which hatch within a short time and develop into females without fertilization. This enables populations to increase very rapidly, to as much as ten times their original numbers in a week.

Under the influence of stimuli which are not fully understood but are thought to involve a combination of temperature, diet and population density, a different type of female producing haploid meiotic eggs develops. If these eggs are not fertilized they produce males parthenogenetically; if they are fertilized they secrete a heavy resistant shell and may not hatch for several months. When they hatch they produce females.

The female reproductive system consists of an ovary and a yolk-producing vitellarium, surrounded by a membrane which is prolonged as an oviduct. The eggs pass through the oviduct to the cloaca and each female can lay only as many eggs as are present in the ovary at birth. Males have a single testis and a ciliated sperm duct opening via a gonopore. Glandular masses, the prostate glands, are associated with the sperm duct whose end is modified as a copulatory organ.

Fertilization takes place by hypodermic impregnation through the body wall into the pseudocoelom. The female can therefore only be fertilized in the first few hours of her life before the cuticle hardens, and once fertilized can produce only fertilized eggs.

The resistant, meiotic eggs are necessary for survival under conditions when normal existence would be impossible. They allow rotifers to survive in areas which are subject to periodic drying and extremes of temperature, and are one of the reasons for the cosmopolitan distribution of the group. The reproductive pattern tends to be cyclic, the meiotic eggs resting over the winter and hatching as ameiotic females in the spring.

Since the nuclei of rotifers do not divide after hatching (hence their restricted cell and nuclear number) they are not capable of regeneration and will die after amputations, or even wounding. This is a point of contrast with forms such as Turbellaria.

Phylum Gastrotricha

The phylum Gastrotricha, with about 150 known species, is one of the smaller pseudocoelomate groups. Like rotifers, gastrotrichs are microscopic and indeed were interpreted as rotifers by early microscopists.

Gastrotrichs are cosmopolitan in distribution, occurring most commonly in freshwater where there is standing vegetation and in seashore sands. Although they bear a superficial resemblance to rotifers, they have no corona or mastax and are generally scaly or spiny. *Chaetonotus* is one of the commonest pond animals and frequently occurs in protozoan cultures.

Chaetonotus: a common gastrotrich

Body form

Chaetonotus has a body divided into four regions: a head bearing several tufts of sensory cilia, a constricted neck, a trunk, and a forked caudal end, made up of a pair of furca. A pair of adhesive glands, characteristic of the phylum, open through the furca and may be compared with the pedal glands of rotifers.

The body wall is covered by a cuticle secreted by a syncytial epidermis, and the dorsal surface is modified as a series of cuticular spines and scales. Circular and longitudinal muscle-blocks lie beneath the epidermis and serve to move the bristles or to contract or curl the body. The pseudocoelom is restricted to a small space between the body wall and the gut.

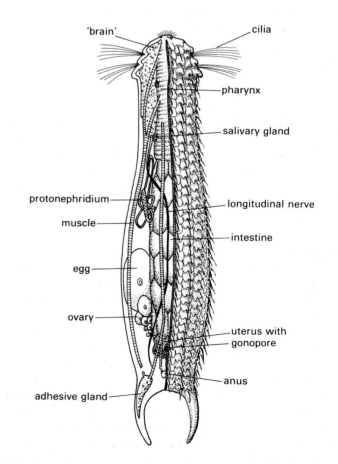

Chaetonotus

Locomotion

The ventral surface is ciliated, the cilia being arranged in two longitudinal bands, and by using these *Chaetonotus* is able to swim or to glide over the substrate. The name of the phylum (*gastrotrichia* meaning 'stomach hairs') is derived from these structures. The animal may also become temporarily sessile, attaching itself by the adhesive glands.

Digestive system

Chaetonotus feeds on organic detritus, algae, protozoans and bacteria which are swept into the terminal mouth by the beat of four ciliated tufts which are arranged in pairs on either side of the head. The mouth opens into a buccal cavity which is lined with cuticle, and thence into a bulbous pharynx with a triangular lumen, cuticular lining and associated salivary glands. In many respects the pharynx may be regarded as comparable with the nematode pharynx. The midgut is a straight tube with a thin external layer of circular muscle. No separate external glands are present, but gland cells are present in the midgut wall and these are thought to produce enzymes for extracellular digestion. The gut is completed by a short rectum leading to a terminal anus.

Osmoregulation and excretion

Two flame bulbs, each containing a single long flagellum, connect with coiled protonephridial tubules, one on either side of the body. Each tubule opens at a nephridiopore near the ventral mid-line. Unlike rotifers, gastrotrichs have no bladder.

Nervous system and sense organs

A large saddle-shaped 'brain' lies dorsal to the pharynx and consists of a pair of lateral ganglia and a broad dorsal connective. A pair of lateral, longitudinal nerves extend the length of the body, and to the body wall and viscera. Sense organs are represented by scattered tactile bristles, tufts of elongated bristles (ciliary tufts) on the head, and ocelli. On the head a pair of ciliated pits lie immediately posterior to the ciliary tufts and are probably chemoreceptors.

Reproduction and growth

Gastrotrichs have a short life span, usually between three and twenty-one days. In contrast to other aschelminths they are generally hermaphrodite. However in forms such as *Chaetonotus* only females are found and they reproduce parthenogenetically. This is presumably a degenerate condition and some traces of testes have been reported. The ovaries are represented by unenclosed cell masses which lie lateral to the intestine. Mature ova move into a space called the uterus which connects to a tiny oviduct opening to the outside by a gonopore. A yolk-producing vitellarium is present. Usually only four or five large eggs are produced, these are released and become attached to the substrate. They may be of two types: a highly resistant thick-walled egg used for overwintering, and a non-resistant thin-walled type which hatches in about four days. Development is direct, and the gastrotrich reaches sexual maturity about two days after hatching.

Phylum Kinorhyncha

The Kinorhyncha are a group of microscopic marine aschelminthes, somewhat longer than rotifers and gastrotrichs but still less than 1 mm in length. They have a segmented body and no external cilia. About 100 species are described but little of their basic biology is known. The majority live in ocean mud in shallow waters, although the earliest discoveries were made amongst algae. (An example of the group is *Echinoderes*.)

Body form

The general body shape is elongated and consists of a head, neck and a body divided into eleven segments (zonites). The cuticle is subdivided into plates, the dorsal plates bearing hollow, movable spines. The head bears 5–6 circles of curved spines (scalids) and an oral cone with a ring of stylets surrounding the mouth. The head can be retracted into the trunk (hence the phylum name Kinorhyncha, meaning 'movable snout') and covered by the neck plates. Like gastrotrichs, kinorhynchs have a pair of ventral adhesive tubes which allow them to become temporarily anchored to a substrate.

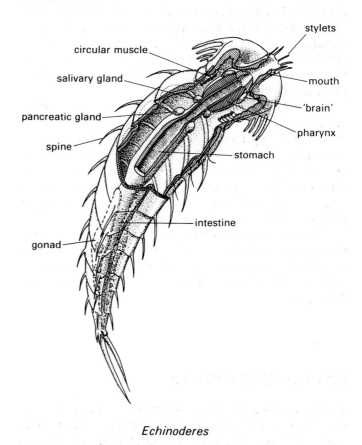

Echinoderes

The epidermis is syncytial and extends into the hollow spines. It also bulges inwards to form three longitudinal cords (mid-dorsal and paired lateral) comparable with the longitudinal cords of the nematodes. Rings of circular muscle are present in the first two zonites, and two pairs of longitudinal muscle bands extend in the dorsolateral and ventrolateral regions of the body. Intersegmental diagonal muscles are present in the trunk. The well developed pseudocoelom is fluid-filled and contains numerous amoebocytes.

Locomotion

Kinorhynchs cannot swim but burrow through the mud. The head is extended into the mud by contraction of the circular muscles and is held in place by the spines. The body is then drawn forwards by contraction of the longitudinal muscles.

Digestive system

The mouth lies at the tip of the oral cone which encloses the buccal cavity. This opens into a tapering tubular pharynx with a triangular lumen, comparable with the pharynx of nematodes and gastrotrichs. The oral cone is protruded during feeding, the stylets penetrate the detritus, and the food is sucked in by the contraction of the pharynx, which is highly muscular and lined with cuticle. The oesophagus is also lined with cuticle and receives secretions from paired dorsal and ventral salivary glands, and posterior pancreatic glands. The stomach lacks a cuticular lining and is believed to be the region of absorption. It is covered by a mesh of longitudinal and circular muscle. A short cuticle-lined mid-gut connects the stomach to the terminal anus on the last zonite.

Osmoregulation and excretion

Two protonephridia are present, one on either side of the intestine in the tenth zonite. These open on to the dorso-lateral face of the eleventh zonite via a nephridiopore. The flame bulbs each contain a long and a short flagellum rather than a ciliary tuft, a structure comparable with that of gastrotrichs.

Nervous system and sense organs

The brain is ring-shaped and encircles the anterior end of the pharynx. A mid-ventral nerve cord with one ganglion in each segment extends from the brain. In addition masses of ganglion cells are present in the mid-dorsal and lateral epidermal chords, though these are not joined by longitudinal nerves. The nervous system is closely associated with the epidermis and may be regarded as an epidermal or sub-epidermal nerve plexus.

Little is known of kinorhynch sense organs which generally restricted to scattered sensory bristles.

Reproduction

Kinorhynchs are dioecious (having the two sexes in separate individuals), and both male and female have a pair of sac-like gonads opening at the last segment. In addition the male normally has two or three penial spines. The eggs hatch to give larvae which undergo successive moults to develop the adult form.

Phylum Nematomorpha

The phylum Nematomorpha, commonly known as the horsehair worms or hair worms, consists of about 230 species of extremely long, slender worms, usually brown or black in colour and with no distinct head. Their common name stems from a fourteenth century myth that they arose spontaneously from horse tails. The phyletic name means 'thread form'.

The juveniles are parasitic in arthropods but the adults are normally free-living in damp soil or fresh water, for example *Gordius*. There is a single pelagic marine genus, *Nectonema*, which parasitizes crabs in its juvenile stages. They may be as much as 100 cm in length, but are usually not more than a milimetre in diameter.

Nematomorphs have a thick outer cuticle, which may consist of as many as 45 fibrous layers, and a cellular epidermis with a ventral, longitudinal chord (or dorsal and ventral longitudinal cords in *Nectonema*). Beneath the epidermis lies a layer of longitudinal muscle fibres and a reduced pseudocoelom. Since the entire function of the adults is reproduction, their internal anatomy is simple. They do not feed; the mouth is therefore small and functionless and the digestive tract degenerate. In the larval stages, whilst the digestive tract is better developed than

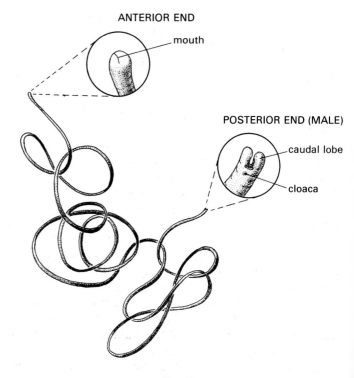

Gordius

that of the adult, the animal absorbs food from the host through the skin. The nervous system consists of an anterior nerve ring surrounding the pharynx, from which extends a single ventral longitudinal nerve cord, attached to the epidermal cord.

The sexes are separate. The ovaries develop lateral diverticula (blind tubes), in which the ova form, and ripe ova are stored in the centre of the ovary which therefore function as a uterus. Paired oviducts join to form a single atrium from which arises a seminal receptacle. The atrium opens via a cloaca. The testes are elongated and tubular and open into the cloaca via separate short sperm ducts.

The males swim or crawl, but the females are normally inactive. The male entwines round the female and deposits the sperm near the cloaca, from where they migrate into the seminal receptacle. The eggs are laid by the female in long strings, stuck together by secretions produced by the cloaca. After hatching, the larva enters an arthropod host either by being eaten in the form of a cyst or by penetration. Development is completed within the host, the nematomorph undergoing several moults and leaving the host as a fully formed adult after a period of weeks or months. A short time after leaving the host the nematomorph attains sexual maturity.

Phylum Acanthocephala

The phylum Acanthocephala, whilst possessing a pseudocoelom, differs from the Aschelminthes in the development of that structure and is not therefore included with them. The 800 known species are all endoparasites, living in vertebrate guts and having an intermediate arthropod host. The majority are around 20 mm in length, but examples up to 650 mm in length are known. In general the longest acanthocephalans occur in mammals, but fish are the most common host. They may be present in large numbers: 1154 have been reported from a seal's gut. A well-known example of the group is *Macracanthorhyncus*, which lives in the gut of pigs.

Acanthocephalans can be recognized by their proboscis which is armed with spiny recurved hooks and used for attachment to the host. The name of the group, which means 'spiny-headed', is derived from this feature.

The acanthocephalan body is divided into an anterior prosoma consisting of the head, neck and proboscis, and a trunk which may also be armed with spines. The proboscis is used for attachment to the gut wall of the host, to which it may cause immense damage, and also enables the animal to move about. Both proboscis and neck can be withdrawn into a muscular proboscis sac by contraction of retractor muscles. Protrusion is by hydraulic pressure, the fluid being stored in lateral diverticula of the body wall, the lemnisci, where the proboscis is withdrawn.

The structure of the body wall shows similarities with that of the aschelminthes. There is a cuticle, secreted by a syncytial epidermis, which is composed of several fibrous layers, a thin dermis, and layers of longitudinal and circular muscle. Acanthocephalans have no digestive system and food is absorbed through the body wall. Spaces (lacunae) between the epidermal layers are thought to be involved in food distribution. Most acanthocephalans lack protonephridia though these are present in one order. The brain is represented by a single ganglion from which arise nerves which lead to the limited tactoreceptors and to the proboscis and muscles. A pair of lateral longitudinal nerves extend to the genital organs.

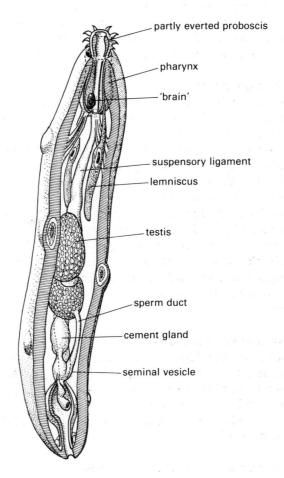

General body form of an acanthocephalan (male)

The main anatomical adaptations of acanthocephalans are associated with reproduction. The sexes are separate and in both males and females the gonopore opens at the end of the trunk. In the female, paired ovaries (or a single ovary) develops in the larva in a sac suspended by a ligament, but these break up into fragments which float within the pseudocoelom as egg balls. As they mature, ova are released from the egg balls. In the male two testes are present, lying one behind the other within a ligament sac attached to the suspensory ligament. Two sperm ducts, one from each testis, run posteriorly through the ligament sac and join to form a common sperm duct with a seminal vesicle at its base. The sperm ducts are surrounded with cement glands which send ducts to the common sperm duct. The duct ends at an eversible penis.

Fertilization takes place within the female. The sperm are injected through the female gonopore, which is then plugged with cement. Development takes place within the female and eventually a larva (acanthor) with an anterior hook-bearing rostellum is passed out of the female and leaves with the host's faeces. The larva is surrounded by a protective shell and can survive for many months. If it is eaten by an appropriate arthropod (often an insect or an aquatic crustacean) the acanthor emerges from the shell and uses its rostellum to bore through the gut wall. In the arthropod body cavity the larva develops to the next larval stage, an acanthella, and then to an encysted cystacanth, which is the infective stage. If the intermediate host is eaten by the primary host, development is complete and the worm leaves the cyst and attaches to the gut wall.

Phylum Annelida

Introduction to the coelomate animals and metamerism

All the remaining members of the animal kingdom possess, at least at some stage in their development, a coelom, and therefore are referred to as coelomates. There are five major coelomate phyla: Annelida, Arthropoda, Mollusca, Echinodermata and Chordata, with many minor phyla. Apart from their possession of a coelom, all these animals show greater complexity than pseudocoelomate and acoelomate forms. For instance, a through gut and a circulatory system are found in forms such as the nemerteans; however such systems are far more complex in the coelomates, elaboration being due to a large extent to their possession of a coelom. Coelomates also show increasing complexity within their nervous system associated with more elaborate behaviour patterns.

An important structural adaptation seen in many coelomates is metamerism. Essentially this means that the animal is divided into a series of similar segments, all of the same degree of development and arranged in a linear series along the antero-posterior axis. Each segment may include subdivisions of the various organ systems, so that in the 'ideal' metamerically segmented animal each segment would be supplied not only with skin, muscles, nerve, and circulatory structures but with excretory and reproductive organs as well. In fact this 'ideal' animal does not exist, metameric animals showing specializations of segments or regions to a greater or lesser extent.

Annelids, arthropods and chordates, with several minor phyla, are all regarded as basically metameric animals although the condition in both arthropods and chordates is so extensively modified in many cases as to be hardly recognizable. The phylum Annelida, however, demonstrates metamerism in its least modified form.

Phylum Annelida

The Annelida is an important phylum of about 9,000 species representing the third major type of worm organization. Their most obvious trait, metameric segmentation, is immediately apparent in the ring-like markings or annuli on the external body surface from which the name of the phylum is derived (*anellus* means 'a little ring').

The general annelid body plan consists of an elongated segmented tube with a straight tubular digestive system positioned in the middle of the coelom. The basic structure of all segments is essentially the same, each consisting of right and left coelomic cavities lined with peritoneum and filled with fluid. These cavities lie against the body wall laterally and against the gut medially. The sheets of peritoneum form mesenteries in the mid-dorsal and mid-ventral line and these hold the gut in place. Anteriorly and posteriorly, marked externally by the annuli, the coelomic peritoneal walls abut those of the adjoining segments to form septa across the coelom. Each septum is formed of two layers of peritoneum with connective tissue and muscle between them. Both the septa and mesenteries may be perforated, allowing for some circulation of coelomic fluid throughout the worm.

There are three main annelid classes: Polychaeta, a large group of 5,500 species of marine worms with segmentally arranged paired appendages and bristles (from which their name is derived: *polychaeta* means 'many bristles'); the predominantly freshwater and terrestrial Oligochaeta with no appendages and few bristles; and the Hirudinea, 500 species of marine, freshwater, and terrestrial leeches.

Archaeannelida, once regarded as representing the primitive members of the phylum and at various times included as a class or subclass within it, is now thought to represent a group of several unrelated families, highly aberrant and with various adaptations to an interstitial mode of life. Many of them are ciliated, and lack parapodia, bristles, and any external segmentation. The polychaetes, as represented by *Nereis*, form the best introduction to the group.

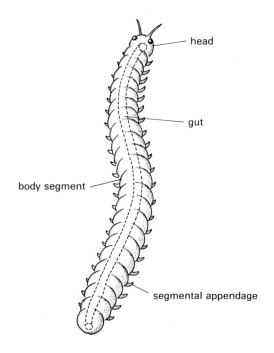

A metameric animal

Class Polychaeta
Nereis: the ragworm

Nereis diversicolor, the ragworm, is one of the commonest intertidal animals, and occurs in large numbers on estuarine mud flats or muddy shores where it is collected extensively by fishermen for use as bait. It lives in burrows and emerges onto the surface to feed.

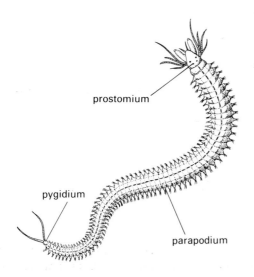

Nereis diversicolor

Nereis is long and slender, variable in colour (hence its specific name), with slightly flattened dorsal and ventral surfaces. The body is divided both externally and internally into about a hundred segments, most of which are identical, but with very distinct head segments and a terminal segment. The initial segment, the prostomium, overhangs the mouth and forms the upper lip, and the terminal pygidium bears the anus.

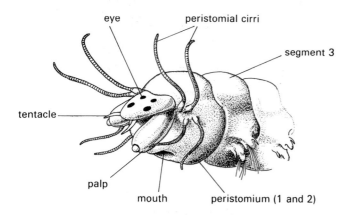

Head of *Nereis*

Each of the body segments bears a pair of parapodia which are muscular hollow lateral extensions of the body wall. Each parapodium is biramous (two branched), with a dorsal notopodium and a ventral neuropodium. These are themselves bilobed structures, and bear specialized bristles, the chaetae, which are used to grip the substrate. The chaetae are formed of chitin and hardened protein (scleroprotein), and each originates from a single cell. They can be protracted, withdrawn, and twisted by means of muscles within the parapodia. An aciculum, a specialized heavy chaeta, supports each chaetal bundle. Muscles arising from the body wall are attached to its base and if these are contracted the entire chaetal bundle is thrust out.

In addition to chaetae and acicula, parapodia have a pair of tentaculate sensory structures, the dorsal and ventral cirri. A pair of similar, though larger, structures, the anal cirri, arise from the pygidium.

Neither the head segments nor the pygidium bear parapodia. The head is formed from two pieces, the prostomium, and a peristomium, which bears the mouth and is derived from two fused segments. Many sensory structures are concentrated in these three segments. The prostomium contains the brain, and also bears two pairs of eyes, a pair of palps, and a pair of tentacles. The peristomium has two pairs of elongated cirri, homologous with those of the body segments, which function as both tacto- and chemoreceptors. In addition a pair of sensory ciliated pits, with an olfactory function, the nuchal organs, lie at the junction of the prostomium and peristomium.

Body wall

The outermost layer of the body wall is a thin cuticle secreted by the epidermis, which is a single layered epithelium of columnar cells with glandular and sensory cells. Two muscle layers lie beneath this and are largely responsible for the external appearance of the worm. The outermost is of circular muscle, which forms a complete ring in the posterior two-thirds of each segment. In the anterior third, the circular muscle is displaced, forming anterior and posterior dorsoventral muscles in association with the parapodia. The inner longitudinal muscle runs within the segments and is divided into four blocks: two dorsolateral blocks lie on either side of the mid-dorsal blood vessel and two ventrolateral blocks lie on either side of the mid-ventral nerve cord. In addition *Nereis* has an elaborate system of extrinsic parapodial muscles associated with the parapodia, lying between the dorsolateral and ventrolateral longitudinal muscles. Three muscle blocks insert on the anterior face, and three on the posterior face of each neuropodium. Oblique muscles extend from the side of the nerve cord to the parapodium base. There is also a complex musculature entirely within each parapodium (the intrinsic parapodial musculature).

The septa are formed of two layers of peritoneum with connective tissue and muscles between them. Some circulation of the coelomic fluid is possible between the segments as the septae are incomplete both dorsally and

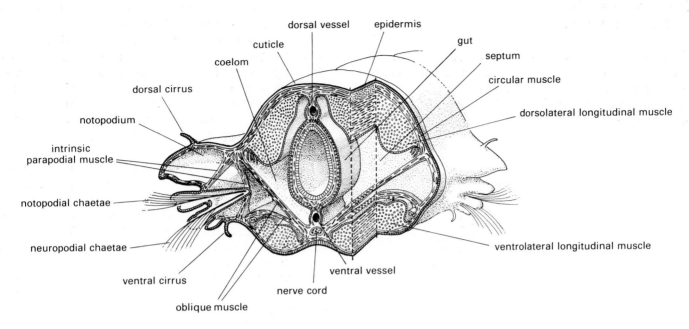

dorsal vessel · epidermis
cuticle · gut
coelom · septum
dorsal cirrus · circular muscle
notopodium · dorsolateral longitudinal muscle
intrinsic parapodial muscle
notopodial chaetae
neuropodial chaetae · ventrolateral longitudinal muscle
ventral cirrus · ventral vessel
nerve cord
oblique muscle

General organization of a segment of *Nereis*

ventrally. However, since the coelom is, to a great extent, divided into compartments, local muscle contractions result in local pressure changes, unlike the diffuse pressure changes in coelenterates and nemerteans.

Locomotion

Nereis generally creeps either slowly or rapidly over the mud but may also burrow and swim. Slow creeping involves the parapodia only, which are used as a series of levers and are moved backwards and forwards in stepwise fashion. Parapodial movement is co-ordinated along the length of the worm, every fifth or sixth parapodium moving in unison. As each parapodium is moved forwards, the acicular muscles contract and the aciculum and chaetae are protruded, providing purchase on the substrate. As the effective stroke ends the aciculum and chaetae are withdrawn, and the parapodium is lifted slightly above the substrate.

Rapid creeping and swimming both involve co-ordination between the parapodia and the longitudinal muscles of the body wall, the two actions serving to amplify each other. Alternate waves of contraction and relaxation pass forwards along the body, throwing it into a series of sideways curves (lateral undulation). At the same time the parapodia on the convex side of the curves are swung forwards to contact the ground and the acicula and chaetae are protruded. Between five and seven parapodia operate together. The position of the undulations is then reversed and the parapodia are moved into the backwards position.

Movement is brought about by a combination of the forward thrust from the parapodia and a forward pull from contraction of the longitudinal muscle when the chaetae are in contact with the ground. Similar movements of the parapodia and body are involved in swimming, but the amplitude of the curves is greater, the muscular waves of contraction being fewer in number.

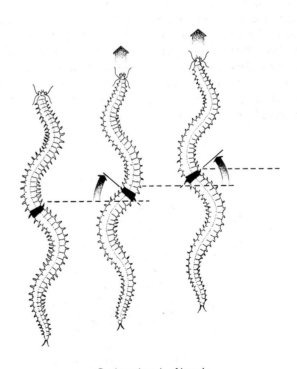

Swimming in *Nereis*

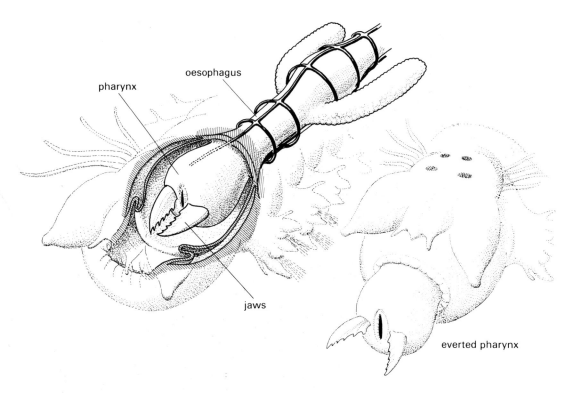

Nereis: anterior digestive system

Nutrition

Nereis feeds on a wide variety of animal and vegetable matter including small corpses and pieces of algae, and may also swallow deposits of mud with a high organic content. Normally it picks up its food by means of its pharynx and jaws, but may secrete a mucous bag at the mouth of its burrow to catch smaller particles by filtration. It is an extremely difficult animal to starve since in the absence of other food it can secrete large quantities of mucus, and feed on bacteria which grow on it.

Nereis spends much of its time within a U-shaped burrow, holding itself in position by means of its chaetae. It creates a water current through the burrow, by dorsoventral body undulations which pass from head to tail and this current brings chemical stimuli from potential food. The worm extends its anterior end from the burrow to capture food and then drags it into its burrow.

The pharynx, which is used to capture food, can be everted to form a proboscis. It is lined with cuticle which has local thickenings forming tooth-like denticles and a single pair of strong jaws with deeply serrated cutting edges. The jaws are moved by means of muscles in the proboscis wall and, with the pharynx everted, these jaws are positioned at the front of the head. When *Nereis* is not feeding the proboscis is withdrawn, the mouth closed by a sphincter muscle, and the peristomium contracted. Eversion is

brought about by a combination of the action of protractor muscles within the pharynx and local increase in coelomic pressure caused by contraction of the body wall muscles. The proboscis is withdrawn by retractor muscles which extend from the body wall to the pharynx. In addition to its role in feeding, the pharynx may be used for burrowing; in this respect it is comparable with the nemertean proboscis. During burrowing the segments posterior to the pharynx are anchored in position by the chaetae.

Immediately posterior to the pharynx the digestive tube narrows to form the oesophagus. Radial muscles in the pharyngeal wall contract and suck food into the gut, and valves between the oesophagus and pharynx prevent its regurgitation. A broad spectrum of digestive enzymes is secreted from the epithelial lining of the oesophagus, from a pair of lateral caecae arising from it, and from the anterior part of the intestine. Digestion is extracellular, and the products of digestion are absorbed in the gut wall by a network of blood capillaries which lie at the bases of the epithelial cells. The walls of the digestive tube have well developed layers of longitudinal and circular muscles which, by alternate contraction and relaxation, produce a series of waves along the gut and push the food through it, a process known as peristalsis. *Nereis* moves backwards out of its burrow for defaecation, waste matter leaving through the anus.

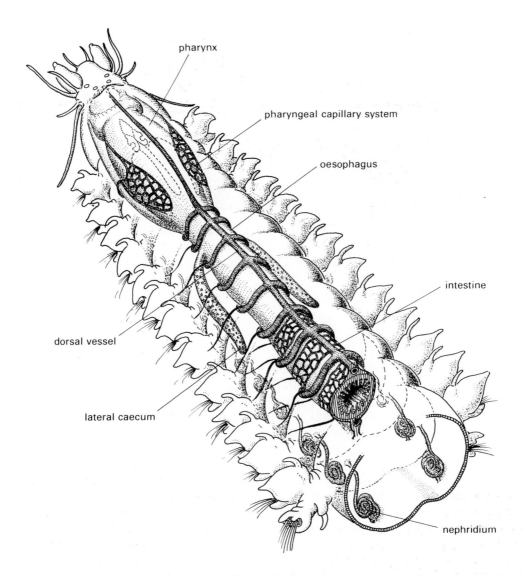

pharynx

pharyngeal capillary system

oesophagus

intestine

dorsal vessel

lateral caecum

nephridium

Dissected *Nereis* in dorsal view

Circulatory system and gaseous exchange

The circulatory system shows considerable advances on those already described, greater efficiency being demanded both by the increased metabolic activity of the annelids and as a consequence of the total separation of the gut from the body wall. The blood, which contains the respiratory pigment haemoglobin in solution in the plasma, is entirely confined to vessels (the 'closed' type of system). The arrangement of the vessels may be regarded as an integration of two systems, a longitudinal system involving a pair of median longitudinal vessels which extend the length of the animal dorsal and ventral to the gut, and a segmental system. Circulation is provided by the action of circular muscle within the main vessels; their action is particularly obvious in the dorsal blood vessel, especially with increased activity of the worm. Blood flows from head to tail in the

ventral blood vessel, and from tail to head in the dorsal blood vessel.

Within each segment a pair of ventral commissural vessels arise from the ventral blood vessel, and extend to the parapodia where they supply the ventral parapodial wall and form a fine network of thin-walled capillaries for gaseous exchange within the dorsal lobe of the notopodium. The movement of the parapodia during locomotion makes them particularly efficient respiratory surfaces. The parapodia also receive blood from a capillary plexus within the gut wall via the circum-intestinal vessels. These vessels are widely looped to permit peristaltic movements of the gut. Blood returns from the parapodia to the dorsal vessel by a pair of dorsal commissural vessels in each segment. There is also a direct linkage between the gut plexus and the dorsal vessel.

The longitudinal vessels are modified anteriorly in

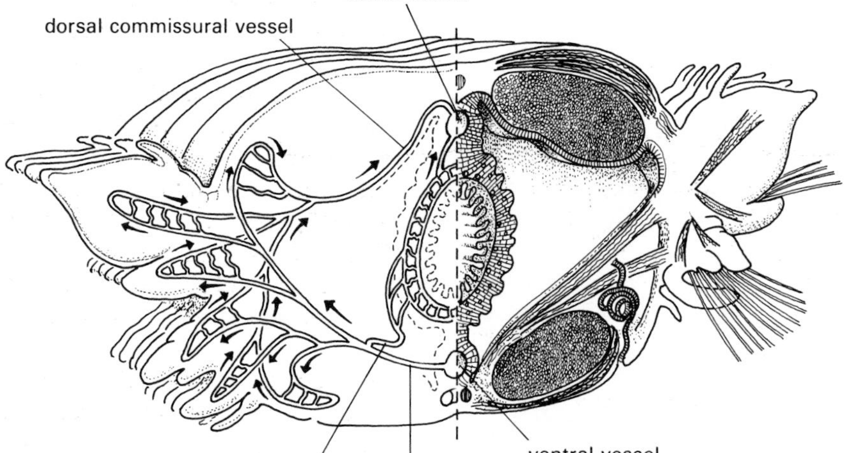

Nereis: segmental blood system

association with the proboscis and brain. The dorsal blood vessel forms a capillary plexus over the cerebral ganglia and the ventral vessel extends to the posterior end of the proboscis. Here two vessels on each side pass through an extensive capillary system over the pharynx, and link with two branches from the cerebral plexus.

Excretion and osmoregulation

A pair of excretory organs, the nephridia, are present in all segments except a few at either end. Each is formed from a long coiled tube with a ciliated lining, and is embedded in connective tissue. At one end of the tube a ciliated funnel, the nephrostome, opens into the coelom. This funnel is suspended from the rim of an opening in the ventral region of the septum. The nephridial tube passes through this opening to the following segment and opens on the ventral surface through a nephridiopore. The nephridium is supplied with blood vessels and bathed in coelomic fluid; waste products are extracted from both types of body fluid.

The nephrostome has well developed cilia, which beat towards its lumen and drive coelomic fluid into it. Larger particles within the fluid are excluded by the cilia, but may be engulfed by the margin of the funnel which is capable of some movement and can ingest particles by means of cytoplasmic extensions. Paired dorsal ciliated organs, triangular in shape and with their bases lying against the dorsal blood vessel, beat to create circulation currents within the coelomic fluid and drive particles of waste products towards the nephrostome.

Water and excretory products, both as particles and in solution, are passed through the tube and to the exterior through the nephridiopore.

Nereis diversicolor is abundant in estuarine conditions, and is able to survive in salinities as low as 0.4% (normal seawater is 3.5%). In this form the nephridial tube is long. However, in *N. cultifera*, which is not able to tolerate reduced salinities, the tube is much shorter. This comparison clearly suggests an osmoregulatory function of the nephridia.

Nervous system

The high level of activity in polychaetes demands a nervous system which is more advanced than that of either the acoelomates or pseudocoelomates, and well developed sense organs. The general arrangement is less diffuse than in these groups, the brain increased in size and in addition each segment has its own nerve centre in the form of ventral segmental ganglia.

The anterior part of the nervous system is formed by an enlarged bilobed cerebral ganglion, the 'brain', lying within the prostomium and connected to the sub-pharyngeal ganglia and ventral nerve cord by a pair of circum-pharyngeal connectives. These have ganglia from which nerves extend from the peristomial cirri. Nerves extend to the 'brain' from the palps, antennae, and eyes; other nerves reach the sub-pharyngeal ganglion from the peristomium. The entire structure, but especially the 'brain' is thus an important centre for the reception of information from peripheral sense organs.

The ventral nerve cord is a fused double structure lying within the body-wall muscle. The segmental ventral nerve cord ganglia occur at the septa, with a small part of each lying anterior to the septum. Four nerves with both motor and sensory fibres arise from each ganglion, one from the anterior part and three from the larger posterior part. Of these the first and fourth run parallel to the septum and are associated with skin receptors and stretch receptors within the longitudinal muscles. They also carry motor fibres to the

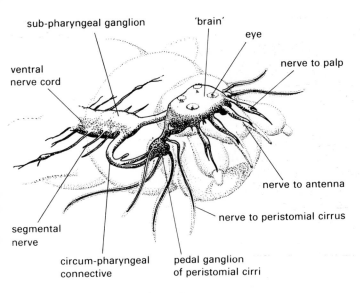

Anterior part of nervous system

The head sense organs have already been mentioned; in annelids the sense of touch is particularly important. The chaetae function as touch receptors, and in addition touch receptive cells are present in the epidermis. Annelid eyes are more elaborate than those of platyhelminths and are of the pigment cup type. The light sensitive cells are arranged in the form of a cup with pigment and supporting cells, and surround a central lens which serves to focus the light. A short optic nerve extends from the back of the cup to the brain. However, like the eyes of platyhelminths, their function is probably restricted to the determination of light intensity and direction, and not the production of an image.

Reproduction, life history, and regeneration

The sexes are generally separate in *Nereis*, and for much of the year no reproductive organs are present at all. During this phase the sexually unripe animal is known as an atoke. However, as the breeding season approaches, sex cells are budded off from the coelomic epithelium throughout the body, in the regions which overlie the principal blood vessels. Maturation takes place in the coelom which eventually becomes packed with gametes, the sexually ripe nereid being known as an epitoke. The majority of species of *Nereis* undergo other changes in developing into an epitoke; they become pelagic in habit, the parapodia are modified to become effective paddles for swimming and the eyes become increased in size. All populations have a high proportion of females. The individual epitokes congregate in swarms, and the females are thought to produce a substance (a pheromone) which attracts the males. Fertilization is external. The epitoke ruptures along the line of the dorsal longitudinal body muscles, which at the time of

muscles and are responsible for co-ordinating muscle action during locomotion. The second nerve enters the parapodium, where it expands into a pedal ganglion from which further nerves arise to supply the parapodial muscles. The third nerve receives fibres from the ventral longitudinal muscles. The segmental ganglia therefore function as co-ordination centres for locomotion.

The ventral nerve cord itself is enclosed by a fibrous sheath with some longitudinal muscle fibres from which the extrinsic parapodial muscles arise. The sheath surrounds a mass of fine nerve fibres and five giant nerve fibres which are specialized for rapid conduction of impulses in either direction. While the fine fibres are associated with the normal activities of the worm, the giant fibres are associated with a rapid escape response.

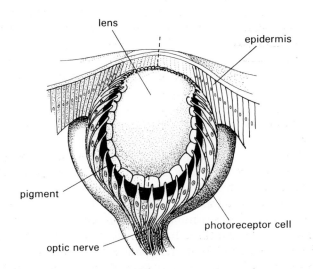

Nereis eye

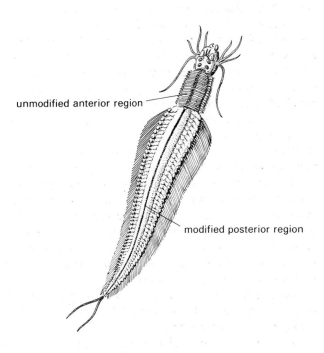

Nereis epitoke

spawning are very thin due to extensive resorption of the body wall as the gametes develop. After spawning the adults die.

The situation in *N. diversicolor* is somewhat different from that described above, perhaps in response to its estuarine habitat. Spawning takes place in burrows or on the surface of the mud. Another species, the freshwater *N. lumnicola*, is hermaphrodite with self-fertilization, the young larvae escaping from the coelom at a fairly advanced stage.

After fertilization the nereid egg hatches to give a small, ciliated, trochophore larva which metamorphoses to the adult form by elongation at its base end. The larva has a supply of yolk so it does not feed and remains on the bottom; among polychaetes this is unusual. Of more general interest is the feeding, pelagic, trochophore larva, which is characteristic of many polychaetes and has the general shape of an old-fashioned spinning-top. Cilia are present, arranged in a band at the equator with tufts at either pole. These are involved in both locomotion and feeding. A gut is present, with oesophagus, stomach, intestine and anus, as well as an excretory organ. The trochophore also has simple sense organs, including a statocyst which responds to changing orientation. After a pelagic period the trochophore elongates at its lower end to develop the adult form.

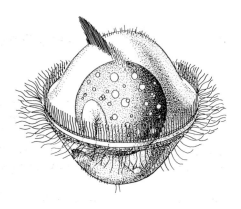

Annelid trochophore larva

The capacity for regeneration is well developed in annelids, although less so than in platyhelminths: tentacles, palps, and even heads may be replaced. It is therefore hardly surprising that polychaetes are capable of asexual reproduction by budding or fragmentation, although the typical method of reproduction is sexual.

Nereis is a representative of one of two groups of polychaetes, the subclass Errantia or free-swimming forms. Representatives of the other polychaete subclass, Sedentaria, are sessile and are often tube dwellers. Both these groups should be regarded as groupings of convenience based on habit rather than true evolutionary relationships. Sedentarians show various degrees of modification away from the basic polychaete plan. Since

they seldom, if ever, leave their tubes, many have become ciliary feeders and the head is frequently highly-modified for collecting, sifting, and sorting detritus or plankton. Often the head is also the principal region of gaseous exchange.

Arenicola marina is an example of a sedentarian polychaete which shows relatively little modification from the basic plan.

Arenicola marina: the lugworm

The lugworm is another common polychaete. It lives between the tidemarks in muddy sand, and like the ragworm is commonly dug up for use as fishing bait. It is a burrowing animal, but in contrast with the U-shaped temporary burrows of *Nereis diversicolor*, the blind-end L-shaped burrow of *Arenicola marina* is a more permanent structure and is lined with mucus to prevent collapse. *Arenicola* remains within the same burrow for many months.

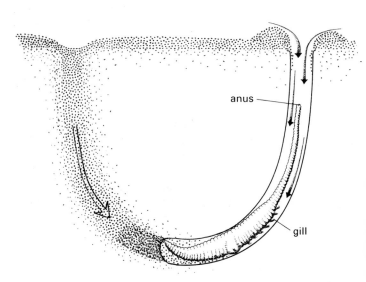

anus

gill

Arenicola marina and burrow

In response to a sedentary life style, the parapodia are reduced, the neuropodia to no more than transverse ridges. The body shows differentiation into an anterior 'trunk', and a posterior 'tail' which consists of a simple, slightly muscular body wall surrounding the gut. The structure of the anterior region is more elaborate with segmentation less apparent than that of *Nereis*. The head is a simple structure, lacking tentacles and palps but with minute eyes (which are not visible externally) and nuchal grooves. The prostomium is small, as are the peristomium and the first trunk segment. Annuli are present but do not correspond to the segment boundaries as in *Nereis*. The peristomium and first trunk segment each have two poorly-defined annuli, and the remaining segments have about four. The peristomium and

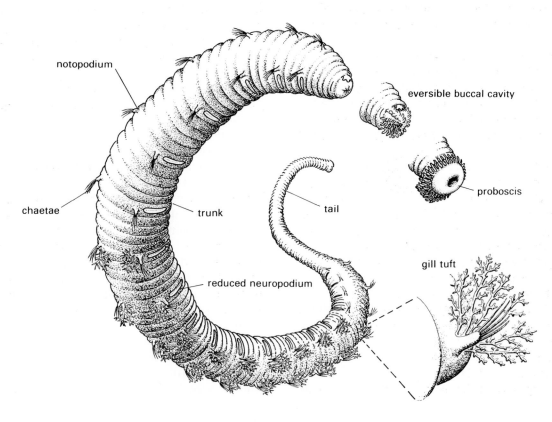

Arenicola marina

first trunk segment lack chaetae as a modification for burrowing. The remaining nineteen anterior segments can be distinguished since each has paired bundles of hook-like chaetae on the reduced notopodium. In addition the posterior thirteen trunk segments each bear a pair of gill tufts, arranged on either side of the mid-dorsal line. With the exception of three very muscular septae at the anterior end, the coelom is not divided.

Locomotion and burrowing

Arenicola is only capable of movement within the burrow: if it is removed from its burrow it cannot crawl back but must re-bury itself. The chaetae are used to grip the burrow walls. The animal moves forwards using waves of muscular contraction which affect both sides simultaneously, rather than alternately as in *Nereis*. Where the circular muscles are relaxed and the longitudinal muscles contracted, the body is fat and the chaetae are extended to grip the burrow and act as a fixed point. A region of the body where the reverse arrangement applies will become elongated and so be pushed forwards. Burrowing involves a combination of these movements with the use of the eversible buccal cavity as a proboscis to loosen the sand. The anterior end of the body is then inserted into the loosened sand, and in turn used as a fixed point whilst the posterior end of the animal is dragged down. Repetition of this sequence creates the burrow. The contraction of the body wall muscles creates a considerable hydrostatic pressure (up to 1 m of water), and

Arenicola is therefore capable of burrowing through hard-packed sand.

Feeding and gaseous exchange

Continuous waves of muscular contraction cause swellings to pass along the dorsal surface of the body and create a water current down through the sand to the blind end of the burrow, and out through the open end at a rate of about 0.1 litres per hour. The current serves for both respiration and feeding.

Organic matter is filtered out of the water by the sand, and the water flow also encourages the growth of colonies of aerobic bacteria on the sand grains. Both sand and organic matter are swallowed by the worm and the sand is expelled from the gut in the form of a compact worm cast. Organic matter forms only about 3.5% of the total material ingested.

The gut consists of the eversible buccal cavity, a short pharynx, a long oesophagus with a pair of oesophageal caecae, and an intestine.

Ingested matter is passed along the gut by muscular contraction and some ciliary action. As in *Nereis*, digestive enzymes are produced both within the oesophageal caecae and in glandular cells of the gut epithelium. In the posterior region of the intestine, water is absorbed from the sand which is then bound together by a lubricating mucus produced by the intestinal epithelium. This prevents damage to the gut wall. Sand is expelled as a worm cast at approximately forty-five minute intervals, the worm crawling backwards to the open end of the burrow to deposit the cast. Tiny mounds of worm casts are a familiar sight on tidal mud flats.

Gaseous exchange takes place over the general body surface, as well as at the gill tufts which have an especially rich supply of blood capillaries. As in *Nereis*, the respiratory pigment haemoglobin is present in the plasma. The dorsal swellings which create the water current also serve to force water through the spaces which surround the gills. As the water current approaches and passes, the gill tufts expand and contract.

Arenicola shows periods of respiratory activity interposed with periods of feeding or rest. When the tide is out and the entrance to the burrow is uncovered, the irrigation current ceases. Instead the worm creeps backwards within the burrow until most of the gill tufts are in the air; since their surfaces remain moist gaseous exchange can still occur.

Circulatory system

The circulatory system is very well developed and in some respects is more elaborate than that of *Nereis*. As in *Nereis*, longitudinal dorsal (anterior flow) and ventral (posterior flow) vessels are present with additional longitudinal vessels extending along the ventral nerve cord. Segmental vessels supply the organs and muscles and link the longitudinal vessels. The principal segmental vessels are arranged as follows: intestinal vessels carry blood from the ventral vessel to the dorsal vessel in the posterior trunk and 'tail'; paired afferent branchial vessels supply blood to the gill tufts, this blood being collected by efferent branchials which join the dorsal blood vessel.

In the mid-trunk region (between gills 1 and 9) the intestine is surrounded by an elaborate system of sinuses. Here the efferent branchial vessels open into a median sub-intestinal sinus. Paired lateral gastric sinuses receive blood both from this sub-intestinal sinus, and from the dorsal blood vessel, by a series of connecting vessels. Each gastric sinus passes blood forwards into a dilated contractile region, the heart, whence it is passed either to the ventral vessel, or to the dorsal vessel via a lateral oesophageal vessel. The hearts have presumably evolved to overcome the problem of a more sluggish circulation consequent on development of a sinus system. The dorsal vessel is also contractile.

Blood is supplied to the three anterior muscular septa and the body wall by branches from the dorsal vessel, and returned to the ventral vessel.

Excretion and osmoregulation

Six pairs of nephridia in segments 4 to 9 excrete water, and probably nitrogenous compounds as well. Each consists of a large nephrostome, a narrow central region, and a posterior bladder which narrows to an external opening, the nephridiopore, in the body wall. Uric acid, an insoluble waste product, has been found in the coelomic fluid and it is probable that soluble nitrogenous waste is lost through the skin and gills.

Arenicola can tolerate wide fluctuations in the osmotic concentration of its body fluid. These seem to follow those of the external medium to some extent. Animals living under conditions of reduced salinity have been shown to have body fluids with an osmotic concentration of half that of those living in sea water.

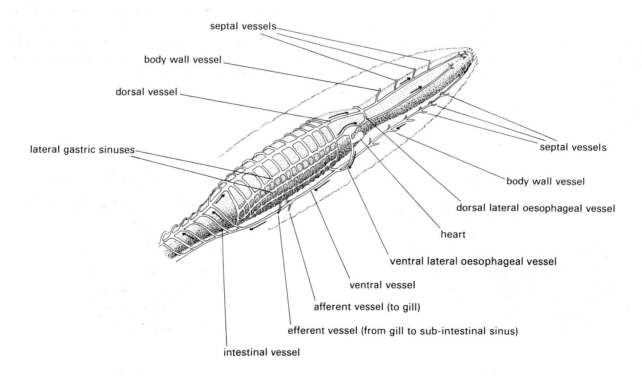

septal vessels

body wall vessel

dorsal vessel

lateral gastric sinuses

septal vessels

body wall vessel

dorsal lateral oesophageal vessel

heart

ventral lateral oesophageal vessel

ventral vessel

afferent vessel (to gill)

efferent vessel (from gill to sub-intestinal sinus)

intestinal vessel

Arenicola circulatory system

Nervous system

The reduced state of sense organs in *Arenicola* has already been mentioned and may be related to its sedentary life style. The nervous system consists of a pair of minute cerebral ganglia linked to the ventral nerve cord by a pair of circum-buccal connectives. No segmental ganglia can be identified, but paired segmental nerves extend to the body wall musculature and to the gut.

Three giant fibres are present in the nerve cord; like those in *Nereis*, they are concerned with rapid conduction of impulses in association with escape reactions.

Reproduction and life cycle

As in *Nereis diversicolor* the sexes are separate. In North temperate seas breeding occurs twice a year, around February and March and in the late summer.

Six pairs of small gonads, which consist of clusters of germ cells without any surrounding epithelium, are present, each attached to the posterior edge of the nephrostome. The developing eggs or sperm detach and complete their development and are stored in the coelom. Mature gametes are shed through the nephridia and transported by the feeding current to the sand surface where fertilization occurs. Trochophore larvae hatch after about four days and remain close to the sand surface, swimming for only short distances. Development is slow, so that young adults resulting from the late summer breeding appear the following spring.

Polychaete diversity

Both free-living and sedentary polychaetes show an immense range of diversity and many of them are extremely beautiful. Free-living surface crawlers, often with the same general body plan as *Nereis*, are found amongst stones, shells and algae. Some show a reduced number of body segments and are short and fat, for example *Aphrodite*, the sea mouse, so-called because of its mat of browny-grey hair-like structures which completely cover the dorsal surface and conceal the animal's external segmentation. Scale worms generally have peculiar plates on their dorsal surface, which are thought to provide a channel for the ventilating current when the animals are burrowing or hidden beneath stones. *Lepidonotus* is a common British example occurring on the lower shore. Contrasting with these are the syllids, which have beautifully-coloured thread-like bodies and dorsal cirri which are dorsally very elongated. The epitoke develops by budding from the posterior end of the atoke rather than by transformation of the atoke as in *Nereis*.

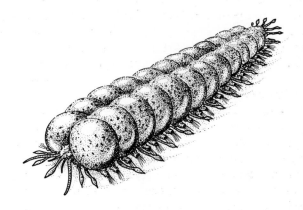

Lepidonotus

Several polychaete families have adopted a pelagic mode of life. Generally they have the same basic body form as surface-dwellers, but many are transparent, a common adaptation of pelagic animals, making them less conspicuous. Tomopterids have no chaetae but membraneous parapodial pinnules for swimming and alcipids are distinguished by large protruding eyes with extremely well-developed lenses.

Aphrodite

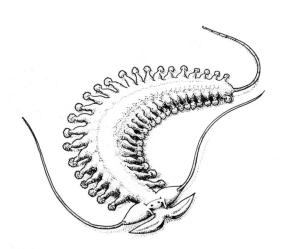

Tomopterus

Burrowing has led to the development of a wide variety of forms, of which *Arenicola* is a comparatively simple example. Many burrowers build tubes as linings to their burrows; other sedentary forms build tubes on substrates. Terebellids use sand grains which they cement together with organic matter; sabellids use mud, and serpulids a hard, calcareous substance. The majority of sedentary polychaetes are highly-adapted, with reduction of

prostomial sensory structures and the development of special anterior structures for respiration and feeding. In sabellids (fanworms) the peristomium bears a crown of many pinnate processes, the radioles, which are normally extended for feeding and respiration but can be rolled up when the animal withdraws into its tube. Sabellids are filter feeders, and the cilia of the pinnules produce a current which causes particles to be trapped by the pinnules themselves. Serpulids show a similar method of feeding. Terebellids, for example *Amphitrite*, have a prostomium bearing large numbers of contractile tentacles which are spread out on the surface of the substrate surrounding the burrow and used to collect detritus which adheres to mucus secreted by each tentacle. The food is moved by ciliary action to the tentacle base and thence to the mouth.

Chaetopterids live in parchment tubes and have a highly-modified body structure. In *Chaetopterus* the notopodia of segments 14 to 16 are greatly expanded as fans which beat and produce a water current. The notopodia of segment 12 are elongated and wing-like, and secrete a mucus net which is supported between them and which is used to trap food particles in the moving water.

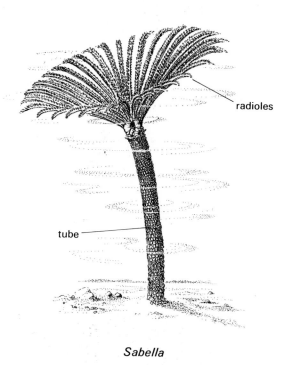

Sabella

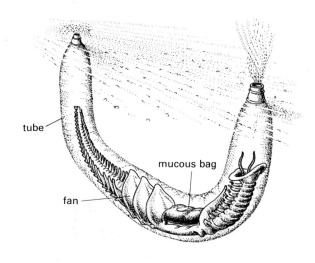

Chaetopterus

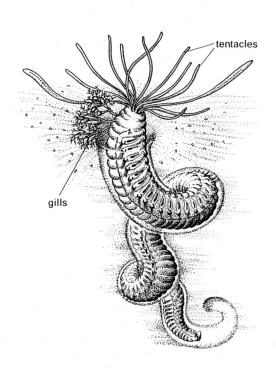

Amphitrite

Both parasitism and commensalism are found in the polychaetes, the latter being particularly common. Some species are known to live with other polychaetes in their tubes or burrows, as well as with crustaceans, sea urchins, sea cucumbers and molluscs. There they feed on some of the food collected by the activities of these animals. Parasitism however is rare.

Myzostomids, sometimes placed in a separate class, Myzostomaria, are parasites or commensals of echinoderms, most being restricted to one echinoderm group, the crinoids. Ichthyotomids are a group of blood-sucking polychaetes which attach themselves to the fins of eels.

Class Oligochaeta

The class Oligochaeta comprises about 3,100 known species and includes the terrestrial earthworms and their marine and freshwater allies. *Tubifex*, used as a fish food and widely distributed in poorly oxygenated and polluted water, is one of the better know aquatic oligochaetes. Like many polychaetes, oligochaetes display almost perfect external metamerism, but may easily be distinguished from polychaetes by their complete lack of parapodia. Chaetae are present, although fewer in number than in polychaetes, and the class name means 'few bristles' (*oligo* means 'few'). Usually eight are present in each segment, arranged in two ventral and two dorsolateral pairs. The size of the chaetae varies according to the mode of life of the animal: they are longer in swimming forms, shorter in burrowing forms. Two other anatomical features are characteristic of the oligochaetes: the head is greatly reduced and lacks sensory appendages, and a clitellum is present in adults. This is a structure formed by several adjacent swollen segments and always occurs in the anterior half of the worm. It is important during reproduction, and will be discussed in more detail in that section.

A superficial inspection of oligochaetes gives the impression that they are simple animals, but in fact they are highly specialised to their mode of life. Many species of earthworms have been described from all over the world; *Lumbricus terrestris* has been particularly well studied and this form will be used as an example of the class.

Lumbricus terrestris

Lumbricus terrestris is one of the more common annelids, occurring frequently in pasture and woodland soils which are not too acid or too waterlogged. It lies within burrows which may extend downwards as much as two metres at times of cold or drought, and which frequently end in a small chamber. Collapse of the walls of the burrows is prevented by a cement lining of defaecated soil. The adult worm can reach a maximum length of 300 mm, and is cylindrical in cross-section for the anterior third of its body, the posterior two-thirds showing some degree of dorsoventral flattening. The body is formed of about 150 segments, with conspicuous annuli which correspond in position to the internal septae. The prostomium is reduced to a small cone, and the peristomium to a simple segment with a ventral mouth. The chaetae are very short, and project only a little way beyond the skin. Like those of *Nereis* they are composed principally of the polysaccharide chitin, held together with scleroprotein fibres. Each chaeta is secreted by a single cell which lies at the base of the chaetal sac. Protractor and retractor muscles associated with the chaetae enable them to be protracted or withdrawn. The clitellum is a conspicuous structure, particularly in sexually mature worms. It involves segments 32 to 36 or 37, and is formed by a highly-developed epidermal glandular thickening.

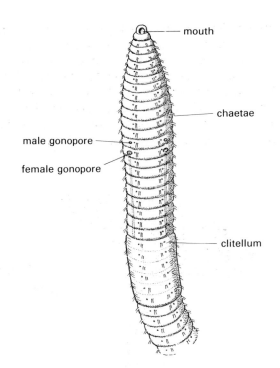

Lumbricus in ventral view

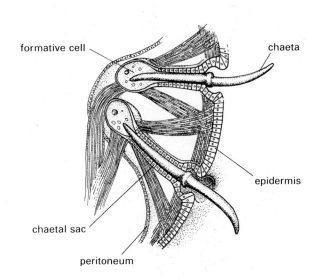

Lumbricus: chaetae and chaetal sacs

Body wall

The structure of the body wall is essentially similar to that of *Nereis*. A thin cuticle, consisting mainly of collagen fibres with polysaccharide and gelatin, is secreted by the epidermis which is a simple columnar epithelium with numerous mucous glands and sensory cells. Mucus secreted by the epidermis passes onto the surface of the worm through perforations in the cuticle. Sensory cells include tacto- and chemo-receptors, which are more or less randomly distributed through the epidermis;

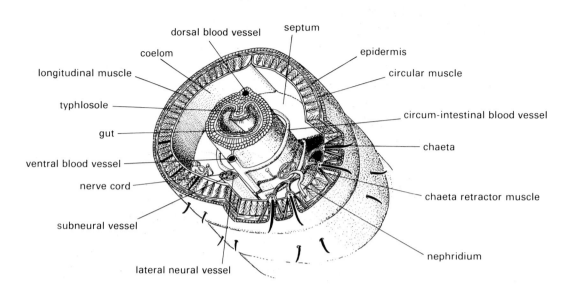

The anatomy of *Lumbricus*

photoreceptors are also present. There is also a sub-epidermal nerve net. The body wall is completed by a well-developed layer of circular muscle and a layer of longitudinal muscle, which is arranged in nine blocks (six ventrolateral, two dorsolateral and one ventral). These muscle blocks have a very elaborate structure, and are strengthened by collagenous lamellae which are well supplied with blood vessels. The muscle fibres are ribbon-like structures, each composed of a single elongated cell, and are grouped in bundles, arranged along the sides of the lamellae. This arrangement has a feather-like appearance in cross-section, and gives the longitudinal muscle considerable strength (approximately ten times that of the circular muscle).

The coelom

The coelom is completely subdivided by well-developed septa with numerous bundles of circular, radial, and oblique muscle fibres. Some communication between segments is possible through a sphinctered aperture lying immediately dorsal to the nerve cord, although this is normally closed. In each segment, a sphinctered mid-dorsal pore opens from the coelom through the body wall to the exterior. Coelom fluid can be exuded through this pore onto the skin, providing additional surface moisture.

Feeding

Lumbricus feeds on decaying fallen leaves and the saprophytic fungi and bacteria which cause their decay. It collects these from the soil surface at night and pulls them into its burrow. During this process it remains anchored to the burrow by the tip of its 'tail'. Considerable quantities of soil are swallowed during feeding and also during burrowing.

The gut is relatively simple. The mouth opens into a thin-walled buccal cavity which is slighty thickened dorsally to form a tongue. This can be protruded through the mouth and is used to pick up small pieces of vegetation. The pharynx, which opens from the buccal cavity, is the principal swallowing organ. It is a thick-walled muscular structure extending to segment 6, and has both intrinsic muscles, and radiating extrinsic radial muscles which are inserted onto the body wall. Alternate contraction and relaxation of these muscles pumps food into the mouth.

The pharyngeal lining is highly glandular, with salivary glands secreting mucus to provide the necessary moisture. (Moisture-secreting glands of some kind are always present in the anterior part of the alimentary canal of terrestrial animals where no liquid is taken in with the food). Glands secreting digestive enzymes, including proteases, are also present in the pharynx, as well as in the crop and intestine.

The pharynx opens into the oesophagus, which is differentiated into three regions: a thin-walled tube extending to segment 13, a crop (segments 14 and 15), and a gizzard (segments 16–19). The crop is thin-walled and used as a storage organ. In contrast the gizzard wall is tough and muscular with a cuticle lining, and here soil grains and leaf particles are ground together. This structure is an adaptation for dry feeding and is greatly reduced in those earthworms which have abandoned a strictly terrestrial mode of life and returned to boggy or aquatic habitats.

Calciferous glands are characteristic of the oligochaete oesophagus, and are related to a burrowing mode of life. They are lateral evaginations of the oesophageal wall (in segments 11 and 12 in *Lumbricus*) which have no direct digestive function, but serve to remove excess calcium and carbonates taken in with the soil which could unbalance the pH of the body fluids. Excess calcium and carbonate ions

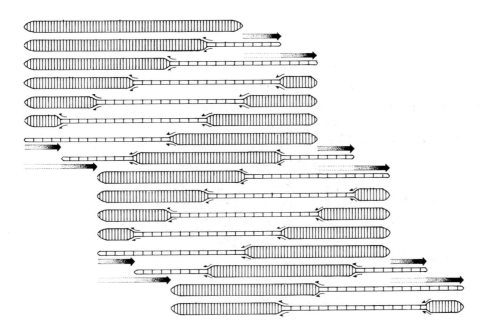

Locomotion in *Lumbricus*

are accumulated in the calciferous glands forming calcite crystals (a crystalline form of calcium carbonate) which then leave with the faeces.

A long, straight intestine completes the gut. A ridge, known as the typhlosole, projects into the lumen from the dorsal surface, serving to increase the surface available for the production of digestive enzymes, (which include cellulase and chitinase), and absorption. Digestion, which is extracellular, is the primary function of the anterior region of the intestine; absorption being the main role of the posterior region. The enzymes produced within the intestine of *Lumbricus* include cellulase and chitinase, so the animal can digest the cell walls of plants and the exoskeleton of soil arthropods. The intestine has an inner layer of circular muscle fibres and an outer layer of longitudinal muscle fibres. Peristaltic waves of contraction of these muscles provide mixing, and conduct food through the gut. The intestinal epithelium is ciliated, the ciliary beat helping to mix the gut contents.

In the region of the intestine the peritoneum gives rise to a dense layer of chloragogenous cells, which gives it a deep yellow colour (the name of the cells refers to this). The function of these cells has been the subject of considerable controversy, but since they contain deposits of glycogen and fat, with varying amounts of ammonia and urea, they are thought to be involved in both nitrogenous excretion and food storage. Certainly their position in relation to both the intestine and the coelomic fluid would suit them for this. They may be shed through the nephridia, and also are found clustered around developing ova. Chloragogenous cells are not restricted to *Lumbricus*, but occur in many oligochaetes and polychaetes in a variety of regions including the oesophagus, intestine, nephridia, and principal blood vessels.

Locomotion

Locomotion in *Lumbricus* shows more similarity to that described for nemerteans such as *Lineus*, or to burrowing polychaetes like *Arenicola*, than to that of *Nereis*. Whether crawling or burrowing, the worm proceeds by extension, anchoring, and contraction rather than by undulation. Each segment becomes alternately short and fat (by contraction of the longitudinal muscle) or long and thin (by contraction of the circular muscle). Alternating waves of contraction pass along the body from anterior to posterior, a new wave beginning as the first passes into the posterior half. The chaetae point posteriorly, allowing segments to move forwards easily, but preventing them from sliding back. Segments in a state of longitudinal contraction have their chaetae protracted and are anchored to the ground; segments in which the circular muscle is contracted have the chaetae withdrawn and are either pulled up or pushed forwards. Burrowing is similar to crawling, but the anchoring segments are pressed firmly against the burrow wall.

Circulatory system and gaseous exchange

The blood is similar in composition to that of *Nereis*, and contains the respiratory pigment haemoglobin in solution. The circulatory system is based on a similar plan, involving a combination of longitudinal and segmental blood vessels. The main collecting vessel is the dorsal blood vessel, which runs for most of its length in close contact with the gut. A ventral blood vessel is suspended in the mesentery ventral to the gut, and in addition three important longitudinal vessels, a median sub-neural vessel and paired, lateral neural vessels are found in association with the nerve cord. Circulation is caused by waves of postero-anterior contraction of the dorsal blood vessels, with dorsoventral waves of contraction of five large commissural vessels in segments 7 to 11, the pseudohearts. Additional control of the blood flow is provided by a system of passive valves in the vessels.

The dorsal vessel receives blood from the body wall via segmental parietals, and from the gut via three vessels per intestinal segment from the typhlosole and two from the gut plexus capillaries. In all segments except the anterior 11 it is connected to the sub-neural vessel by a pair of dorso-subneural vessels which run in the septa. Longitudinal extensions from each of the anterior dorso-subneurals, the lateral oesophageals, collect blood from the oesophagus and pharynx.

The ventral blood vessel is the principal route of distribution, giving off five main vessels in each segment. A pair of ventro-parietals extend to the body wall, and three

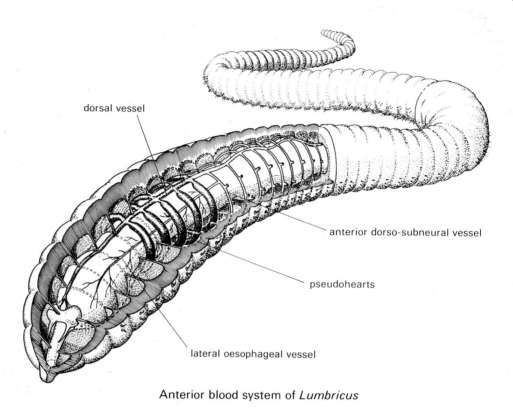

dorsal vessel

anterior dorso-subneural vessel

pseudohearts

lateral oesophageal vessel

Anterior blood system of *Lumbricus*

dorso-intestinal

body wall plexus

dorso-subneural

nephridial plexus

ventroparietal

subneural

gut plexus

ventro-intestinal

dorsal

pseudoheart

ventral

Principal segmental blood vessels of *Lumbricus*

ventro-intestinals to the gut plexus. The nephridia and reproductive organs are supplied with blood by small branches from the ventro-intestinal vessels. Blood is supplied to the nerve cord by branches of the lateral neural vessels, and collected from it by the sub-neural vessel.

Gaseous exchange takes place by diffusion over the general body surface, the necessary moisture being provided by exudations of the coelomic fluid through the dorsal pores, by the secretions of epidermal mucous glands, and by excretions from the nephridia. The extensive sub-epidermal capillary network is supplied with blood from the ventral vessel via the ventro-parietals and drained into the dorsal vessel via the segmental-parietals.

Excretion and osmoregulation

A pair of nephridia, closely comparable with those of *Nereis*, are present in every segment except the first three

and the last. However the nephridial tube is extremely elongated, and divided into distinct regions; an anterior narrow tube, a middle tube and a wide tube. An elongated, posterior, muscular bladder leads to the nephridiopore which opens on the ventrolateral surface. Considerable resorption of proteins, water and certain essential salts and ions (including potassium, sodium, and chloride) is thought to take place in the middle and wide tubes.

Nervous system

This is built on the same general plan as that of *Nereis*. A bilobed cerebral ganglion or 'brain' lies dorsal to the anterior end of the pharynx and is connected by a pair of circumpharyngeal commissures to the bilobed sub-pharyngeal ganglion, and thence to the paired ventral nerve cords. In *Lumbricus*, however, the 'brain' lies slightly more posteriously than that of *Nereis*, in the third segment. Segmental ganglia, associated with the ventral nerve cord, are present in approximately the centre of each segment, those at the posterior end being particularly prominent. Three pairs of nerves are normally given off in each segment. Exceptions are the prostomium which is supplied by nerves from the cerebral ganglion, and the peristomium which receives nerves from the sub-pharyngeal ganglion.

As in the polychaetes, giant fibres are present in addition to the arrangement of nerves just described. *Lumbricus* has five giant fibres, three of which are large and medially placed on the dorsal side of the nerve cord, and two of which are smaller, ventrally placed, and more widely separated. They innervate the longitudinal muscles through giant motor neurons arising in each ganglion, and cause all the longitudinal muscles to contract simultaneously, resulting in a sudden rapid shortening of the worm and providing an effective escape reaction.

The relationship between the nervous system and behaviour has been extensively studied in *Lumbricus*. If the ventral nerve cord and body wall are cut, the muscles behind the cut do not contract. However if the body wall is then sewn together, waves of contraction are able to pass the cut in a co-ordinated manner. This is possible because of local reflexes between the segments. If the longitudinal muscles are stretched, sensory cells within the muscle blocks (proprioreceptors) are stimulated, and these in turn stimulate the circular muscle to contract in the succeeding segment and so stretch the longitudinal muscles in that segment. In this way co-ordinated waves of contraction can pass along the length of the worm.

Although oligochaetes lack sensory appendages and obvious eyes, they show a range of responses to environmental conditions. Their sense organs are relatively simple in structure, consisting of single epidermal sensory cells, or groups of cells. The commonest type are touch receptors, which consist of single cells, each with a short hair-like projection through the cuticle; these may be extremely numerous (up to 1,000 per segment) and are highly concentrated on the prostomium and last segment. Photoreceptors are concentrated on the prostomium, but

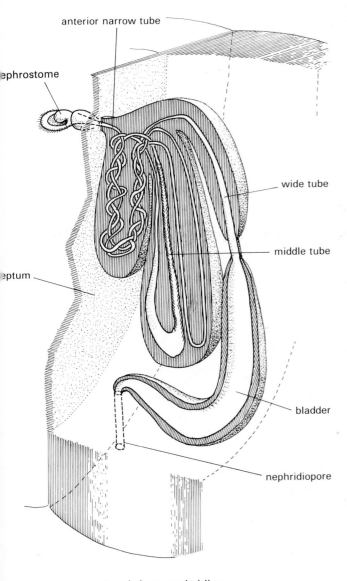

anterior narrow tube

nephrostome

wide tube

septum

middle tube

bladder

nephridiopore

Lumbricus nephridium

also occur over the dorsal surface of the body generally. Each is derived from a single cell and has a small transparent lens which focuses light onto a neurofibril network.

Chemoreceptors are formed from clusters of sensory cells. They appear as raised tubercles and occur in groups forming three rings round each segment. In *Lumbricus* they show a particular concentration on the prostomium where as many as 700 per mm^2 may be present.

Earthworms generally move away from light, and emerge onto the soil surface at night; the early bird may catch the worm, but the worm should have returned to its burrow long ago! They also tend to move towards moisture, burrowing more deeply in times of drought.

Reproduction and growth

Oligochaete reproductive systems show important points of contrast with those of polychaetes. The oligochaete is hermaphrodite, with distinct gonads and gonadial ducts, and the reproductive organs are restricted to a small number of segments in the anterior part of the body.

In *Lumbricus* the reproductive process is highly adapted for a terrestrial mode of life. There is an annual reproductive cycle, with the gonads maturing in early summer. One pair of ovaries and two pairs of testes are present as small, pear-shaped structures suspended from the lower part of the posterior face of the relevant septa near the mid-ventral line. The testes lie within large testis sacs in segments 10 and 11, and these open into three seminal vesicles in segments 9, 11 and 12. The germ cells develop to mature sperm within the seminal vesicles which are filled with nutrient fluid. Mature sperm return to the testis sacs and then enter either the anterior or posterior vasa efferentia through funnel-shaped openings. The vasa efferentia of each side join to form a vas deferens which opens by the ventral mid-line in segment 15. A pair of seminal grooves extend from the openings to the clitellum. The ovaries, in segment 13, are rather more elongated than the testes, and are not enclosed in sacs. Instead the ova are shed and develop within the coelomic fluid before passing into the ovisacs. A single pair of oviducts opens to the ventral surface of segment 14.

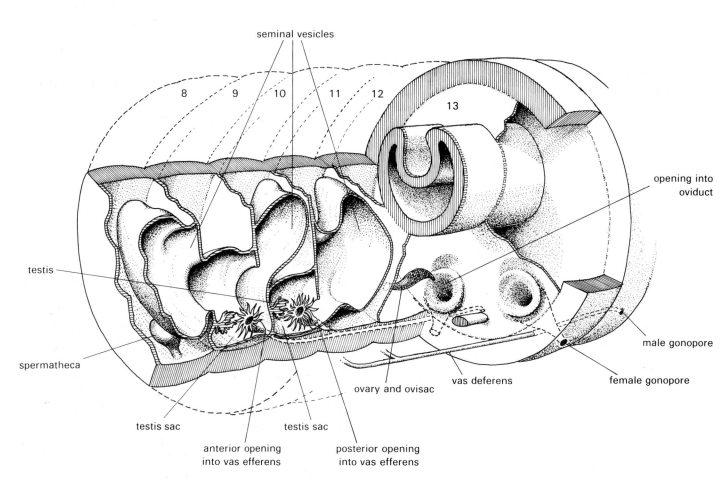

The reproductive organs of *Lumbricus terrestris*

Copulation takes place on wet warm nights on the soil surface, with a mutual exchange of sperm. The worms come into contact, with their ventral surfaces pressed together. The anterior end of each partner is directed towards the posterior end of the other, so that the clitellum (segments 32 to 36) of one lies opposite segments 9 to 11 of the other. Each worm is then surrounded by a tube of mucus produced by the epidermis and extending from segment 8 to the clitellum. In addition the clitella themselves are held together initially by their chaetae, which are dug into the skin of the opposite worm, and later by a tube of mucus surrounding both worms. Seminal fluid is released into the seminal grooves and transported along them to the anterior margin of the clitellum; it is prevented from escaping by the mucous tube. The seminal fluid is moved by the action of acriform muscles which contract in sequence, pulling the epidermis inwards and forming a sequence of pits. The seminal fluid accumulates in the space between the worms before it enters the spermathecae, where it is stored, through the spermathecal openings in segment 9. The worms then separate.

The eggs are laid in cocoons, formed by secretions from the clitellum and preceding segments. A mucus tube is formed from segment 6 to immediately posterior to the clitellum; the clitellum also secretes a membrane and an albuminous lining. Between 5 and 16 eggs are passed into the cocoon from the female gonopore. A wave of expansion from the posterior end then forces the cocoon forwards and as it passes the spermathecal openings, seminal fluid is discharged into it from the spermathecae. A number of cocoons are formed until all the eggs and sperm have been used. Fertilization takes place within the cocoons, which dry and shrink to dark-brown, ovoid, pea-sized structures which are deposited in the earth or under stones. The eggs develop to an advanced stage within the albumen produced by the clitellum, and the larval stage is suppressed. In *Lumbricus* only a single worm survives to hatch from each cocoon.

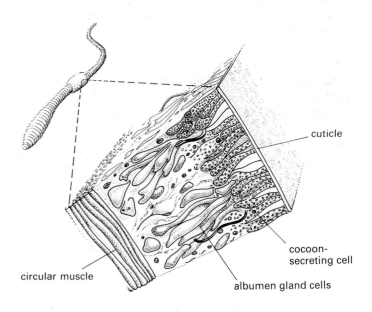

Lumbricus: histology of the clitellum

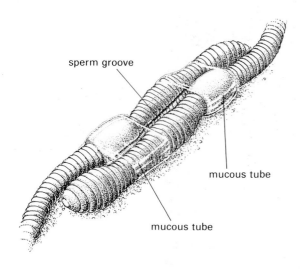

Anterior segments of copulating worms

Earthworms and the soil

Earthworms are of considerable economic and agricultural importance, chiefly because of the activities of the many species of earthworms within the soil. The value of their contribution in soil formation and improving soil texture, and also in recycling plant nutrients, was first recognised by Charles Darwin in 1837. He estimated that 50,000 worms per acre were present and that they produced between 10 and 18 tons of worm casts annually. More recent measurements of worm populations suggest that in the soil of average pasture (which is base rich) there are 50–120 g of earthworms per m^2, with 120 g per m^2 in deciduous forests. The enchytraeids, a family of minute earthworms about 10 mm in length, are particularly numerous in acid soils of the moorland type, where up to 50 g per m^2 have been recorded, and in coniferous forest with 11 g per m^2).

Many earthworms, like *Lumbricus terrestris*, are herbivores feeding on plant matter they drag down into their burrows. In this way they facilitate the breakdown and decay of plant material which is completed by organisms such as bacteria, fungi, nematodes, soil arthropods and molluscs. Other earthworms, particularly those which live in pastures, feed largely on dung. Many enchytraeids feed on fungi or bacteria present in the litter and soil organic matter.

It is difficult to assess the precise importance of earthworms in decomposition and nutrient recycling although they certainly are important. In soils where larger burrowing species of earthworms are absent, for example in acid heathland, there is usually a large accumulation of undecomposed organic matter at the surface, and such soils are often very infertile. Equally, their burrowing is of value in increasing the stability of the soil and improving its drainage.

There are many genera of earthworms of which two, *Lumbricus* and *Allolobophora* are particularly common in British soil. *Allolobophora* forms definite worm casts on the surface where *Lumbricus* lines its burrows with its faeces. *Lumbricus* can tolerate a wide range of pH conditions and is common in drier soils. *Allolobophora* is characteristic of compact, dry soils and of wetter soils of lower pH, although it cannot survive the high acidities tolerated by the enchytraeids. It is rather difficult to distinguish the many different species of earthworms and a careful examination of characters such as the number of segments and the position of the clitellum is necessary.

Class Hirudinea

The hirudineans, or leeches, are a relatively homogeneous group of highly-specialized annelids, including about 500 described species. The majority live in freshwater, but a few are marine and some have adapted to a terrestrial mode of life in warm moist regions.

Whilst many of the characteristics of leeches are absent in other annelids, they resemble the oligochaetes in their lack of parapodia and head appendages, and in their possession of a clitellum and a hermaphrodite reproductive system. They are therefore thought to share a common ancestor with the oligochaetes, and the two classes are sometimes placed together in a subphylum, Clitellata.

The popular view of leeches as blood-sucking parasites is misleading. Most cannot be regarded as parasites at all since they make only infrequent visits to other animals for blood meals. Some verge on true ectoparasitism by living more or less permanently on one individual; others are ordinary free-living predators or scavengers.

Like nematodes, leeches vary little in body form and most are dorsoventrally flattened and taper anteriorly and posteriorly. In contrast with the other annelid classes the body is formed from a fixed number of 34 segments, which is constant through a leech's life. Each segment is usually superficially subdivided by 2 to 5 annuli, and chaetae are absent.

Suckers for attachment to prey or substrate are characteristic of the leeches; a ventrally directed anterior sucker is always present which surrounds the mouth and is formed from the prostomium and the next two segments. A posterior sucker, which may be absent, is also directed ventrally and is formed from the eight posterior segments.

A century ago the name leech was sometimes used to describe a physician, and referred to the common method of treating conditions such as swellings and fevers by using leeches to draw off blood from the patient's veins. Control of the amount of blood taken was simple, with small leeches being used for small amounts. For example as many as eleven leeches were once applied to the leg of George IV in an attempt at treatment. Several million leeches were sold annually for medicinal purposes, all belonging to a single species, *Hirudo medicinalis*. Although it has no place in modern medicine, *H. medicinalis* is a common laboratory animal and will be used as an example of the group.

Hirudo medicinalis

Hirudo medicinalis is one of the longer leeches, up to 120 mm long, and living in marshes, ponds and streams. The body may be divided into five regions: the head, formed by the prostomium and first six segments, a pre-clitellar region of four segments, a clitellar region of three segments, a trunk of thirteen segments, and a posterior sucker of eight fused segments. The mouth is surrounded by a conspicuous anterior sucker, and since *H. medicinalis* has a posterior sucker, the anus opens dorsally.

Body wall
The body wall is highly muscular, with a layer of oblique muscle fibres lying between the longitudinal and circular muscles. In addition bands of dorsoventral muscle extend through all the muscle layers of the body wall, and it is these muscles which remain contracted and give the leech its characteristically flattened appearance. The epidermis is covered by a thin cuticle, and a well-developed connective tissue, dermis is present between the epidermis and body wall musculature. In the region of the suckers the body wall musculature shows modification, the circular muscles being arranged concentrically around each sucker.

The coelom, circulation, and gaseous exchange
The structure of the coelom is highly unusual; there are no septa and the coelomic cavity is largely filled by connective tissue. Remnants of the coelomic cavity are found in the sinus system, which is a series of longitudinal spaces filled with coelomic fluid containing haemoglobin in solution. This has taken on the role of the blood vascular system, as found in other annelids, and true blood vessels are therefore absent. Fluid is circulated through this system by the contraction of well-developed lateral sinuses. Gaseous exchange takes place over the general body surface; haemoglobin is present in solution and is responsible for about half the oxygen transported, the remainder being carried in physical solution.

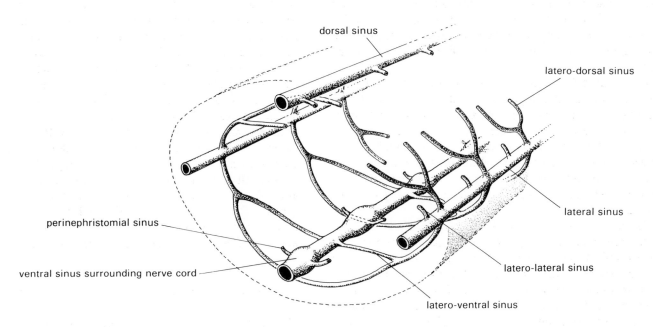

Coelomic sinus system of *Hirudo*

Locomotion

Leeches move either by swimming, or by a looping type of walk, in some ways reminiscent of that of *Hydra*. Swimming is by dorsoventral undulation caused by waves of contraction along the longitudinal musculature. It is not the normal method of locomotion and is used only if the leech becomes detached from the substrate. Crawling is more common. With the posterior sucker attached to the substrate the animal extends its body by a wave of contraction of the circular muscles. It then attaches by the anterior sucker, releases the posterior sucker, and by contraction of the longitudinal muscle shortens its body and draws it up into a loop. These movements are repeated with alternate attachment of the posterior and anterior suckers.

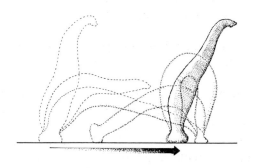

Hirudo crawling

Feeding

Hirudo medicinalis is a blood sucker, feeding on a variety of vertebrate hosts. The preferred host is a mammal, but *Hirudo* also attacks amphibians and reptiles. *H. medicinalis* is capable of ingesting blood equivalent to five times its own body weight at a single meal, another meal being unnecessary for several months. During feeding it attaches to the host by the anterior sucker and cuts into the skin of the host using three, saw-like chitinous jaws, one dorsal and two lateroventral. The jaws lie immediately within the mouth cavity and are operated by muscles attached to their bases. As they move together and apart they form a characteristic three-pronged wound. Masses of unicellular salivary glands open between the jaws and these secrete a substance called hirudin which prevents coagulation of the blood, as well as other substances of unknown origin with an anaesthetic effect. As a result of these secretions a leech can feed for many hours whilst the host remains entirely unaware of its presence; only after the animal has left is an irritation noticed.

A buccal cavity behind the jaws opens into a muscular pumping pharynx from which a short oesophagus opens into a large thin-walled midgut region, the crop, which is greatly expanded by eleven pairs of caecae. Blood is pumped by the pharynx into the crop, where some of its water is absorbed to be excreted by the nephridia, thus reducing the bulk of the meal by 40%. When the crop is full, the greatly-distended leech drops off the host.

The alimentary canal is completed by a small stomach, a straight intestine and a short rectum leading to the anus. Digestion is unusual in that it is chiefly carried out by a symbiotic gut bacterium, *Pseudomonas hirudinis*. The bacteria break down proteins of high molecular weight,

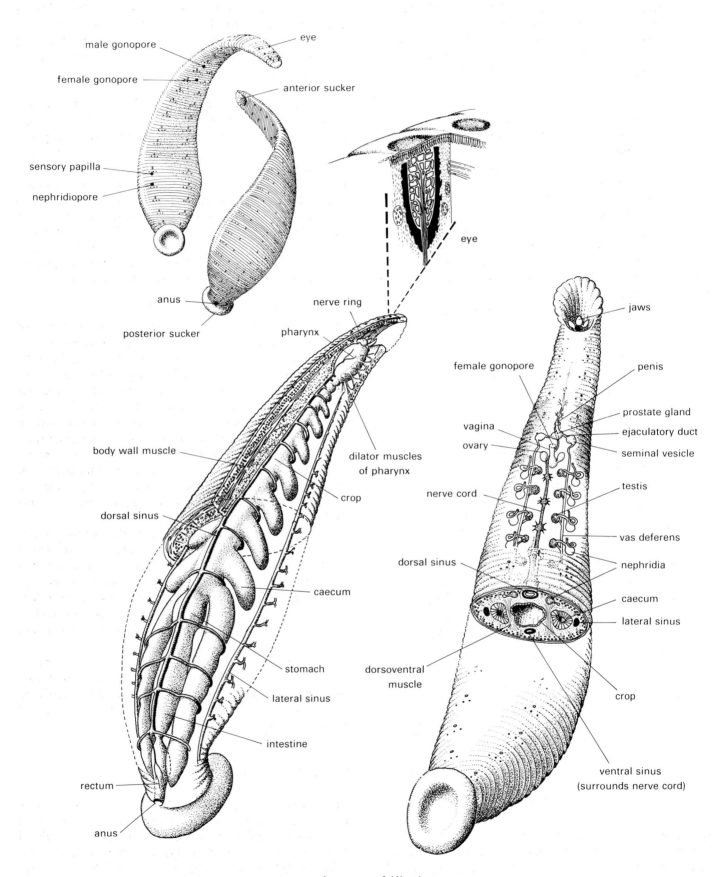

male gonopore

eye

female gonopore

anterior sucker

sensory papilla

nephridiopore

eye

anus

posterior sucker

nerve ring

pharynx

jaws

penis

female gonopore

prostate gland

vagina

ejaculatory duct

ovary

seminal vesicle

body wall muscle

dilator muscles
of pharynx

crop

testis

nerve cord

dorsal sinus

vas deferens

dorsal sinus

nephridia

caecum

caecum

lateral sinus

stomach

crop

lateral sinus

dorsoventral
muscle

intestine

rectum

ventral sinus
(surrounds nerve cord)

anus

Anatomy of *Hirudo*

carbohydrates and fats. The intestinal cells of the gut produce only limited amounts of exopeptidase enzymes. The process of digestion is very slow, and it may take *H. medicinalis* up to 200 days to absorb a single meal. No wonder physicians needed millions of them!

Excretion

Hirudo medicinalis has 17 pairs of segmentally arranged nephridia (in segments 6 to 22) which are structurally similar to those of the oligochaetes though with some specializations related to the reduction of the coelomic cavity. In each nephridium the coelomic chamber is represented by a non-ciliated capsule into which open clusters of nephrostomes, the whole structure forming a ciliated organ. A long, coiled tubular region from the ciliated organ opens into a dilated muscular bladder which discharges through a ventrolateral nephridiopore. The structure of the tubule is unusual, in consisting of a cord of cells enclosing a non-ciliated intracellular canal.

The principal waste product is ammonia, which is removed from the leech's body fluid by the nephridial tube. A considerable amount of resorption of ions is thought to occur within the tube so that the nephridia control the chemical composition of the body fluid. They also have an osmoregulatory function.

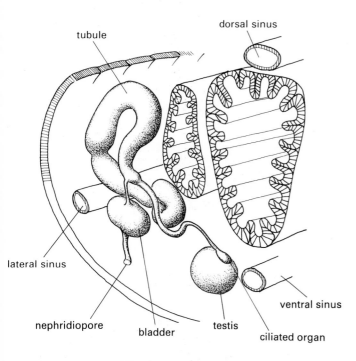

Hirudo: nephridium

Nervous system and sense organs

The nervous system of leeches reflects their specialized body structure. The ventral nerve cord is superficially single, but in fact double, and bears a series of fused ganglia, each corresponding to a segment and giving off branch nerves. A large nerve ring surrounding the pharynx represents the cerebral and sub-pharyngeal ganglia, the circum-pharyngeal connectives, and the ganglia of the anterior segments, all of which are more discrete in other annelids. The posterior ganglion also represents eight fused ganglia.

The epidermis contains touch, photo- and chemo-receptors. In addition two specialized types of sense organs are present. Sensory papillae in the form of projecting discs of sensory cells occur in each segment, and eyes, consisting of clusters of photoreceptor cells surrounded by a pigment cup, are present on the dorsal surfaces of the anterior segments. The sense organs are remarkably effective and *Hirudo* can apparently detect a man standing in water, even at a considerable distance. In addition to its acute sensitivity to chemical stimuli it also appears to respond to slight increases in temperature. Generally leeches avoid the light, but tend to move towards it when seeking a meal.

Reproduction

Hirudo medicinalis is hermaphrodite: it has nine pairs of spherical testes, each enclosed in a coelomic pouch and closely associated wth the nephridia in segments 12 to 20, and a single pair of coiled ovaries in segment 11. Each testis discharges into a vas deferens (one on each side) which coils at its anterior end to form a seminal vesicle. From each seminal vesicle ejaculatory ducts extend to join medially at the prostate gland from which an eversible muscular penis opens through the median male gonopore in segment 10. The entire female reproductive system lies in segment 11; it consists of two short oviducts which emerge from the ovaries to join at a common duct and a short vagina leading to the female gonopore.

Mating normally occurs on land in the summer and is comparable with the process in oligochaetes. The ventral surfaces of the two leeches are brought together, the partners lying head to tail, so that the female gonopore of one partner is opposite the male gonopore of the other. Exchange of sperm takes place, and the penis of the male is everted into the vagina of the female where the sperm are stored.

The fertilized eggs are usually laid two or three weeks after copulation. The clitellum becomes conspicuous and secretes a cocoon into which albumen and the fertilized eggs are deposited. The cocoons are then left in damp soil and young leeches hatch directly from them.

Hirudinean diversity

In comparison with polychaetes hirudineans show relatively little diversity. The class is divided into four orders of which one, Acanthobdellida is regarded as primitive and restricted to a single North European species which is parasitic on salmonid fish. It is the only leech in which chaetae are present. The Rhynchobellida includes the piscicolids, a large group of fish leeches, and the glossiphonids which are also ectoparasitic on both vertebrates and invertebrates. This group includes the giant Amazon leech, *Haementeria ghiliani*, which can reach 300 mm in length, though the majority of hirudineans are between 20 mm and 50 mm long. The jawed leeches are members of the order Gnathobdellida, which includes *Hirudo*, and many other forms which feed on mammals, and the family Haemadipsidae, limited in distribution to Australasia. *Erpobdella*, a common pond leech which feeds on insect larvae and other invertebrates, has no jaws and is a representative of the order Pharyngobdellida.

Synopsis of the phylum Annelida

Numerous annelid families are recognized, and the following list is restricted to those mentioned in the text.
Class Polychaeta
  Subclass Errantia
    Family Aphroditidae: (*Aphrodite*)
           Polynoidae: (*Lepidonotus*)
           Syllidae: (*Syllis*)
           Tomopteridae: (*Tomopteris*)
           Alciopidae: (*Alciopa*)
           Nereidae: (*Nereis*)
           Ichthyonomidae:(*Ichthyonomis*)
           Myzostomidae: (*Myzostoma*)
  Subclass Sedentaria
    Family Chaetopteridae: (*Chaetopterus*)
           Arenicolidae: (*Arenicola*)
           Sabellidae: (*Sabella*)
           Terebellidae: (*Amphitrite*)
           Serpulidae: (*Serpula*)
Class Oligochaeta
  Order Lumbriculida
  Order Tubificina
    Family Tubificidae: (*Tubifex*)
           Enchytraeidae: (*Enchytraeus*)
  Order Hoplotaxida
    Family Lumbricidae: (*Lumbricus, Allolobophora*)
Class Hirudinea
  Order Acanthobdellida
        Rhynchobdellida
    Family Glossiphoniidae: (*Haementeria*)
           Piscicolidae
  Order Gnathobdellida
    Family Hirudinidae: (*Hirudo*)
           Haemadipsidae
  Order Pharyngobdellida
    Family Erpobdellidae: (*Erpobdella*)

Phylum Arthropoda

Introduction to the Arthropoda

The phylum Arthropoda is the largest group within the animal kingdom approached in species number only by the Protozoa and Nematoda, and with over 800,000 known species. The phylum includes 80% of the entire animal kingdom. Within the group there is an amazing range of diversity of both structure and mode of life, with representatives being found in every known habitat. However the total biomass is less than might be expected as the majority of species are very small. The group includes many familiar forms such as lobsters, crabs, spiders, centipedes, bees and houseflies.

As a group, arthropods probably affect man more directly than any other phylum, and have therefore been the subject of intensive study. Their activities can be either detrimental or beneficial. They can destroy crops, the Old Testament plague of locusts being an early recorded example, and also food, clothing and entire buildings.

> Some primal termite knocked on wood,
> And tasted it and found it good,
> And that is why your cousin May
> Fell through the parlor floor today.
>
> Ogden Nash.

Many diseases are transmitted by insects, including parasitic infections such as malaria and filariasis, as well as typhus and plague. The Great Plague of London, transmitted by fleas, was responsible for the deaths of 68,596 people in London in 1665, out of a population of about 460,000 of which two-thirds are said to have fled in an effort to escape the disease. Insects also contribute minor irritations to daily life in the form of bites or stings.

> The Lord in His wisdom made the fly,
> And then forgot to tell us why.
>
> Ogden Nash.

and

> The Honey Bee
> The honey bee is sad and cross
> And wicked as a weasel
> And when she perches on you boss
> She leaves a little measle.
>
> Don Marquis.

On the other hand man benefits from the arthropods. Products obtained from insects include shellac, dyes such as cochineal, beeswax and honey. Crabs, shrimps and lobsters are an enjoyable addition to man's diet and, of greater significance, arthropods are important components of many food chains. Aquatic insects are a major element in the diet of river fish, their place being taken by marine crustacean larvae in the sea. Mammals as large as the whale are able to survive entirely on a diet of crustaceans. A wide range of soil arthropods (including mites, springtails, woodlice) are important in the process of plant decay. Arthropod scavengers remove organic matter such as faeces or dead animals. The pollination and hence sexual reproduction of many plants entirely depends on insects and very often there is a close symbiotic relationship between plant and insect.

Some insects have been used by man for the biological control of pests, which are themselves very often insects. Finally insects have contributed largely to scientific research: much of present day knowledge of the mechanisms of inheritance is based on the use of the fruit fly, *Drosophila melanogaster*; studies of insect flight have contributed to the development of aviation, and insect larvae (mealworms) have been used to study population dynamics. The position of man with regard to arthropods is therefore highly ambivalent: many are beneficial, some are harmful, but in either case they cannot be ignored.

Biological success is difficult to define. However a successful group can be regarded as one that is numerous, both in terms of species number and number of individuals; that is widespread both in terms of habitat and geographical distribution; that is capable of exploiting the greatest variety of food sources, and can defend itself against attack. By all of these criteria, arthropods rank as the most successful animals.

Arthropods are thought to have evolved from annelids, with which they show certain similarities. Both groups show metameric segmentation, the segments of arthropods bearing appendages which may be compared with the parapodia of polychaetes. However in many arthropods the basic metameric pattern in which every segment performs every function is considerably obscured: groups of segments are specialized to perform functions for the entire animal. These groups are known as tagmata (singular, tagma) and the whole process is tagmatization. Some degree of tagmatization is found in all annelids, but it occurs to a far greater extent in the arthropods and is an important evolutionary trend. Indeed, one of the advantages of metamerism is the opportunity it provides for specialization of this kind.

Other points of similarity between annelids and arthropods include the structure of the central nervous system, essentially the same in both groups, and to some extent the arrangement of the circulatory system. Arthropods have a dorsal tubular heart, which may be compared with the annelid contractile dorsal blood vessel. Paired segmental coelomic compartments develop in the young arthropod but are greatly reduced in the adult by the

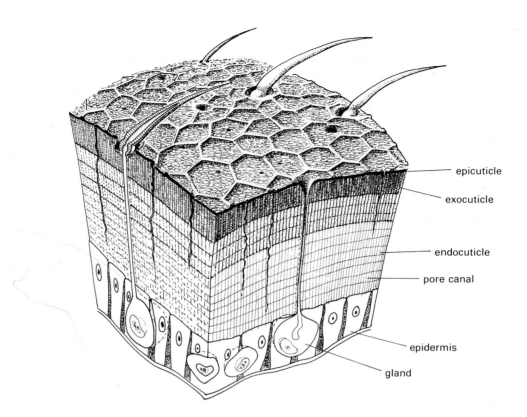

epicuticle
exocuticle
endocuticle
pore canal
epidermis
gland

Arthropod integument

development of a blood-filled space, the haemocoele, as the major body cavity.

The main distinguishing feature of arthropods, immediately visible to the casual observer, is a tough, semi-rigid cuticular exoskeleton secreted by the epidermis and consisting basically of chitin, a polysaccharide, enmeshed with a protein element, arthropodin. The cuticle is a layered structure, with a thin, non-wettable outer surface. Whilst the cuticle is always built on a foundation of chitin, this may be invaded or overlaid by different materials in the various arthropod groups. In forms such as insects, the cuticle is rather thin and light but is strengthened by impregnation with tanned proteins and waterproofed by an outer layer of wax. In forms like crabs the cuticle is thick and hard due to the deposition of calcium salts in the endocuticle; a degree of water-proofing is provided by lipids in the outer epicuticle, and the whole structure is protected by an outer layer of cement.

The cuticle is of great significance in the success of the arthropods, since it serves as a skeleton, providing both support and an efficient means of locomotion, and also provides protection from desiccation. It is laid down in a series of plates, the sclerites, joined together by flexible regions, the arthrodial membranes. Separate plates or groups of plates cover the head, thorax, and abdomen and their limited flexibility means that progress by any form of undulation is no longer practical. Arthropods therefore depend on a series of jointed legs for locomotion and it is from these structures that the name of the phylum is derived (*arthropoda* means 'jointed legs'). Arthropod legs are always hollow, and composed of a number of tubular sclerites, the podomeres, which are able to move both

relative to the body and to each other. Movement is brought about by this system of levers moved by antagonistic muscles attached within the sclerites. The continuous muscle layers characteristic of animals with a hydrostatic skeleton are split and the muscles arranged in discrete blocks either as flexors (bringing sclerites together) or extensors (moving them apart). In one respect walking in many arthropods retains features comparable with locomotion in polychaetes: each leg performs a series of movements in which it is lifted, swung forwards, and then placed on the substrate. All the legs on one side of the body carry out these movements in sequence, in a locomotor wave, and walking is the result of alternating locomotor waves on the right and left hand sides of the body. Appendages are also necessary as sense organs, (antennae and antennules) and for food manipulation.

In many respects arthropod cuticle may be compared with a suit of mediaeval armour, and like such suits, it cannot be increased in size. A young knight had new armour made as he grew, and the same applies to the growing arthropod. Arthropods increase in size by periodically moulting or shedding their cuticles, a process known as ecdysis. Arthropod growth is not a gradual process, but takes place in a series of steps, each size increase taking place in the short period after the old cuticle is shed and before the new one is secreted. The period between moults is called the instar.

An exoskeleton can be a very effective protection against water loss, and this has been an important factor in the evolution of the many groups of terrestrial arthropods. However, it also restricts the availability of suitable permeable surfaces for excretion and respiration and arthropods have therefore developed more elaborate respiratory and excretory organs than the animals described so far. Indeed all the features described as diagnostic of arthropods can be traced directly to the development of an exoskeleton; even the modification of the circulatory system and coelom is related to the loss of hydrostatic function of the body fluids.

The recognition of Arthropoda as a phylum implies that all members of the group are related to each other. In fact this assumption is no longer generally accepted and the term arthropod can more appropriately be applied to a grade or level of organization, probably reached by several separate evolutionary lines following more or less parallel paths. Zoologists traditionally divided living arthropods into two taxonomic groups: the subphylum Chelicerata, which lack antennae and have feeding appendages called chelicerae; and the subphylum Mandibulata, which have antennae and whose feeding appendages are called mandibles. The former group includes the scorpions and spiders; the latter such diverse forms as crabs, centipedes, and insects. Modern classifications tend, however, to divide present-day arthropods into three distinct taxonomic groups on the basis of their general limb anatomy, and particularly that of the appendages used for cutting and preparing food. The mandibulates are regarded as an artificial assemblage and two groups – Crustacea and Uniramia – are recognized in its stead. There are now believed to be at least four main lines of arthropod evolution, represented by the subphyla Trilobitomorpha (entirely extinct), Chelicerata, Crustacea, and Uniramia. The crayfish *Astacus* (Crustacea) will be used as an introduction to the phylum.

Synopsis of the phylum Arthropoda

Subphylum Trilobitomorpha (an entirely fossil group of primitive arthropods)
Subphylum Chelicerata
    Class Merostomata
    Class Arachnida
    Class Pycnogonida
Subphylum Crustacea
    Class Cephalocarida
    Class Branchiopoda
    Class Ostracoda
    Class Mystacocarida
    Class Copepoda
    Class Branchiura
    Class Cirripedia
    Class Malacostraca
Subphylum Uniramia
    Class Insecta
    Class Chilopoda
    Class Diplopoda
    Class Symphyla
    Class Pauropoda

Subphylum Crustacea

The crustaceans can be regarded as highly successful arthropods, and include shrimps, lobsters and crabs as well as numerous smaller forms of no gastronomic interest. The majority of species are marine but some live in fresh water. Aquatic crustaceans are extremely important; in addition to including many of the larger arthropods they are of great ecological importance in marine and freshwater food chains. Several species of crustaceans live above the high tide line on beaches but rather few have invaded the land. As terrestrial animals they have not been very successful because they have retained a characteristically aquatic physiology and are therefore restricted to damp environments.

Astacus

Astacus, the crayfish, is an excellent introduction to the group since it is fairly readily available, less highly modified than many of the advanced larger crustaceans such as the crabs, and large enough for convenient dissection. *Astacus fluviatilis* is a European representative of the genus, with a wide distribution, especially calcareous streams, though it is less common now than formerly. It is a bottom dweller and lives under stones or in burrows which it makes in the banks, emerging at night to feed.

Exoskeleton
The external covering is formed by a segmented exoskeleton which is extremely thick and hard except at the arthrodial membranes, and which is secreted by a single-layered epidermis. The surface layer, the epicuticle, is thin and composed of tanned lipoprotein, giving impermeability to the entire structure. The endocuticle is thick and composed of three layers; there is an outer pigmented layer of sclerotized protein, a middle layer of deposited calcium carbonate and an inner uncalcified layer.

Body organization
Astacus can be divided into three regions or tagmata, the head (consisting of six segments), thorax (eight segments) and abdomen (six segments and a posterior telson); the head and thorax, however, are almost indistinguishably fused as a cephalothorax covered dorsally by one large rigid exoskeletal plate, the carapace. A cephalic groove approximately indicates the division between head and thorax dorsally, and ventrally some signs of external segments are visible. An accurate assessment of the number and position of the segments can be made from the appendages: one pair is present on each segment, modified in response to specific feeding or sensory functions. The head is prolonged anteriorly as a jointed rostrum, with a stalked eye lying on either side.

It is thought likely that ancestral crustaceans had a series of identical appendages, arranged one per segment along the whole length of the body. In *Astacus* the least specialized appendages are found on the first five abdominal segments: these are the pleopods or swimmerets. The exoskeleton of the abdomen also shows the least modification, with each segment consisting of a ring formed from a dorsal tergite (tergum) and a ventral sternite (sternum). The tergites overlap dorsally whereas the sternites are joined by a wide region of flexible membrane. The pleopods are ventrally directed biramous appendages attached to the abdomen by a basal protopodite composed of two pieces, a coxopodite and a basipodite. The two branches attach to the basipodite, an outer exopodite and an inner endopodite, and both are fringed with bristles or setae. The appendages of the head and thorax, as well as the other abdominal appendages show considerable modification of this basic structure, either by loss of parts (the appendage becoming secondarily uniramous) or by addition of parts. Both the exopodite and the endopodite may be composed of several segments, and the basipodite and coxopodite may be fused, lost, or expanded.

In female *Astacus* the first five abdominal appendages are all similar to the basic structure described except that the first pair shows some reduction in size. In the male the first two pairs of abdominal appendages are modified for sperm transfer, the first being reduced to an unjointed rod, and the endopodite of the second being stiffened. The sixth pair of

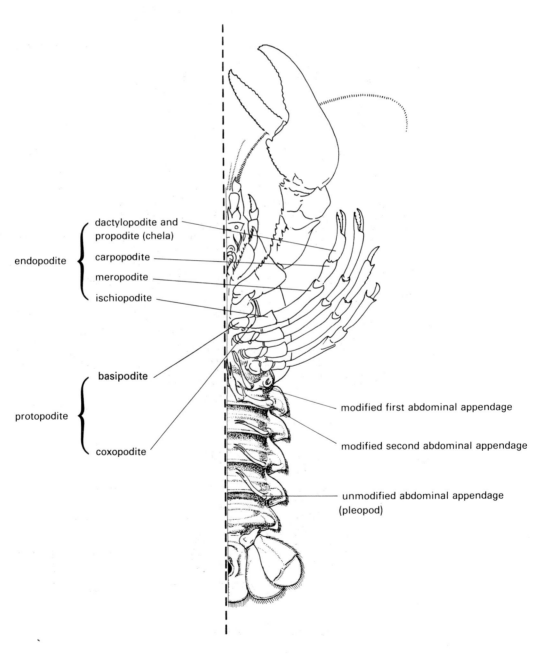

endopodite {
  dactylopodite and propodite (chela)
  carpopodite
  meropodite
  ischiopodite
}

protopodite {
  basipodite
  coxopodite
}

modified first abdominal appendage

modified second abdominal appendage

unmodified abdominal appendage (pleopod)

Astacus (male): ventral surface

abdominal appendages are modified in both sexes to form broad fan-line structures, the uropods. The protopodite consists of a single segment only and both the endopodite and exopodite are greatly expanded, the latter having a transverse hinge. In combination with the telson they form an effective rudder and swimming appendage: sudden contractions of the abdominal muscles produce fast sweeping movements of the uropods and telson enabling the crayfish to swim rapidly backwards to escape danger.

The head bears five pairs of appendages. The most anterior are the first antennae, or antennules which are relatively short and have two filaments borne on a peduncle of three segments. The second antennae lie posterior to the stalked compound eye and originate posterior to the mouths. They have two functions: the endopodite is represented by a long segmented filament and serves as a sense organ, the exopodite is a basal plate which controls the angle at which the animal dives. The possession of two pairs of antennae is a diagnostic feature of the subphylum Crustacea.

The head is completed by three pairs of feeding appendages surrounding the mouth, which is ventral in position. A pair of short heavy mandibles (segment 4) with opposing surfaces for grinding cover the mouth, which has a non-flexible labrum at the front. Behind these are two pairs of accessory feeding structures, the first and second

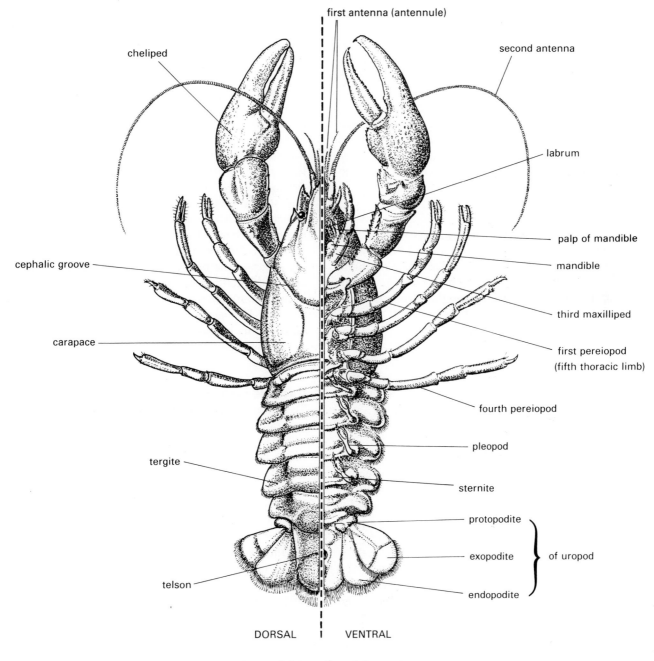

Astacus (female)

maxillae. The first maxillae (segment 5) are small and flattened and have two basal joints which are expanded and fringed with setae, and a reduced, unjointed endopodite. The second maxillae, also known as maxillules (segment 6), have both the coxopodite and basipodite expanded as bilobed structures, and the exopodite is also greatly enlarged as a scaphognathite, which functions as a paddle to create the respiratory current.

Five pairs of thoracic appendages are present. The most anterior of these are the first maxillipeds, which have the coxopodites and basipodites extended inwards; the endopodites form a broad plate and a tiny, two-jointed structure, and the exopodites are expanded into a long, many-jointed filament. The second and third maxillipeds each bear a pair of gills on the coxopodite. The exopodite is slender, with many segments, while the endopodite is large and consists of five joints (named, from the base: ischiopodite, meropodite, carpopodite, propodite, and dactylopodite) each bearing setae. The same joints and gills are found in the posterior thoracic appendages, though these all lack exopodites. The first pair, the chelipeds, are the longest and most conspicious limbs of *Astacus*, with the propodite and dactylopodite modified as a pair of massive pincers or chelae. The four posterior thoracic appendages, the walking legs or pereiopods, are similar to each other except that the anterior two pairs bear small pincers while the others have a pair of simple spines.

Locomotion

Astacus normally moves by walking using its walking legs and holding its chelipeds out in front to counter-balance the weight of the abdomen. Movement is helped by a rhythmic anterior-posterior beat of the pleopods. Unlike many crustaceans *Astacus* cannot swim using the pleopods alone as these are disproportionately small for its size. If startled, *Astacus* retreats by an abrupt dart backwards which is caused by sudden flexure of the abdomen with the uropods and telson acting together as a paddle. The animal cannot dart forwards. The rapid nervous co-ordination necessary is provided by a giant fibre system, similar to that described in the annelids.

Feeding

Astacus feeds at night, when it comes out from its hiding place to seek food. This consists of almost anything organic, animal or plant, alive or dead; it will eat snails or insect larvae, and frequently feeds on shells or calcareous algae for their calcareous content. During the coldest times of the year *Astacus* retreats into a burrow and does not feed.

The process of transferring food into the mouth is complicated, and involves six pairs of appendages, the mandibles, maxillae, maxillules, and first, second and third maxillipeds. The third maxillipeds are the largest and function both to protect the other appendages when the animal is not feeding and to pick up and tear the food. During tearing larger food pieces are grasped by the mandibles and pulled by the second maxillipeds. Food is then passed forwards to the mandibles via the second and first maxillipeds. It is crushed by forwards and backwards movements of the mandibles and pushed into the large mouth by the maxillae.

Astacus has a straight gut with an anterior end highly modified for grinding. There are no cilia, and food is moved entirely by peristalsis. The mouth opens into a stomodaeum, lined with chitin, and consisting of a short oesophagus and a capacious proventriculus. The proventriculus has two chambers. In the first (cardiac) chamber there is a series of chitinous teeth and articulating plates moved by muscles, which together form a gastric mill where food is crushed and ground. The second (pyloric) chamber forms a sieve with rows of chitinous setae. The food therefore has to be ground into extremely fine particles before it is able to pass down the alimentary tract into the midgut. This protects the midgut from damage by abrasion and also allows *Astacus* to have a very broad and unselective diet.

In addition to the mechanical process of the gastric mill, chemical digestion also occurs in the stomodaeum. A large bilobed digestive gland, the hepatopancreas, consisting of numerous finely-branched tubules, lies in the haemocoele and opens into the midgut. A watery dark-brown mixture of protease, carbohydrase and lipase is produced and sucked forwards into the cardiac chamber by dilator muscles, through paired ventral canals protected by a setal sieve. Digestion therefore starts in the gastric mill, and when it is sufficiently advanced, partly digested food and fine particles are passed back along the ventral canals into the pyloric chamber. This is tubular, the most anterior region acting for storage, and the next as a press in which partially digested and liquid matter is squeezed out. Digested food is absorbed in the hepatopancreas which is also a site for storage of food reserves. Undigestable matter is passed along to the midgut and eventually leaves the anus via the cuticle-lined hindgut (proctodaeum). The anus opens in the ventral mid-line of the telson.

Gaseous exchange

To compensate for the restriction of permeable surfaces imposed by an exoskeleton, *Astacus* has developed a series of outgrowths (epipodites) which function as gills and have only a thin cuticle separating the blood from the respiratory current. Eighteen pairs of gills are present, crowded into a branchial chamber which is formed by lateral extensions of the carapace, known as branchiostegites. The ventral surface of the branchial chamber is provided by the bases of the limbs, from the second maxilliped to the posterior end of the thorax.

The epipodites are of two types and are arranged in three layers. The pleurobranchs and arthrobranchs arise from the side of the thorax and from the articulatory membrane between the leg and the body. They both have a feather-like structure, with a main stem (which carries the afferent and efferent blood vessels) bearing closely set branchial filaments on each side. The podobranchs arise from the

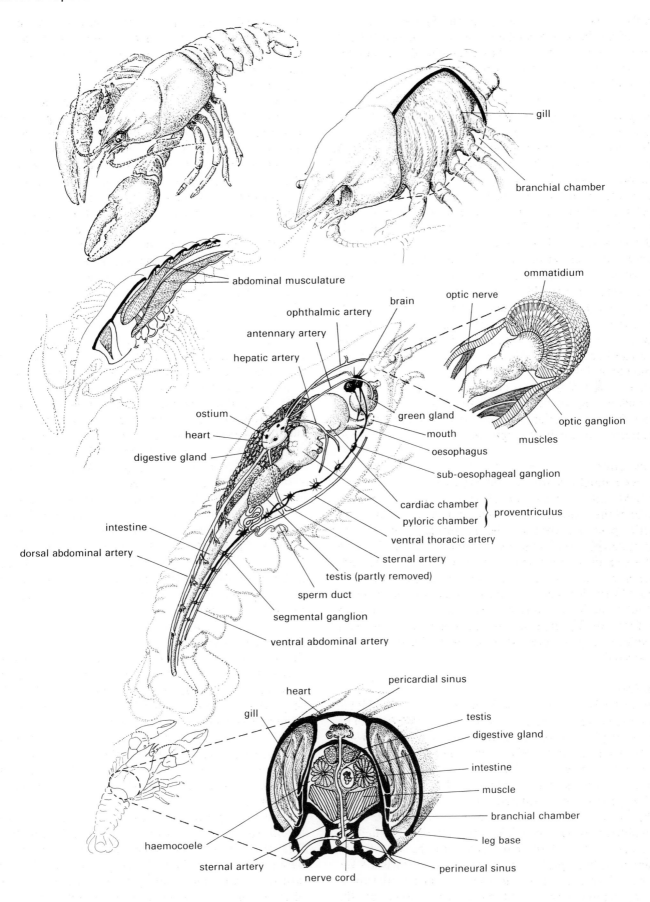

gill

branchial chamber

abdominal musculature

ommatidium

optic nerve

brain

ophthalmic artery

antennary artery

hepatic artery

optic ganglion

muscles

ostium

heart

digestive gland

green gland

mouth

oesophagus

sub-oesophageal ganglion

cardiac chamber ⎫
pyloric chamber ⎬ proventriculus

ventral thoracic artery

sternal artery

testis (partly removed)

intestine

dorsal abdominal artery

sperm duct

segmental ganglion

ventral abdominal artery

pericardial sinus

heart

gill

testis

digestive gland

intestine

muscle

branchial chamber

leg base

perineural sinus

haemocoele

sternal artery

nerve cord

The anatomy of *Astacus*

coxopodites of the appendages and are double structures; the anterior branch being feather-like and the posterior plate-like. Considerable variation of gill arrangement is seen within the crustaceans. For example the freshwater genus *Cambarus* has 17 gills on each side and the pleurobranchs are replaced by a double arthrobranch row; in other genera as many as four rows of epipodites may be present.

For the gills to function successfully a water current is essential. The current flows forward and is created by rhythmic sculling of the scaphognathites (second maxillae) which has a pumping effect. Water enters between the limb bases and at the posterior margin of the carapace; the gills are arranged to curve towards the current. Periodically the direction of flow is reversed, so that debris is washed off the gill surfaces.

Gaseous exchange takes place at the gill filaments which are very thin-walled and together provide a large surface area. Haemocyanin is present in solution in the blood, giving it a characteristic blue-green colour. This pigment functions in a similar way to haemoglobin, but contains copper rather than iron. Oxygen is taken up and transported, both in combination with the haemocyanin and in physical solution.

The effect of the water current is to cause oxygen-rich water to pass over gills whose blood supply has a low oxygen complement; in this way a high diffusion gradient is maintained. Approximately 60 to 70% of the oxygen available in solution in the water current can be taken up by the crayfish.

Circulation

Like all arthropods, *Astacus* has a haemocoele. Thus in contrast to the closed system of vessels and sinuses found in annelids, the arthropod circulatory system is open, and many of the main organs lie in blood-filled space. On the dorsal side of the thorax the haemocoele forms a large pericardial sinus surrounding the heart. The heart is a muscular structure, polygonal in cross section and held in position in the sinus by six strands of elastic fibres. After each contraction of the heart muscles these fibres cause it to expand again. Blood enters the pericardial sinus laterally from the gills, and passes into the heart itself by three pairs of ostia (antero-dorsal, lateral, and ventral). The ostia are valves which open passively when the heart expands to admit blood. Blood leaves the heart through seven arteries which convey it to the various haemocoeles and organs. At the entrance to each artery there is a pair of valves which ensure that blood is not sucked back into the heart when it expands.

A median anterior artery (the ophthalmic artery) extends forwards to supply blood to the brain, antennules, and eyes; near its anterior end it has a contractile dilation which functions as an accessory heart. In many small crustaceans this may be the only artery present. A pair of lateral antennary arteries supply the antennules, antennae, the green glands, and the anterior region of the alimentary

canal. A pair of hepatic arteries leave the heart latero-ventrally, to supply the digestive gland and gonads. From the posterior end of the heart a median dorsal abdominal artery extends along the abdomen supplying the abdominal muscles and intestine. The sternal artery also leaves the heart posteriorly and descends to the ventral region of the body where it divides into two branches: one of these, the ventral thoracic artery, extends anteriorly, and the other, the ventral abdominal, posteriorly. Both branches give off a series of blood vessels which supply haemocoeles within the appendages. Blood passes from the various spaces into large ventral sinuses and thence to the gills. From there it returns to the heart.

Excretion and osmoregulation

The principal osmoregulatory organs in *Astacus* are the paired green (antennary) glands which lie in the haemocoele anterior to the mouth. They are of particular interest because they contain a rudimentary coelom and coelomoduct. The green glands consist of a ventral glandular region over which lies a large, thin-walled bladder opening through a duct in the base of the coxopodite of the antenna. The glandular region has an inner sac with a cavity subdivided by partitions; this leads to a green labyrinth of cells from which a coiled nephridial canal leads to the bladder. Blood is supplied to capillaries within the inner sac via the antennary artery, and it is thought that ultra-filtration occurs between the inner sac and the blood: salts are resorbed in the nephridial canal. Thus the green glands are able to regulate the salt/water balance of the animal. The urine, which collects in the bladder, is hypotonic with the blood. Nitrogenous waste, principally in the form of ammonia, is eliminated through the gills. Small amounts of amino-nitrogen (10%) and urea are also excreted.

Nervous system and sense organs

There is a close resemblance between arthropod and annelid central nervous systems although arthropods show a higher degree of cephalisation. *Astacus* has a double ventral nerve cord; the strands are separate over most of their length but are linked by segmentally arranged ganglia, which show some degree of fusion. The brain, which forms a ring round the anterior end of the alimentary canal, is formed by cerebral ganglia (which lie anterior to the mouth) connected to the sub-oesophageal ganglion (formed by the fusion of the ganglia segments 4 to 8) by a pair of circum-oesophageal commissures. The remaining thoracic and abdominal ganglia are distinguishable as separate structures in segments 9 to 19. The sternal artery passes between the two nerve strands between the ganglia of segments 12 and 13. The cerebral ganglia receive nerve fibres from the antennae, antennules, and eyes, and the sub-oesophageal ganglion controls all mouthparts except the third maxillipeds.

There is a well-developed system of giant fibres comparable with that of the annelids: two median fibres extend from the brain to the telson, and two laterals arise segmentally and are unconnected to the brain. The giant fibres innervate the longitudinal abdominal muscles causing sudden rapid flexure, and also bring about sudden movements of the abdominal and thoracic appendages and antennae. The whole system provides a very effective escape reaction.

Astacus is an active animal, and has extremely well-developed sense organs. The eyes are of the compound type found throughout the arthropods, and capable both of detecting different light intensities and of forming images. These are large and conspicuous, and made up of about 2,500 units called ommatidia, each of which points in a slightly different direction. An ommatidium is essentially a self-contained light-sensitive unit capable both of refracting light and of initiating nerve impulses which are transferred to the cerebral ganglion via the optic nerve. It is elongated

and cylindrical in shape, and the outermost region is a biconvex lens composed of transparent cuticle which is secreted by specialized epidermal cells lying beneath it. Like other parts of the cuticle, the lens is shed and replaced when the animal moults. Beneath the lens lies the crystalline cone secreted by a group of vitrellar cells, which focuses light onto light sensitive cells, the retinulae. These have a fibrillar fringe of closely-packed microtubules arising from their inner layers, and this is the region where the light sensitive pigment is deposited. The microtubules of each retinula form a rhabdomere, and these together form the rhabdome which appears under the light microscope like a solid rod. The bases of the retinulae are prolonged as nerve fibres to synapse with the cells in a well-developed optic ganglion from which the optic nerve passes to the brain. Each ommatidium signals the average intensity of light falling on its rhabdome; the image formed by the compound eye is therefore a compound of dots of light and may be compared with a photograph printed in a newspaper.

The ommatidia are separated from each other by layers of pigment, arranged in two regions: the proximal pigment surrounds the inner (retinular) region of the ommatidium, and the distal pigment the outer part. The position of the pigment layers can be changed, thus increasing or decreasing the degree of isolation of the ommatidia. In bright light the pigment moves to screen the retinulae, so that light which has passed down each ommatidium is not scattered onto the neighbouring cells. This improves the detail of the image formed. Under dim light the pigment moves to expose the retinulae, which can then receive light from several ommatidia. Although this reduces the detail of the image and makes it rather blurred it enables the animal to see in poor light.

Balancing organs or statocysts are of considerable importance in aquatic forms such as *Astacus*. These take the form of simple sacs, which open to the exterior through a minute pore opening on the base of the antennules. Each sac is lined with sensory hairs and contains the statolith, a clump of sand grains which shifts around on the sensory hairs, sending impulses to the brain. The statolith is lost with the cuticular lining and has to be replaced at each moult, suitable grains being scooped up by the antennae and introduced through the statocyst openings. Rather nasty experiments have been conducted in which the only grains available have been iron filings and the animal has then been disorientated by holding a magnet near it and drawing the statolith to one side.

Setae which respond to touch, and chemoreceptors are present on the mouthparts as well as on the antennules and antennae. The antennae are particularly important as they can be held forwards or bent posteriorly to gain information from both in front and behind as the animal moves.

Reproduction and growth

In *Astacus* the sexes are separate and the reproductive organs are simple. The two gonads, whether ovaries or testes, are fused to give a Y-shaped structure with two

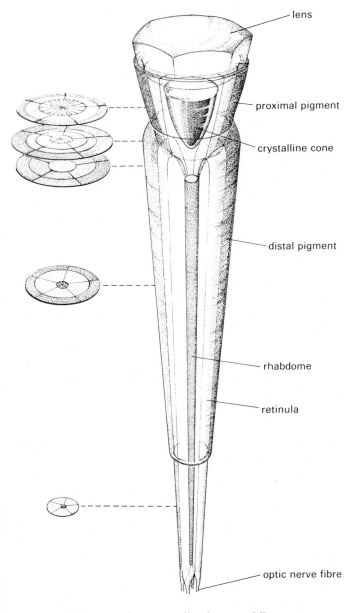

lens

proximal pigment

crystalline cone

distal pigment

rhabdome

retinula

optic nerve fibre

Diagram of a generalised ommatidium

anterior canals which extend beneath and to the sides of the pericardial sinus, and a single posterior canal. In the female a pair of short thin-walled ducts arise in the region of the pericardium and extend to openings near the bases of the second pair of walking legs. The walls of the oviducts secrete a protective chitinous shell over the eggs. In the male the sperm ducts also arise near the pericardial sinus, but are long and highly convoluted, and extend to the bases of the fourth walking legs. The glandular lining secretes a sticky substance which surrounds the sperm while they are stored in the sperm ducts. The sperm have an unusual shape with no tails and a star-shaped structure provided by numerous tangential spines.

Mating takes place in the autumn, shortly after the female has undergone her pre-adult moult. The partners lie sternum to sternum, the male holding the female, who lies on her back, by his chelipeds and walking legs. Packets of sperm (spermatophores) are deposited by the modified first two pairs of abdominal appendages of the male onto the posterior sterna and pleopods of the female, to which they stick. When the spermatophores come into contact with the water, the sticky secretion hardens and they may survive for a considerable period of days or even a month.

Fertilization is external and takes place later, after the partners have separated. The female lies on her back with her abdomen curved forwards to form a space, roofed by the abdominal terga with sides formed by long setae on the lateral margins of the terga. Eggs are discharged into this space, and moved about by a current produced by the beating pleopods. The sperm, released from the spermatophores, stick to the eggs by their spines and explode, firing the nuclei into the eggs. The fertilized eggs are attached to special setae on the pleopods by a cementing fluid produced by specialized glands, which forms a protective coat round each egg.

Development takes place entirely within the egg, which has a large supply of yolk. The eggs are carried by the female until the spring and she is described as being 'in berry'. At hatching the young lack some of their abdominal appendages and remain attached to the female's pleopods by specially modified, hooked chelipeds. In most respects however they are similar in form to the adult although very much smaller.

This type of life history is highly unusual and should not be regarded as typical of the Crustacea, or even the Decapoda, the order to which *Astacus* belongs. The majority of crustacean eggs hatch to give larvae which are totally unlike the adult and go through a succession of different larval forms to develop as an adult. The lack of larval stages in crayfish life history is undoubtedly associated with its freshwater habit, and made possible by large-yolked eggs. In a freshwater environment larvae would be unlikely to find a sufficient source of food in the limited freshwater plankton. Furthermore they would be under considerable predation pressure and would be at risk of being carried by water currents to totally unsuitable surroundings.

Crustacean diversity and classification

Astacus illustrates most of the distinguishing features of the Crustacea. The members of this major arthropod group all have the following characters: a series of appendages which are fundamentally biramous; a pair of mandibles, formed by the expansion of the limb bases (gnathobasic) and with an action based on a forwards and backwards swing; and two pairs of antennae. In addition gills are characteristically present, associated with the thoracic appendages, but may be absent in some very small species. The crustaceans are the only major group of aquatic arthropods, and are normally given the status of either a subphylum or a superclass. If the Arthropoda were not regarded as a phylum, the Crustacea would have phylum status.

About 31,300 species of crustacea are known, and most modern classifications divide these into seven classes: Cephalocarida, Branchiopoda, Ostracoda, Mystacocarida, Copepoda, Branchiura, Cirripedia and Malacostraca. The latter is the most advanced group of crustaceans and includes more species (20,000) than the rest put together. *Astacus* is a highly specialized member of the class Malacostraca, and differs in many important respects from other Crustacea.

The Cephalocarida is a very small group of crustaceans, only discovered in 1955 and comprising seven known species. They are tiny shrimp-like forms (the root *carid-* means 'a shrimp') and are thought to represent some of the most primitive living crustaceans.

The Branchiopoda ('gill-feet') also represent a relatively primitive stage of crustacean development. There are 800 species which are restricted to freshwater where they may be quite abundant.

Class Branchiopoda

Branchiopods show a considerable degree of structural variation but all have two things in common: their appendages have a flattened leaf-like structure and the coxopodite bears a flattened epipodite or gill, hence the name of the group. In addition to their role in respiration the appendages function for filter-feeding and frequently for locomotion as well. They are mostly small in size, the longest being about 100 mm.

Branchiopods occur in two common forms, either a short compact body with few limbs and obscure segmentation, or an elongated form with many segments. *Daphnia*, the water flea, is an example of the first type. It is highly cosmopolitan and one of the commonest animals in ponds and ditches where it provides an important item in the diets of animals such as *Hydra*, as well as many fish. It is the basis of much of the fish food sold by pet shops.

Daphnia is similar in size to rotifers and often found in freshwater plankton with them. It has a very different shape from *Astacus* with a short stocky body and a well-developed carapace which completely covers the limbs so that there is no external evidence of segmentation. The head protrudes anteriorly and bears a pair of disproportionately large and powerful second antennae, which the animal uses to row itself upwards through the water in a series of jerks. A period of upward movement is followed by a period of sinking, in which the antennae are used to slow the animal's movement, functioning rather like a parachute. The head also bears the first antennae (antennules) and a single compound eye in the median mid-line formed from the fusion of paired eyes and known as a nauplius eye.

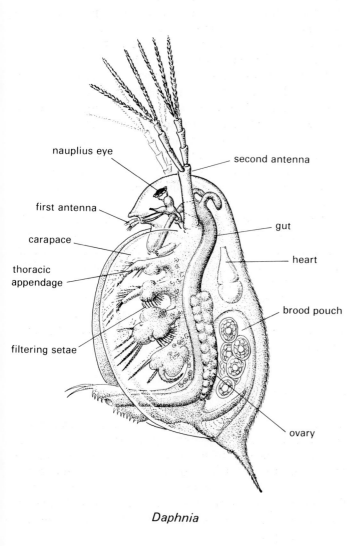

Daphnia

The remainder of the body is visible through the transparent carapace and is greatly abbreviated. It bears five pairs of trunk appendages, each bearing fine filtering setae and used for both feeding and respiration.

Reproduction in the branchiopods is unusual, since many of them reproduce parthenogenetically. For example, most of the individuals in a *Daphnia* population are likely to be parthenogenetic females. In many respects the process parallels that described for freshwater rotifers. Unfertilized thin-shelled eggs pass into a brood pouch, which is a large chamber lying between the body and the roof of the

carapace where the eggs develop. They are prevented from escaping by a projection at the posterior end of the body which is opened only when the young *Daphnia* are fully developed.

Under adverse conditions *Daphnia* may lay thick-shelled, resistant eggs which require fertilization. This normally happens towards the end of the summer when male *Daphnia* appear in the population, but may also occur in the spring.

The eggs are produced two at a time and fertilized in the brood pouch, where they remain. At the next moult the brood pouch, by this time greatly thickened, is shed and forms a box in which the eggs are protected until better conditions return. Females hatch from the eggs and reproduce parthenogenetically so that the population is able to undergo a sudden rapid increase.

Chirocephalus, the fairy shrimp, is an example of the elongated form of some branchiopods. It lacks a carapace and so its segmentation is clearly visible. It normally swims upside down using its limbs and these are also used for feeding and gaseous exchange. *Chirocephalus* has both a median (nauplius) eye and paired lateral eyes. It is occasionally found in temporary pools in southern England.

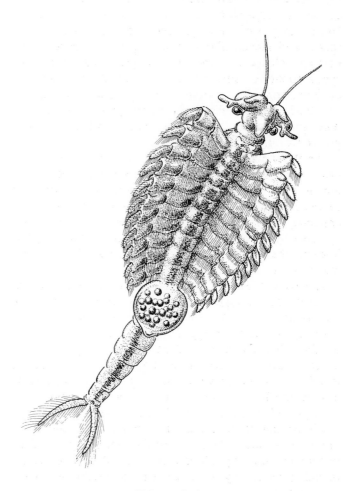

Chirocephalus

Class Ostracoda

The Ostracoda (the name is derived from the Greek *ostracon*, 'a shell' or 'tile') are a group of over 2,000 crustacean species and with a wide distribution in marine (two-thirds) and freshwater habitats.

They are small, most about 10 mm in length, with the largest species attaining a length of 30 mm. The majority are bottom dwellers, but a few are excellent swimmers. The ostracod body is short and entirely enclosed in a bivalved carapace (the shell mentioned above), the two halves being hinged together at a dorsal ligament. Only the tips of the antennae and antennules, and of a short abdomen can be protruded at the vertical edge of the carapace. The body is short, and indistinctly segmented with a total of seven pairs of appendages including the antennae. *Cypris* is a common example found on the mud at the bottom of ponds. Like *Daphnia* it uses its antennae for swimming, but it also uses both antennae and trunk appendages for crawling.

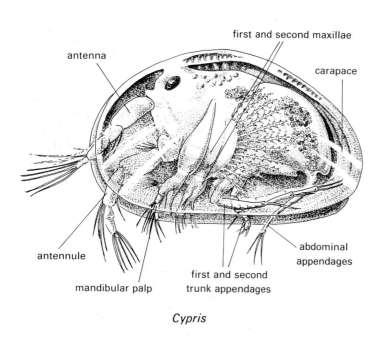

Cypris

Class Copepoda

The copepods, with 7,500 species, are the most highly successful and abundant group of small crustaceans. They are abundant in marine and freshwater habitats, and are important in aquatic food chains, feeding on microscopic organisms and providing food for a number of carnivores. They are, indeed, the principal link between the phytoplankton and the higher trophic levels of the sea where they sometimes occur in vast quantities. For example at certain times of the year *Calanus finmarchicus* is the main food of the herring and many other fish feeding near the surface. Several groups of copepods are parasites; for example *Penella* is parasitic on fish and whales, the caliguloids are parasitic on marine and freshwater fishes, and the larvae of the monstrilloids are parasitic on polychaete worms. As a group, copepods show three different types of life history: some are free-living forms at all states while others have parasitic larvae, or the adults are parasitic.

The free-living forms may be planktonic or bottom dwellers: one order, the cyclopoids, includes both planktonic and benthic species. *Cyclops*, which belongs to this group, is perhaps likely to be most easily available for laboratory study, as it is a cosmopolitan genus, occurring in fresh and brackish water and being particularly common in ponds. It feeds on diatoms and protozoa, and may even attack small fish. The general body organization resembles that of a crayfish, with a cephalothorax formed from the head and first two thoracic segments, a thorax, and an abdomen. The genus is instantly recognizable by the large nauplius eye, for which it is named. Cyclops was one of a fabled race of giants with one eye in the middle of the forehead, and the word has passed into English as a descriptive term for one-eyed forms. Similarly the term copepod is descriptive; it means 'oar-footed' and, refers to a series of biramous thoracic appendages. However, like *Daphnia*, *Cyclops* swims chiefly by rowing with well-developed head appendages, in this case the elongated antennules are the primary swimming organs though the thoracic appendages and the antennae may assist to some

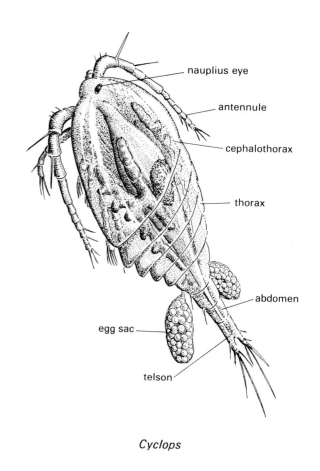

Cyclops

degree. The abdomen of *Cyclops* has no appendages on any segment except the telson which bears two long caudal rami.

The females of *Cyclops* are larger and more abundant in the population and are easily recognized by a pair of large egg sacs containing fertilized eggs which are attached to the first abdominal segment. Breeding normally takes place in spring, summer and autumn, though species which live in temporary pools breed in winter and early spring. During copulation the male fastens two spermatophores to a mid-ventral spermatheca which lies in the seventh thoracic segment of the female immediately anterior to the point of attachment of the egg sacs. Two lateral oviducts from a single median ovary also open in this segment. Fertilization is internal, the spermatophores providing sufficient sperm for several batches of eggs. Each batch of fertilized eggs is then covered by a secretion which hardens on contact with water forming the ovisacs. The eggs hatch as nauplius larvae, in between 12 hours to 5 days, and a new batch of fertilized eggs is produced immediately. This kind of larva is characteristic of crustaceans, and has a single median nauplius eye (the term for the median eye of adults is derived from the larval form), three pairs of appendages (first and second antennae and mandibles) and no apparent trunk segmentation. Five or six nauplian larval stages or instars follow during which development proceeds from anterior to posterior and increasing numbers of appendages develop. After this the larva develops as a copepodid larva which shows the general adult features, although the abdomen is unsegmented and the set of thoracic appendages is not complete. Five copepodid instars are necessary before the full adult form is attained.

Class Cirripedia

Class Cirripedia, the barnacles, also show a world-wide distribution. They are the only non-parasitic crustaceans to have adopted a sessile mode of life in the adult stage, and show extensive modification for this. Their structure is so strange, with the exoskeleton modified as a series of flat calcareous plates, that they were originally described as molluscs and it was not until their larval stages, which look rather like ostracods, were discovered in 1830 that their true crustacean affinities were recognized. The neatest description of a barnacle comes from a famous French zoologist of the last century, Agassiz: 'nothing more than a little shrimp-like animal standing on its head in a limestone house and kicking food into its mouth.' It is difficult to understand the process by which barnacles developed their peculiar evolutionary line; they are unique in the manner in which they stand on their heads throughout their adult lives, attached firmly to the ground by cement produced by glands at the bases of the antennules. Both head and abdomen are virtually absent, the body consisting mainly of the thorax.

Barnacles usually live in groups attached to algae, rocks, piers, and ships, as well as to animals such as whales, turtles, and even decapod crustaceans. Two genera are particularly well known; *Balanus*, a common barnacle on coastal rocks, and *Lepas*, the goose barnacle, which is often found attached to driftwood and ships. Many barnacles are parasitic, and have undergone extensive degeneration, with a reduction in their segmentation and a loss of the calcareous plates. Other species bore into wood, and these also lack calcareous plates.

first antenna   second antenna

mandible

nauplius eye

Nauplius larva

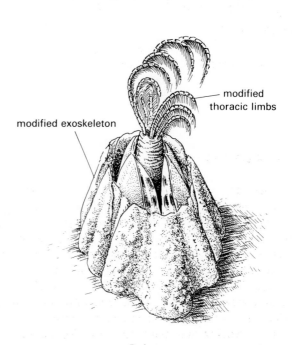

modified exoskeleton

modified
thoracic limbs

Balanus

Both *Lepas* and *Balanus* are filter feeders, and create a water current by beating their fringed thoracic limbs which are then used to sieve out planktonic animals and plants. In both cases the thoracic limbs are highly modified, both the exopodite and endopodite being represented by long flexible cirri with many setae. *Lepas* has a retractable stalk and a skeleton of 5 translucent plates and is one of the larger barnacles with a stalk length of between 100 mm and 200 mm and a body length of about 5 mm. *Balanus* is smaller; six of the plates are arranged symmetrically round the body and four others form a valve which can be closed over the withdrawn cirri.

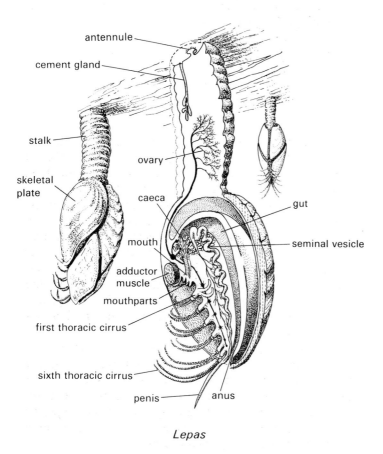

Lepas

Classes Mystacocarida and Branchiura

Both mystacocarids and branchiurans are small classes of crustacea, related to the copepods and barnacles. The mystacocarids are minute (5 mm in length or less) with elongated bodies and relatively long mouth appendages with setae. They live between the sand grains of the littoral zone and are thought to feed on detritus. The class was first discovered in 1943, and includes a single genus *Derocheilocaris*.

The branchiurans are a group of about 75 species of blood-sucking ectoparasites, living on fish and some amphibians. They are found attached to the skin or within the gill chamber of the host and show various methods of

attachment, including modification of the first pair of antennae into claws, or of the bases of the first maxillae as suckers. They have well developed thoracic appendages, which they use when swimming between hosts.

Class Malacostraca

Almost 75% of lower crustaceans, including all the largest and most successful forms, are malacostracans. The group as a whole shows great diversity; the majority are entirely aquatic and marine, but many (including *Astacus*) have invaded freshwater, and a very few have adapted successfully to the land. Terrestrial malacostracans are confined to two orders, Amphipoda and Isopoda, both of which also include marine and freshwater forms.

One of the most familiar and readily available genera of amphipods is *Gammarus*, found in large numbers under stones on the lower and middle shores, as well as in freshwater streams. *Gammarus* has a laterally compressed body and is usually seen crawling or jumping about on its side. The posterior three abdominal appendages are modified for jumping. Amongst the Isopoda, the familiar woodlouse is one of the most successful terrestrial crustaceans, found living in rotten wood, decaying vegetation or in large numbers under stones. They feed mainly on vegetable matter, and are able to survive fairly dry conditions.

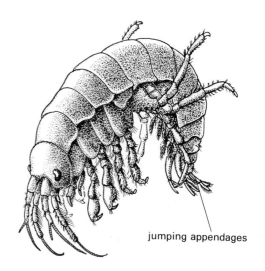

Gammarus

There are several small orders of marine malacostracans regarded as being amongst the more primitive members of the group. Among these, *Nebalia* (Order Nebalacia) is common throughout the world in shallow water, buried in sand and mud and feeding on particles of organic matter. The Stomatopoda eg. *Squilla*, are the mantis shrimps, so called from a superficial resemblance between them and the insect, the praying mantis. They are fairly large (*Squilla* can be up to 340 mm in length), and are aggressive predators, attacking prey of any suitable size including fish. The second pair of thoracic appendages is modified for grabbing and crushing and has several large spines on the last joint. Normally they lurk inside burrows and pounce on likely prey with their grabbing legs.

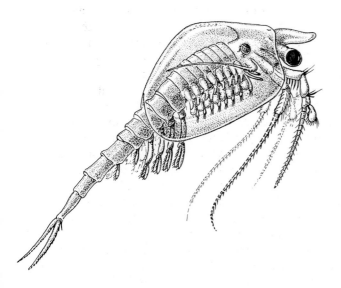

Nebalia

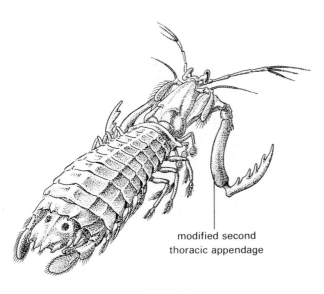

modified second
thoracic appendage

Squilla

The Mysidacea are swimming forms, with a thin carapace and long abdomen. *Mysis relicta* has a very wide distribution, in cold and temperate seas as well as lakes in North America, Europe, and Asia. It is small (20 mm) and like many other pelagic forms, protected from discovery by its glassy transparent structure.

The most highly-specialized and successful crustaceans are found within the super-order Eucarida, a group

Mysis

characterized by a highly developed carapace. It includes the order Euphausiacea, the krill, a pelagic group of great commercial importance since they constitute the chief food of several species of whale and economically important fish, as well as the order Decapoda to which *Astacus* belongs. Decapods show a great variety of habitat, and considerable diversity of form. They have two main lines of specialization as swimmers (suborder Natantia) and crawlers, (suborder Reptantia), although the distinction between these two methods of locomotion is not always clear. However the division does serve to distinguish between the shrimps and prawns (which are basically swimmers) and the bottom-dwelling crabs, crayfish, and lobsters.

Natantians have light exoskeletons, and many are laterally flattened. Prawns and shrimps may be roughly distinguished on the basis of the rostrum, which is well developed in prawns, for example *Leander*, a common inhabitant of rock pools, and greatly reduced in shrimps, for example *Crangon*, commonly found in shallow water and in estuaries, as well as burrowed in sand.

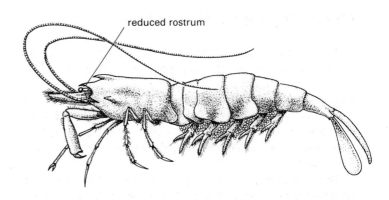

reduced rostrum

Crangon vulgaris

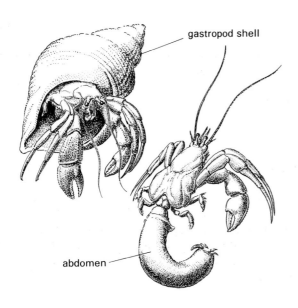

Eupagurus

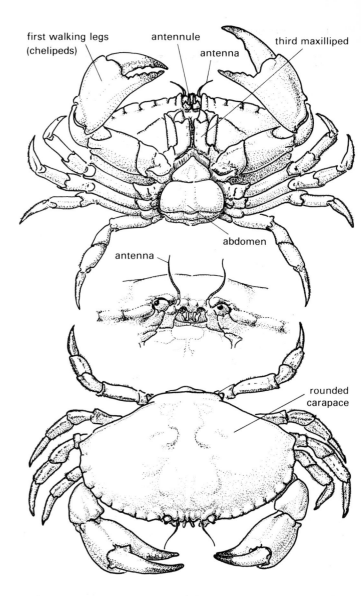

Cancer

Reptantians all have heavy skeletons, and are not flattened laterally. They are considerably more diverse than the Natantia, with three divisions, Macrura (the true lobsters and crayfish), Anomura (the squat lobsters and hermit crabs) and Brachyura (the true crabs). The anatomy of *Astacus* is fairly representative of the Macrura, with a large abdomen and a conspicuous swimming fan formed from the uropods and telson. The abdomen in the Anomura is considerably smaller and there is no tail fan. In the hermit crabs, for example *Eupagurus*, the abdomen is a skinned structure, without any form of calcareous exoskeleton, but protected by an empty gastropod shell which the animal drags around with it. It often happens that sessile animals, such as sea anemones or sponges, may come to live on the gastropod shell in commensal association with the hermit crab. Other anomurans have an abdomen which is secondarily hardened and carried bent forwards underneath the thorax. In the true crabs (Brachyura) the abdomen is very reduced and bent forwards ventrally. In addition the carapace is flattened and rounded to give a typical crab shape; the antennae are short, and the first pair of walking legs have well-developed, powerful pincers. In some species, such as fiddler crabs, these pincers are unequally developed and the animal advances sideways with the longer of the pair held forwards. There is a considerable range in size of crabs from the minute pea crabs such as *Pinnotheres* to the edible crab *Cancer pagurus* which is found amongst rocks on the lower shore and in deep water. Contrasting with the solid rounded form of *Cancer* are the spider crabs, some of which are very large indeed. The largest living arthropod is a Japanese spider crab, *Macrocheira kaempferi*, at over three metres in length with its chelipeds outstretched.

It is impossible to give a complete coverage of the crustaceans in a book of this size, and the above discussion is an attempt only to describe their very great diversity including forms likely to be encountered in the British fauna. A complete list of the main crustacean groups is given in the synopsis.

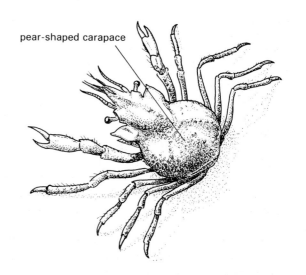

A spider crab, *Hyas araneus*

Synopsis of the subphylum Crustacea

Class Cephalocarida
Class Branchiopoda (*Daphnia, Chirocephalus*)
Class Ostracoda (*Cypris*)
Class Copepoda (*Calanus, Penella, Cyclops*)
Class Mystacocarida (*Derocheilocaris*)
Class Branchiura
Class Cirripedia (*Balanus, Lepas*)
Class Malacostraca,
   Order Nebalacia
   Order Euphausiacea (*Euphausia*)
   Order Mysidacea (*Mysis*)
   Order Isopoda (*Ligia*)
   Order Amphipoda (*Gammarus*)
   Order Decapoda
    Suborder Natantia (*Crangon*)
    Suborder Reptantia
      Division Macrura (*Astacus*)
      Division Anomura (*Eupagurus*)
      Division Brachyura (*Cancer, Pinnotheres, Macrocheira*).

SUCCESSFUL LAND ARTHROPODS

The centipedes and millipedes, the scorpions and spiders, and the insects, are all successful land arthropods. However it is important to emphasize that these are not closely related groups of animals; in particular scorpions and spiders almost certainly have a different origin from the others. Between them, these land arthropods represent the two main kinds of arthropod organization still to be considered.

The crustaceans have already been described as a group of arthropods with biramous limbs, mandibles formed from an expanded basipodite, and two pairs of antennae. The spiders are representatives of another major group, the chelicerates, which also have biramous limbs, and gnathobasic mandibles, but only one pair of antennae. The insects, millipedes and centipedes are all members of the third major group, the uniramiate arthropods, so called because their limbs are fundamentally uniramous (single-branched) but also recognizable by mandibles which are derived from the entire limb, and a single pair of antennae, believed to be appendages of the second segment. All land arthropods have faced problems of survival in the terrestrial environment and, in fact, many of the same problems will be discussed in the context of other terrestrial animals, such as land snails and vertebrates. Support is an example; water is a dense medium which provides buoyancy, but land animals have to support themselves in a non-dense atmosphere. Osmoregulation, respiration and reproduction all present particular problems in a drying environment. It is hardly surprising that groups of arthropods at a similar level of organization solved these problems in similar ways, and that various examples of convergence are found between the two main land-living groups.

Subphylum Uniramia

The uniramiate arthropods are a vast assemblage dominated by the class Insecta which includes more than 750,000 species. It is therefore impossible to give a short and detailed account of this group within a book of this size, or to select one form which may be regarded as representative of the group.

Class Insecta

Insects can be distinguished instantly from all other arthropods by their possession of three pairs of legs, a characteristic referred to by the alternative name for the group, Hexapoda. In addition many insects are adapted for flight, with one or two pairs of wings attached to their thorax, and the head bears a pair of well-developed compound eyes, and a pair of antennae. The migratory locust, *Locusta*, and the cockroaches *Blatta* and *Periplaneta* are all readily available and of a suitable size for dissection and study. The following introduction to the insects is based on *Locusta*, with a brief account of the cockroach being given to assist those who prefer to use that animal.

Locusta, the locust

General body form
In most respects, the locust is a fairly typical insect and a good representative of its order, Orthoptera, which also includes the grasshoppers and crickets. The body is somewhat laterally flattened, and like that of *Astacus* it is divided into three regions: head, thorax and abdomen. In the insects these three regions are clearly distinguishable. The head is connected to the thorax by a soft slender neck, and is held at right angles to the body so that its dorsal surface is pointed forwards and the mouthparts directed downwards.

A study of the embryonic development of insects shows that the head is formed from six segments which correspond to the first six segments in Crustacea. However the segments are not distinguishable in the adult head, which is enclosed in a conical head capsule and bears only four pairs of appendages, namely a pair of antennae (segment 2) and three pairs of mouthparts (segments 4–6). The antennae, with the other sensory appendages (well-developed compound eyes and three simple eyes called ocelli) are all borne on the lateral or dorsal surfaces of the head. The head

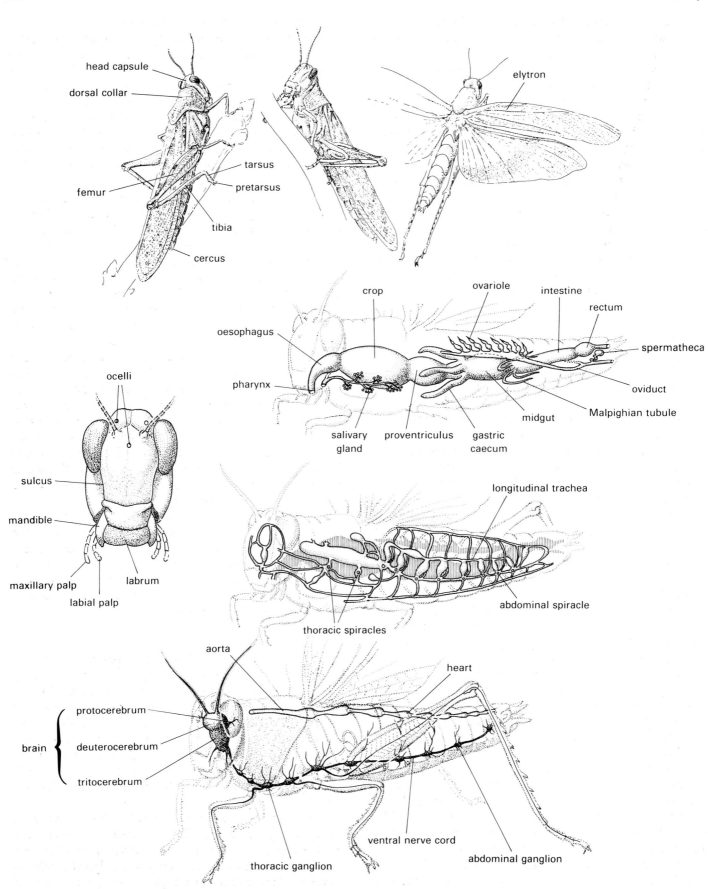

Anatomy of the locust

is marked by a series of grooves which do not represent segments. Some, the sulci, represent strengthening ridges on the inside of the head capsule, others are lines of weakness along which the capsule splits at moulting.

The thorax is formed from three segments, generally referred to as the prothorax, mesothorax, and metathorax, and each bears a pair of legs. In the locust the hind legs are highly modified and elongated for jumping. In addition both the mesothorax and metathorax bear a pair of wings which are folded backwards over the abdomen when at rest. A posterior extension of the prothorax forms a dorsal 'collar' over the mesothorax.

The abdomen is rounded in cross-section, and formed of eleven segments, though the eleventh is only distinguishable in embryos, being fused to the tenth in adults. A vestigial telson completes the posterior abdominal segment. There are no abdominal appendages apart from a pair of sensory structures, the cerci, on the last segment.

The exoskeleton
Like that of other arthropods the exoskeleton is composed of sclerites: the dorsal tergites, lateral pleurites and ventral sternites; these are joined together by arthrodial membranes. The head sclerites fuse to form the head capsule.

The cuticle is thin and of typical insect structure. The outer epicuticle is three-layered with a thin outermost layer of lipoprotein cement, secreted after moulting and poured onto the surface through pore canals which extend from the epidermis. This layer, and the underlying wax layers, give a high degree of impermeability to the cuticle and greatly reduce water loss in dry conditions. The epicuticle is completed by an inner layer of cuticulin. The inner procuticle consists of a sclerotized exocuticle and a non-differentiated endocuticle.

Nutrition
Locusts are plant eaters with strong mandibles for biting and chewing foliage. The destructive nature of their feeding habits is well known; famines caused by 'plagues' of locusts were recorded in the Old Testament and are still a problem today. The mouth is bounded anteriorly by a large plate-like upper lip, the labrum, and posteriorly by the labium, which though a single structure, represents the paired second maxillae. At the sides of the mouth are a pair of extremely powerful mandibles which are simple blades. A pair of maxillae with an extremely complicated structure lie between the mandibles and labium and are used for manipulating food. Copious saliva, produced by a pair of large salivary glands is discharged through an opening in the labium and spread onto the food by the insect before it picks it up and chews it. This serves to moisten the food and, since it contains starch digestive enzymes, also begins the process of digestion. Food is seized and cut up by the mandibles and maxillae and pushed into the mouth by the mandibles, maxillae, and labium.

The alimentary canal consists of foregut, midgut, and hindgut regions, with most absorption occurring in the midgut since the fore- and hindgut are lined with cuticle. Food is pushed into the pharynx and down a narrow oesophagus into a large crop where it is mixed with more saliva, and also with digestive enzymes produced by the midgut. From there it passes into the proventriculus which is modified as a highly muscular grinding region with a cuticular lining thickened in places to form teeth. These break up the food into fine particles, and also act as a filter, preventing overlarge particles passing into the midgut. The midgut is separated from the foregut by a valve.

The insect midgut, or stomach, is tubular and has a loose chitinous lining, the peritrophic membrane. This is produced by delamination over the surface of the midgut epithelium, and protects it from abrasion. Digestive enzymes secreted by the midgut epithelium diffuse through the peritrophic membrane, which is also permeable to the passage of digested food. The gut wall is completed by muscle layers and connective tissue. Six elongated gastric caecae, each with a forward-pointing tube and a shorter backward-pointing one, extend from the anterior end of the midgut. These caecae produce digestive enzymes and absorb the products of digestion. The hindgut consists of the pylorus, intestine and rectum. It is lined with cuticle, though this is thinner and more permeable than that of the foregut and some absorption can occur. The hindgut serves to transport waste matter to the anus and in addition is an important region for the resorption of salts, amino acids and in particular water. The conservation of water in this way is an important function and is a significant factor in the survival of insects as land forms.

The fat body is a diffuse gland lying within the haemocoele dorsal to the gut. It functions as one of the principal food storage areas of the insect, and contains reserves in the form of glycogen, protein and fat. In some insects, such as cockroaches, the fat body may include urate cells, in which uric acid accumulates, which may act as a source of nitrogen for the animal.

Locomotion
Locusts show three methods of locomotion: walking, leaping, and flight. The legs consist of six articulating regions (coxa, trochanter, femur, tibia, segmented tarsus and pretarsus) with associated levator and depressor muscles. They are elongated structures and so the animal can walk without scraping the ground with its body which is slung between the legs. This arrangement permits more powerful movement with long strides and greater speeds than is possible in animals which must drag their bodies over the ground.

During walking most insects move their legs in groups of three, each group being a triangle formed from the first and third legs of one side, and the second leg of the other. Six legs is a small number in arthropod terms, but the reduced number of legs has two advantages: it allows the elongation already mentioned, and also reduces the chance of interference between legs as the animal moves. The legs are set closely together attached to a rigid plate, which eliminates any possibility of body undulation as they move.

During walking no leg is raised until that behind or in front of it is in contact with the ground, and the legs in each pair move alternately. For each step the leg is raised and extended forwards to contact the ground. It then pushes backwards, first shortening and then lengthening again as the body is pushed forwards past the point of contact with the ground. At any time the three legs on the ground function as props, while the other three function as levers and push the insect forwards rather as an oar functions in rowing.

Locusta uses short jumps as a normal mode of progression but is also capable of considerable leaps as an escape mechanism. The femur and tibia of the hind legs are greatly elongated, and the necessary power is provided by the extensor tibiae muscles, lying within the femur, which are extremely well developed.

Insects are the only invertebrates which can fly. *Locusta* has two pairs of wings, which develop as outgrowths of the meso- and metathorax and therefore are flattened integumentary lobes. They consist of a double layer of cuticle which is strengthened by veins, each formed by a lumen surrounded by heavy rings of cuticle. The veins follow a distinct pattern of main longitudinal veins connected by smaller cross veins. They connect with the haemocoeles, and contain circulating blood, and the main veins also contain sensory nerves and extensions of the tracheal system.

The front wings of *Locusta* are elongated and straight and consist of relatively thick cuticle. In contrast the posterior edges of the hind wings are expanded and thin to give a broadly triangular structure with a large surface area. The

forewing is more highly sclerotized than the hind wing, giving it a tough, leathery structure. This type of modified forewing is known as an elytron. The elytra are folded over the hind wings and used to protect them when the animal is at rest. The name of the order, Orthoptera, is derived from these conspicuously straight-edged forewings (Greek *orthos* meaning 'straight', *pteron* meaning 'wing'). An alternative name for the order which is becoming increasingly popular in modern entomology (Saltatoria) refers to their jumping locomotion (Latin *saltare* meaning 'to leap'). Both names are highly appropriate to the group.

Each wing articulates with the edge of the tergite at a membraneous joint and is positioned with its ventral surface resting on an extension of the pleurite, near the articulation point. The pleural process acts as a fulcrum on which the

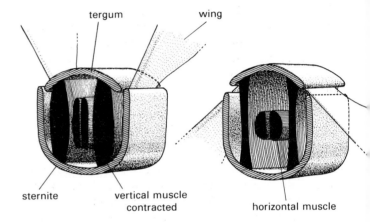

Mechanism of the basic wing stroke

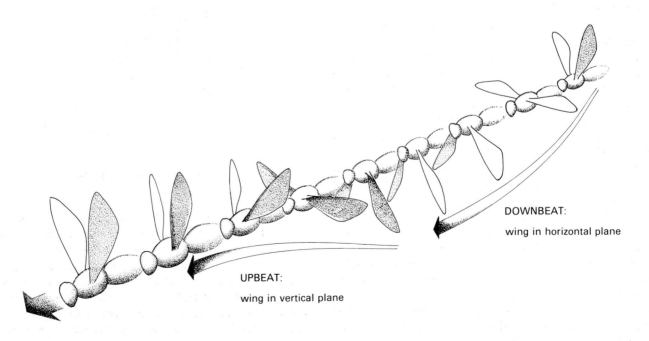

Wing movement during flight

wing moves like a seesaw although the insect wing is a long way off centre. The position of the wing on the fulcrum means that a small amount of muscle movement can produce a very large movement in the wing. Upward movement of the wing results indirectly from the contraction of vertical muscles between the tergites and sternites. The powerful downward movement is produced directly by basalar muscles attached to the wing bases, and also indirectly by transverse horizontal muscles which contract and raise the tergites. These sets of muscles are antagonistic and contraction of one set serves to stretch the other which can then contract. In addition, the elasticity of the thoracic exoskeleton is able to contribute to wing movement, accelerating the beat. An uplift caused by the downstroke keeps the insect in the air.

Up and down movement is not the only action needed for flight, which also needs forward movement. The wing is able to move extremely freely on the thorax since it has a membraneous base with four auxiliary articulatory sclerites. During flight each wing tip passes through an ellipse, the basalar muscles twisting the wing during each stroke and thus providing the backward thrust necessary for forward movement. Locusts begin to fly by jumping into the air, and the wings begin to beat when the hind-leg tarsi lose contact with the ground. The beat frequency is between 4 and 20 beats per second which is relatively slow in insect terms; it contrasts with a figure of 190 beats per second in houseflies and honey bees, and an amazing 1000 beats per second in gnats. Locusts are very powerful fliers and migrate long distances. When the young locusts have developed wings they may gather together in huge swarms and migrate, for example, right across parts of Northern Africa. (As a form of locust control these swarms may be detected by radar, and then tracked and sprayed with pesticides from the air).

Gaseous exchange

As a terrestrial animal, *Locusta* is dependent on atmospheric oxygen for respiration. Like other animals with a highly impermeable cuticle, gaseous exchange through the general surface is impossible. Indeed in a land animal it is advantageous to keep the permeable region to a minimum. The respiratory system of *Locusta* depends on a system of air tubes, which ramify through the body and transport oxygen directly to the individual cells. The circulatory system is not involved in gaseous exchange.

The air tubes, or tracheae, are invaginations of the body lined with a thin cuticle which is continuous with that of the body surface. These tubes are supported by cuticular spiral thickenings, the taenida, which protect the tracheal lumen from collapse. Oxygen enters the tracheae through openings called spiracles. In *Locusta* ten pairs of spiracles are present, two on the sides of the thorax and eight on the first eight abdominal segments. The thoracic spiracles are slits in the pleurites, but the abdominal spiracles are smaller and lie in the arthrodial membrane between the tergites and sternites. Each is surrounded by a ring-shaped sclerite and leads into an atrium and then to the trachea. The spiracles

have a closing mechanism which may consist either of valves in the spiracle opening, or an internal system cutting off the atrium from the trachea by constriction. The thoracic spiracles of *Locusta* are closed by valves under muscle control: in the second thoracic segment two semicircular valves are moved into the closed position by a muscle which pulls down their bases, causing them to rotate towards each other and so close the spiracle. Relaxation of the muscle allows the spiracle to open. The first thoracic spiracle has muscles to both open and close the aperture. The abdominal spiracles are closed by constriction. The closing mechanisms enable the spiracles to be kept open for the minimum time necessary for gaseous exchange and are essential for reducing unnecessary water loss.

The tracheae take the form of a pair of longitudinal trunks with cross connections in the thoracic and anal regions. The head is supplied by longitudinal anterior extensions from the atria of the first thoracic spiracle and first thoracic cross connection. Branches from the main tracheae anastomose throughout the body, and finally terminate in clusters of tracheoles derived from tracheole cells and with a very fine cuticular lining which is not shed when the animal moults.

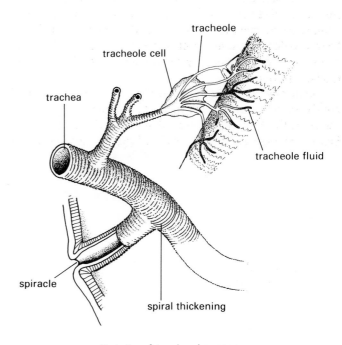

Details of tracheal system

(The cuticular lining of the tracheae is shed at moulting). The tips of the tracheoles branch and are filled with fluid which transports oxygen in true tissue solution into the cells.

When a small insect is inactive gaseous exchange takes place by simple diffusion through the tracheal system, oxygen passing in and carbon dioxide passing out. In more active insects and in large forms like *Locusta* this is supplemented by active ventilation, produced by compression and expansion of some of the tracheae

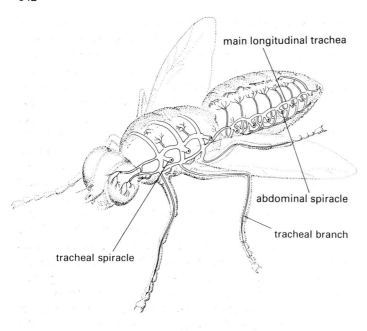

Tracheal system of a generalised insect

(particularly the longitudinal trunks) and the tracheoles. This results indirectly from muscular contraction of the abdomen, and produces a flow of air from front to back.

Circulation

Locust blood (haemolymph) is a colourless fluid containing cells (haemocytes) and with a high concentration of organic molecules, as well as inorganic ions. Insects have an open blood system with circulation being produced by the activity of an elongated tubular heart. This lies in the dorsal mid-line within the pericardial sinus and extends anteriorly as the aorta, which opens into the haemocoele of the head. The heart is made of thirteen segmentally arranged chambers, each with a pair of incurrent ostia with internal valves through which blood enters. It can be distended to allow the inflow of blood by the contraction of twelve pairs of triangular alary muscles which extend laterally to the body wall. The alary muscles form part of the dorsal diaphragm which separates the pericardial sinus from the perivisceral sinus which surrounds the gut.

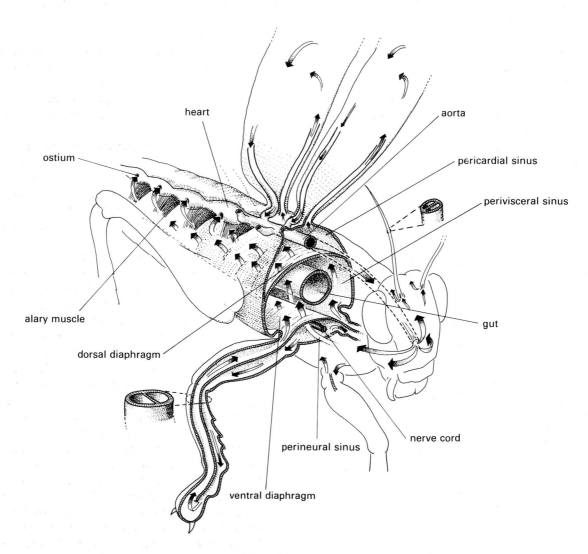

Insect blood supply

Blood is pumped forwards through the heart, leaving it through the aorta and the ventrolateral excurrent ostia. It passes backwards through the perivisceral sinus (surrounding the gut) and percolates through the ventral diaphragm to the perineural sinus (surrounding the nerve cord) and thence to the limbs. In many insects pulsatile organs supply blood from the aorta to the wings, and from the pericardial sinus to the antennae. Contraction of the alary muscles depresses the dorsal diaphragm, causing blood to re-enter the pericardial coelom and thence pass into the heart.

Excretion and osmoregulation

Excretion is a potential source of considerable water loss, since like the respiratory surface, the excretory surface has to be permeable. Furthermore many common nitrogenous products, including the primary nitrogenous waste product, ammonia, are poisonous except in very dilute solution. Insects excrete most of their nitrogenous waste as uric acid, a common excretory product in animals for which water conservation is important, as it is relatively harmless and in any case crystallises out in high concentrations.

The insect excretory system is based on numerous, blind-ending unbranched structures, the Malpighian tubules, which arise from the gut near the junction of the midgut and hindgut. The hindgut also plays a part in the excretory process as the site for reabsorption of water. The Malpighian tubule wall is one cell thick, and is supported by a tough basement membrane in which strands of muscle are arranged in wide spirals, and an outer sheath of peritoneum. The tubules lie freely in the perivisceral cavity and writhe continuously so that they are brought into the maximum possible contact with the haemolymph, and the fluid in the

lumen is moved to some extent, Cilia, used by animals such as annelids to drive fluid through their nephridia, are absent throughout the phylum and insects use osmotic gradients to achieve the same result. Potassium and probably sodium ions are actively secreted across the tubule membrane from the haemolymph. It has been shown that the concentration of potassium ions in the tubule is never less than six times that of the haemolymph. This sets up an osmotic gradient across the tubule wall which draws water into the lumen. Solutes, including uric acid, can then pass across the membrane; it is possible that some of these may be actively transported. This method could prove extremely wasteful of both water and other metabolically useful structures, and the rest of the process concerns reabsorption of these. Some reabsorption occurs at the base of the tubules but most takes place in the rectum of the hindgut. The wall of the rectum is thickened, and the cells have their surfaces greatly expanded by microvilli beneath the cuticle. An osmotic pump based on potassium also functions in the rectum, but in the opposite direction, returning water to the haemocoele. Uric acid is precipitated out and leaves with the faeces, but other solutes are reabsorbed.

Nervous system and sense organs

The basic plan of the nervous system is similar to that of *Astacus* (page 128). A double ventral nerve cord, which includes giant fibres, is present, with ganglia in each segment; the three thoracic ganglia are larger than the abdominal ganglia and insects show varying degrees of ganglia fusion.

The anterior end of the nervous system forms a ring round the oesophagus composed of a dorsal bilobed brain, circum-oesophageal connectives, and a ventral sub-oesophageal ganglion. The brain is an elaborate structure, divided into three regions: protocerebrum, deuterocerebrum, and tritocerebrum. The compound eyes and ocelli are associated with the protocerebrum, antennae with the deuterocerebrum, and the circum-oesophageal connectives with the tritocerebrum. The lobes of the tritocerebrum are connected by a commissure which passes behind the oesophagus to a small frontal ganglion and are also associated with nerves to the labrum. The sub-oesophageal ganglion, which controls the mouthparts and salivary glands, is formed from three fused ganglia.

Locusta, like all insects, has extremely well-developed sense organs. The compound eyes are large, with a basic structure similar to that already described (page 129). In addition two lateral ocelli and a median ocellus are present which detect degrees in light intensity and are very sensitive to low light. Other sense organs are scattered generally over the body, and are particularly numerous on the appendages. They are thought to be derived from setae, and consist of a receptor cell associated with modifications of the cuticle. In chemoreceptors, which are concentrated on the mouthparts, antennae and legs, extensions of the receptor cells pass through fine pores in the cuticle, which is modified either as a peg or a hair. Chemical stimuli of various kinds

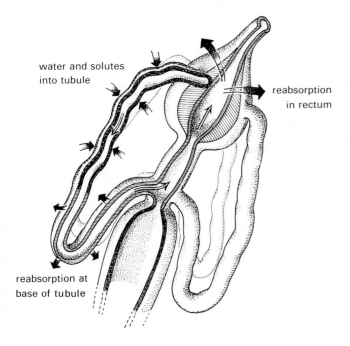

water and solutes
into tubule

reabsorption
in rectum

reabsorption at
base of tubule

Malpighian tubule

play a very important part in the lives of insects. They help them to recognize and locate their food. Insects also recognize their own species for mating or social behaviour by means of chemical signals called pheromones. Chemoreceptors are almost incredibly sensitive to the presence of particular chemicals such as pheromones and the animals can respond to the most minute traces (literally a few molecules), of such substances. One method of capturing insects is to attract them by a volatile chemical to which they respond and a good deal of research on insect pheromones is directed towards this end. There are various kinds of touch receptor such as a tactile hair, or a campaniform sensillum, a thin, domed area of cuticle above a single receptor cell. Campaniform sensilla are concentrated in groups in areas of the body subject to mechanical stress, such as the leg joints.

Locusts, like many other insects, are capable of producing sounds which they make by a process known as stridulation. This involves scraping a series of ridges or pegs with some other part of the body, causing it to vibrate. In the male locust a ridge on the inner surface of the femur of the hind leg is rubbed against a row of pegs on the elytra producing a surprisingly loud noise. The wing pegs are reduced in the female, which is silent. In order to detect sound, hearing organs are required. Two tympanal organs are present in locusts, one on either side of the first abdominal segment. These consist of a thin area of cuticle, the tympanic membrane, backed by an air sac. The tympanic membrane is able to vibrate and is associated with specialized organs which respond to changes of tension, the chordotonal organs. These are formed of groups of scolopida, which are receptors stretched between two points beneath the cuticle. Apart from their function in detecting sound similar chordotonal organs may be found in the region of joints where they function as proprioreceptors.

Reproduction and development

In insects the sexes are separate. The female has paired ovaries, which lie in the abdomen above the gut. Paired oviducts join to form a short median oviduct leading to the vagina, which opens into a genital chamber and then to the exterior by a ventral median slit at the tip of the abdomen. A long (35–45 mm) coiled spermatheca, where sperm are stored, opens from the vagina.

The ovaries are elongate structures, formed from a number of tubules, the ovarioles, which open independently into the oviduct and in which the eggs develop. Viewed externally each tubule has a number of swellings which represent the developing eggs or oocytes. The wall of the ovariole is made up of two layers: an outer sheath rich in glycogen and lipids, and an inner elastic membrane, the tunica propria, which has an important role in ovulation, as well as acting as a supporting structure to the ovariole. At its distal end the ovariole is extended as a terminal filament, while proximally it is connected by a fine duct to the oviduct. Oocytes are produced by oogonia in the distal region (germinarium) of the ovariole and grow in the more proximal vitellarium which forms the greater part of the tubule.

The male reproductive system consists of a single median testis (formed from the union of two) with paired sperm ducts and a median ejaculatory duct. The testis is composed of numerous testis follicles, bound together by a peritoneal sheath which also binds the two halves of the testis to each other. Each follicle opens separately into the sperm duct via a short vas efferens. Sperm are produced at the distal end of each follicle in the germinarium, development taking place in the rest of the follicle. Mature sperm are enclosed in gelatinous protein capsules (spermatophores) before transfer to the female during copulation. Fifteen pairs of accessory glands including white glands, hyaline glands, opalescent glands and seminal vesicles open into the ejaculatory duct. They produce a variety of secretions which mix with the sperm in the seminal fluid and contribute to the formation of spermatophores. The sperm are flagellate.

During mating the spermatophore is placed in the genital opening of the female; as the partners separate it ruptures and the sperm escape. They then migrate to the spermatheca where they are stored until fertilization, which occurs as the eggs pass through the oviduct.

The fertilized eggs of locusts are laid buried in sand and surrounded by a frothy material produced by secretions of the female accessory glands. This hardens to form an egg pod. Appendages of abdominal segments eight and nine are modified in the female to form an ovipositor which is used to deposit the eggs well below the surface. The eggs have a large yolk and the young locusts emerge as nymphs or hoppers, which look similar to adults but lack wings, which are represented by pads. The adult form is reached by a series of moults, the number of which is not constant but is normally around four. The wings grow larger with each moult until the adult form is reached. This type of development from a larva with a body structure similar to that of an adult insect is known as incomplete metamorphosis; it contrasts with complete metamorphosis, also found in insects, in which the adult and larva bear little resemblance to each other.

Periplaneta

Cockroaches, such as *Periplaneta* and *Blatta*, which belong to the order Dictyoptera, are commonly used for laboratory purposes. These two differ in size, *Periplaneta* being the larger, and also in wing development since the female *Blatta* has only vestigial wings. They are tropical in origin and live as nocturnal scavengers, often associated with human habitation. They have been introduced into temperate regions and are common in kitchens, bakehouses and similar warm places, living on food and waste matter left by man.

Cockroaches show a similar body construction to the locust, but are dorsoventrally flattened. The appendages

follows a series of moults leading to the development of the adult. The cockroach, therefore, provides another example of incomplete metamorphosis.

Metamorphosis

Incomplete metamorphosis (hemimetabolous development) has already been described in two examples, locusts and cockroaches. It is always characterised by three forms, egg – nymph – imago (adult), in which the nymph and adult are alike in many respects. Other groups showing incomplete metamorphosis include the dragonflies (Odonata) whose larvae can be amongst the most ferocious and aggressive invertebrate predators in ponds and streams. In contrast complete metamorphosis (holometabolous development) is a four stage process, including an additional resting stage, the pupa, between the last larval moult and the adult. Furthermore the larval form normally has a body form and way of life which is totally unlike that of the adult. Larvae are often worm-like, showing segmentation but with reduced wings, an indistinct division between thorax and abdomen and no wing buds. Complete metamorphosis, therefore, may be summarised as egg – larva – pupa – imago (adult), and is characteristic of several insect orders. The life history of butterflies and moths (order Lepidoptera) provides an excellent example of complete metamorphosis. The egg hatches to give a caterpillar, a chewing animal and very often a plant eater which bears no resemblance at all to the adult. Lepidopteran caterpillars are immensely varied, but they all have the same external structure consisting of a head with ocelli, antennae and mouthparts, a three segmented thorax bearing three pairs of jointed walking legs, and a segmented abdomen with four pairs of abdominal prolegs and posterior and prolegs (claspers). They have a weakly sclerotised cuticle, and are relatively inactive, crawling rather slowly about on their food. They crawl with a characteristic humping movement, the thoracic limbs and claspers being attached and moved forwards alternately. This is seen most clearly in loopers where the body is drawn upwards into an arch before the claspers grip the substrate. In this way the caterpillar can travel a distance equivalent to a body length in one loop. Lepidopteran caterpillars are normally herbivores, and some are serious pests of crops and forest plantations. A notable exception is the clothes moth, which feeds on keratin and can damage wool and silk clothes, and carpets. Since the adult diet does not include significant amounts of protein, the amino acids necessary for egg production have to be acquired in the larval stages; indeed many lepidopteran adults do not feed at all.

After a series of larval moults the larva enters a quiescent pupal stage in which the adult organs are formed. In temperate regions this often occurs at the end of the summer, so the pupa also serves as the overwintering stage. With the exception of the nervous system all the larval organs are broken down and reformed in the pupa, so that

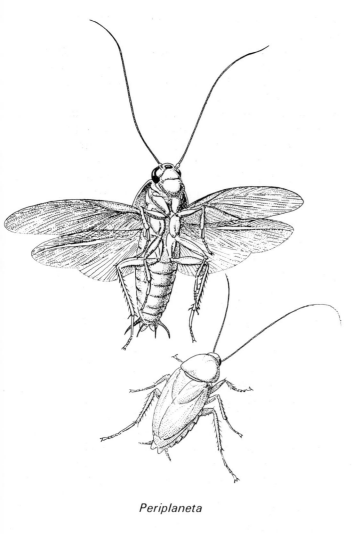

Periplaneta

are also generally similar though the legs are slender and elongated with five tarsal segments (compared with two in the locust), and the antennae are larger. They do not have the greatly developed hind legs of locusts and therefore cannot jump. Instead they depend for escape on their ability to run very fast. Although *Periplaneta* has limited powers of flight, it uses them rather seldom.

The alimentary tract of cockroaches is similar to that of the locust, but eight caecae arise at the juncture of the fore- and midgut. The tracheal system, circulatory system and excretory system are also closely similar.

The reproductive organs differ in detail from those of the locust. During mating, spermatozoa are transferred into the female genital pouch in which fertilization occurs and the eggs become enclosed in a tanned protein egg case. This is a purse-shaped structure containing two rows of eight eggs. Along the top of the egg case there is a system of cavities connected to the exterior by small pores which allow the eggs to 'breathe'. The female carries the egg case for a time between her terminal abdominal segments but eventually she desposits it, and development is then completed. The young cockroaches emerge as nymphs, which like those of the locust, resemble the parents but lack wings. There

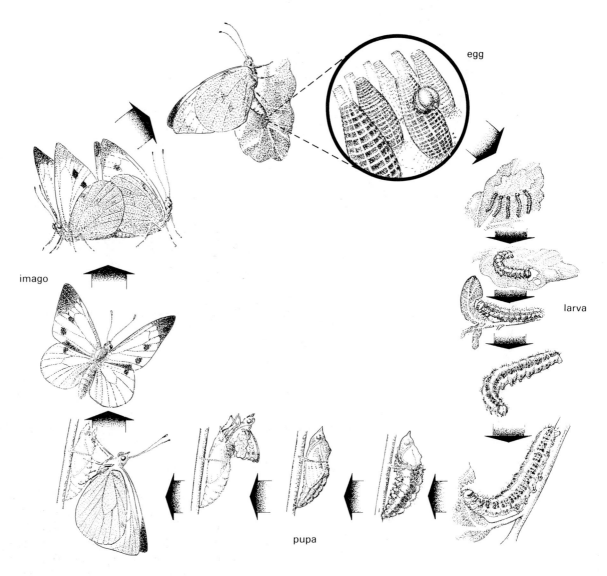

egg

larva

imago

pupa

Complete metamorphosis: the cabbage white butterfly

the adult shows complete structural contrast with the larva. The wings of the adult develop internally and are not visible externally until the pupal stage. The maxillae of the adult are modified as a sucking proboscis for feeding on fluids such as nectar which are drawn up as through a straw.

The Trichoptera (the caddis flies) provide another example of complete metamorphosis. In their case the larvae are aquatic and build houses or portable cases of various materials including leaf fragments, wood fragments and sand or mud bound together with silk. Trichopteran larvae, and the many other insect larvae which are aquatic, form an important link in the food chains of fish and water birds.

Other examples of complete metamorphosis are provided by beetles (Coleoptera), lacewings (Neuroptera), flies (Diptera), bees (Hymenoptera), and fleas (Siphonaptera). Ichneumon flies (Hymenoptera) show complete metamorphosis with a parasitic larval stage; the egg is 'injected' into the larva of other insects, commonly butterflies and moths, by means of a long pointed ovipositor in the female. The larva feeds on the less essential organs of the host, which therefore continues to live and feed. The essential organs are only killed when the ichneumon larva is ready to pupate, which may be before or after the host reaches that stage.

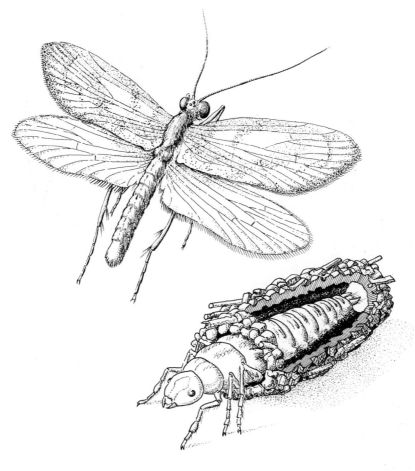

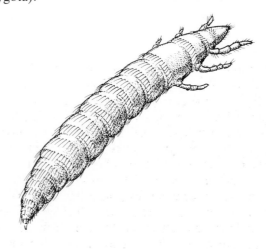

A trichopteran larva and adult

Insect diversity

The insects show an immense diversity of form and habit, but the most important orders of insects can usually be recognized without too much difficulty. A major subdivision is made in the class between the primitively wingless insects (subclass Apterygota) and insects which either have wings or have lost them secondarily (class Pterygota).

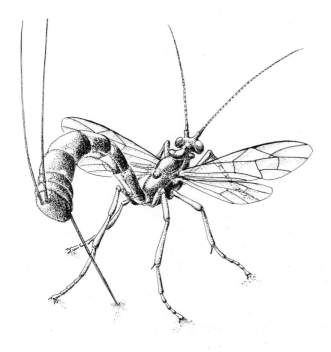

An ichneumon fly, *Rhyssa,* depositing eggs

A proturan

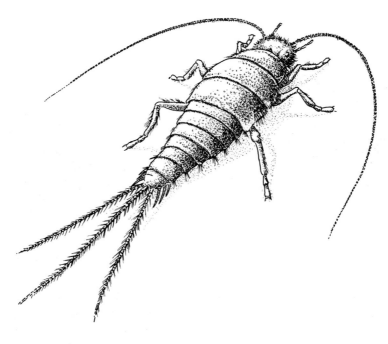

A thysanuran

Apterygote insects are found in three orders, Collembola (springtails), Protura, and Thysanura (silverfish). All are small and usually live in moist places under stones, in soil or rotting wood. In addition some thysanura species (*Lepisma, Thermobia*) have invaded houses and warehouses, where they are regarded as pests living on starchy food, books or clothing. *Petrobius*, another species of thysanuran, lives on the shore at the high tide mark where it feeds on algae, hiding in rock crevices when the tide is in. The springtails are of great ecological importance for their role in litter decomposition in the soil. It has been estimated that one acre of grassland supports about 230,000 of these tiny (less than 5 mm) inconspicuous animals. They have biting mouthparts and begin the process of litter breakdown by chewing dead plant remains. They in turn provide a food source for other soil animals at higher levels in the food chain. The common name, springtail, refers to the springing adaptation on the fourth abdominal segment which the animal can release causing it to leap into the air when disturbed.

The subclass Pterygota includes the majority of insect species, as widely differing as the butterfly and the flea, the locust and the midge. Their success as a group can probably be related to three major organizational achievements: the ability to fly which improves their dispersal, enables them to escape from predators, and allows them to reach new potential habitats and food; the development of complete metamorphosis, which in several more advanced orders allows the adult and the larva to exploit different food sources; and the development of a wide range of modification of the mouthpart structure, which allows the insects to exploit a wide variety of diets. In addition several groups have evolved parasitism, either in the juvenile stages

(for example blowflies and ichneumon flies), adult stages (fleas) or throughout the life cycle (lice). Two orders, the termites (Isoptera), and the ants, bees, and wasps (Hymenoptera), have developed highly-structured social organisation although some hymenopterans retain a solitary mode of life.

Flight

The ability to fly has already been discussed in the context of the locust (Orthoptera), in which two pairs of wings are present. Two pairs of wings are found in the majority of flying insects although these may be modified in various ways. The wings of the Ephemeroptera (mayflies) and Odonata (dragonflies and damselflies) are both long, transparent structures with a complex network of veins. The wing of a recently discovered fossil dragonfly was 300 mm in length, making it amongst the largest ever known. Many species of thrips (Thysanoptera), a group of minute flower-dwelling insects, are wingless, though the structure of the wings when present is highly unusual, distinguishing them from all other insects. They are feather-like, with a narrow central axis and a long fringe of bristles on either side. (The order name means 'fringe-wing' from the Greek *thysanos* which means 'a fringe').

The Trichoptera (caddis-flies) have wings with hairs (the Greek *trichos* means 'a hair') and those of the Lepidoptera are covered with scales (the Greek *lepides*, means 'scale'), thought to have evolved from trichopteran-type hairs. It is the possession of these scales and hairs which cover the wings and body which is responsible for the brilliant colouration possessed by many species of butterflies and moths making them some of the most beautiful of insects. A common question asked about butterflies and moths is

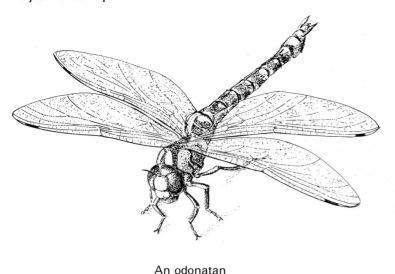

An odonatan

A thysanopteran

'What is the difference between them?' This is not easy to answer precisely, though in general moths are dull night-fliers and butterflies are brightly-coloured day-fliers. Butterflies generally have knobbed or clubbed antennae, whereas those of moths may have a variety of shapes but are never knobbed or clubbed. Generally butterflies fold their wings together over their backs whilst the moths hold them outstretched and flat. In fact the division is an artificial one, and most of the distinctions break down somewhere when applied outside the European fauna.

In several groups of winged insects, the forewings show various degrees of modification. The role of the forewings as protective elytra has already been mentioned in the Orthoptera, although in this case they are also used for flight. However in the beetles (Coleoptera) the forewings are modified entirely for the protection of the hind wings when at rest and are thick stiff structures which do not flap in flight. Like many other names of insect orders the Coleoptera refers to the wing structures: *koleos* is the Greek for 'a shield'. This adaptation enables beetles to combine flight with a crawling habit under stones or in litter and appears to have been highly successful. The Coleoptera are the largest insect order with over 300,000 species. Like the butterflies and moths, they have a long history of interest by naturalists of all ages, some very young indeed!

'I found a little beetle so that Beetle was his name,
And I called him Alexander and he answered just the same.
I put him in a matchbox, and I kept him all the day. . . .
And Nanny let my beetle out – yes, Nanny let my beetle out
She went and let my beetle out
    And Beetle ran away.'

                                        A. A. Milne

In the true flies, Diptera (the name means 'two wings'), the forewings are the only functional wings, and the hind wings are reduced to knob-like structures called halteres. In flight the halteres vibrate with the wings, and because they have heavy heads they continue to vibrate in the original plane even if the fly changes its orientation. This produces mechanical stress in the cuticle at the base of the halteres which is detected by a small group of campaniform sensilla and by chordotonal organs. These react to the forces at the base of the haltere and send impulses to the brain. In response to these the fly can correct its flight back onto a level course. The Diptera are another very large and successful order, chiefly because of the diversity of their mouthparts, and their range of diets.

Feeding

Pterygotes exploit a wide range of diets, but modifications to their mouthparts fall into two broad categories: biting and chewing, and piercing and sucking. Both the locust and the cockroach provide examples of biting mouthparts, but biters are also found in the Coleoptera, Plecoptera (stoneflies), Dermaptera (earwigs) and Trichoptera, amongst others. The Coleoptera are unusual in that biting mouthparts of a fairly generalised type are found in both the larvae and the adults. The majority of beetles are plant feeders, but some are predators. A ladybird, *Rodolia cardinalis*, preys on aphids and was involved in the first successful case of biological pest control. In the late 19th century an Australian hemipteran, *Icerya purchasi*, introduced accidentally into California, was posing a serious threat to the citrus fruit industry. *Rodolia*, a natural predator of *Icerya purchasi*, was deliberately introduced to California, and the problem solved.

The beetles also include many economic pests. More than 600 species are found commonly associated with stored food products; stored cereals and other food stuffs may be infested with meal worms (the larvae of *Tenebrio*) and flour beetles (*Tribolium*). Many weevils (a very large family of beetles, Curculionidae) are found in cereals, for example the 'rice' Weevil, *Sitophilus oryzae*, and the grain weevil, *S. granarius*. The larvae of woodworm (*Anobium punctatum*) and deathwatch beetle (*Xestobium rufovillosum*) bore into dead wood, and are well known for damaging furniture as well as floorboards and structural timbers. The adults and larvae of *Dermestes* feed on dry substances of animal origin

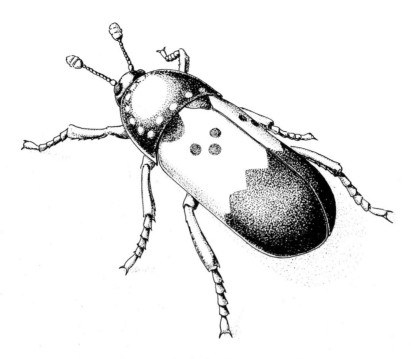

Dermestes lardarius (larder beetle)

(feathers, fur, bones, glue, leather, dried meat) and the larvae may damage wood or even mortar by burrowing into it when the time comes for them to pupate. On the credit side a minor use of *Dermestes* is to clean vertebrate skeletons for preparation of zoological specimens. Wireworms (click beetle larvae) are common garden and crop pests, living on the ground and feeding on roots. The Colorado beetle *(Leptinotarsa decemlineata)* is one of the most notorious of crop pests and the discovery of one specimen is enough to cause extreme alarm to potato farmers in Britain.

The Psocoptera (booklice) are minute insects with chewing mouthparts which feed on moulds, algae and scraps of organic materials. They derive their common name from the fact that they may be found amongst old books which have become mouldy, though they may also occur in warehouse stocks. Their natural habitats are among vegetation and particularly on the bark of trees.

The mouthparts of the Hymenoptera (bees, wasps and ants) are basically of the biting type with mandibles always present. However they can show considerable modification, particularly of the labium, which is divided into paired lateral paraglossae and a glossa. An extreme example of this is found in the honey bee, *Apis mellifera,* whose mandibles have completely lost their biting function and are spatulate structures, used for moulding wax. The glossa is prolonged as a sucking tube with a motile tip and sheathed by the labial palps; the paraglossae are very small.

The Hemiptera, the true bugs, are a large order (about 50,000 species) of insects with piercing and sucking mouthparts which feed by extracting juices from plant or animal material. The mandibles and maxillae are modified

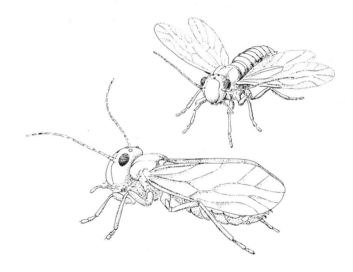

Ectopsocus, a winged psocopteran

to form needle-like stylets used for piercing and held when not in use along the ventral surface of the body in the grooved labium which is known in these forms as a rostrum, and modified as a sheath. The rostrum also has a sensitive tip, and is used to select the feeding site, and guide the stylets into position. The puncture is made by the mandibles which have sharp, saw like edges. The maxillae and mandibles are curved and arranged concentrically with the maxillae inside. Two grooves on the inner surfaces of the maxillae form two tubes when the maxillae are held tightly together: saliva is passed into the wound down one canal and food sucked up by the other. The pharynx is modified as a pump to produce the necessary suction.

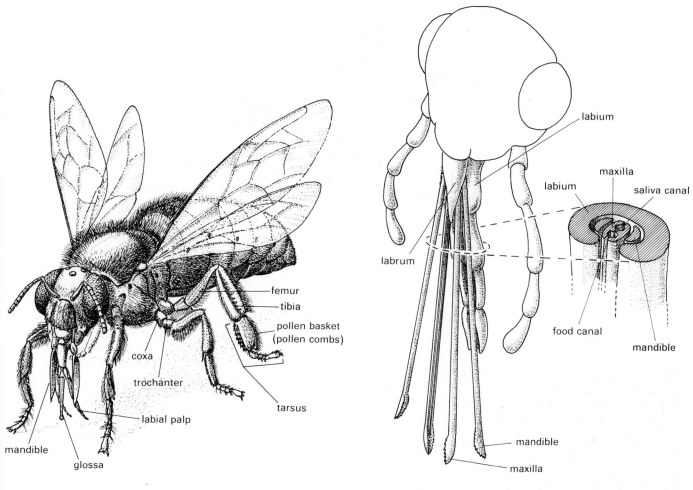

Apis mellifera (worker)

Hemipteran mouthparts

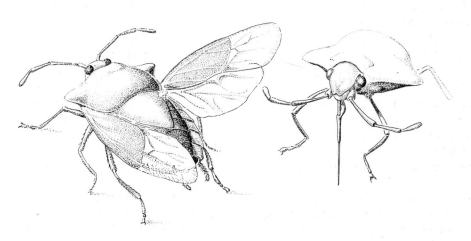

A representative hemipteran

Some of the commonest pests are hemipterans. The aphids (which belong to the suborder Homoptera, sometimes regarded as a separate order) include some of the commonest garden pests, the ubiquitous blackfly and greenfly. They feed by inserting their stylets into the phloem cells of the plant and sucking out the nutrient-rich phloem sap. In addition to damage caused by the aphids themselves they may transport viral diseases between plants and are therefore serious agricultural pests. Two thirds of known plant viruses are transmitted by aphids.

Some hemipterans feed on the blood of vertebrates or insects. *Rhodnius* (Reduviidae), a South American genus and one of the best known experimental insects, sucks the blood of mammals and in doing so transmits trypanosome parasites. The aptly nicknamed assassin bugs belong to the same family, but feed mainly on arthropods though they may also feed on vertebrates. The family Cimicidae, which includes *Cimex*, the bed bug, are collectively blood suckers of birds and mammals. Sucking mouthparts are also found in the Lepidoptera, although many species do not feed in the adult stage. The proboscis is formed from the maxillae which are grooved and hooked together on the inner side forming a tube through which nectar and liquid food can be sucked. When not in use the proboscis is held coiled beneath the insect's head. In some Lepidoptera the proboscis is very long to enable nectar to be taken from deep within the spur of a flower like the colombine (*Aquilega*). One tropical orchid has nectar at the bottom of a spur 300 mm long and is visited by a moth with a correspondingly long proboscis. Included in the Lepidoptera are many examples of very close co-evolution of insects and flowering plants; this is of selective advantage to both since the flower is pollinated and the insect obtains food. There are countless other examples, and close association with flowering plants is a recurring theme of insect evolution and may have contributed to their success.

Whilst true flies (Diptera) are all liquid feeders in their adult stage, they manage to exploit a very wide range of food materials which either are liquid or can be made so by the addition of saliva. There are 70,000 known species of which many are scavengers of decaying organic matter such

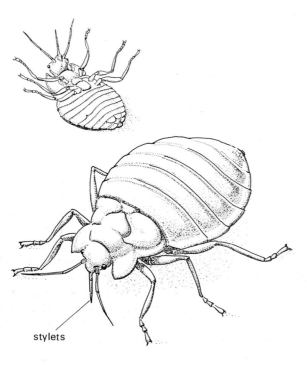

stylets

Wingless aphid

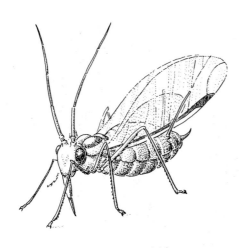

Cabbage aphid

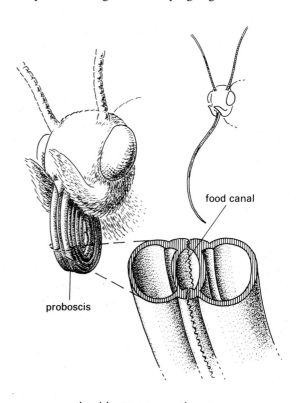

food canal

proboscis

Lepidopteran mouthparts

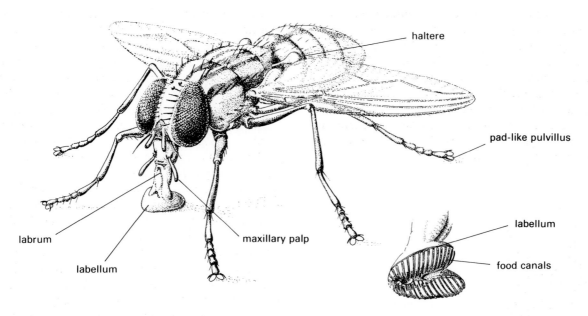

Musca domestica, dipteran

as dung, while others feed on nectar or on blood. Although dipterans show considerable variation in their mouth parts they all have certain features in common. The feeding canal (proboscis) is formed from the labrum, a projection from the floor of the mouth which carries the salivary duct (the hypo-pharynx) and the labium which is expanded into two lobes, the labella. Maxillary palps are always present.

In some blood sucking forms such as mosquitoes, mandibles and maxillae are present as piercing organs and the labella are small. In the stable fly (*Stomoxys calcitrans*), also a blood feeder, the labellum is well developed and has numerous small teeth which are used to make the wound into which the proboscis is pushed. In contrast to these, the housefly (*Musca domestica*) has large and fleshy labellae with numerous tubes with which it mops up food.

The Ephemeroptera (mayflies) whilst having quite a long nymphal life, do not feed at all as adults and have greatly reduced mouthparts; their adult life may be less than a day. The name of the order is derived from the Greek *ephemeros* which means 'living a day'. Mayfly adults are of great importance in the diet of many surface-feeding fish and are the models on which many of the 'flies' made by fishermen are based.

Parasitic insects

Parasitic insects are found in several orders, but notably in the Mallophaga (the chewing lice), Anoplura (sucking lice), Strepsiptera and Siphonaptera (fleas). Parasitic pterygotes generally show a reduction or loss of wings and other adaptations towards their mode of life. For example, both the chewing lice and the sucking lice are wingless, and dorsoventrally flattened. They have reduced antennae and compound eyes and no ocelli, and their legs are short with long, sharp claws which are used to cling to the host. The Mallophaga are generally ectoparasitic on birds, although

two families infect mammals, feeding on skin, feathers, hair, scales or dried blood around wounds. The largest are about 6 mm in length. The eggs are cemented to the host by the female, and the young pass through all their larval instars attached to the host's body. The group includes an important pest on domestic poultry, *Menopon gallinae*. The Anoplura are very similar to the Mallophaga but have mouthparts adapted for sucking, and are ectoparasites of mammals. From the point of view of man one of the most important families is the Pediculidae, which includes the human louse, *Pediculus humanus*. This occurs in two distinct races, (*P.h. capitis*), the head louse and (*P.h. humanus*), the body louse. Both keep very much to their respective parts of the human anatomy; *P.h. capitis* attaches its eggs to the hair while those of *P.h. humanus* are laid on

A representative anopluran, *Pediculus humanus*

the body and clothes. One of the problems of lice is that they are involved in the transport of diseases such as typhus.

The Strepsiptera are minute, and spend their early larval stages as parasites on other insects. The adult males are free-living with reduced forewings and broad hind wings, while adult females are grub-like and normally remain on the host. The fleas are blood-sucking forms, feeding off birds and mammals. Whilst the entire life cycle is spent in the vicinity of the host, only the adult flea lives on the host and even this drops off its host at intervals. The rabbit flea, *Spilopsyllus arniculi*, is unusual in that, like the sand fleas or jiggers common in Africa and South America, the adult female remains more or less permanently attached to the host.

Fleas have no wings, but are adapted for jumping with greatly elongated back legs, and are easily recognisable by their considerable lateral flattening which enables them to crawl between the fur or feathers of the host. They find a host by detecting heat radiated by it to which they respond by jumping. *Pulex irritans*, the human flea, is capable of leaps of up to 300 mm.

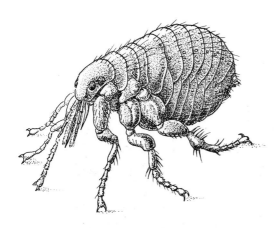

A siphonapteran, the rat flea

The breeding cycle of fleas is normally linked with that of the host and the presence of reproductive hormones in the blood of the host may be necessary to stimulate the reproductive processes. The eggs may be laid on the host, but are not attached. They fall into the nest or bedding, where they develop and the young hatch. The young fleas feed on detritus and later infest the young of the host. Bird fleas are normally restricted to breeding when the bird is breeding and has a nest: a limitation which probably accounts for their rarity. Only about 5% of the flea species have bird hosts.

Fleas show varying degrees of host specificity; *Pulex irritans*, for example, is found on man but also on certain other mammals including foxes and badgers. Fleas, like lice, may be carriers of disease organisms. Bubonic plague is

transmitted by the rat flea which may also feed on other rodents, and on man as well; myxomatosis is carried by the rabbit fleas.

It should be remembered that the majority of insects are parasitic only as larvae. Such insects are referred to as parasitoids and the habit may be exploited as a form of biological control, for example using the many wasp parasitoids to control plant pests.

Insect social organization

Highly structured social organizations are found within both the Hymenoptera (ants, bees and wasps) and the Isoptera (termites). The manner in which these colonies are organized provides some of the most fascinating studies in all biology. The honey bee, *Apis mellifera,* is probably the most familiar and best known and has been taken as an example. In the late summer three types of bees are found in the hives; there are a number of mature males, the drones, and numerous worker bees which are sterile females, with only one mature female, the queen. A normal bee colony has about 50,000 bees of which the great majority are workers.

Both the drones and the queen are solely concerned with reproduction; all other necessary activities, which includes collecting nectar and pollen, producing wax and building the combs, tending the young and cleaning the hive are carried out by the workers. The mouthparts of the workers are modified for collecting nectar and moulding wax, the hind legs are modified for collecting pollen, and the epidermis for producing wax. These modifications are vestigial or absent in the queens and absent in the drones.

Honey bee cells are formed from wax, and are hexagonal in cross section. They are formed in vertical sheets (combs) and are used both for storing pollen and honey, and for rearing the larvae. The queen is fertilized away from the hive on her 'nuptial flight' by the drones, and afterwards she returns to the hive to spend all her time in egg laying. Eggs are laid in the normal brood cells, in drone cells which are slightly larger, and also in irregular cone-shaped cells at the edge of the comb called the queen cells. Drone cells are normally produced towards the end of the summer, and the queen lays non-fertilized drone-producing eggs into them. Fertilized eggs are laid into the brood cells, and into the queen cells; the latter only appear in the hive when the colony is about to swarm or when the queen is failing. Fertilized eggs produce either queens or workers, the future of these eggs being controlled by the worker bees. For the first three days after hatching all bee grubs are fed on 'royal jelly', which is a secretion produced by the salivary glands of the workers, and is very rich in protein. Queen larvae are fed continuously on 'royal jelly', but drone and worker larvae are fed subsequently only on a mixture of pre-digested pollen and nectar. If a colony loses its queen the workers immediately enlarge ordinary worker cells and set about feeding the larvae with 'royal jelly', to produce an emergency queen. Apart from her role of egg production, the queen produces 'queen substance', a pheromone produced by the mandibular glands which is passed from worker to worker and, by regulating behaviour, keeps the colony together.

When the first new queen emerges from her cell she stings and kills all other developing queens whilst they are still in their cells. She then goes off on her nuptial flight during which the drones compete to fertilize her in mid-air. On her return to the hive the workers may kill the old queen, or the new queen may leave with a swarm of workers to set up a new colony. A queen normally lives for several years, but worker bees survive for only a few weeks in the summer, or they may overwinter. The drones live until the autumn when they are killed or thrown out and the depleted colony settles down to rest and feed on stored pollen and honey for the winter.

A requirement of any kind of social behaviour is some form of communication between members of a colony. The pheromone produced by the queen is a chemical method of communication, but the honey bee can communicate in other ways. During the summer worker bees inform other workers where they have found nectar or pollen. The remarkable way in which this seems to be done is by a series of symbolic dances and was the subject of a classic study by Karl von Frisch. The returning worker brings samples of food to its fellows in the hive and dances on the face of the combs to show the location of the food source. If the food is near the hive, the bee dances in circles, reversing the direction at intervals. If the distance is greater than about 100 m away the dancer makes a short straight run, waggling its abdomen as it goes, and then returns to its starting point by a semicircle. The process is repeated, semicircles being made alternately to the left and the right of the straight run. The length of the waggle run indicates the distance of the food from the hive, measured as the time taken for the bee to get there. The direction of the run indicates the food's position relative to the sun. For example, food 40° to the left of the sun is indicated by a waggle run 40° to the right of the vertical. Even when the sky is overcast bees can detect their position, possibly by the use of ultra-violet light or by the pattern of polarised light.

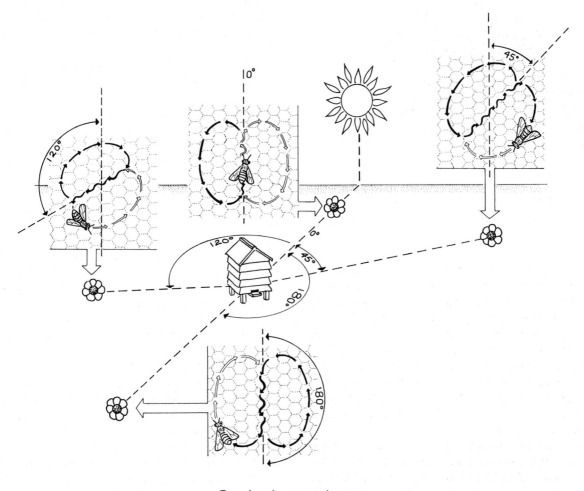

Bee dancing: waggle runs

Synopsis of the main insect orders

Subclass Apterygota
  Orders Protura
         Thysanura (silverfish *Lepisma* also *Thermobia*, *Petrobius*)
         Collembola (springtails)
Subclass Pterygota
  Orders Ephemeroptera (mayflies)
         Odonata (dragonflies and damselflies)
         Orthoptera (grasshoppers and locusts, for example *Locusta*)
         Dictyoptera (cockroaches, for example *Periplaneta*, *Blatta*, and mantids)
         Isoptera (termites)
         Plecoptera (stoneflies)
         Dermaptera (earwigs
         Psocoptera (booklice)
         Mallophaga (chewing lice such as *Menopon*)
         Anoplura (sucking lice, for example *Pediculus*)
         Thysanoptera (thrips)
         Hemiptera (bugs, for example *Rhodnius*, *Cimex*, *Aphis*)
         Neuroptera (lacewings)
         Coleoptera (beetles, for example *Rodolia*, *Anobium*, *Xestobium*)
         Strepsiptera
         Trichoptera (caddis-flies)
         Lepidoptera (butterflies and moths, clothes moth)
         Diptera (true flies, for example *Musca*, *Stomoxys*)
         Hymenoptera (bees such as *Apis*) wasps, ants,
         Siphonaptera (fleas, such as *Pulex*, *Spilopsyllus*)

THE MYRIAPODOUS ARTHROPODA

The Chilopoda (centipedes), Diplopoda (millipedes) and two minor groups, Paurapoda and Symphyla, are arthropods whose bodies are divided into two tagmata instead of three as in the crustaceans and insects. There is a complex head separated from an elongated trunk consisting of many, similar, leg-bearing segments. At one time these animals were included in a single class, Myriapoda, which included about 10,500 species, however the tendency in modern classifications is to regard the four main divisions as classes in their own right, and to use 'myriapoda' as a convenient collective name rather than a strict taxonomic term. They are all terrestrial arthropods, though unlike the insects they are mostly restricted to humid regions since they lack a waxy epicuticle to protect them against desiccation. They are commonly found in soil or humus, or under wood or stones.

The head is derived from five or six segments, and bears a single pair of antennae. Ocelli are normally present, but compound eyes are generally absent. The mouthparts are held directly forwards and include a labrum, mandibles and one or two pairs of maxillae. The structure of the mandibles and legs is essentially similar to those of the insects. Like insects they have a tracheal respiratory system, and an excretory system of Malpighian tubules. In fact many authorities regard the myriapods and insects as having evolved from a common stock and have proposed that those arthropods with uniramiate appendages should constitute a separate phylum, Uniramia.

The two most familiar groups of myriapods are the centipedes and millipedes, which exhibit two contrasting modes of life. Centipedes are fast-moving carnivorous predators, whereas millipedes are burrowing or litter-dwelling vegetarians.

Class Chilopoda: the centipedes

Lithobius is a widespread and common centipede, occurring under stones in vegetation or under the bark of trees throughout temperate and tropical regions. It is dorso-ventrally flattened and has a head with six segments and two pairs of maxillae, and eighteen trunk segments each bearing a pair of walking legs. As in all centipedes the appendages of the first trunk segment are modified as a pair of powerful recurved maxillipeds or poison claws, these project forwards under the head, and virtually conceal the mouthparts. The maxillipeds are used to catch and kill small animals and are associated with poison glands which open at

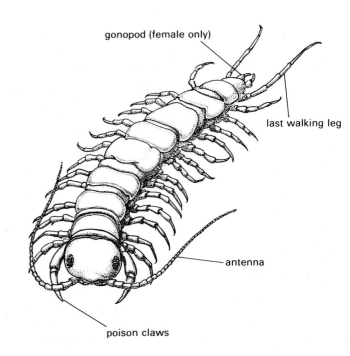

gonopod (female only)

last walking leg

antenna

poison claws

Lithobius (female)

their tips. The effect of the bite from some species can be extremely painful to humans and the large *Scolopendra gigantica*, a tropical American form which may reach 260 mm in length, is recorded as having killed people.

After it has been caught and killed, the prey is held by the poison claws and second maxillae, and then cut and manipulated into the mouth by the mandibles and first maxillae. Centipedes have a pair of well-developed mandibular glands which produce saliva. The digestive tract is a straight tube with a short oesophagus and hindgut.

Each walking leg has seven joints: coxa, trochanter, femur, tibia, and three tarsi. The legs are generally relatively long and spread out laterally and the animal is therefore capable of making long strides and rapid movement. When the animal walks, the legs are moved in a regular metachronal rhythmm.

Lithobius normally breeds in the summer. Sperm is transferred to the female indirectly in the form of a spermatophore which she picks up, with courtship behaviour taking place both before and after the spermatophore is deposited. Gonopods, present in males and females, are used to handle the spermatophore. The eggs are laid singly and hatch to give a larva which is similar in form to the adult but which has a reduced number of segments and only seven pairs of legs. It takes several years for *Lithobius* to reach sexual maturity.

Class Diploda: the millipedes

Millipedes are herbivores and normally lead a secretive mode of life so that they are rather seldom seen. They are less agile than the centipedes and easily distinguished from them. The head is composed of five segments, with a single pair of maxillae; the trunk consists of diplosegments, each formed by the fusion of two originally separate somites and bearing two pairs of walking legs. Internally each displosegment has two ganglia and two pairs of heart ostia.

Diplopods vary considerably in size: the smallest are only 2 mm in length while the biggest, which are tropical, may be as much as 280 mm long. They also show a variation in the number of diplosegments, from 11 to more than 100.

Some of the commonest millipedes are found in the family Iulidae, a cosmopolitan group of closely related genera, which includes some large tropical species. *Iulus*, the wireworm, occurs in Britain in dark damp places under stones, or under the bark of dead trees. It has a small head and an essentially cylindrical body composed of rounded shining segments. Whilst it is only capable of fairly slow crawling, it can exert a very powerful push, which is sufficient to force it into humus or loose soil. As in the centipedes, a metachronal wave passes along the length of the body though it is more apparent because of the larger number of legs. When the animal runs, as few as 12 legs may be involved in each wave of movement, but during pushing as many as fifty may be involved. Since *Iulus* cannot avoid predators by running away it protects its more vulnerable

ventral surface by coiling up when at rest and can also emit a noxious evil-smelling liquid in a spray, produced by a pair of repugnatorial glands in each segment.

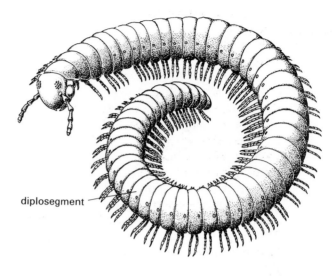

diplosegment

Iulus

Iulus feeds on decaying organic matter which it moistens with saliva and scrapes up and chews with its mandibles. Like centipedes, millipedes have a straight alimentary tract with a short foregut and midgut.

Iulus normally reproduces in spring and early summer. Sperm transfer is indirect and involves courtship behaviour. The female digs a spherical nest in the soil, into which she deposits about 100 eggs, which she guards until they hatch. At first the young animals have only three or four leg-bearing segments. Moulting takes place in moulting chambers, similar in construction to the nest, and the discarded exoskeleton is generally eaten, possibly to recover its calcium content.

Class Symphyla

The Symphyla is a small class of about 120 species, which superficially resemble centipedes and live among leaf mould and soil. Their mouthparts are of particular interest to zoologists since they share some characteristics with the insects, such as the fusion of the second maxillae to form a labium.

Class Pauropoda

Pauropodia are soft-bodied and grub-like animals, about 0.5 to 2 mm long. They are common inhabitants of the forest floor, where they feed on humus, fungi, or animal corpses. They bear some similarities to the diplopods, having large terga which overlap two segments. Perhaps because they are so small, they have no heart or tracheal system.

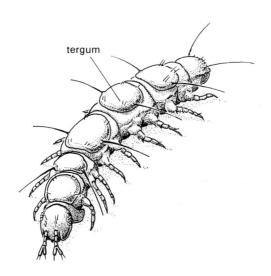

tergum

A pauropod

THE CHELICERATE ARTHROPODS

The subphylum Chelicerata is the third major arthropod taxon, it represents another evolutionary line, and includes the well-known spiders, scorpions and ticks as well as some lesser-known aquatic forms. All chelicerates have limbs which are basically biramous. They have no antennae and the first pair of appendages are feeding structures called the chelicerae, from which the group is named. The second pair of appendages are the pedipalps. The body is divided into two regions, a cephalothorax or prosoma, and an abdomen or opisthosoma.

The chelicerates are divided into three major groups. The great majority are terrestrial and belong to the class Arachnida. The class Merostomata is an ancient group of primitively aquatic animals, and the class Pycnogonida comprises some strange marine species.

The majority of the merostomates are only known as fossils and there are only four living species in three genera. One of these, *Limulus*, the horseshoe or king crab, forms a valuable introduction to the chelicerates as a whole and is large enough for convenient study.

Class Merostomata: *Limulus*

Limulus polyphemus lives in shallow water on the coast of North America, and another species lives on Asian coasts. It is a bottom dweller which spends much of its time partly buried in sand or mud. It ploughs through the surface mud in search of the invertebrate prey on which it feeds.

General body form

Limulus is a relatively large arthropod which can grow to about 600 mm long. It is dorsoventrally flattened with a body divided into a broad horseshoe-shaped prosoma (hence its American common name), a smaller opisthosoma, and an elongated caudal spine.

The dorsal surface of the prosoma is completely covered by a large convex carapace with two posterolateral projections and a median row of spines. The carapace is hinged to the opisthosoma, and serves both to protect the ventral prosoma appendages since its anterior rim curves ventrally, and to facilitate pushing through the mud. The mouth is on the ventral surface and is surrounded by the prosomal appendages: these are the labrum, a pair of three-jointed chelicerae with distal joints forming pincers, and a pair of pedipalps. The pedipalps each consist of a well developed coxa projecting towards the mouth as a spiny gnathobase, and five other segments (trochanter, femur, tibia, and two tarsal segments). Five pairs of walking legs are present of which the first four pairs bear pincers and have their coxae modified as gnathobases for chewing and moving food. The fifth pair do not have pincers, instead there are four leaf-like processes attached to the first tarsal segment which are used for sweeping away mud and sand and for pushing during burrowing. A pair of spiny chilaria, which serve to chew and manipulate food, complete the prosomal appendages. They are thought to represent modified coxae, and are similar to the gnathobases of the first four walking legs.

The opisthosoma consists of 9 fused segments and fits into the posterior border of the prosoma, to which it is joined, in effect, by a hinge. Its posterior edges are bordered by six short spines. It has six pairs of opisthosomal appendages of which the first pair is fused to form the genital operculum, and the remainder are modified as gills. The posterior two abdominal segments have no appendages. The caudal spine, which articulates with the opisthosoma, is highly mobile and used for pushing during burrowing. The animal can also right itself with its caudal spine if it is turned onto its back.

Nutrition

Limulus is a scavenger and predator, feeding mainly on molluscs and polychaetes. The food is caught and picked up by the chelicerae, and passed to the chilaria and gnathobases where it is broken up and passed forwards to the mouth. The gut includes a dilated crop for food storage, and a gizzard which is modified for grinding and has cuticular denticles and strong muscles. Two large glandular hepatic caecae open into the midgut. Enzyme production and digestion take place in the midgut, and also in the hepatic caecae which are the principal sites for food absorption. Undigested matter leaves through the anus which opens just anterior to the caudal spine.

Gaseous exchange and circulation

Each of the 5 posterior pairs of opisthosomal appendages, which are modified as gills, has an inner narrow endopodite

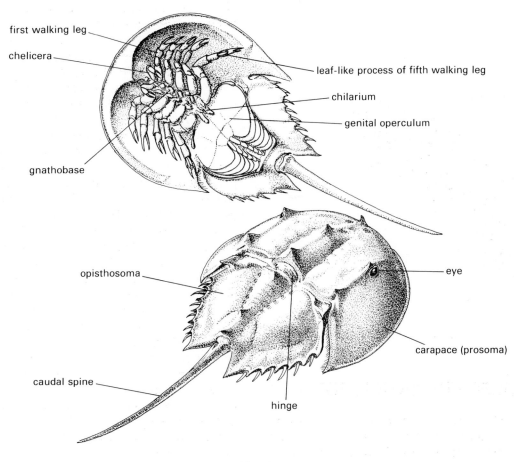

first walking leg

chelicera

leaf-like process of fifth walking leg

chilarium

genital operculum

gnathobase

opisthosoma

eye

carapace (prosoma)

caudal spine

hinge

Limulus

and a broad outer exopodite, bearing 150 to 200 leaf-like lamellae which form the surface for gaseous exchange. The lamellae have a superficial resemblance to the leaves of a book and are sometimes called book gills. To maintain a water flow over the lamellae the appendages are moved. Small individuals can swim upside-down using the gills as paddles.

The circulatory system is well developed and of the typical arthropod pattern. A dorsal heart with 8 ostia pumps blood into an arterial system. Blood eventually passes into two longitudinal ventral sinuses from which it flows into the book gills. Oxygenated blood returns to the heart via the pericardium. The blood contains a greenish respiratory pigment, haemocyanin.

Excretion and osmoregulation

The excretory organs are two pairs of coxal glands (which may be compared to the green glands of *Astacus*), on either side of the gizzard. Excess water and waste excreted from the blood passes first into a single chamber, and then through a coiled tubule where some resorption occurs. The urine passes into the bladder which opens through an excretory pore at the base of the last pair of walking legs.

Nervous system and sense organs

The nervous system shows a large degree of fusion which partly obscures the underlying segmental arrangement. The brain forms a ring around the oesophagus; the anterior part represents the protocerebrum, and the lateral parts are formed from a fusion of the tritocerebrum and the ganglia of the remaining first seven segments. A ventral nerve cord with five ganglia extends posteriorly from the brain. Paired lateral longitudinal nerves extend parallel to the ventral nerve cord and control the heart beat.

The prosoma bears a pair of simple eyes in the form of invaginated cups near its anterior margin and a pair of compound eyes which are more laterally placed. Although the compound eyes are composed of a number of ommatidia, their detailed structure differs from those of other arthropods and they are not directly comparable. Each ommatidium consists of between 8 and 14 photosensitive cells (retinula cells) grouped round a rhabdome. In addition there are one or two eccentric cells which synapse with neighbouring eccentric cells and transmit information to the brain. Pigment is present but it does not provide a moveable screen as it does in *Astacus* (page 129). King crabs can differentiate between light and dark, and may be able to detect movement with their eyes, but they have insufficient ommatidia for image formation.

The anterior rim of the carapace bears a frontal organ, which is thought to be a chemoreceptor.

Reproduction and growth
Limulus polyphemus mates in the spring. The females and males migrate inshore where the male clambers onto the female's back, clinging on with the pedipalps which are modified to bear a single curved claw. The female digs a series of shallow holes on the beach in which the large eggs are laid, fertilized, and buried; between 200 and 300 eggs are laid by each female. The young *Limulus*, when it hatches, is known as a trilobite larva from its resemblance to a fossil arthropod group, the trilobites (page 166). The larva has a very short caudal spine and only two pairs of book gills but in other respects it is rather like the adult. The young king crab, which swims and burrows, takes approximately thirteen months to attain the adult form, and becomes sexually mature after three years.

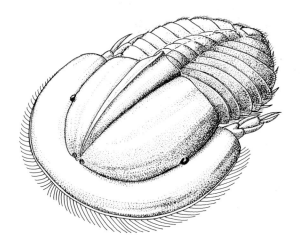

Trilobite larva

Class Arachnida

Next to the insects, the arachnids are probably the best known arthropod group and like the insects they show an immense diversity of form and habit. Altogether there are about 71,000 known species ranging in size from microscopic animals to giant spiders the size of a soup plate, and including predators, parasites, and secondarily aquatic forms. With such diversity it is difficult to find a typical arachnid to introduce the group. Spiders must be among the most familiar arachnids, though they have a notoriety that is generally undeserved, and a common spider genus, *Araneus*, will be described.

Araneus, a spider

General body form
Araneus has the typical spider body form with a prosoma bearing conspicuous appendages, separated by a narrow waist (the pedicel) from the opisthosoma which superficially appears to have no appendages. The body and limbs have a thick covering of hairs, which are thought to be mainly tactile sensory organs. The dorsal surface of the prosoma is formed by a carapace, which bears a group of eight eyes anteriorly near the mid-dorsal line and is marked by a series of depressions which correspond with internal points of muscle attachment. The prosoma bears six pairs of jointed appendages: chelicerae, pedipalps, and four pairs of walking legs. The chelicerae are moderate in size, and each consists of a large basal joint which encloses a poison gland, and a fang through which the gland opens. The pedipalps are six jointed, with well developed coxae which project forwards and inwards as gnathobases (the maxillary lobes). The legs are formed from eight segments (coxa, trochanter, femur, patella, tibia, metatarsus, tarsus, pretarsus) with claws present on the pretarsi which are used to hook into the strands of the web.

The globular abdomen appears superficially to be unsegmented and bears lung books and the spinning organs, or spinnerets. Anteriorly, on the ventral surface of the abdomen, the reproductive ducts open into the epigastric furrow.

Locomotion
In arachnids the coxa is almost immovably fixed to the body, and only a limited amount of rocking movement is possible at this joint. Spiders walk by a kind of punting action in which the posterior limbs push and the anterior legs pull. The legs are rotated around the coxa, the limb is bent dorsally and then straightened.

Feeding
Araneus feeds on small insects such as flies which it catches in a web made of silk. The web is suspended in the vertical plane, and flying insects collide with it and are trapped. The spider then runs up to the prey, injects poison into it using its chelicerae, wraps it in silk, and carries it off.

The silk is a protein and is very similar to that produced by caterpillars. It is produced by large silk glands on the posterior part of the opisthosoma, and is at first a liquid but hardens as a result of being drawn out. The ducts from the silk glands terminate at the ends of six small spinnerets, two anterior and four posterior. The silk is of two types: the supporting frame of the web is made from dry silk. A spiral of adhesive silk coated with droplets of a glue-like substance superimposed on the frame is used to catch the insects. The result is the very familiar orb web. Since the adhesive silk loses its stickiness fairly rapidly, the web has to be re-made, usually daily. The old silk is generally eaten, and therefore the protein is re-used. Once the prey is safely caught, the spider pours enzymes secreted by the midgut into it. Partly

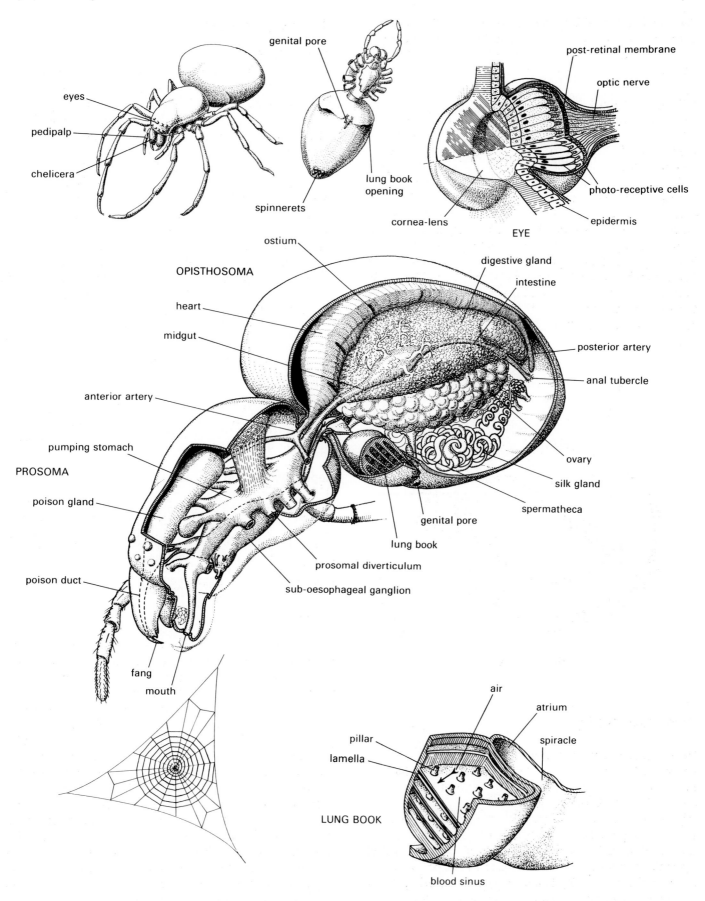

eyes

pedipalp

chelicera

genital pore

spinnerets

lung book
opening

post-retinal membrane

optic nerve

photo-receptive cells

epidermis

cornea-lens

EYE

ostium

OPISTHOSOMA

heart

midgut

anterior artery

pumping stomach

PROSOMA

poison gland

poison duct

fang

mouth

digestive gland

intestine

posterior artery

anal tubercle

ovary

silk gland

spermatheca

genital pore

lung book

prosomal diverticulum

sub-oesophageal ganglion

air

atrium

spiracle

pillar

lamella

LUNG BOOK

blood sinus

Anatomy of *Araneus*

digested food is sucked in through the mouth by pumping movements of the pharynx and a highly muscular posterior enlargement of the oesophagus, the pumping stomach. The food passes into the midgut which consists of a central tube with numerous lateral diverticula which extend through the prosoma and opisthosoma. Further enzymes are mixed with the partially digested food in the midgut which is also the site of absorption; the products of digestion are stored in the surrounding interstitial cells. A short intestine (hindgut) connects the midgut to the anus which opens through a small anal tubercle at the end of the opisthosoma.

Gaseous exchange and circulation

Araneus has two types of respiratory organ, tracheae and book lungs. The tracheae are very similar to those of insects and open through a median ventral spiracle, near the posterior end of the opisthosoma. A pair of book lungs opens more anteriorly on the ventral surface of the opisthosoma. These structures are regarded as a more primitive kind of respiratory organ. Although they are similar to the book gills of *Limulus* they are internal structures, and consist of a chitinised pocket which represents an invagination of the abdominal wall and opens to the exterior through a slit-like spiracle. On one side the wall is folded into lamellae, held apart by pillars. Gaseous exchange takes place between blood circulating in the lamellae, which have a very thin layer of cuticle, and the air in the cavity. Gas moves in and out of the book lung mainly by diffusion.

The main body cavity is a haemocoele which in the opisthosoma contains a heart with ostia. A number of major arteries extend from the heart: there is a large anterior artery to the prosoma, a small posterior artery to the posterior opisthosoma, and abdominal arteries from each heart segment. These transport blood into tissue spaces and then into the ventral sinus which supplies the book lungs. Blood is transported from these to the pericardial chamber by venous channels. The respiratory pigment, haemocyanin, is present.

Excretion and osmoregulation

Araneus has two types of excretory organs: there are branched Malpighian tubules which are believed to function like those of insects, and a pair of coxal glands comparable with those of *Limulus*, which open at the bases of the first pair of walking legs. The most abundant nitrogenous waste product is guanine, which is excreted but may also be stored in the body wall.

Nervous system and sense organs

Like that of *Limulus*, the nervous system shows an extensive degree of concentration. The brain, composed of the protocerebrum and tritocerebrum, lies above the oesophagus; it is connected by a ring to a sub-oesophageal ganglion which represents a fusion at the ventral nerve ganglia. The brain gives rise to nerves to the eyes and chelicerae, while the sub-oesophageal ganglion supplies the

other appendages. A single posterior nerve extends to the abdomen.

Three types of sensory organs are present: eyes, sensory hairs and slit sense organs. *Araneus* has eight eyes placed in two rows of four in the anterior dorsal mid-line of the prosoma. Each is composed of a combined cornea and lens which is continuous with the cuticle. A layer of epidermal cells provides the vitreous body and below this is a retinal layer of photoreceptive cells, and a post-retinal membrane. The anterior eyes have the photoreceptive cells orientated directly to the light source (direct type); the remaining eyes have light reflected onto the photoreceptive cells by the post-retinal membrane which is modified as a mirror-like tapetum. Spider eyes can detect movement and changes in light intensity but cannot form static images.

The sensory hairs are scattered all over the body and are stimulated by slight vibration or air currents. They may be simply innervated setae, or finer, longer structures, and are probably the most important arachnid sense organs.

There are numerous slit organs on the appendages and body which respond to changes in tension of the exoskeleton, and are therefore proprioreceptors and sound detectors. Each consists of a slit in the cuticle covered by a thin membrane which is in contact with a process from a sensory cell. In *Araneus* they are important for detecting prey in the web, particularly those receptors in the joint between the tarsus and metatarsus which can detect vibrations in the supporting strands of the web.

Reproduction and life history

Araneus has a simple reproductive system though the processes of mating are elaborate. The female has a pair of elongated hollow ovaries in the abdomen in which the eggs are formed and stored. The lining epithelium of the ovaries also secretes an adhesive substance. Paired oviducts extend from the ovaries to a median vagina which opens at the genital pore in the epigastric furrow. A pair of spermathecae open to the exterior, and are connected to the vagina by short ducts. The male has two large, tubular testes from which convoluted sperm ducts extend to the genital pore in the epigastric furrow. The distal tarsi of the pedipalps of the male are enlarged as knobby hollow intromittent organs for use in mating.

During mating the male places a drop of sperm on a sperm web which he weaves with silk from special silk glands. He then fills his intromittent organs with the sperm and finds a female. The pedipalps are then inserted into the external spermathecal openings and the spermathecae filled with sperm. Obviously in spiders, which are predatory, the male is at great risk as he approaches the female and courtship is important to ensure recognition and acceptance. *Araneus* attaches a safety line of silk above the female's web and swings towards and away from the web on this. He then plucks the radial thread held by the female, and sends a distinctive message in a kind of morse code.

Egg-laying normally takes place soon after copulation, in the autumn. The female lays several hundred eggs and fastens them with silk to a solid surface. Soon after this she dies, and the eggs remain in the cocoon, hatching the following spring. The young spiders, or spiderlings, go through a series of moults, becoming adult after about 18 months in their second summer.

Arachnid diversity

The living arachnids are divided into eleven orders of which five, Palpigradi, Uropygi, Schizomida (sometimes included in the Uropygi), Amblypygi, and Ricinulei, are small groups with fewer than a hundred known species. The larger groups show a broad superficial diversion of body shape between the elongated scorpions, pseudoscorpions, and whip scorpions, and the shorter and rounder spiders, harvestmen, ticks and mites.

Scorpions (order Scorpiones)

The scorpion body consists of a prosoma and an elongated opisthosoma, which may be subdivided into a mesosoma or pre-abdomen of seven segments and a narrow five-segmented metasoma or post-abdomen. This arrangement of the abdominal segments is generally regarded as primitive. At the end of the abdomen is the stinging apparatus, with a well-developed barb. The pedipalps are greatly enlarged, and form a pair of pincers used for capturing prey. The group is entirely carnivorous, and prey is caught by the pedipalps and then killed or paralysed by the sting. The abdomen with its sting is curled forwards and over the head and is used in defence as well as for food capture. The venom, produced by glands in the base of the stinging apparatus, normally acts on the nervous system and varies in toxicity according to the species. The majority of scorpions produce a venom which, whilst being fatal to invertebrates, is more unpleasant than fatal to man. There are exceptions to this, however: one particularly dangerous species, *Androctonotus australis* of the Sahara desert, has venom which can apparently kill a dog in seven minutes, and a human in about seven hours. *Buthus* is a Mediterranean form whose venom can cause fever, considerable pain, and in children even death.

Scorpions are most common in the tropics and subtropics, and the different species are adapted for desert conditions, or for humid environments such as tropical rain forests. Generally they are nocturnal, hiding under wood or stones or in burrows during the day. Scorpions may come into houses, and in certain countries the practice of shaking out one's shoes in the morning before putting them on is probably wise.

Scorpions generally have elaborate courtship behaviour surrounding the transfer of spermatophores, thus ensuring that the male is not killed before it can mate. Even so, in many species the male is often killed and eaten by the female. The eggs hatch inside the female reproductive tract and after hatching the young are carried on her back for some time.

Pseudoscorpions (order Pseudoscorpiones)

Scorpions are among the largest arachnids; the pseudoscorpions on the other hand are very small (up to 8 mm long), and are harmless to humans. They occur in leaf mould, soil, or under stones or in rock crevices in the upper shore, and feed on small arthropods such as collembolans or mites. The group gets its name from a superficial resemblance between these animals and true scorpions since in this group also the pedipalps are modified as pincers; however the abdomen is short and does not have stinging apparatus. Pseudoscorpions do have poison glands but these are associated with the pedipalps.

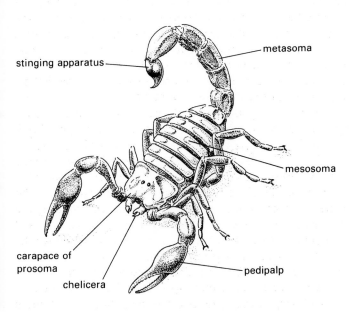

A scorpion (*Androctonus*)

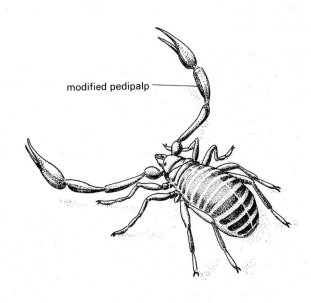

A pseudoscorpion

Solifugids (order Solifugae)

The wind scorpions or sun spiders (Solifugae) are tropical or semi-tropical arachnids, commonly found in deserts. They are instantly recognisable by two distinctive characters: the chelicerae are very large, modified as pincers, and used to kill prey and solifugids run extremely fast on three pairs of walking legs, the anterior pair being reduced in size and used as tactile organs. The pedipalps are leg-like in form but have specialised adhesive organs at their ends which are used to capture prey which is then killed and torn apart by the chelicerae. The prosoma is divided into two regions which are articulated to each other, and the opisthosoma is broad and visibly segmented.

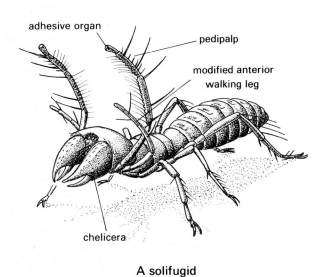

A solifugid

Spiders (order Araneae)

Spiders form one of the largest arachnid orders with 32,000 described species. This is, however, unlikely to represent their full number as many are small (0.5 mm length) and inconspicuous and some almost certainly remain to be discovered. Spiders are also extremely abundant in terms of numbers of individuals, it has been estimated that as many as 5,600,000 may be present in a single hectare of grassy meadow.

Araneus gives a fairly good idea of spider organisation as well as an example of a web-building form. The use of silk is not restricted to webs, and many spiders build silken nests, or use their silk as a safety line when climbing (in the same way as mountain climbers use a rope), or to lower themselves to the ground.

The majority of spiders feed on insects although the larger forms may feed on small vertebrates. Some have an extremely dangerous bite, which can even prove fatal to humans; for example, the black widow (*Latrodectus*) secretes a neurotoxin which may cause death by respiratory failure, and the brown recluse spider (*Loxosceles reclusa*) produces a haemolytic venom which can cause severe ulceration.

Spiders hunt in a variety of ways. Wolf spiders and tarantulas, amongst others, stalk their prey and catch it by a sudden pounce. The trapdoor spiders burrow in the ground, forming silk-lined burrows with a lid of moss or soil beneath which they wait to jump out on passing prey.

Harvestmen (order Opiliones)

The harvestmen are familiar long-legged forms, widely distributed in humid habitats. They have small chelicerae and leg-like pedipalps. The legs are generally long and slender and the body is held high above the ground so that they can move extremely rapidly if alarmed. Like the majority of arachnids they are predators, but may also feed

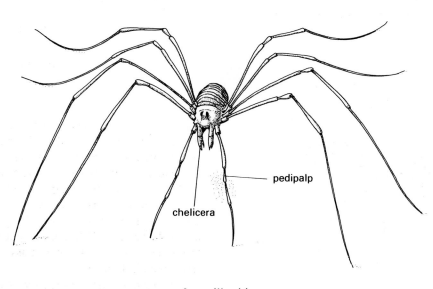

An opilionid

on dead organic matter including vegetables and fruit. Unlike spiders they have a permeable cuticle and are very vulnerable to desiccation.

Mites and ticks (order Acarina)

From the economic point of view the mites and ticks (order Acarina) are by far the most important arachnids. They are highly cosmopolitan and extremely numerous both in terms of recorded species (over 32,000) and in terms of individuals. The majority are small, mites being normally not more than 1 mm long, and the largest ticks reaching 30 mm. Like the insects, the Acarina are both useful and harmful to man, although the harmful effects are more insignificant. Mites are extremely abundant in soil and leaf litter and are important in litter decomposition, particularly in drier areas. For example there may be 160,000 in a square metre of undisturbed grassland soil. They are also likely to be present wherever there is dead organic matter such as in dust in houses or among rubbish where they feed as scavengers. Many arachnid mites feed on stored food stuffs such as cheese, dried fruit, or flour. *Dermatophagoides* is commonly associated with house dust and may precipitate allergic reactions similar to hay fever in many people.

The herbivorous spider mites (family Tetranychidae) have their chelicerae modified for piercing plant cells and are thus able to suck out the cell contents. The gall mites (Tetrapodili) also feed on plant cells. They are usually extremely small (0.1 mm long), blind, without respiratory or circulatory systems, and with legs reduced to two parts. The mouth parts are adapted for sucking plant juices. They induce the plants on which they feed to form galls by producing chemicals which stimulate abnormal cell division.

Many acarinids are parasites of reptiles, birds and mammals. Some are parasitic on man, some on his domestic animals. Like other parasitic animals they can be dangerous because of the disease organisms they carry. Ticks attack all vertebrate groups, and not only take blood but transmit diseases such as Tick Typhus. *Ixodes*, the sheep tick, is a cosmopolitan genus feeding on sheep but also man,

hedgehogs, rabbits and dogs. Some mites also spread disease, for example some species of *Trombicula* spread scrub typhus. Some mites spend all their life on the host and are true parasites. The female of *Sarcoptes scabiei*, the human itch mite, burrows into the epidermis and causes scabies. The eggs are laid and the young hatch within the host, spending their developmental stages in the hair follicles.

Class Pycnogonida

The pycnogonids are a group of about 500 highly aberrant marine chelicerates, commonly known as sea spiders. Whilst not well known animals, they are very common on the middle or lower shore under stones, or amongst hydroids and bryozoans. *Nymphon gracile* and *Pycnogonum littorale* both occur commonly on British shores. In addition to the littoral pycnogonids some species occur at great depths in the oceans. Shore-living pycnogonids are generally small animals, but deep sea forms may have a leg span of as much as 750 mm.

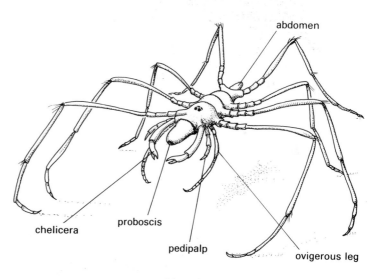

Nymphon

The body is narrow and separable into three regions: an anterior proboscis, a cephalothorax which bears four pairs of relatively long walking legs composed of eight joints, and a posterior abdomen of a single segment. *Nymphon* has both chelicerae and pedipalps whereas *Pycnogonum* has neither. In some pycnogonids there is an additional pair of large, ovigerous legs which in *Nymphon* consist of ten joints and a claw and are held beneath the body. They are used by the males to carry cemented masses of eggs and are poorly developed in females.

Pycnogonids are normally carnivorous, and suck extracted juices and small particles from soft-bodied forms

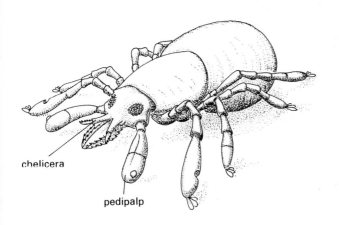

Ixodes

such as sponges, hydroids and sea anemones. Their long legs enable them to stand over their prey. They have a typically arthropod circulatory system, but no respiratory or excretory organs. The central nervous system is of a simple but typical chelicerate form, with sense organs in the form of simple eyes and sensory hairs.

The eggs hatch to give a proto-nymphon larva which has three pairs of appendages, chelicerae, palps, and ovigerous legs, each composed of three segments. The adult form develops through a series of moults with gradual addition of appendages.

Synopsis of subphylum Chelicerata

Class Merostomata (*Limulus*)
Class Arachnida
    Order Scorpiones (*Androctonotus, Buthus*)
    Order Pseudoscorpiones
    Order Solifugae
    Order Palpigradi
    Order Uropygi
    Order Schizomida
    Order Amblypygi
    Order Araneae (*Araneus, Latrodectus, Loxosceles*)
    Order Ricinulei
    Order Opiliones
    Order Acarina (*Dermatophagoides, Ixodes, Trombicula, Sarcoptes*)
Class Pycnogonida (*Pycnogonum, Nymphon*)

PRIMITIVE ARTHROPODS

Subphylum Trilobitomorpha

The most primitive arthropods are the 4,000 species included in the Trilobitomorpha. The group is an ancient one, living between 300 and 600 million years ago. Trilobites reached their peak in the Lower Cambrian and, with many other marine groups, became extinct in the Permian. The reason for this mass extinction is unknown and none of the explanations given for it is entirely satisfactory.

Trilobites are believed to have lived in shallow seas, generally crawling over the muddy parts of the sea bottom. The majority were small, about 20 to 100 mm in length, but specimens 800 mm long have been recorded. Although whole trilobites are found comparatively rarely, the group is known in remarkable detail. Even development has been described for two genera.

Trilobites had a conservative, and easily recognizable, body form. The majority of fossils are preserved with the dorsal side uppermost, probably because they were dorsoventrally flattened, and the dorsal surface was covered by a hard shell. The body was oval in outline and divided into three regions: head (cephalon), thorax, and pygidium. Of these, the head always included five segments, with both thorax and pygidium having a variable number of segments. The head, and, in the majority of trilobites, the pygidium, were covered by a single dorsal plate. The dorsal skeleton of the thorax consisted of a series of overlapping dorsal plates, one per segment. The dorsal armour was divided by two longitudinal grooves into three lobes, characteristic of the group and the basis of their name.

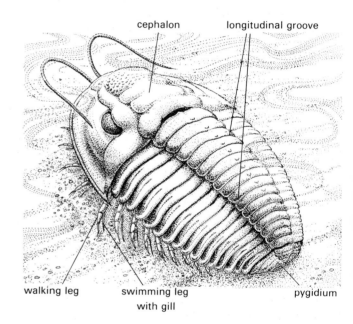

A trilobite

Ventrally, trilobites had a series of similar biramous appendages on every segment except the first, with limited specialization in the head and pygidial regions. The first segment bore a pair of long sensory antennae regarded as homologous with the first antennae of Crustacea and the antennae of insects. A labrum, which partly covered the mouth, lay between the antennae.

The typical trilobite appendage consisted of an outer walking or swimming leg, and an inner branch which bore a series of filaments and is generally regarded as gill-bearing. Both branches were jointed. The appendages of both head and pygidium showed some modification; those of the head may have shown reduction in size or modification of their bases as jaws, while those of the pygidium showed progressive reduction in size.

Apart from the antennae already mentioned, trilobite sense organs consisted of a pair of dorsally-placed eyes on each side, in the middle of the head shield. These varied in size, but had a basic structure similar to the compound eyes of other arthropods. X-ray studies have shown trilobites to have had a pear-shaped stomach surrounded by a digestive gland, and a long, straight intestine.

Trilobite development involved three distinct larval forms: protaspis, meraspis and holoaspis, each with several instars. The protaspis larva was planktonic and could be interpreted as representing the head segments only, the meraspis as head and pygidium. Gradual development of thoracic segments at the anterior border of the pygidium eventually gave a holoaspis; this was small but otherwise had the typical adult structure.

Trilobites show diversity of size, shape, eye position and filament structure, and are therefore presumed to have occupied a variety of marine habitats. The filaments could be stout or feather-like and could have been used for digging or swimming as well as for gaseous exchange. Many trilobites are thought to have been bottom-dwelling scavengers; others may have lived on the organic matter in mud, in the manner of many annelids. Some are thought to have been predators, hiding in burrows and pouncing on passing prey. Yet other trilobites had an elongated, narrow body shape and lateral eyes, while others had the dorsal shield modified as long, radiating spines. These are interpreted as having been swimming or pelagic forms. Many of the known fossils are rolled up, and it is thought that these forms protected their ventral surface by curling up, a method used by modern woodlice.

The evolutionary position of the trilobites is unclear. It is tempting, on the basis of their biramous appendages, to regard them as a link between the polychaetes and Crustacea; the occurrence of a seven-jointed walking leg in both trilobites and cephalocarid crustaceans supports this view. However, the evidence generally is slim and, since the position of the gills in Crustacea is highly variable, the limb homologies are doubtful. For the present it seems better to regard trilobites as an independent arthropod division.

Phylum Onychophora

Whilst the phylum Onychophora is small, with only about 70 existing species, it is a very ancient one and is of considerable importance in considering the origin of the arthropods; in fact onychophorans have been described as the 'missing link' between the annelids and the arthropods. Many zoologists believe that they are similar to the early forms on an evolutionary line which led to the uniramiate arthropods: the myriapods and insects. Indeed some workers regard them as primitive arthropods and include them as a subphylum within that group. It is therefore convenient to discuss them at this point.

Modern onychophorans live in damp terrestrial habitats in the tropics and subtropics where they are fairly common in leaf litter, in crevices in the soil and in rotting logs.

However it is interesting to note that *Aysheaia*, a fossil onychophoran, very similar in appearance to modern forms, was marine.

The majority of onychophorans are about 20 mm in length, but the group ranges in size from 14 to 150 mm long. *Peripatus*, which includes some of the larger species and is probably the best known, will form the basis of the description which follows.

Peripatus

Peripatus is found in equatorial rain forests. It is nocturnal, hiding during the day under leaves, rotting wood, or stones and creeping out at night to hunt and feed.

General body form
Peripatus is superficially rather caterpillar-like, with a cylindrical body and short stubby legs. Anteriorly it has a pair of big antennae and a pair of knob-like oral papillae. The antennae are tactile and are the chief sense organs; the oral papillae are also tactile and in addition have openings at their tips through which an adhesive substance, secreted by a pair of large glands, can be extruded. This secretion can be shot out a considerable distance (up to 0.5 m) and then hardens, forming sticky threads. This is used in defence. There is a ventral mouth and a pair of mandibles, which, as in the arthropods, are regarded as modified segmental appendages. The roof of the mouth is thickened to form a rasping tongue, with a row of small chitinous teeth.

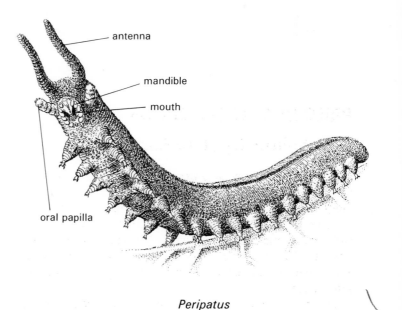

Peripatus

Peripatus has around 17 pairs of walking legs which are all identical. Each superficially resembles an annelid parapodium, but the internal musculature is closer to that of an arthropod limb. The unjointed legs are roughly cone-shaped, and ringed by tubercle-studded ridges. Each terminates in a foot with a pair of claws and a series of transverse pads.

Peripatus has a chitinous exoskeleton which is tanned in its outer layers like that of an insect. However it is thin and soft and the animal can therefore change its outer shape to squeeze into small cracks. It is also very permeable and provides rather little protection against desiccation. Like arthropods, onychophorans moult. The body wall is completed by three muscle layers: circular, diagonal, and longitudinal, and is therefore built on the typical annelid plan. However the principal body cavity is a haemocoele, markedly similar to that of the arthropods.

Locomotion

Peripatus crawls rather slowly, using its legs, but also using extension and contraction of the body itself. the legs are kept extended by hydrostatic pressure in the body cavity, and support the body away from the ground. Waves of contraction pass along the body in an antero-posterior direction. As a segment is extended, the legs are lifted above the ground and moved forwards; legs in contact with the ground are used to push.

Feeding

Peripatus feeds on small invertebrates which it searches for at night. The food is cut by the mandibles, salivary excretions are passed into it, and a semi-liquid mass sucked into the mouth. The gut is a simple straight tube.

Gaseous exchange and circulation

Both the respiratory and circulatory systems are arthropod-like. *Peripatus* has a dorsal heart with ostia, and a tracheal system opening through numerous spiracles. Unlike insects it has no means of closing these, which is another factor restricting it to damp places.

Excretion and osmoregulation

Peripatus has segmentally arranged nephridia which open at the inner limb bases. Each has a ciliated funnel and nephrostome lying in a sac which represents a vestigial coelom. The tubule enlarges to form a bladder before opening at the excretory pore. While the nephridia are clearly similar to those of annelids, they are in many respects also comparable with the green glands of Crustacea. Modified nephridia form the salivary glands and female gonoducts.

Nervous system and sense organs

The nervous system is formed of a bilobed brain lying dorsal to the pharynx, a pair of ventral nerve cords, and a series of segmental ganglia. Apart from the antennae and oral papillae, which are tactile organs, *Peripatus* has a simple eye at the base of each antenna with a well-developed lens and a retina. Sensory cells are also distributed more generally in the skin, with particular concentrations at the tubercles of the legs.

Reproduction and growth

The sexes are separate. The male system consists of paired testes and ducts, uniting to form a median duct in which spermatophores are formed. Each spermatophore consists of a number of spermatozoa in a chitinous envelope. The duct opens at the genital pore between the posterior pair of walking legs.

In the female there are paired ovaries, each of which is connected to an oviduct. During copulation the male deposits spermatophores on the female's body and the spermatozoa make their own way to the oviducts; fertilization is internal. The eggs develop within the uterus and the young *Peripatus* are fed on secretions from the uterine wall. They finally leave through the genital pore; further growth follows, with frequent moults.

Phylum Mollusca

The phylum Mollusca is the second largest in the animal kingdom including about 100,000 living species and with a long fossil record of some 35,000 more. The group is extremely diverse and includes such well known forms as the squid, octopus, slugs, snails and oysters. Like other large phyla, molluscs are not only represented by a large number of species but many of the species are very abundant. The vast majority are marine and the group originated in the sea. Of the four most abundant classes of living molluscs, Polyplacophora (chitons or coat-of-mail shells), Bivalvia (bivalves), Gastropoda (slugs and snails) and Cephalopoda (squids and octopuses), only two, the gastropods and bivalves, have invaded freshwater and in one, the gastropods, there are terrestrial forms. In addition to these important groups there is one class, Monoplacophora, believed for a long time to be extinct and now known from a single genus, and two classes, Aplacophora and Scaphopoda with only a few living species.

The history of man's interest in the group is particularly lengthy. Molluscs have been used as food, and their shells as currency, jewellery, and tools since prehistoric times, and are still used in these ways today. The evidence for their early importance as food comes from the middens of limpet and mussel shells which litter many of the most ancient archaeological sites. In the houses of the neolithic village of Skara Brae on Orkney, tanks were found in which mussels could be kept alive until they were needed. Today, marine molluscs such as mussels, clams, squids and octopus are an important part of the diet in almost all parts of the world close to the sea. The farming of certain shellfish, in particular oysters, has developed into a major industry and a great deal of research is carried out to find the most productive methods of growing them. Oysters, of course, are also farmed for their pearls which are produced in response to foreign bodies such as grains of sand introduced into the shell. Natural pearls are considered more valuable but artificially seeded or 'cultured' pearls are in reality almost indistinguishable. Terrestrial species are less important as food but certain genera such as *Helix* are highly esteemed as delicacies.

Many shells are extremely beautiful and because they are also durable they have been used in many societies as currency. In some tribes of New Guinea a large and rare cowrie shell is worn as a sign of status or as a badge of office. Everywhere shells are used for decoration and ornament. It is hardly surprising that shell collecting (conchology) is a popular hobby with a long history. The 'collections of the curious' made by wealthy amateurs in the seventeenth century included many shells brought back by seamen on trading ships. Some of these collections were extremely valuable. Sir Hans Sloane, whose collections, bequeathed to the nation in 1753, founded those of the British Museum, is said to have spent fifty pounds on shells alone, a fabulous sum in his day. Today there is a serious danger that uncontrolled collecting will cause the disappearance of the more desirable and rarer species.

The scientific study of molluscs (malacology) also has a long history. Aristotle knew of the phylum and recognized the relationships between the cephalopods and the shelled molluscs (Testacea) which he subdivided as univalves and bivalves. The term 'Mollusca' was introduced by Jonston in 1650 as a substitute for 'Mollia' which had been introduced by Pliny; his classification included the cephalopods and also the unrelated barnacles. The word was also adopted by Linnaeus in his system of classification and so stands to the present day. It is ironic that a group of animals which are chiefly famous for their hard shells should have a name derived from the Latin *mollia* meaning 'soft things'. This is due to misapprehensions, both on the part of Jonston, who did not recognize the shelled forms as molluscs, and on the part of Linnaeus who included a mixture of soft-bodied forms (for example sea anemones, medusae, polychaetes and sea cucumbers) as well as pteropods, slugs, and cephalopods, in his group, retaining the term 'Testacea' for the shelled molluscs with the barnacles and serpulid worms. Through the eighteenth and nineteenth centuries the molluscs gradually lost their extraneous members and took shape as the group we know today.

Molluscs can be regarded as a group of bilaterally symmetrical animals with a reduced coelom. They have well-developed heads, indistinguishably joined to a muscular structure, the foot, which extends both anteriorly and posteriorly. The head bears sensory structures (typically eyes and tentacles) and the muscular region is modified for a range of functions including movement, burrowing and digging.

The body organs form a dorsal visceral mass or hump, and are covered by an epidermis. In molluscs the epidermis (referred to as the mantle or pallium) extends ventrally like a skirt towards the head and foot forming a cavity, the mantle cavity, between them and the body wall. Structures such as gills are found within this cavity which is also modified as a lung in terrestrial molluscs.

In shelled molluscs the mantle secretes the shell which consists of a protein (conchiolin) matrix reinforced by crystalline calcium carbonate in the form of either calcite or aragonite. Mother of pearl (nacre) consists of layers of tiny aragonite blocks and forms an inner lining to most shells. Pearls are also made of nacre deposited around particles, such as sand grains, which get into the oyster's mantle cavity.

The molluscs are a particularly difficult group to study since many of the living groups have little apparent resemblance to each other. Their relationships only become apparent by reference to the long fossil record extending back to the pre-Cambrian period. Unfortunately, soft body structures are very rarely preserved as fossils. It has long been common practice to introduce the group by using a generalized description of an 'ancestral mollusc' based on a comparative study of the more primitive living forms and regarded as the way a 'hypothetical ancestor' may have looked.

The molluscan plan: the hypothetical ancestor

'Ancestral mollusc' is thought to have lived in shallow pre-Cambrian seas, crawling over rocks and feeding on algae. It was probably small in size, bilaterally symmetrical and basically oval in shape. Protection was provided by a shell, probably little more than a tough cuticle, which could be clamped down onto the rocks by retractor muscles between its inner surface and the head and foot as an effective defence against would-be predators. The animal crawled by means of a muscular foot with ciliated cells and mucous glands on its ventral surface. A combination of ciliary action and rippling muscle contractions within the foot brought about a slow gliding movement.

'Ancestral mollusc' is thought to have fed on small particles of vegetable matter, probably algae, which it scraped off the rock surface with a rasping action of its radula, a characteristic feeding structure of modern molluscs. The radula in present day forms is derived from the buccal cavity and can be projected out of the mouth. It consists of a membrane bearing longitudinal rows of cartilaginous teeth and supported by a tongue-like odontophore. Salivary glands produce mucus which both lubricates the radula and entangles the food particles forming long food-mucus strings. Certain modern molluscs, regarded as primitive, exhibit a very curious method of getting food into the stomach and an ancestral form may well have used a similar method. The anterior part of the stomach, the style sac, has a ciliated lining, with the cilia all beating at right angles to its longitudinal axis. This causes the strings of food and mucus to rotate, forming a central mass known as the protostyle. Once formed, the protostyle acts as a spindle on which the mucous strings are rolled and pulled down the narrow oesophagus and into the stomach. A digestive gland is associated with the stomach, and here

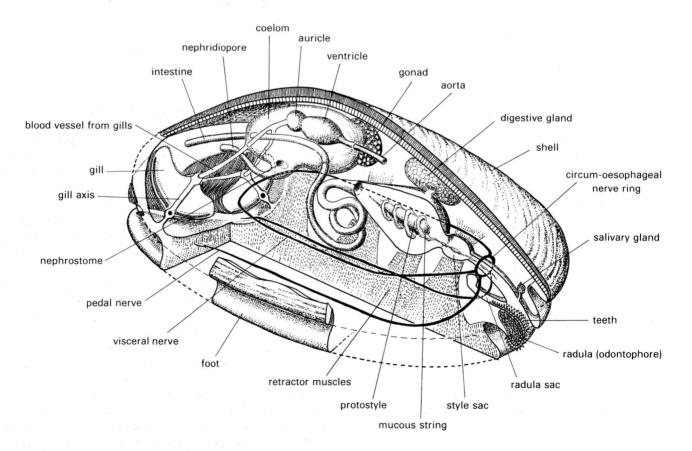

'Ancestral mollusc'

food is digested intracellularly. The alimentary canal is completed by a long intestine, in which faecal pellets are formed. These are swept away by the respiratory current.

The respiratory organs of many molluscs are gills (ctenidia). Those of our hypothetical friend 'ancestral mollusc' were probably similar to those of primitive living gastropods and bivalves. These have a central flattened axis projecting from the body wall and rows of broad, wedge-shaped, ciliated filaments on either side (bipectinate gills). The ciliary beat creates a current which enters below the gills, passes between them, and leaves the cavity dorsally. A side effect of the current is to sweep away faecal pellets as they leave the anus.

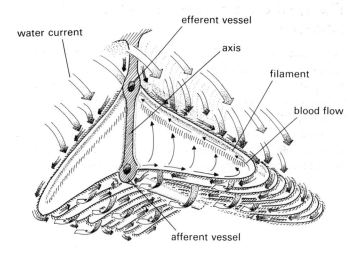

Bipectinate gill

The excretory organs of molluscs are commonly referred to as kidneys and open from the body cavity into the mantle cavity. Those of 'ancestral mollusc' were probably similar in many respects to annelid nephridia, with a nephrostome, tubular duct, and nephridiopore. In 'ancestral mollusc', eggs or sperm produced by the gonads may well have been released into the body cavity to leave via the nephridia.

Today's molluscs have an open blood system. The coelom is restricted to a region surrounding the heart dorsally and part of the intestine ventrally, in other words a pericardial and perivisceral structure. The typical molluscan heart is a more elaborate structure than any yet described. It consists of a ventricle with a pair of auricles opening into it. Blood enters the auricles from the ctenidia and passes into the ventricle. It leaves the heart by a single aorta which branches to supply sinuses in which the organs are bathed directly. 'Ancestral mollusc' may well have been organized in a similar way.

The basic plan of the molluscan nervous system consists of a circum-oesophageal nerve ring from which two pairs of longitudinal nerves extend posteriorly. The more ventral pedal cords innervate the foot while the dorsal visceral cords extend to the viscera and mantle. 'Ancestral mollusc' is

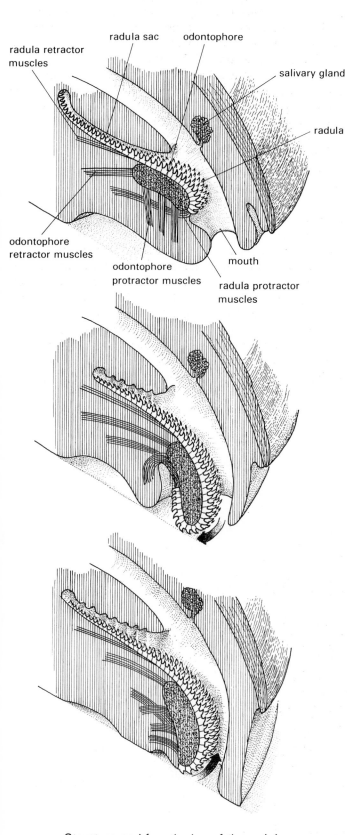

Structure and functioning of the radula

thought to have conformed to this plan. The sense organs of 'ancestral mollusc' are a subject of pure conjecture though they may well have been similar to those of primitive living molluscs: normally eyes, statocysts and osphradia (chemoreceptors positioned on the posterior margin of the gill membranes).

Molluscan larvae

The earliest stages of embryonic development in invertebrates are described elsewhere (page 192). In molluscs the gastrula develops into a free-swimming trochophore larva, similar to that found in many other marine invertebrates including the annelids. However in the majority of molluscs this quickly develops to a more elaborate veliger stage and in some cases the trochophore stage may be suppressed or may occur before hatching. The veliger is free-swimming, and has a ciliary girdle (the prototroch) which develops into a velum by putting out lobes at each side. The velum is a bilobed swimming organ, edged with powerful cilia. The lobes are held forwards and the cilia beat to bring about movement. The veliger has many of the characteristic structures of adult molluscs including a head, foot, and shell. There is also a mantle cavity into which the velum can be withdrawn. The inner walls of the mantle cavity have sensory, mucous, and ciliated cells, and ctenidia are present. Internally the veliger has a simple tubular gut, with an anterior odontophore and radula and a posterior anus which opens into the mantle cavity. A pericardial cavity surrounds the heart and coelomic cavities are also present in the gonads and excretory organs. The nervous system has four pairs of ganglia (cerebral, pleural, pedal, and visceral) linked by connectives and commissures.

The veliger provides the common link in the great range of adult molluscan forms, and all the basic mollusc shapes can be derived by simple transformations of this structure: the polyplacophorans are formed by a dorsoventral flattening; the gastropods (for example snails) develop by twisting the visceral mass and mantle cavity through 180° bringing it above the head; the lamellibranchs (bivalves) form by lateral flattening and the cephalopods by elongation and fusion of the head and foot.

Primitive living molluscs

Class Monoplacophora: *Neopilina*

In 1952 ten living specimens of a peculiar mollusc were discovered from a deep ocean trench in the Pacific Ocean off Costa Rica. They were small, between 3 mm and 30 mm long, with a single symmetrical shell (Monoplacophora means 'one plated') and were identified as *Neopilina*, the sole surviving genus of a group regarded as long extinct. Since then other species of *Neopilina* have been found in deep water and their survival as a genus is undoubtedly related to their ability to live at great depths (2,000 to 7,000 m) where there is less competition from more modern forms.

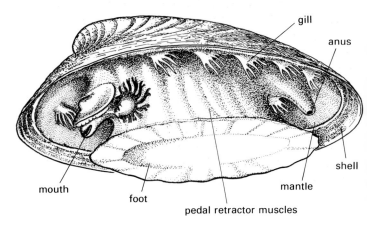

Neopilina

In many respects *Neopilina* resembles the hypothetical form which malacologists have found it necessary to invent. The mantle cavity is represented by a pallial groove separating the edge of the foot from the mantle and containing 5 to 6 pairs of gills. The head is small, and the foot is a broad, flat structure with 8 pairs of pedal retractor muscles. The structure of the radula and digestive system are like those already described, however the intestine is elongated and coiled. Its stomach contents suggest *Neopilina* feeds chiefly on diatoms and Foraminifera, although sponge spicules have also been found. Six pairs of nephridia open via nephridiopores into the pallial groove.

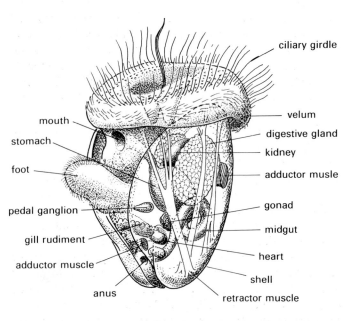

Veliger larva

The heart has two auricles and also two ventricles (one is more usual in molluscs); each ventricle gives rise to an aorta and these fuse to form a single anterior vessel. A paired pericardial coelom surrounds the heart. The nervous system is similar to the basic mollusc plan described. Little is known about the reproduction of *Neopilina* but the sexes are separate and fertilization must be external. Two pairs of gonads are present, connecting by a single gonoduct to a nephridium.

Repetition of structures such as retractor muscles and nephridia is apparently characteristic of Monoplacophora; in fossil species multiple muscle scars can be seen in the shells. This has been interpreted as evidence of metamerism with a consequent close evolutionary relationship between molluscs, annelids, and arthropods. An annelid mollusc relationship is also supported by their early embryology. On the other hand, some aspects of the general body plan, nervous system, method of locomotion and the use in both groups of intracellular digestion suggest a relationship with platyhelminths, although the latter are at a much simpler level of organization. The present conclusion regarding molluscan ancestry tends to regard them as sharing a remote common ancestor with the annelids, but developing independently at the coelomate level of organization.

Class Polyplacophora

Among common living molluscs the type of organization described as the hypothetical ancestral mollusc is most nearly approached by the chitons (class Polyplacophora). These are frequently found in the intertidal and sublittoral zones and are highly adapted for clinging to the underside of stones. They are oval and dorsoventally flattened, and have a shell made up of 8 articulated transverse calcareous plates from which the class name, meaning 'many plated' derives. If dislodged, the chiton rolls up into a ball in a similar way to a woodlouse; this is a highly unusual reaction in a mollusc and is thought to enable the animal to right itself easily as well as to protect its undersurface.

A chiton

Chitons have a broad, flat, foot, which is used both for clinging, and for locomotion in the same way as that described for 'ancestral mollusc'. They are microphagous, and the feeding process is similar to that already described. The respiratory system is also similar to that described in the hypothetical ancestral form.

The polyplacophoran nervous system is of a primitive type, with a circum-oesophageal nerve ring from which nerves supply the buccal cavity. Well-developed pedal and lateral nerve cords are present. Chitons have no eyes, statocysts or tentacles. A sensory structure, the sub-radula organ, is associated with the radula and innervated from the circum-oesophageal nerve ring. Apart from this, the chief sense organs are the aesthetes, which are unique to chitons. They are specialized mantle cells with sensory endings lying within minute vertical canals in the shell, and can be extremely numerous (1,750 nerve endings per square mm in *Lepidochitona cinereus*). Although their functions are not fully understood they are believed to be photoreceptors and in one family (Chitonidae) are modified as single 'eyes'.

Chitons have separate sexes, external fertilization, and a larval stage lasting about a week.

The major mollusc classes

Class Gastropoda

The class Gastropoda is the largest molluscan class with about 75,000 living species, and a long fossil record. Gastropods ('belly-feet') show considerable adaptive radiation and are found in all types of marine environment, both at the surface and on the bottom, as well as in freshwater and on dry land. The group is divided into 3 subclasses: Prosobranchia, Opisthobranchia, and Pulmonata. Generally, gastropods have a well developed head with tentacles and an asymmetrical body. The visceral mass has been twisted through 180° so that the mantle cavity, ctenidia, anus, and nephridiopores lie immediately behind the head, a modification known as torsion. The shell is adapted as a 'house into which the animal can retreat rather than as a simple protective shield'.

The gastropod larva is at first bilaterally symmetrical and undergoes twisting quite suddenly in response to asymmetric growth of the retractor muscles. The evolutionary significance of the development of this torsion in gastropods is not fully understood, although a number of malacologists have suggested that it evolved as a means of protecting the head, either of the larva or of the adult. The position of the anus and ctenidia relative to each other and to the head has certainly created fouling problems for the group. Much of gastropod diversification can be interpreted as attempts to solve these problems. Torsion should not be confused with simple coiling of the shell, which evolved in response to an increase in the volume of the visceral mass.

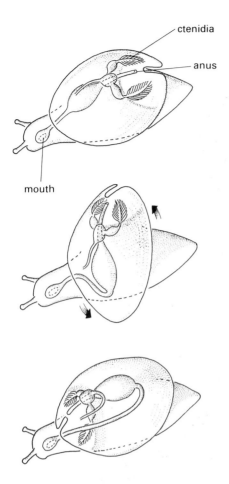

Torsion in gastropods

The genus *Helix*, a member of the subclass Pulmonata, is a gastropod which shows molluscan adaptations to a terrestrial mode of life. *Helix pomatia*, the Roman snail, is by far the largest British pulmonate and has a long history as a gastronomic delicacy. For example, snail shells have been found associated with the cave dwellings of prehistoric man in France. The Romans fattened them in nurseries called 'cochlearia', feeding them on bran soaked with wine. They are recorded (rather improbably) as having grown so large that the shell had a volume of 10 quarts. Pliny regarded 2 snails, 2 eggs, a barley cake and a lettuce as an adequate meal. In Europe, 'escargots' are still popular; they are fasted in 'escargotoires' for several days before they are eaten to rid them of any leaves which might taint them and could be poisonous. Alternatively they may be fed on flour, lettuce or herbs during this period. The Romans are popularly assumed to have introduced *H. pomatia* to Britain, hence its common name; however, both this species and *Helix aspersa* (the common snail), which is also edible, have been found associated with pre-Roman habitations in this country and must, therefore, have been enjoyed by the indigenous population.

Both species of *Helix* grow to a reasonable size and are readily available for dissection. A description of *Helix* will therefore form the basis for a discussion of the gastropods.

Helix

Helix pomatia lives in open woodland, quarries, banks, and hedges on calcareous ground, but not usually on cultivated land such as gardens or parks. *H. aspersa*, a rather smaller species, is extremely common in gardens and may also be found in quarries, hedges, and banks. It often has a fixed 'home' in a sheltered place to which it returns after each foraging period; these homes may be used by many generations of snail and, in limestone areas, give the impression that the animal has bored into the rock. Both species feed at night and after rain. In Britain they hibernate during the winter. *H. pomatia* buries itself just below the surface of the soil, while *H. aspersa* congregates in groups in crannies or may bury itself. During hibernation the shell aperture is sealed with a membraneous diaphragm of hardened mucus impregnated with calcium carbonate and small quantities of silica, phosphates and iron.

Body form: the shell

The most conspicuous external feature is the shell, which, as in most gastropods, is a spiral coil. Its structure may be most easily understood by regarding it as an elongated, hollow cone, wound round a central hollow axis, the columella. The columella opens at a small ventral pore or umbilicus and the sutures between the coils are closely fused. The shell itself is made up of three layers: an outer horny pigmented periostracum of conchiolin, a middle calcareous cross-lamellar layer formed of long strips of aragonite in a conchiolin matrix and an inner layer of nacre. Obtaining sufficient calcium for shell formation is a particular problem for terrestrial snails. As a result snails are largely restricted to regions where the soil is fairly calcareous. *Helix* stores calcium carbonate in cells in the digestive gland whenever it is available in excess of its immediate needs. When reared on a diet which is lacking in calcium it forms a shell which is thin and transparent and may form only an incomplete cover to the visceral hump.

The body

When the snail is resting the head and foot are pulled, by contraction of a columellar muscle, into a large chamber created by the lowest whorl of the shell. This muscle originates on the columella, runs down the side of the visceral mass, and splits into bundles extending to the head and anterior and posterior foot. In *H. pomatia* the columellar muscle can contract to a tenth of its relaxed length.

When the snail is active the head and foot protrude anteriorly and posteriorly from the aperture of the shell. The foot is broad and highly muscular with a flat under-surface (the sole), and the head bears two pairs of tentacles

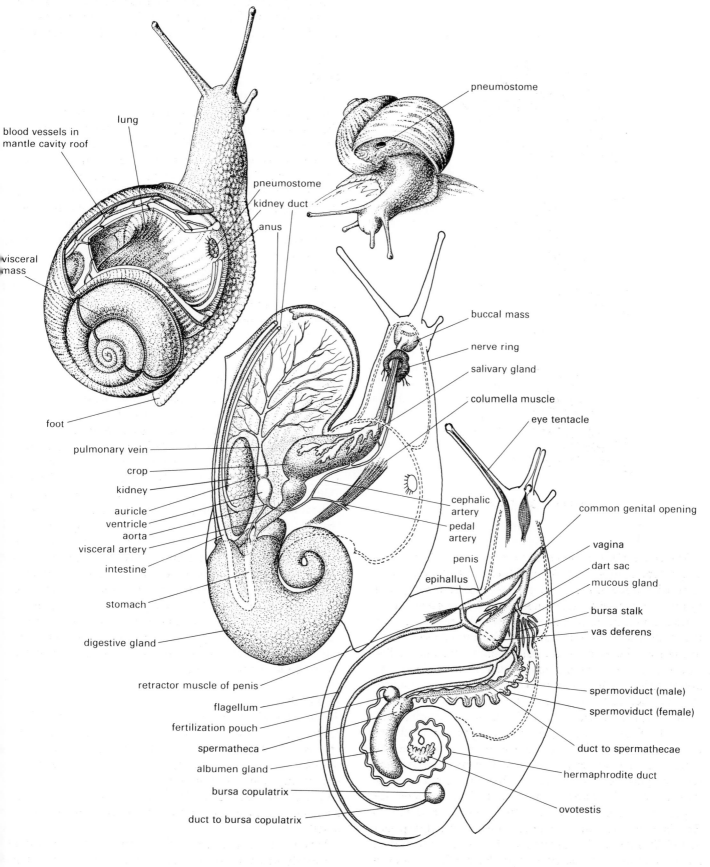

pneumostome

blood vessels in
mantle cavity roof

lung

pneumostome

kidney duct

anus

visceral
mass

foot

buccal mass

nerve ring

salivary gland

columella muscle

eye tentacle

pulmonary vein

crop

kidney

auricle

ventricle

aorta

visceral artery

intestine

stomach

digestive gland

cephalic
artery

pedal
artery

penis

epihallus

common genital opening

vagina

dart sac

mucous gland

bursa stalk

vas deferens

retractor muscle of penis

flagellum

fertilization pouch

spermatheca

albumen gland

bursa copulatrix

duct to bursa copulatrix

spermoviduct (male)

spermoviduct (female)

duct to spermathecae

hermaphrodite duct

ovotestis

Anatomy of *Helix*

and a slit-like mouth. The remainder of the soft parts, the visceral hump, is always enclosed by the shell. The lower margin is thickened and fused with the mantle edge to form a collar, and this region secretes all but the innermost layers of the shell. These last are secreted by the mantle. Fusion of the mantle margin to the head and foot modifies the mantle cavity as an enclosed space, the lung. Air passes into the lung through a conspicuous opening on the right hand side, the pneumostome.

Locomotion

In many ways locomotion in *Helix* is comparable with that of the gliding movements of larger free-living platyhelminths. The flattened ventral surface of the foot (sole) is pulled into a series of minute ridges by waves of contraction of the pedal musculature, clearly visible through a sheet of glass as the snail crawls up the other-side. The musculature is a complex web (including dorsoventral, longitudinal, transverse and oblique fibres) whose contraction is controlled through a nerve net by the pedal ganglion. The edges of the ridges are directed backwards and waves of contraction pass from the hind end towards the anterior. About eight waves may pass over the sole in sequence at any one time. As a result the animal is pushed forwards.

For gliding movement of this kind it is necessary for the substrate to be lubricated; the necessary lubrication is provided by mucus produced by epidermal glands in the sole as well as by an anterior glandular mass, the pedal gland. After the animal has passed over it the mucus dries, leaving the well-known shiny snail trails.

Feeding

Helix is herbivorous and feeds on a wide variety of plants. *H. aspersa* can be particularly damaging to young plants and soft fruit in the garden. The mouth opens into a buccal cavity with a tongue-like odontophore extending from its floor. The odontophore is supported by structures called cartilages (because they show close structural similarities with vertebrate cartilage), and moved by a complex array of muscles. A thick cuticle, to which the radula is fused, covers the odontophore and the whole structure is called the buccal mass. The radula consists of chitin and protein, and bears a series of highly regular recurved teeth, formed of chitin and hardened protein impregnated with inorganic materials (iron and silica).

The buccal mass is protruded through the mouth and the radula used like a rasp to tear off pieces of food. It is withdrawn by a well-developed retractor muscle. The radula is continuously worn away through use, and is renewed by specialized cells (odontoblasts) which lie at the inner end of the radula sac into which the posterior end of the radula fits. The floor of the radula sac secretes a ribbon of cuticle onto which the teeth are fused. As more rows of teeth are secreted the entire structure moves forwards until it appears on the dorsal surface of the odontophore.

The particles of food are taken into the buccal cavity which opens at its posterior end into a narrow tubular oesophagus. A pair of elongated salivary glands open into the roof of the buccal cavity. These secrete a watery mucus, containing amylase and other enzymes, which are used to lubricate the moving radula and begin the process of digestion.

The oesophagus has a thin muscular wall with a glandular, ciliated lining, and widens posteriorly to form a large thin-walled crop. The cilia beat to give a strong, backwardly-directed current which, with peristaltic movements of the oesophageal wall, forces the food towards the crop. Enzymes produced by the oesophageal glands are mixed with the food. In the crop further mixing takes place with a brown digestive fluid produced by the digestive gland and also containing enzymes. In addition to these enzymes symbiotic bacteria living in the crop and intestine are thought to produce a cellulose-splitting enzyme, cystase, which may also be secreted by the digestive gland.

The digestive gland opens into the stomach, which follows the crop. It is a voluminous organ, brown in colour and forming packing tissue throughout the visceral hump. In *Helix* it is composed of three kinds of cells: enzyme secreting cells, absorptive cells, and calciferous cells (responsible for calcium carbonate storage). The stomach has a ciliated lining and particles of digested food are swept into the digestive gland for absorption. Indigestible matter is carried to the distal end of the stomach and thence to the coiled intestine. A style and style sac are absent in *Helix* but are both typically present in gastropods. The intestine extends to the rectum which opens, via the anus, into the mantle cavity just behind the pneumostome.

Gaseous exchange

The mantle cavity is modified to form a lung by fusion of the mantle edge to the back and neck in front of the visceral hump. The lung is roofed by the mantle which is extensively supplied with blood by the pulmonary vein. The floor of the lung is formed by the head and foot which, in their relaxed position, arch up into the lung but can be flattened by contraction of intrinsic longitudinal muscles. Gaseous exchange takes place over the surface of the mantle, with air being drawn into the mantle cavity through the pneumostome by alternate arching and flattening of the lung floor. The pneumostome can be closed by a valve, reducing water loss to a minimum.

Circulation

In pulmonates the heart consists of a single auricle and a ventricle lying within the pericardial cavity. Blood from both the general body surface and the pulmonary vein is received by the auricle and pumped into the ventricle. The single aorta, arising from the ventricle, forks to form a cephalic artery which supplies the anterior part of the body and the foot (as the pedal artery), and a visceral artery which supplies the visceral hump. The arteries branch and enter a series of irregular cephalic, pedal or visceral haemocoelic spaces or sinuses in which the organs lie bathed in blood. The spaces are small and the tissue resembles platyhelminth parenchyma. Blood is collected by two veins which supply the lung, and thence returns to the heart. A copper-containing respiratory pigment, haemocyanin, is present, giving a distinct green colour to the blood.

Excretion and osmoregulation

Helix has a single kidney, representing that of the post-torsion left-hand side. It is a sac of coelomic origin connected to the pericardial cavity by the renocardial canal and drained by the kidney duct, or ureter, which extends along the rectum to a slit-like opening at the right edge of the pneumostome close to the anus. Both the renocardial canal and the ureter are ciliated; in the former the cilia beat towards the kidney creating a flow of fluid, though in *Helix* the canal is minute and the flow of fluid small. The kidney is yellow and has folded glandular walls which secrete uric acid; this can be discharged in a solid form and therefore with little water loss. Furthermore uric acid can be stored in the kidney over a long period and this is an important factor in survival through periods of hibernation. The greater part of the fluid which enters the kidney through the renocardial canal is reabsorbed.

Nervous system and sense organs

In *Helix*, the ganglia, commissures and connectives characteristic of molluscs are concentrated near the anterior end to form a nerve ring enclosed in a connective tissue envelope around the oesophagus and salivary ducts. Nevertheless some distinctions can be made between the component structures of this ring. The cerebral ganglia lie dorsally and are linked by a pair of cerebro-buccal connectives to the small buccal ganglia which lie lateral to the salivary gland ducts. The cerebral ganglia give off nerves to the body wall, mouth tentacles and eyes, and the buccal ganglia supply nerves to the buccal mass and cavity.

Paired cerebro-pedal and cerebro-pleural connectives extend ventrally from the cerebral ganglia to the pedal and pleural ganglia respectively. The latter are joined by fused visceral ganglia to form a ring through which the cephalic artery passes. The ganglia supply nerves to the various body organs and the foot.

The sense organs are mainly concentrated on the head. The first pair of tentacles (cephalic tentacles) are chemoreceptors and also have a well-developed tactile sense. Each of the second pair of tentacles (or eye stalks) bears a single eye at its tip. This structure consists of an invagination of the epidermis lined with pigment and photoreceptor cells, and has a large spherical lens and a vitreous layer covered by a cornea of thin, unpigmented translucent cells. It can detect light and light direction but has only limited powers of image formation.

A pair of statocysts, each consisting of a spherical cavity with 11 to 13 giant cells and a number of syncytial cells, lies embedded in the foot near the pedal ganglia. These statocysts are innervated from the cerebral ganglia and contain fluid, secreted by their lining, with calcareous particles. In addition to the special sense organ the entire body surface is sensitive to touch and chemical stimuli, especially in certain areas such as the lips and sides of the foot. As a modification associated with its terrestrial mode of life *Helix* has no osphradia.

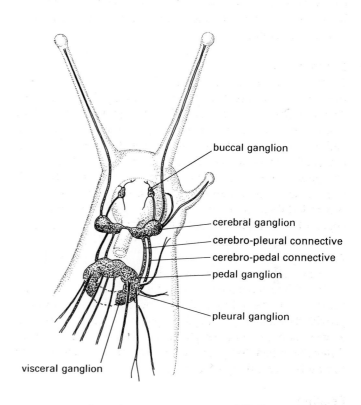

buccal ganglion

cerebral ganglion
cerebro-pleural connective
cerebro-pedal connective
pedal ganglion

pleural ganglion

visceral ganglion

Anterior nervous system of *Helix*

Reproduction

Helix is hermaphrodite. Copulation takes place with mutual exchange of sperm in the form of spermatophores, and fertilization is internal (a common adaptation of terrestrial animals).

A single gonad (the ovotestis) produces both eggs and sperm simultaneously and is a small white structure embedded in the digestive gland. Sperm are produced continuously and passed from the ovotestis into a short coiled hermaphrodite duct which extends towards the albumen gland. This duct functions as a seminal vesicle and appears to be full of sperm throughout the year. Just before it reaches the albumen gland the hermaphrodite duct dilates forming a fertilization pouch which is connected to the albumen gland chamber by a narrow duct; the pouch also receives short narrow ducts from between three and five sacs called spermathecae, in which sperm transferred from another snail during copulation (foreign sperm) is stored. The albumen chamber is joined to the remainder of the reproductive system by a spermoviduct which, as its name implies, serves for transfer of both sperm and eggs and is functionally separated into two grooves by paired longitudinal folds. The wall of the albumen chamber is also deeply folded with one fold forming a groove which joins the fertilization pouch to the sperm groove of the spermoviduct. At the beginning of copulation sperm pass from the hermaphrodite duct through the albumen chamber to the sperm groove, thence to the vas deferens and epihallus where the spermatophore is formed. This process is extremely rapid (taking less than one minute) due to the beating of long cilia which line the entire sperm duct.

The spermatophore is an elongated structure with distinct head, neck, body and tail regions. Sperm is stored in its body. In *Helix aspersa* (but not in *H. pomatia*) the spermatophore tail is formed in a long caecum or 'flagellum' which extends from the epihallus.

In *H. pomatia* mating occurs immediately after hibernation in Britain, in April or May and possibly also in the late summer. *H. aspersa* mates throughout the summer. In either case copulation is preceded by an elaborate courtship in which the partners circle round each other, make tentacular contact and entwine their bodies. In *H. pomatia* this phase may last for as long as one and a half days. When the animals are intertwined, a dart sac (part of the vagina) secretes a calcareous spicule which is driven into the partner's body wall and which stimulates copulation. The intromittent organ or penis is everted and inserted into the bursa stalk (a direct continuation of the vagina) of the partner. After the spermatophore head has been placed in position, the penis is withdrawn and the rest of the spermatophore is pushed out by the donor and drawn in by the recipient. Partners transfer their spermatophores simultaneously and the whole process is long and slow, taking several hours.

Sperm are released from the spermatophore and migrate to the spermathecae. The remains of the spermatophore and any excess sperm are transported along an elongated duct by peristalsis to the bursa copulatrix where they are digested. The snails then resume their normal activities.

Eggs pass down the hermaphrodite duct and are fertilized in the fertilization chamber by foreign sperm transferred from the spermathecae. The eggs are large and yolky; after fertilization they are surrounded by albumen secreted by the glandular lining of the fertilization chamber, and by a leathery calcareous shell as they pass down the oviduct.

Helix lays its eggs in humid places in hollows which it digs with its foot. A typical clutch size is between 19 and 96 eggs. The larval stages develop inside the shell of the egg which enlarges as the developing *Helix* grows. The veliger is modified from its basic planktonic form by the development of a large cephalic vesicle in the velar region which provides an increased surface for gaseous exchange.

On hatching, the young snails are complete apart from their reproductive system. Maturation is normally reached in the second year, and mature snails live for several years.

Helix has high regenerative powers, particularly of its shell, although almost any part of the body except the central nervous system can be regenerated if damaged.

Gastropod diversity

There are three gastropod subclasses: Prosobranchia are chiefly marine, and have an anterior mantle cavity and gills. Opisthobranchia show detorsion and have a reduced shell and mantle cavity; Pulmonata (the group to which *Helix* belongs) have no gills and a mantle cavity modified as a lung.

The majority of gastropods are prosobranchs and the other two subclasses are believed to derive from this group. Some classifications, therefore, regard all gastropods as prosobranchs.

Subclass Prosobranchia

The chief adaptive variation in prosobranchs concerns gaseous exchange and the direction of the water current generated by the animal. Torsion produces unusual problems of sanitation in that the snail defaecates on its head, and gastropod radiation shows various solutions to this.

The least modified prosobranchs (order Archaeogastropoda) are typically algal browsers and are restricted in distribution to rocky surfaces since their gills present a large surface area for fouling (for example from sand or mud). Typical examples include the limpets (*Patella*, *Emarginula*) and ormers (*Haliotis*) which have conical or flattened shells, and the whelks (*Gibbula*). The simplest form of gill arrangement is seen in forms such as *Emarginula* or the keyhole limpet *Diodora*, which have bipectinate gills (of the primitive type described for 'ancestral mollusc') and a slit or cleft in the shell for the outgoing water current. Both the anus and the kidneys open near the aperture so that waste products are swept away. In *Haliotis* the excurrent leaves through several small holes, an arrangement which does not weaken the shell and provides greater protection than a single larger aperture. Many archaeogastropods (such as *Acmaea*) reduce the gills to a single bipectinate structure, usually on the left hand side, resulting in an oblique water current entering on the left and leaving on the right. Again the anus opens into the excurrent. *Patella* shows loss of the original gills and has evolved secondary gills (folds of the mantle) in association with a similar oblique current.

Most of the above forms are adapted to live in the intertidal zone and parallel the chiton in that they can be held clamped down hard against rocks by powerful pedal retractor muscles. Their shells are typically very strong and modified as protective shields rather than houses, having a shape which gives minimum resistance to water currents and battering by waves. In contrast the whelks have conical shells and (like *Helix*) can retreat into an enlarged shell aperture which can be closed by a flat plate of conchiolin, the operculum. Like *Acmaea* whelks have a single gill and an oblique water current across the mantle cavity. In shape they are less well adapted for life on exposed rocks and

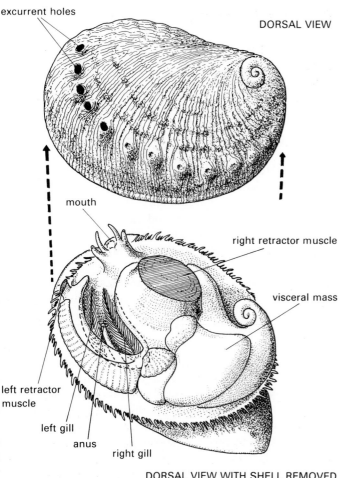

excurrent holes

DORSAL VIEW

mouth

right retractor muscle

visceral mass

left retractor muscle

left gill

anus

right gill

DORSAL VIEW WITH SHELL REMOVED

Haliotis

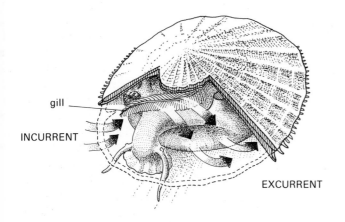

gill

INCURRENT

EXCURRENT

Water current of *Acmaea*

the possibility of fouling by detritus in turbulent water or on soft bottoms. The filaments on one side have been lost to give a unipectinate structure which attaches directly to the mantle. In addition many neogastropods have developed a tubular structure for water intake (the siphon), by extending the mantle edge and rolling it inwards. The siphon has proved an extremely valuable adaptation because it can be extended above a soft substrate in burrowing forms or used as a sense organ. To support the siphon, animals such as *Buccinium*, *Murex* and *Nucella* have expanded the shell to form a siphonal canal.

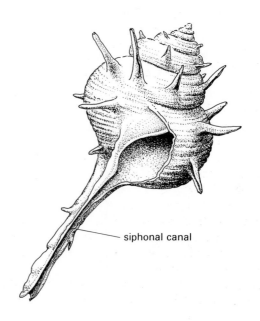

siphonal canal

Murex branderis

Mesogastropods feed on algae and organic sediments and are easily recognizable as their shell has a nacre lining. Common examples include species of the genus *Littorina* (periwinkles) which feed on seaweeds or lichen. *L. saxatilis* is particularly interesting as it is an air-breathing form living high up on the shore and has its mantle cavity modified as a lung. *Xanthina exigua* is also unusual as a pelagic species

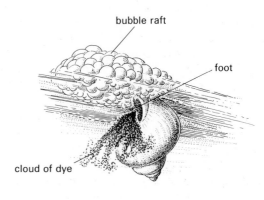

bubble raft

foot

cloud of dye

Xanthina

usually shelter in crevices, under stones or seaweed, or buried in sand.

The remaining prosobranchs (mesogastropods and neogastropods) live in a wide range of marine habitats. Undoubtedly much of their success is due to modifications of the gill which have reduced its surface area and therefore

which floats upside down on a raft of air bubbles and preys on siphonophoran cnidarians. It is a beautiful and conspicuous violet colour, and confuses would-be predators by emitting a cloud of purple dye.

Most neogastropods feed on organic deposits or are predators. Among the latter is *Ocenebra erinacea* which is a pest of oyster beds; it drills a hole through the oyster's shell by means of a modified radula, and then sucks out the body contents.

Subclass Opisthobranchia

The opisthobranchs are a marine group. The primitive members of the group such as *Actaeon* have the typical gastropod body form with a coiled shell but the overall trend is for shell reduction or loss, a secondary detorsion and development of bilateral symmetry, and a reduction of the mantle cavity with loss of the original gills. Forms such as *Bullaria* or *Scaphander* cannot withdraw completely into the shell, and in many species the shell is completely enfolded in the mantle. Many have become adapted for a free-swimming or pelagic mode of life. *Aplysia penctata*, the sea hare, crawls around on seaweed but may also swim using flaps (parapodia) which are lateral expansions of the foot. When alarmed it releases a copious violet dye which obscures its location. In *Aplysia* the shell is represented by a horny structure embedded in the mantle. The pteropods or sea butterflies may be shelled (order Thecosomata) or unshelled (order Gymnosomata). In the unshelled forms the mantle cavity is absent, and gaseous exchange occurs over the entire body surface. Pteropods are pelagic. They are suspension feeders and trap food in mucus on the parapodia.

The nudibranchs ('naked gills') are beautiful and brightly coloured. They have no shell or mantle cavity, and have achieved a total secondary bilateral symmetry; they have secondary gills which are external and may form a ring round the anus. The second pair of tentacles are modified as rhinopores for chemoreception. *Facelina auriculata* and *Aeolidia papillosa* are common British examples. *A. papillosa* has numerous papilla-like appendages (cerata), which contain extensions of the digestive gland, along its back. It feeds on sea anemones and may transport the nematocysts of its prey to its own appendages and then use them for defence. The Opisthobranchia includes one parasitic order (Parasitica) of worm-like endoparasites of sea cucumbers.

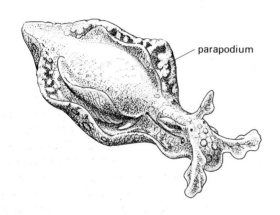

Aplysia

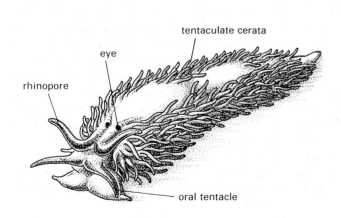

Aeolidia papillosa

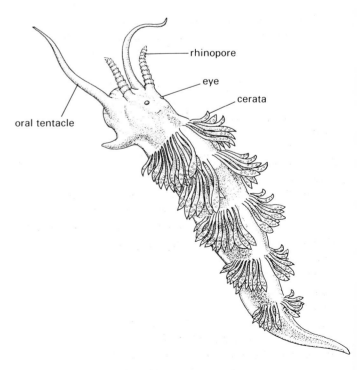

Facelina auriculata

Subclass Pulmonata

This group includes the highly successful land snails (such as *Helix*) and slugs (for example *Limax*) as well as many snails of fresh and brackish water (for example *Limnaea*); the few marine representatives live in estuaries or intertidal water. Pulmonates can be distinguished from land and freshwater prosobranchs by their lack of an operculum. Pulmonates, as their name suggests, always have the mantle cavity modified as a lung although some aquatic forms develop secondary gills. *Limnaea* has the mantle cavity edges extended into a long tube, the pneumostome, which can be extended to the water surface.

Slugs have perhaps undergone the most conspicuous modification away from the basic gastropod form. *Limax* is widely distributed on cultivated land and has a similar mode of life in many ways to *Helix*. There is a shell but it is a small asymmetrical structure, partly horny and partly calcareous, buried in the mantle.

A particular significance of many pulmonates is their involvement in parasitic life cycles as intermediate hosts. For example snails in the family Limnaeidae are hosts to the larval stages of liver flukes; and the family Planorbidae are hosts to the larval stages of *Schistosoma mansoni* though *S. japonicum* uses a prosobranch intermediate host. Slugs may be hosts to the cysticercus stage of tapeworms. Land pulmonates act as intermediate hosts to many strongiloid nematodes which infect mammals in the mature stage of their life cycle.

Synopsis of class Gastropoda

Subclass Prosobranchia
    Order Archaeogastropoda: limpets, top shells and
        ormers (*Diodora, Patella, Emarginula, Haliotis,
        Gibbula, Acmaea*) and the terrestrial family
        Helicinidae)
    Order Mesogastropoda: periwinkles, tower shells,
        cowries (*Littorina, Xanthina, Ocenebra*, and the
        operculate land and freshwater snails.
    Order Neogastropoda: whelks
        Buccinium, Murex, Nucella)
Subclass Opisthobranchia
Various orders including:
    Order Tectibranchia: sea slugs and bubble shells
        (*Actaeon, Bullaria, Scaphander, Aplysia*)
    Order Thecosomata: sea butterflies (shelled pteropods)
    Order Gymnosomata: naked pteropods
    Order Nudibranchia
    Order Parasitica
Subclass Pulmonata
Various orders including
    Superorder Basmmatophora (aquatic)
        (*Limnaea, Planorbis*)
    Superorder Stylommatophora (terrestrial)
        (*Helix, Limax*)

Class Bivalvia

Bivalves are easily recognized; they are compressed laterally and, as their name suggests, have a shell composed of two parts (valves) one on each side of the body joined at a dorsal hinge. The foot is also laterally compressed and has no flattened crawling surface, which explains the term Pelycopoda (meaning hatchet foot) as an alternative for the group.

The vast majority of bivalves are sedentary or burrowing animals which have no tentacles and may have either a rudimentary head or no head at all. Most are particulate or filter feeders and the radula is absent. *Mytilus* (the mussel) is an intertidal form, attaching itself to rocks by means of threads secreted by the foot. In the oyster (*Ostrea*) the foot is lost altogether and the left valve cemented to the underlying substrate. There are many edible bivalves including mussels, scallops, oysters and cockles.

Mytilus will be described as an introduction to the group, as it is readily available and large enough (up to 10 cm) for convenient dissection.

Mytilus edulis, the edible mussel

Mytilus edulis is common on shores or rocks near the low tide mark where it is found in tightly packed masses, sometimes in large quantities. It attaches itself to the substrate by a tuft of fine byssal threads, which are secreted by a byssal gland on the foot and protruded between the ventral edges of the valves. *Mytilus* usually remains in one place, but can move very slowly by breaking and resecreting the byssal threads. Retractor muscles can shorten the foot and so tighten the byssal threads.

Body form

Like gastropods, *Mytilus* secretes its shell from the mantle but has two centres of calcification rather than one. The valves are joined at a dorsal hinge or ligament, secreted by a special region of mantle called the isthmus. The shell consists of a dark, horny periostracum, a middle layer of aragonite and calcite deposited in prisms, and an inner nacre layer. The isthmus secretes conchiolin only. As the ligament is elastic and is strained when the valves are closed they will tend to spring open unless held together by adductor muscles (and are always so in dead mussels). *Mytilus* has anterior and posterior adductors, each composed of two types of muscle fibres. Quick muscle fibres are adapted for rapid contraction to close the shell against a predator; catch muscle fibres, which are smooth, are used for periods of slow sustained contraction when the animal is left exposed by the tide.

The mantle edge projects slightly beyond the shell and is highly sensitive. The dorsal margins are joined, except at the posterior end where the mantle forms an exhalent siphon. Postero-ventrally and ventrally the edges are separate, and it is in this region that water enters the mantle cavity. The most conspicuous features of an open *Mytilus* are greatly expanded ctenidia (gills) which are used for both gaseous exchange and feeding. The foot is small and the head reduced to a few labial palps surrounding the mouth.

Feeding

Mytilus is a filter feeder, living on detritus, protozoans and diatoms. The ctenidia are greatly expanded curtain-like structures which act as strainers. Additional filaments have been added to the series so that they extend anteriorly as far as the mouth and posteriorly to the anus. In addition the filaments are increased in length and folded upwards in a V-shape. They are suspended from the ctenidial axis which is fused along the dorsal margin of the mantle; their ends turn up close to this region so that in cross-section the axis and its filaments form a W-shaped structure. A supra-branchial cavity lies within each V, this is exhalent and is connected to the exhalent siphon. The ventral infra-branchial cavity connects to the inhalent region of the mantle edge. A water current, created by lateral cilia on the gill filaments which beat with a metachronal rhythm, passes from the infra-branchial to the supra-branchial chambers. Filtering is provided by the filaments themselves and by bundles of fused cilia (the latero-frontal cilia) with a pinnate (feather-like) structure which flick particles onto the filament surfaces where they are trapped by mucus. The latero-frontal cilia are also sufficiently close to form a very effective sieve. Trapped particles and mucus are moved by the frontal cilia into food grooves; there are five of these which lie ventrally at the ctenidial axis, the distal ends of the filaments, and the angles of the filaments. Food and mucus are moved anteriorly along the grooves and then transferred to the mouth by the labial palps.

Once it has entered the mouth the food is propelled along the oesophagus to the stomach by cilia and by the action of a structure known as the crystalline style. Cilia are generally used for food propulsion in *Mytilus* which does not have a muscular gut wall, probably because it is a particulate feeder. The stomach is capacious and globular and lies entirely embedded in the digestive gland which opens into it by ducts: the surface area of the stomach may be further increased by a ventral diverticulum. The crystalline style arises from a style sac at the posterior end of the stomach and projects dorsally forwards to rest against an area of cuticle on the stomach wall, the gastric shield. It consists of a column of mucus containing starch-digesting enzymes which is rotated by the beat of style sac cilia towards its long axis. The style is secreted continuously by the style sac and slowly dissolves away, releasing mucus and enzymes into the stomach. Its rotation stirs the stomach contents and pulls mucus-food strings into the stomach by a kind of winching action.

Partly digested food passes into the ducts of the digestive gland; digestion is completed and food absorbed in the digestive gland. Food particles which do not enter the digestive gland pass into the intestine with indigestable matter where they may be taken up by numerous amoebocytes within the intestinal lumen. The intestine is long and looped, presumably to allow sufficient time for this process. Undigested matter leaves via the anus which opens near the exhalent siphon.

Gaseous exchange and circulation

Gaseous exchange takes place as water moves across the ctenidia through folded membranes at their bases. The amount of oxygen removed from the water is low; this is because the water current generated by the animal contains oxygen greatly in excess of the respiratory needs of the animal. The blood has no respiratory pigment. Mussels uncovered at low tide remain closed and inactive and the ctenidia are kept moist by water retained within the shell.

The heart in *Mytilus* is an elongated structure with two auricles which receive blood from the gills and empty into the ventricle. During its development the ventricle comes to surround the intestine. Blood is discharged from the ventricle via an aorta which immediately divides, one branch extending to supply the head, viscera, and foot and the other the mantle and siphon. Many bivalves have large visceral and pedal haemocoelic spaces comparable with those of gastropods but in *Mytilus* only a pericardial sinus is present.

Excretion and osmoregulation

The excretory system of bivalves consists of two U-shaped kidneys, within the pericardial cavity and functionally very like those of gastropods. One arm of each kidney is glandular and lies alongside the heart opening into the pericardium (the pericardial gland), the other is a bladder with an external opening at the anterior end of the suprabranchial cavity.

Nervous system and sense organs

The nervous system is bilaterally symmetrical and relatively simple, with three pairs of ganglia (cerebro-pleural, visceral and pedal) and two longitudinal nerve cords. The cerebro-pleural ganglia, which are joined above the oesophagus by a short commissure, each give rise to two nerve cords. One extends to the visceral ganglion (the cerebro-visceral connective), the other ventrally to the pedal ganglion. The cerebro-pleural and pedal ganglia control the foot and the anterior adductor muscle, while the visceral ganglion controls the exhalent siphon and the posterior adductor muscle.

The sense organs are simple and concentrated mainly at the mantle edge which is very sensitive both to touch and to changes in light intensity. The innermost fold of the mantle edge is thick and has a large number of short mantle tentacles. Statocysts are present near the pedal ganglion and also a sensory patch known as the osphradium in the exhalent chamber. The functions of this structure are uncertain, though it may act as a chemoreceptor or detect particles in the exhalent current. Its position within the exhalent current makes it extremely unlikely that it is

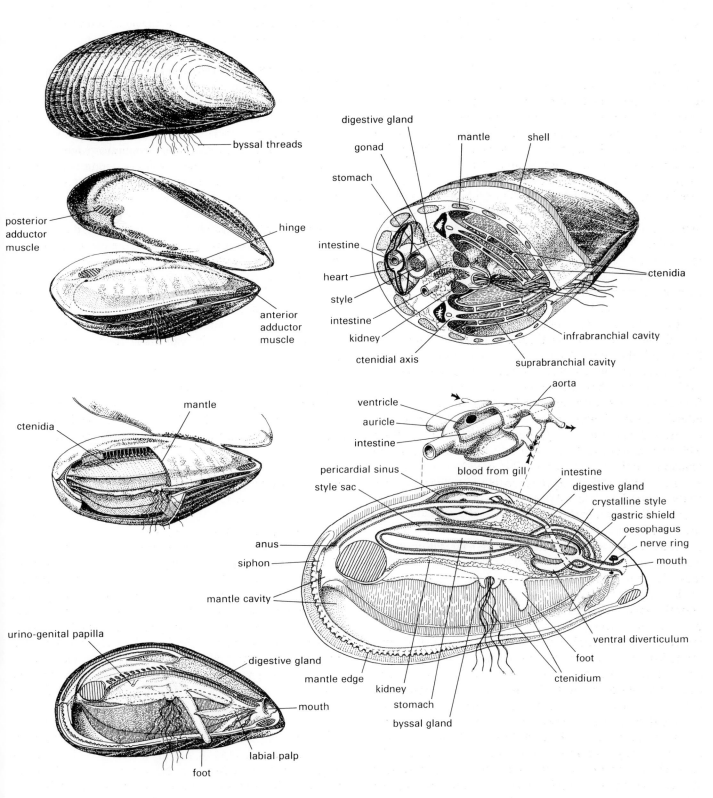

byssal threads

posterior adductor muscle

hinge

anterior adductor muscle

digestive gland

gonad

stomach

mantle

shell

intestine

heart

style

intestine

kidney

ctenidial axis

ctenidia

infrabranchial cavity

suprabranchial cavity

mantle

ctenidia

ventricle

auricle

intestine

aorta

blood from gill

intestine

digestive gland

crystalline style

gastric shield

oesophagus

nerve ring

mouth

pericardial sinus

style sac

anus

siphon

mantle cavity

urino-genital papilla

digestive gland

mantle edge

kidney

stomach

byssal gland

ventral diverticulum

foot

ctenidium

mouth

labial palp

foot

Anatomy of *Mytilus*

homologous with gastropod osphradia which are in an inhalent position.

Reproduction and growth

Reproduction in *Mytilus* may occur at any time between mid-winter and early autumn. The sexes are separate, and fertilization is external. The reproductive system is far simpler than that of *Helix* and consists of paired gonads, and gonoducts which open alongside the kidney ducts at the urinogenital papillae. The ovaries are cream coloured and the testes reddish. Gametes are shed into the supra-branchial cavity and leave with the exhalent current.

After fertilization, the larva develops, first as a trochophore and then as a symmetrical veliger very much like that described in the introduction. The veliger is planktonic and is important as a dispersal phase for an otherwise sedentary animal.

Bivalve diversity

There are two classifications of bivalves in common usage. One, the older and simpler, recognises three subclasses: Protobranchia (the least modified bivalves), Lamelli-branchia (of which *Mytilus* is an example) and Septibranchia. The alternative, which was proposed initially by palaeontologists and probably gives a more accurate picture of relationships within the group, subdivides these groups to give several subclasses. The relationship between the two classifications is as follows: Protobranchia is divided into two subclasses: Palaeotaxodonta and Cryptodonta; Lamellibranchia into three: Pteriomorphia, Palaeo-heterodonta and Heterondonta, and Septibranchia replaced by subclass Anomalodesmata (which also includes some lamellibranchs).

The following discussion uses the older classification, the other being summarised to help those using additional texts.

Subclass Protobranchia

Most protobranchs, for example *Nucula nucleus*, the common nut shell, live buried in clay, sand or gravel in shallow water. Unlike other bivalves they have a single pair of bipectinate gills and therefore are not filter feeders. Instead they are selective deposit feeders and pick small animals, such as other molluscs or ostracods, using their well developed labial palps and tentaculate proboscoides. The foot is relatively large and muscular and can be used both to move the animal over the bottom in a series of leaps, or for burrowing. An unusual feature in comparison with other bivalves is that the respiratory current passes from front to back (the animal is buried with its anterior end downwards). As in other bivalves the current is created by lateral cilia; any sediment deposited on the gills' surfaces being removed by the frontal cilia.

Subclass Lamellibranchia

Lamellibranchs ('plate gills') are all filter feeders. The ctenidia in *Mytilus*, although they may appear elaborate, are in fact comparatively simple; in the more specialised forms individual filaments are joined by sheets of tissue to form a more solid structure. In these forms, water enters the ctenidium through pores (ostia) and is transported through water tubes. Many lamellibranchs burrow into soft substrates, often obtaining their water by means of elongated siphons which can be extended upwards to the surface. *Cardium edule*, the common cockle of the fishmongers, burrows to fairly shallow depths, using its foot to dig. Others may burrow much deeper and make semi-permanent burrows. They may use their siphon for feeding. *Tellina*, which is one of the commonest shells found on beaches, has a long inhalent siphon which it uses like a vacuum cleaner to pick up detritus from the surface. *Ensis*, the razor shell, has greatly elongated valves and burrows vertically but comes up to the surface to feed.

Lamellibranchs which live fixed in one place at the surface such as the mussel and oyster have already been mentioned. In addition there are a few forms which are not attached in any way and can move freely. For example, although juvenile scallops (*Pecten*) are anchored to a substrate by byssal threads, the adults are not; indeed scallops are among the very few bivalves which are effective

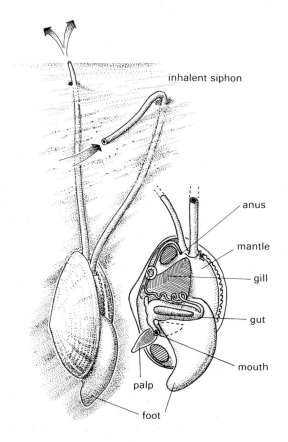

inhalent siphon

anus

mantle

gill

gut

mouth

palp

foot

Tellina

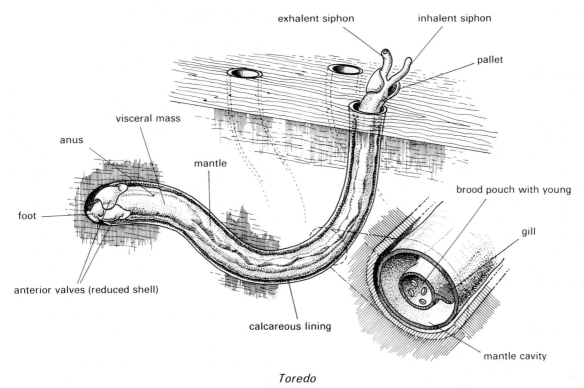

Toredo

swimmers. They move rapidly and spasmodically by quickly opening and closing their valves, thus forcing water out of their shells. They have flaps of tissue at the shell edges which can be used to direct the water jets and therefore control the direction of movement. Swimming is used principally as a means of escape from predators such as starfish, but may also be used if the animal needs to change its feeding site. Like other lamellibranchs, *Pecten* is a filter feeder.

The shipworms (*Toredo*) are highly-specialised bivalves which bore into submerged wooden structures such as driftwood, boats and pilings and can therefore be serious pests. They are considerably elongated and worm-like and have their shell reduced to two small anterior valves which they use as a drill. The mantle covers the body behind the shell and secretes a hard calcareous tube which lines the cavity the shipworm produces. The animal has long delicate siphons which can be extended to the surface to facilitate gaseous exchange and *Toredo* can also close itself off in its burrow with special hard calcareous pallets at the posterior end. It feeds on the sawdust it excavates and has symbiotic bacteria living within its gut which digest the cellulose. The digestive gland is also specialized for this unusual diet.

Subclass Septibranchia

In septibranchs the ctenidia are modified as perforated muscular septae to form a pumping system between the inhalent chamber and the supra-branchial cavity. Septibranchs are carnivores or scavengers. The force of the current created by the pump is sufficient to bring small animals such as Crustacea and various kinds of worm into the mantle cavity where they can be seized by the labial palps and passed into the mouth.

Freshwater bivalves

Several bivalves live in freshwater. For example the freshwater mussel, (*Anodonta*, a lamellibranch) is a filter feeder which lives buried in the bottom of streams, rivers or lakes. The chief adaptations of freshwater bivalves concern their development. Thus in many freshwater forms including *Anodonta* the veliger is not free swimming but is

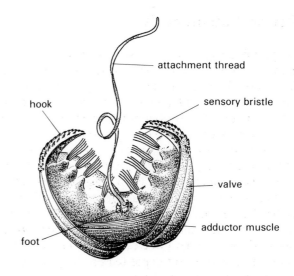

Anodonta: glochidium

highly specialized as a parasite (a glochidium) of fish. The *Anodonta* glochidium has a hook on each valve, which it clamps into the fish, and an adhesive attachment thread. Later the fish surrounds the glochidium with a cyst. Another variant of larval life history is found in the sphaerids, in which the young are retained in the gills of the adult until their development is complete.

Synopsis of Class Bivalvia

Subclass Protobranchia (*Nucula*)
Subclass Lamellibranchia (*Mytilus, Ostrea, Cardium, Tellina, Ensis, Pecten, Toredo, Anodonta*)
Subclass Septibranchia

Two minor mollusc classes

Class Scaphopoda

The scaphopods or elephant's tusk shells have a long tubular tapering shell as their common name suggests. *Dentalium* may be found buried in sand or mud with the tip of its shell protruding in shallow water but the majority are found at greater depths.

The body shows considerable elongation along the anteroposterior axis and the mantle cavity is large. The head

is reduced and proboscis-like and there is a reduced trilobed foot used for burrowing. Scaphopods are detritus feeders, using threadlike feeding tentacles (captacula), and have a fairly typical molluscan alimentary canal. They have no sensory tentacles, eyes or osphradia, and their circulatory system is chiefly noteworthy for the absence of the heart.

The larval development is similar to that of bivalves and they are therefore regarded as a specialized off-shoot of that group.

Class Aplacophora

Aplacophorans ('without plates') are a strange group of worm-shaped molluscs commonly known as solenogasters. They lack the typical molluscan features of foot, mantle, and shell (although calcareous spicules are generally present in the skin) and have a poorly developed head.

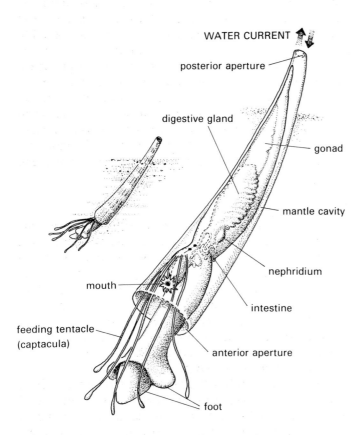

Dentalium

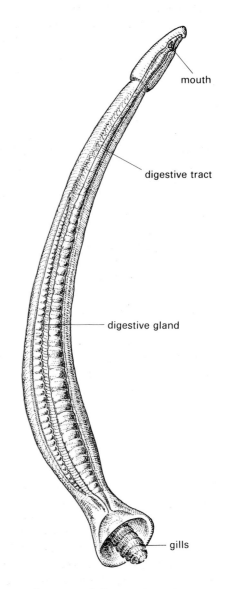

A representative apalacophoran

Their worm-like shape is derived by inrolling at the margins of the body. However, since they have the molluscan characters of a radula and a style sac in their alimentary canal, there is no doubt about the phylum to which they belong.

Solenogasters live either as burrowing scavengers or amongst corals and hydroids on which they feed. At one time they were considered as members of a class Amphineura with the chitons, although the two groups have little in common.

Class Cephalopoda

The cephalopods (cuttlefish, squid, octopodes and *Nautilus*) are the most organized and specialized molluscs. While these animals have retained much of the basic mollusc plan they appear completely different in both external features and mode of life from the other classes of the phylum. They have adapted the typically bottom-living slow-moving form of other molluscs to become fast-moving, often free-swimming, and highly effective predators with acute sense organs and a much higher level of intelligence. As a group they demonstrate the extreme plasticity of the molluscan form which has contributed greatly to the success of the phylum as a whole.

Cephalopods are a diverse group with several grades of organization, even amongst their limited number of about 650 species. However all have certain major features in common. The anterior region of the foot is divided into a number of large prehensile tentacles or arms which are attached to the head and surround the mouth (Cephalopoda means 'head-foot'). The visceral hump has moved to a posterior position and the coelom is more spacious than in other molluscs. Associated with their active and predaceous behaviour, all have well developed and conspicuous eyes. External shells are rare in cephalopods; in the majority the shell is an internal structure or may be absent altogether. However the group has a long fossil history and several shelled forms are known.

Cephalopods have a different development from other molluscs, with no free swimming larval stage; instead large-yolked eggs are laid from which the young hatch directly. *Sepia*, the cuttlefish, has been selected for more detailed description.

Sepia

Sepia is fairly common in British coastal waters, particularly where there are sandy bottoms, and they may also be found in estuaries or bays. In some years large numbers of dead cuttlefish are washed ashore after they have migrated to inshore waters for reproduction. The internal shell or 'cuttlebone' is commonly found on beaches and is used as a source of grit for cage birds. *Sepia* has two methods of escaping detection by predators. When threatened it can discharge large quantities of dye containing melanin, a black pigment. This is released from an ink sac as an inky cloud which confuses the predator. In addition it is the chameleon of the invertebrates, and is capable of rapid colour changes to blend with its surroundings, using special pigment cells (chromatophores) which it can expand or contract to give a deeper or lighter colour.

During the day *Sepia* normally half buries itself in sand; it hunts at night, swimming over the surface in search of suitable prey.

The shell

Like that of other molluscs, the shell (cuttlebone) of *Sepia* is formed from calcium carbonate and protein. It is concealed within the mantle and is secreted by it, layers being added throughout the animal's life. However the organization of the shell is totally unlike that of other mollusc classes; it is formed of a series of thin walls making a stack of gas and liquid filled chambers less than 1 mm deep. As a result the structure has a low specific gravity (about 0.6), and functions as a buoyancy organ, which counter balances the weight of the animal and gives it neutral buoyancy in water. The proportion of gas to liquid in the chambers can be varied by the animal, increasing or reducing the lift provided by the cuttlebone. To increase the cuttlebone density and reduce its lift the concentration of the fluid is increased by active ionic secretion by the living tissue; the fluid in the chambers becomes hypertonic to the blood and water enters the cuttlebone by osmosis, thus displacing some of the gas. Ion uptake from the fluid causes water to leave the cuttlebone and gas diffuses in from the blood to replace it. The buoyancy organ is an extremely valuable adaptation since it enables the animal to hover almost motionless whilst searching for prey, or to rest on the bottom, thus conserving energy expended during swimming. However it is not a strong enough structure to withstand great pressures and so restricts the distribution of *Sepia* to shallower waters. In addition to its function as a buoyancy organ, the shell serves for muscle attachment and for support.

Body form

Sepia is relatively broad and slightly dorsoventrally flattened, thus it is oval in cross section. It is somewhat elongated along the antero-posterior axis, and has paired fins which extend horizontally along the body. At the anterior end the mouth is surrounded by eight short tentacles and two which are greatly elongated and have expanded spatulate ends. These can be extended or retracted and are used to grasp prey. The inner surface of the short tentacles and the ends of the elongated ones are covered by short round suckers.

The mantle cavity is on the ventral surface of the body, it is closed by a thick muscular wall, apart from a well developed funnel developed from the mantle skirt which allows water to be both sucked into, and expelled from, the mantle cavity.

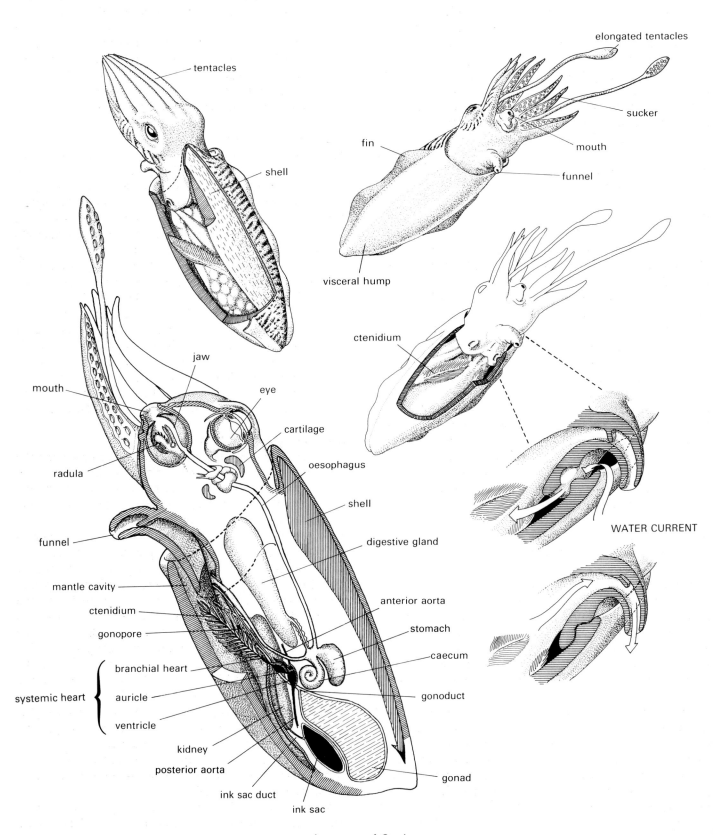

tentacles

shell

elongated tentacles

sucker

fin

mouth

funnel

visceral hump

ctenidium

WATER CURRENT

jaw

mouth

eye

cartilage

radula

oesophagus

shell

funnel

digestive gland

mantle cavity

anterior aorta

ctenidium

stomach

gonopore

caecum

branchial heart

gonoduct

systemic heart

auricle

ventricle

kidney

posterior aorta

ink sac duct

gonad

ink sac

Anatomy of *Sepia*

Locomotion

Sepia is a powerful, graceful, and highly controlled swimmer. It can hover, or cruise slowly, by waves of contraction which pass along the fins. Alternatively it may swim rapidly by jet propulsion when actively chasing prey or avoiding predators. When this happens, water is sucked into the funnel from both the right and left sides, and is expelled in sudden powerful jets by simultaneous contraction of the mantle musculature. Since the funnel is flexible and can be pointed in many directions, *Sepia* can dart forwards or backwards or move up or down. Obviously neutral buoyancy contributes considerably to the versatility of the animal's movements.

Feeding

Sepia is carnivorous and feeds chiefly on crustaceans such as prawns. The prey is caught between the ends of the elongated tentacles, which are moved rapidly forwards, and is then drawn in towards the mouth.

Sepia has a radula and it also has a pair of hard chitinous jaws shaped like a parrot's beak and frequently called the beak. Two pairs of salivary glands open into the buccal cavity; when *Sepia* bites it injects a venom (a neurotoxin, tyramine) secreted by the second pair of salivary glands into the prey. The beak is used to tear up the food which is pulled into the buccal cavity by tongue like movements of the radula. It passes down the oesophagus, which is dilated at its lower end forming a crop, to the stomach. *Sepia* has a large stomach and its distal region gives rise to a caecum, with its surface increased by leaf-like inward projections. The digestive gland secretes enzymes into the stomach and caecum where most of the digestive processes occur. Absorption takes place principally in the digestive gland but also in the caecum and intestine. Undigested matter passes down the intestine to the anus which opens into the mantle cavity close to the opening of the ink sac.

Gaseous exchange and circulation

There is a single pair of ctenidia in the mantle cavity. They have a surface area which is greatly expanded by folding, reflecting the increased respiratory needs of these active animals. The ctenidial cilia have been lost, and the respiratory current is provided by gentle rhythmic contractions of the mantle, similar to those used in jet propulsion.

The circulatory system is elaborate and entirely closed with fine capillaries connecting the arteries and veins, and no large haemocoelic spaces. Large networks (plexi) of capillaries are present in the ctenidia, increasing their efficiency considerably. The blood contains a respiratory pigment haemocyanin, in solution. All those features may be associated with the higher metabolic rate and vigorous activity of cephalopods compared with other molluscs.

Three hearts are present. The main systemic heart, corresponding with that of other molluscs, lies within a spacious pericardium. It receives blood from the ctenidia into two auricles and despatches it from the ventricle to the body through an aorta which gives off branches to the viscera and continues forwards to supply the head. There are also two branchial hearts which lie within the coelom at the bases of the gills. These collect deoxygenated blood from an elaborate venous system (anterior vena cava from the head: abdominal veins from the viscera; pallial veins from the mantle) and pump it through the gills. In addition many of the main blood vessels have contractile walls.

Excretion and osmoregulation

Sepia has two kidneys. Each consists of a large renal sac which opens into the pericardium through a reno-pericardial canal and into the mantle cavity through a nephridiopore. The main vessels to the branchial hearts (afferent branchial veins) pass across the renal sacs where they produce large evaginations (the renal appendages) which project into the sacs. Blood is forced into and out of the renal appendages as the branchial hearts pulsate. Each branchial heart also has an evagination or appendage which lies in close contact with the pericardium and is thought to correspond to the pericardial gland in other molluscs. Ultra-filtration occurs from the branchial hearts and renal appendages into the pericardial cavity and the filtrate is transported down the reno-pericardial canal to the renal sac. Some reabsorption occurs in the canal, and glucose and amino acids also appear to be reabsorbed by the appendages. Urine leaves through the nephridiopore.

Nervous system and sense organs

The brain is greatly enlarged and forms a ring around the oesophagus. It is enclosed by a cartilage-like box which, in its protective function, may be compared with the vertebrate skull. The homologies between the brain and the central nervous system of other molluscs are not entirely clear as there is a high degree of fusion. The supra-oesophageal region is probably equivalent to the cerebral (possibly with the pleural) ganglia. It gives rise to a pair of optic lobes and a pair of buccal nerves extending to paired superior and inferior buccal ganglia associated with the buccal mass. The sub-oesophageal region appears to correspond with the visceral and pedal ganglia. The pedal ganglia have anterior subdivisions (brachial ganglia). The funnel is supplied from the pedal ganglia directly, the tentacles by the brachial ganglia. The visceral ganglia give off three pairs of posteriorly directed nerves: one pair supplies the internal organs and, via the branchial ganglia, the ctenidia; one pair supplies the stomach, and one pair the mantle. A series of dorsal lobes of the brain are the seat of memory and learning capabilities as well as tactile and visual discrimination; together these make the cephalopods exceptional among invertebrates.

Sepia has a well developed system of giant nerve fibres which permit rapid synchronous contraction of the mantle muscles for fast movement and escape.

The sense organs of all cephalopods are well developed. The eyes show striking functional similarities with those of vertebrates although their construction is very different.

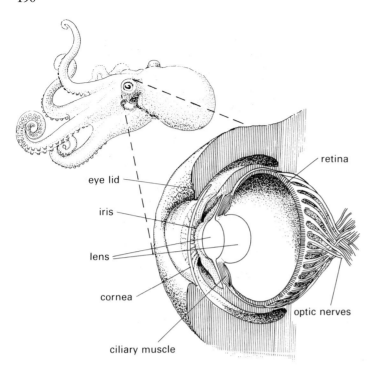

Structure of a cephalopod eye

They lie within sockets of a cartilaginous material associated with that which surrounds the brain. Structures which function as a lens, retina, iris, cornea and eyelid are all present. The lens is a rigid sphere and is suspended by a ciliary muscle, so that it can be moved forwards and backwards and therefore focused. An iris diaphragm in front of the lens, controls the amount of light admitted to the eye. The retina is formed of a layer of rod-like photosensitive cells, arranged so that they are directed towards the light source; these are connected to retinal cells which send fibres to the optic ganglia at the other end of the optic nerves. In *Sepia* one area of the retina is particularly sensitive, with 100,000 retinal cells per square millimetre concentrated in this region.

Sepia has well developed statocysts close to the ventral surface of the brain. There are olfactory pits on the side of the head close to the mantle cavity opening, in addition tacto- and chemoreceptors are scattered over the body surface and are particularly concentrated on the tentacles and suckers.

Reproduction and growth

The sexes are separate, a single gonad being located at the posterior end of the body. The gonads are saccular and the cavity is believed to be coelomic in derivation. In the male a long coiled vas deferens transports the sperm to a seminal vesicle where spermatophores are formed. These are stored in a sac which opens into the mantle cavity. The ovary produces large yolked eggs which pass down the oviduct. Fertilization is internal, and preceded by a courtship in which the male develops a striped display pattern and swims about above the female. One of the short tentacles of the male, the heterocotylus, is specialized as an intromittent organ with several rows of smaller suckers. The male seizes the female head on, picks up a spermatophore with his heterocotylus and deposits it in a fold below the female's mouth. Here the spermatophore bursts, and the sperm swim into the mantle cavity, enter the oviduct and fertilize the eggs which are then surrounded by protective membranes in the oviducal gland. The fertilized eggs are blown out of the funnel singly and attached by a stalk to seaweed or corals. Development is direct, and miniature *Sepia* hatch. At certain times of the year large numbers of *Sepia* move inshore to reproduce and many then die.

Cephalopod diversity

All living cephalopods belong to one of two subclasses, the subclass Coleoidea which includes cuttlefish, squids and octopodes and the Nautiloidea. These are mainly extinct and known only as fossils, except for *Nautilus* which is the only cephalopod with an external shell. Indeed *Nautilus* may be regarded as a relict of a fairly primitive level of cephalopod organization. It is now a rare animal found only in the Pacific and is highly valued by shell collectors. It has a pearly shell, composed entirely of nacre, which is coiled over its head in a flat spiral. Inside, the shell is divided by septae into a number of chambers which represent stages of growth, the animal living in the largest and most recent one. The shell functions (as in *Sepia*) as a buoyancy organ and the amount of buoyancy can be controlled. It is a thick structure and *Nautilus* can therefore survive at considerable depths.

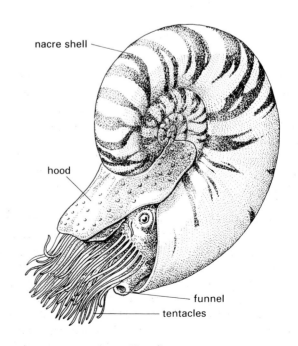

Nautilus

Nautilus differs from other cephalopods in many ways apart from its shell; for example it has no suckers, numerous slender tentacles, four gills, four auricles to the heart and four kidneys. The sense organs, brain and locomotor organs are all comparatively simple. Like *Sepia*, *Nautilus* swims by jet propulsion. It is a bottom-living scavenger and preys on small crustaceans.

Squid (for example *Loligo*) are long-bodied and specialized for rapid movement. The body is shaped rather like a torpedo with fins restricted to a diamond structure at the posterior end which functions as a stabilizer and hydrofoil. The shell has no calcareous matter and consists only of a horny 'pen'. In other respects the body has become more rigid with the development of cartilage-like plates in the mantle and neck region in addition to those which surround the brain.

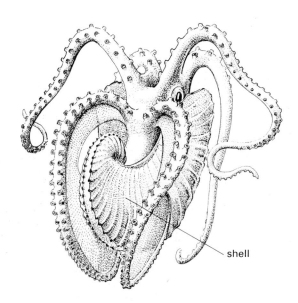

Argonauta: female

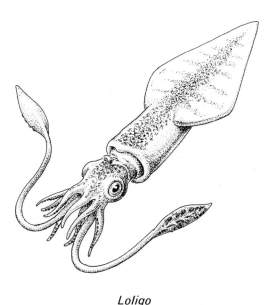

Loligo

Most squids live in mid-water, but there are deep-water dwellers (for example the giant *Architeuthis*, which can reach a length of about 6 m when fully extended) as well as pelagic forms. In some of the smaller pelagic squids the coelom is as much as two thirds of the total body volume and functions as a buoyancy chamber. Squids are capable of the fastest swimming of any invertebrates, and some species which live near the surface may even shoot out of the water and glide for short distances.

Octopodes have abandoned the mid-water in favour of a secondary bottom-dwelling mode of life. They are generally rather slow moving and often sedentary although they can swim by jet propulsion. They mostly clamber around over the rocks using eight greatly elongated arms which have well developed suckers. They are territorial and live in dens which they defend and from which they make feeding excursions. They may also lurk in their dens and pounce on passing crustaceans, snails or fish. Once prey is caught it is

bitten and paralysed, and then filled with digestive enzymes. The partly-digested contents are then sucked up by the octopus. *Argonauta* (the paper nautilus) shows considerable sexual dimorphism: the female is partly enclosed by a strange horn-shaped paper-like shell which she clasps with two specialized arms, while the male is much smaller and has no shell.

Some of the deep-water octopodes and also some deep-water squids have developed large webs between their tentacles which they use as nets to catch detritus and similar particles.

Synopsis of class Cephalopoda

Subclass Nautiloidea (*Nautilus*)
Subclass Ammonoidea (extinct ammonites with complex shells and sutures)
Subclass Coleoidea
    Order Decapoda: ten-armed forms (squids and cuttlefish such as *Sepia*, *Loligo*, *Architeuthis*)
    Order Octopoda: eight-armed forms (octopodes such as *Octopus*, *Argonauta*)

Early embryonic development: the protostomes and deuterostomes

A consequence of sexual reproduction in Metazoa is that a multicellular adult has to develop from a single cell. This is an extremely complicated and hazardous process involving repeated cell divisions and cell movement to permit organ development and growth. Embryonic development, that is the development of the individual within the fertilized egg, may be a short or lengthy process. A short embryonic stage, which is more usual in invertebrates, results in a very small juvenile unable at first to follow the adult mode of life. A larval period forms part of the development process, the larvae frequently looking totally unlike their parents and following a different mode of life. Alternatively the period of embryonic development may be prolonged, in which case the embryo has to be provided with a source of food (commonly yolk) and hatches as a miniature replica of the adult, adopting the same mode of life.

In the embryonic development of all metazoans the earliest stage is a process of repeated divisions of the zygote, known as cleavage. The egg divides repeatedly, to produce successive doublings in number (2-cell stage, 4-cell stage, 8-cell stage etc.) without any increase in size.

Two different patterns of cleavage are seen in the Metazoa. The simplest is radial cleavage, which begins with two vertical divisions, at right angles to each other. This is followed by a horizontal division, the resulting cells lying directly above each other. The process continues in a simple pattern of alternating vertical and horizontal divisions, so that the cells produced always lie directly above or beside each other. Spiral cleavage is rather more complex. As in the radial cleavage egg the first two divisions (4-cell stage) take place vertically and at right angles to each other. However in the third (8-cell stage) the top four cells come to rest obliquely on the lower four, and at an angle to the longitudinal axis of the fertilized egg. Subsequent divisions are also orientated obliquely, the inclination from the vertical being alternately to one side of the axis or the other.

Whatever the pattern of cleavage, the cells tend to round up after division and to remain in contact with each other. Eventually the number of cells becomes too great for all round contact and the egg forms a multicellular hollow ball of cells, one cell thick. This stage is the blastula, the cavity being known as a blastocoele, and the individual cells as blastomeres.

In certain metazoans (for example platyhelminths, annelids and molluscs) the ultimate destination in the adult of the blastomeres is fixed at a very early stage. If the embryo is taken at the four-cell stage and the blastomeres separated, each will develop into a fixed quarter of the later

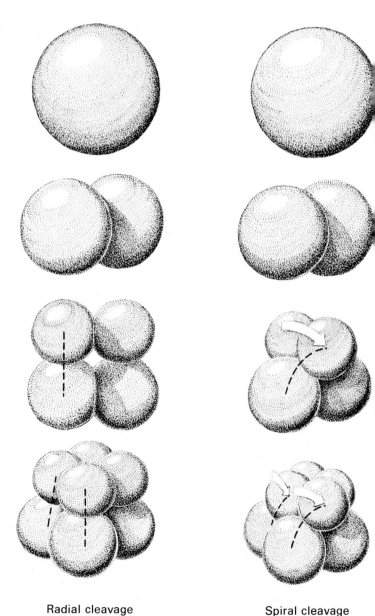

Radial cleavage   Spiral cleavage

embryonic and larval stages. This condition is referred to as determinate cleavage, contrasting with indeterminate cleavage in which the ultimate fate of the blastomeres is established at a later stage. A form that shows indeterminate cleavage, for example a sea urchin, can be separated at the four-cell stage and each blastomere will give rise to a complete larva.

The next stage of embryonic development is one of cell movement (gastrulation), during which the developing embryo is changed from a single layered blastula into a two-layered gastrula. The cells of one half of the embryo come to lie enclosed within those of the other half. In the simplest type of development gastrulation occurs by a process of invagination not unlike the effect gained by sticking one's thumbs into a hollow, flexible ball. Normally during gastrulation the blastocoele is eliminated, but in one type of animal organization seen in the pseudoceolomates, it persists as the adult body cavity, the pseudocoele. The new cavity created during this process of invagination is called the archenteron, and the region of invagination, the blastopore, may be modified in one of two ways. In some metazoans (for example platyhelminths, annelids and molluscs) it becomes the mouth, while in others (for example echinoderms and chordates) it becomes the anus, and the mouth is formed by a new penetration. In either case the new body cavity, the archenteron, ultimately forms the gut cavity.

The process of gastrulation may be modified in various ways, depending on the presence or absence of yolk. Yolk is a biologically inert substance, consisting of proteins and fats used to feed the developing embryo. Generally the more yolk that an animal has, the more it approaches the adult form on hatching. The simplest form of gastrulation of the kind just outlined is found in animals with very little yolk in which cleavage occurs equally throughout the egg, giving a blastula formed of cells of an almost equal size. This gives total equal cleavage. When yolk is present it tends to concentrate in one part of the egg, where it impedes the process of cell division; as a result that region of the blastula is formed from fewer and larger yolky cells giving total unequal cleavage. A large yolk makes invagination difficult and the gastrula is formed by a process of invagination with rapid growth of small cells around the yolky cells termed epiboly. In animals which have an enormous amount of yolk, total cleavage is impossible; instead cleavage is confined to a small area of living protoplasm at the surface of the egg and the embryo develops as a flattened blastodisc floating on the yolk mass.

Another important aspect of the embryology of metazoans is the manner in which the coelomic mesoderm (an embryonic tissue from which structures such as the muscles, gonads and excretory organs develop) arises. In forms like annelids and molluscs all the mesoderm arises from a single blastomere which proliferates to form a mass of cells. In metameric forms such as the annelids this develops as a linear series of cell blocks, in which a split appears which enlarges to form the coelom (schizocoely). A

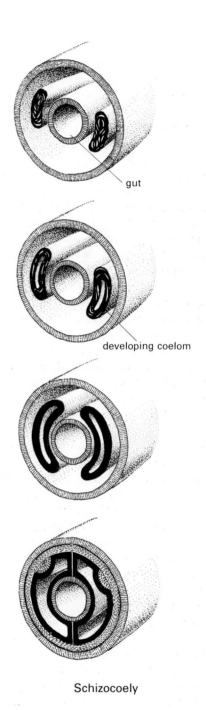

gut

developing coelom

Schizocoely

different method of coelom and mesoderm formation is seen in echinoderms and primitive chordates; in these the mesoderm and coelom arise as a series of pouches from the archenteron, which later round up and separate from it (enterocoely).

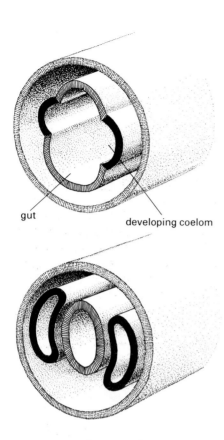

gut

developing coelom

Enterocoely

These aspects of development have been used by many zoologists as a basis for recognizing two major taxonomic groups, Protostomia and Deuterostomia. Protostomia have a mouth formed from the blastopore ('protostome' means 'first mouth'), determinate and usually spiral cleavage, and schizocoely. Deuterostomia ('deuterostome' means 'second mouth') have an anus developed from the blastopore, radial indeterminate cleavage, and in primitive chordates, but not vertebrates, enterocoely. The amount of yolk present is highly variable within each group and depends on ecological factors. Major protostome groups are the molluscs, annelids and arthropods; major deuterostomes are the echinoderms and chordates. Although this classification is widely used, many zoologists argue that the use of such supertaxa is of doubtful value, and is possibly misleading.

The minor coelomate phyla

Introduction to the minor coelomate phyla

Three major protostome phyla, Annelida, Arthropoda, and Mollusca, have already been discussed. However not all the invertebrate adaptive lines resulting from the development of a coelom are represented by these groups. There are many minor phyla, which, whilst they may include a few extremely common species, are generally not abundant in terms of either species or individual number. The majority have neither economic importance nor ecological significance and the study of these animals has been largely neglected. However they do represent evolutionary divergences at the annelid-arthropod level of organization and as such are of interest to students of animal diversity. In ecological terms these minor phyla are not successful, and the animals tend to be found in the poorer and less accessible habitats. They may be conveniently considered under the following headings: the minor protostome coelomates, the lophophorate phyla, and the minor deuterostomes. The minor protostome coelomates include six phyla: Sipunculoidea, Echiuroidea, Priapuloidea, Tardigrada, Pentastomida, and Pogonophora. The lophophorates are a group of four phyla (Phoronida, Brachiopoda, Entoprocta, and Bryozoa) which all share the same type of food collecting organ, the lophophore. The minor deuterostomes will be considered later, after Echinodermata, the major phylum of invertebrate deuterostomes.

THE MINOR PROTOSTOME COELOMATES

The inclusion of six phyla (Sipunculoidea, Echiuroidea, Priapuloidea, Pentastomida, Tardigrada and Pogonophora) under a single heading is entirely one of convenience, since none of the phyla shows a close relationship with any of the others. The phylogenetic relationships of the priapulids, tardigrades and pogonophorans are particularly uncertain, indeed it has been suggested that both priapuloids and tardigrades are really pseudocoelomates. In both cases, the true nature of the body cavity is uncertain.

Four of the phyla, Priapuloidea, Echiuroidea, Sipunculoidea and Pogonophora, are elongated and worm-like in body form. In contrast the tardigrades are short and rounded and bear a superficial resemblance to the mites. The pentastomids are all parasitic, and live in the lungs of vertebrates, especially reptiles. Some indication of probable relationships will be given at the end of each section.

Phylum Priapuloidea

Priapuloids owe their name to a fanciful resemblance seen between *Priapulus* and the human penis. Nine species are known and all are marine burrowing animals which live in sand or mud in cool or cold waters. They may be found buried on the lower shore and also occur in deeper waters, down to 7,200 m. *Priapulus caudatus* is one of the best known priapuloid species, about 80 mm in length and widely distributed on cold or cool water coasts from the lower shore downwards. Priapuloids range in size from *Priapulus*, which is up to 200 mm long, to *Tubiluchus*, which is microscopic with a length of 0.5 mm.

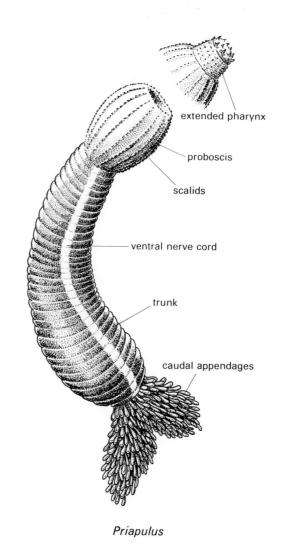

extended pharynx

proboscis

scalids

ventral nerve cord

trunk

caudal appendages

Priapulus

Body form

Priapuloids are fleshy, gherkin-shaped animals, with an anterior retractable proboscis (or prosoma) which is barrel-shaped and covered with spines or scalids arranged in longitudinal rows. The trunk is segmented externally (but not internally) with 30–40 rings and may be covered with spines or warts. In some species, for example *Priapulus caudatus*, the trunk bears paired tassel-like appendages which are thought to function either as respiratory organs or chemoreceptors.

Body wall and body cavity

The priapuloid body wall has an outer cuticle composed partly of chitin, which is moulted at intervals. There is a single-layered epidermis and layers of circular and longitudinal muscles.

The innermost layer of the body wall is a cellular membrane which forms the mesenteries supporting the internal organs. The existence of such a structure strongly suggests that the priapuloid body cavity is a true coelom (and it is so regarded here) but this interpretation is questioned by some zoologists. The coelom is filled with a fluid which includes amoebocytes.

Locomotion

Priapuloids burrow through the mud by means of muscular contractions of the body wall, with the coelom functioning as a hydrostatic skeleton. During this process the proboscis is alternately withdrawn and protracted. *Priapulus* may also lie at rest, buried in mud, and with its mouth, which is at the tip of the proboscis, level with the surface.

Feeding

Priapulus is a carnivore feeding on slow-moving, soft-bodied invertebrates such as polychaetes. The prey is captured, using the pharynx, which can be everted, and swallowed whole. The pharynx has a lining of cuticle, with thickenings in the form of small teeth which are used to seize prey. The alimentary canal is simple, consisting of a straight intestine leading to a short cuticle-lined rectum and an anus at the posterior end of the trunk.

Gaseous exchange and circulation

With the possible exception of the tassel-like organs, *Priapulus* has no specialized organ for gaseous exchange. In some priapuloids, such as *P. caudatus*, the fluid-filled coelom provides a kind of circulatory system. *P. caudatus* also has corpuscles containing a respiratory pigment, haemerythrin.

Nervous system and sense organs

The nervous system is closely associated with the epidermis. A nerve ring surrounds the anterior end of the pharynx and gives rise to a ganglionated mid-central nerve cord from which peripheral nerves and commissures arise. The proboscis and trunk bear numerous small papillae which may be sensory in function.

Excretion and reproduction

Priapulus has a protonephridial excretory system which is closely associated with the gonads, thus forming a urinogenital system. On each side of the intestine is an elongated urinogenital duct from which arises a mass of solenocytes on one side, and a single gonad (formed from a mass of tubules) on the other. The duct opens as a pore at the posterior end of the trunk. The sexes are separate and fertilization is external. The egg undergoes radial cleavage and develops to give a priapulid larva which also lives in the mud, surrounded by a cuticular case, the lorica, into which it can withdraw. It grows through a series of larval moults and finally takes up the adult form and habit. Like so much of the biology of this group, the details of development, and particularly of the body cavity formation, are not known.

Relationships

The precise phylogenetic status of the priapuloids is open to question. Their moulting, and the structure of the proboscis, could relate them to the aschelminths. However if the body cavity is interpreted as a coelom, a question which will only be answered when we have a full knowledge of their embryology, their relationship with other coelomates is obscure. It is interesting that haemerythrin, a rare respiratory pigment, is also found in the sipunculoid worms.

Phylum Sipunculoidea

The sipunculoids are a widespread group of about 350 species of marine animals of rather sedentary habits. The name means 'little pipe', and presumably refers to their rather elongate shape. Some, for example *Sipunculus*, are active burrowers, while others bore their way into coral reefs. Others, for example *Phascoloin*, live in abandoned annelid tubes or gastropod shells. Sipunculoids are generally of moderate size (*Sipunculus nudus* may be as long as 200 mm), but some attain lengths of up to 720 mm and others are as small as 2 mm. One tiny sipunculoid has been found in discarded foraminiferan shells.

Body form

Sipunculoids are cylindrical, with no external segmentation. They have a short protrusible proboscis, which ends in an oral disc of small frilly tentacles (*S. nudus* has four) surrounding the mouth. This proboscis is often studded with spines or papillae. They have grid-like markings on the trunk, due to the longitudinal and circular muscles which are visible through the epidermis.

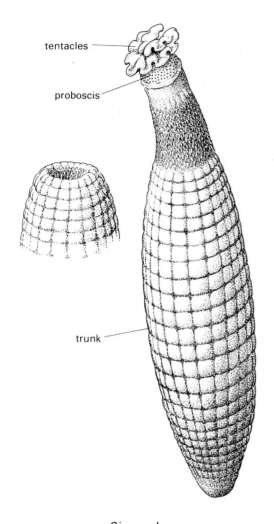

tentacles

proboscis

trunk

Sipunculus

Body wall and coelom

The sipunculoid body wall is similar to that of the annelids, with layers of cuticle, epidermis, dermis and circular and longitudinal muscle. The innermost layer of the body wall is peritoneum, and a true undivided coelom is present.

Locomotion

Sipunculus burrows through mud or sand. With the trunk held fixed, the proboscis is pushed forwards, the coelomic fluid acting as a hydrostatic skeleton. The proboscis is then retracted and the body pulled up into the space which has been created.

Feeding

Sipunculus ingests the sand and mud through which it burrows, digesting associated organic detritus. Other sipunculoids may capture small organisms, such as protozoans and larvae, by means of their tentacles. The proboscis can be easily and rapidly everted and the oral tentacles have a separated hydraulic system of tentacular canals associated with a ring canal and compensation sacs. Fluid is forced into the tentacles when the muscular walls of the compensation sacs contract. Contraction of the tentacular musculature forces the fluid back into the sacs. The mouth opens into a muscular pharynx, from which extends a long, coiled U-shaped intestine. Food is moved along the intestine by a conspicuous ciliated groove. The intestine terminates at a short rectum which opens at a mid-dorsal anus at the anterior end of the trunk.

Gaseous exchange and circulation

The coelomic fluid is involved in both circulation and gaseous exchange and contains amoebocytes as well as corpuscles containing haemerythrin, a respiratory pigment. Since this pigment has also been found in the fluid of the tentacle canals, the tentacles are probably an important site of gaseous exchange.

Excretion

Sipunculoids have a pair of large sac-like nephridia which open through a pair of nephridiopores beside the anus and discharge nitrogenous waste in the form of ammonia. In addition, many species have some extremely unusual excretory structures, the coelomic urns. These are vase-shaped clusters of peritoneal cells topped by a ciliated cell. These clusters collect waste particles from the coelomic fluid. In due course they become detached and, together with a trailing mass of waste particles, may be excreted through the nephridia or deposited on the intestinal peritoneum in the form of chloragogenous tissue, comparable with that of annelids.

Nervous system and sense organs

The nerve cord is similar to that of annelids, with an anterior ring including cerebral ganglia and a single ventral nerve cord. However the nerve cord is not ganglionated. The sensory cells are particularly concentrated on the proboscis, and some genera have a pair of nuchal organs which are thought to be chemoreceptors. Pigment cup ocelli may be present, embedded in the brain.

Reproduction and growth

The sexes are separate. The gonads arise from the anterior peritoneum, and the gametes are shed to mature in the coelom. Fertilization is external. Development is similar to that of the annelids, leading to a trochophore larva. In *Sipunculus* the larva is somewhat elongated and has a free-swimming period of about a month before taking up a benthic existence.

Relationships

In spite of a lack of any evidence of metamerism, the sipunculoids are generally regarded as related to the annelids. Their trochophore larva provides strong support for this affinity.

Phylum Echiuroidea

The echiuroids are a group of 100 species of marine worms which show some similarities to the sipunculoids. Like them, they are sausage-shaped animals, about 100 to 150 mm in length, living in sand, mud, or in crevices. However they are readily distinguished by their long non-retractable proboscis with which they feed on small animals or organic deposits. Several genera including *Bonellia* and *Echiurus* have males which are parasitic on the females. *Echiurus* lives in a U-shaped burrow in the sand and *Bonellia* inhabits holes in rocks. One tiny form, *Thelasemma melitta*, lives in sand dollar (echinoderm) tests which it enters when it is very small. Generally echiuroids are sedentary, remaining in their holes or burrows and extending the proboscis to feed.

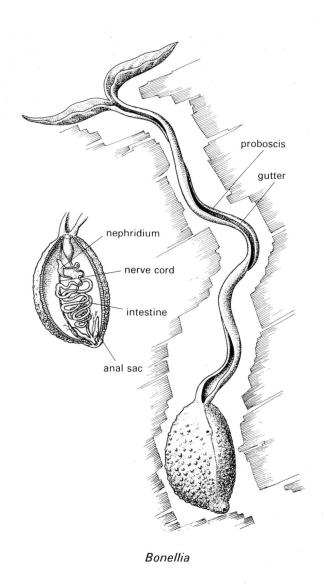

Bonellia

Body form

The proboscis is the most conspicuous feature of the body. It is an extension of the head and is probably homologous with the annelid prostomium, its mouth opening at its base.

It is a flattened structure, but the edges are rolled ventrally giving it an inverted gutter shape. Sometimes, as in *Bonellia*, the tip is forked. The trunk may have well-developed chaetae, which in *Echiurus* are arranged as an anterior pair and several circlets. The proboscis is a highly glandular structure, and the ventral gutter is ciliated. Otherwise the body wall and the structure of the coelom are similar to those of sipunculoids.

Feeding

Small organisms and particles of organic detritus stick to the mucus-covered surface of the proboscis and are transported along the ciliated gutter to the mouth. The proboscis can be extended or contracted and is used to explore and exploit the substrate around the animal's retreat. The alimentary canal is long and coiled and food passes along it inside a distinct ciliated groove. The first part of the intestine is presumed to be the region of enzyme secretion. Running parallel to this is a separate narrow tube, the siphon, which may serve as a bypass for excess water taken in with the food.

Respiration and circulation

There are no specialized organs of gaseous exchange. The coelomic fluid contains amoebocytes, and also haemocytes which contain haemoglobin and may be used for oxygen storage when the tide is out. A closed blood vascular system, comparable with that of the annelids, is present.

Excretion

Nephridia with a structure similar to those of the sipunculoids, are present, and open near the anterior end of the animal. In *Bonellia* one nephridium is present, and in *Echiurus* there are two. In addition, a pair of simple or branched structures, the anal sacs, arise from the rectum. These have numerous ciliated funnels similar to nephrostomes and probably function in the same way as nephridia, the waste products leaving via the anus.

Nervous system and sense organs

The sedentary habit of the echiuroids is reflected by their lack of special sense organs. The nervous system is like that of the sipunculoids.

Reproduction and growth

The sexes are separate. Gametes arise from the peritoneum and mature in the coelom. These are normally shed through the nephridia and fertilization is external. In forms such as *Bonellia*, where the male lives as a parasite in the female nephridium, the eggs are fertilized in the nephridia and pass through the early stages of development there. Echiuroids undergo spiral cleavage and develop a free-swimming trochophore larva similar to that of annelids.

Relationships

Like the sipunculoids, the echiuroids are regarded as being related to the annelids.

Phylum Pogonophora

The pogonophorans are a group of deep water animals, of widespread distribution in marine waters below a depth of about 100 m. They are a fairly recent scientific discovery, the first specimen having been found in 1900. They are thread-like animals, up to 350 mm long but usually no more than 1.0 mm in diameter. They generally have a closely packed tuft of tentacles with cilia and pinnules at their anterior end and the name Pogonophora (meaning beard-bearers) refers to these structures. Some species have only a single tentacle.

The worms are sessile and live in stiff chitinous tubes which hold them upright in the mud, or in decaying wood or debris. They are highly unusual amongst free-living metazoans in their total lack of a mouth or a digestive tract; in fact early workers thought the specimens were incomplete! The feeding process in pogonophorans is not understood, though they are thought to collect detritus in their tentacles and to digest it extracellularly, absorbing the products of digestion through the body wall. The tentacles are also believed to function as the region of gaseous exchange. The main part of the body is a long trunk with two girdles of chaetae in the mid-region; a short segmented opisthosoma also bears chaetae and probably functions as an anchor since it extends below the tube into the mud. There is a coelom in each of the main body divisions.

The female has paired cylindrical ovaries and paired oviducts. The male has paired cylindrical testes and a median sperm duct in which sperm are packaged into spermatophores. The process of spermatophore transfer is not understood; possibly they are moved to the mouth of the tube by the tentacles and float away to be collected by the female. The process of development is also unknown though it has been shown that some pogonophorans brood their eggs in their tubes. The young larvae develop ciliated girdles, however it is not certain whether young pogonophorans disperse or sink directly to the bottom.

Relationships

Prior to the discovery of the opisthosoma (which was broken off in the first specimens collected) pogonophorans were regarded as deuterostomes related to the hemichordates. Nowadays some zoologists regard them as intermediate between protostomes and deuterostomes, but others relate them to annelids on the basis of the opisthosoma.

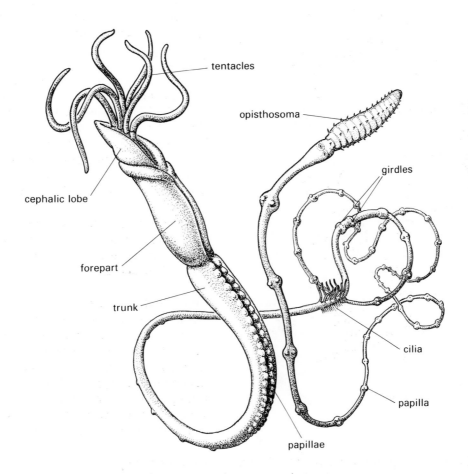

A representative pogonophoran

Phylum Tardigrada

The tardigrades, commonly known as the water bears from a somewhat bear-like appearance (!), are a group of 400 species of tiny animals (generally less than 0.5 mm) found in a variety of specialized habitats. They are cosmopolitan in distribution and not uncommon, but because of their minute size they have to be looked for. Some are marine, living amongst sand particles, and some live in freshwater amongst the bottom detritus or on water plants. The majority are semi-aquatic, living in the water film in lichens, mosses, liverworts, or in the soil, a habitat they share with some of the rotifers. Like the rotifers they have a remarkable ability to withstand desiccation and low temperatures, as well as other stringent environmental conditions.

Body form
Tardigrades have short cylindrical bodies, with four pairs of stubby legs each ending in a bunch of four to eight claws. The head is not clearly distinct from the trunk.

Body wall and coelom
The outermost layer of the body wall is a cuticle formed from chitin and polysaccharide which may, as in *Echiniscus*, be divided into regularly arranged plates. The structure is similar to that of gastrotrichs. The cuticle is periodically moulted and the epidermis secretes a new one. The musculature consists of muscle bands, each derived from a single cell and extending between sub-cuticular attachment points, and may be compared with that of arthropods in this respect. Like the rotifers the tardigrades show some regularity of cell number, probably related to their small size. The main cavity is a haemocoele, and the coelom is confined to the gonadial cavities.

Locomotion
Tardigrades use their legs to crawl clumsily and slowly about, grasping the substrate by means of the terminal hooks. The name, Tardigrada (from the Latin meaning 'slow to step') refers to their ungainly gait.

Feeding
Plant-dwelling tardigrades feed on plant cell contents which they obtain by piercing the plant with a pair of sharp stylets similar to those of plant-dwelling rotifers and nematodes. The mouth opens into a cuticularized buccal cavity, and food is sucked up by a pumping action of the heavily muscular pharynx. Digestion and absorption take place in the intestine which opens via a short rectum at a terminal anus. In *Echiniscus* defaecation is associated with moulting, the faeces being left behind with the cuticle.

Other tardigrades feed on detritus or are predators of small organisms such as soil nematodes.

Respiration and circulation
Like other very small animals, tardigrades have no respiratory or circulatory organs.

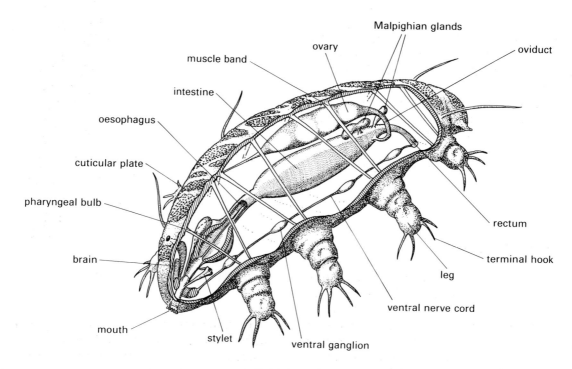

Echiniscus

Excretion

Three large glands, thought to be excretory organs and sometimes called Malpighian tubules, are present at the junction between the intestine and the rectum. Nitrogenous waste is also deposited in the cuticle and shed at moulting.

Nervous system and sense organs

The nervous system is comparable with that of annelids and arthropods. A dorsal brain is present, composed of three median lobes and a pair of lateral lobes. This is attached by connectives to a metameric ventral double nerve cord and four segmental ganglia. Several lateral nerves extend from the ventral ganglia to each leg, one of which terminates in a pedal ganglion in each leg. The body and head surfaces bear numerous sensory bristles and spines, and simple eye spots are also present.

Reproduction

The sexes are separate and a simple ovary or testis lies in the mid-line above the intestine. The male has two sperm ducts which pass to a single median gonopore just in front of the anus, the female has a single oviduct which opens either via the rectum or at a female gonopore. Mating normally occurs at the time of a moult; the sperm may be deposited in a discarded female cuticle which contains eggs or may be deposited in the oviduct or in a small seminal receptacle in the female in which case fertilization is internal. As in rotifers, some species produce thick- or thin-shelled eggs in response to adverse or favourable conditions; terrestrial tardigrades always lay thick shelled eggs. Also as in rotifers, parthenogenesis is common. Development is direct and rapid, but the embryonic stages are not fully understood.

Relationships

The phylogenetic position of tardigrades is particularly difficult to understand. Studies of embryonic development are needed to determine the nature of the body cavity. Tardigrades show points of similarity with both the gastrotrichs and rotifers, but these may in part be related to their small size and specialized habitats. Some zoologists believe that the tardigrades will prove to be related to aschelminths. An alternative interpretation relates them to the mites, though the similarities are probably superficial.

Phylum Pentastomida

The Pentastomida (this means 'five-mouthed') are a group of about 90 species of parasitic, blood-sucking animals (commonly called tongueworms) which live in the respiratory organs of reptiles, and also in mammals and birds. *Linguatula* is a parasite of dogs. They have an elongated worm-like body (20 to 130 mm long) with five protuberances at the anterior end. The name of the phylum is derived from these structures, though in fact four of them are legs bearing claws and are used for attachment to the host, and only the fifth, median, structure bears a mouth.

Pentastomids have a chitinous cuticle which is shed at intervals, longitudinal and circular muscles, and a haemocoele. They feed on blood from the host which they pump into a simple alimentary canal. They have no circulatory, respiratory or excretory organs.

The reproductive system is well developed, as in many parasites. The sexes are separate and fertilization is internal. Fertilized eggs are passed into the digestive tract of the host, and leave with the faeces. The life cycle normally requires an intermediate host (a rabbit in the case of *Linguatula*) where the larval development takes place through a series of moults. The larva has four or six legs, and if the intermediate host is eaten, the larva climbs up the oesophagus from the stomach to reach the lungs and nasal passages. Since the intermediate host has to be the prey of the primary host, common intermediate hosts of pentastomids include fish or small mammals.

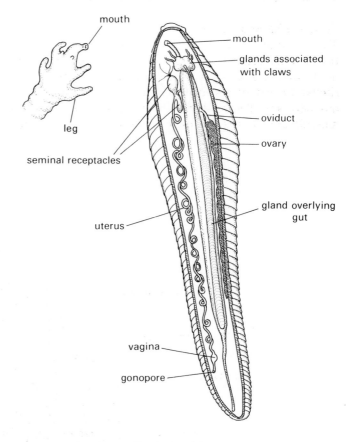

Pentastomida

Pentastomids are regarded as being derived from the arthropods and various arthropod groups have been suggested. The most likely hypothesis suggests that they originated from a group of crustacean fish parasites, the Brachyura.

THE LOPHOPHORATE PHYLA

Introduction

The lophophorate phyla, Bryozoa, Brachiopoda and Phoronida, are three relatively minor groups of coelomates which have sufficient features in common to be regarded as a natural group, sometimes termed the superphylum Lophophorata. They are sessile animals, and probably their principal adaptive feature, a tentaculate food collecting organ called the lophophore, is related to this habit. The body is divided into the following three sections: a small anterior epistome which overhangs the mouth dorsally and is surrounded by the lophophore; a ring-shaped middle section with a coelomic compartment, the mesocoele, which surrounds the anterior region of the gut and the nerve ganglia, and which bears the lophophore; and finally the main part of the body with a metacoele surrounding the viscera. The gut is characteristically U-shaped and the anus opens outside the lophophore. A U-shaped gut, bringing the anus to the upper surface, is a common adaptation in sessile animals.

Phylum Bryozoa (also known as Ectoprocta or Polyzoa)

Phylum Bryozoa (the sea mosses), with about 4,000 species, is the largest of the lophophorate phyla. Bryozoans are widely distributed and are among the commonest animals on stony and rocky shores. For example, colonies of hornwrack (Flustra) looking like bleached seaweed are very commonly found washed up on the strand line. They are also economically important because some genera (including Bugula) foul underwater structures such as ship hulls. A few bryozoans have invaded freshwater, including one class (Phylactolaemata) which are restricted to this environment. Although there are only fifty or so species in the class, the group has large numbers of individuals and a wide distribution. In general, bryozoans have been neglected by zoologists because of their small size (individual animals are microscopic and the colonies normally less than 10 mm in height) their typically neutral colours, and their sessile colonial habit, and superficially plant-like appearance (the name, derived from the Greek, means 'moss animals').

Bryozoans occur in colonies which are derived by asexual budding from one individual. The individual animals or zooids grow within non-living cases of cuticle (zooecia) with chitin and/or calcium carbonate which are secreted by the epidermis. The zooid body wall is thin and has no muscle layer, and pores in the zooecium walls allow some contact between the living tissue of adjacent zooids. An opening in the zooecium, the orifice, allows the lophophore to be protruded.

Characteristically the colony shows polymorphism, individual zooids being specialized to fill one of a variety of roles such as feeding, defence, creation of water currents, attachment or reproduction. The mature stage is invariably sessile and is achieved by a free-swimming larva. Many interesting comparisons can be made between the Bryozoa and the colonial hydroids discussed earlier though these do not reflect any close relationship. Instead similarities of adaptation can be traced to the small size of the animals, their sessile colonial organization, and their outer skeletal cases.

The phylum Bryozoa is divided into three classes: Stenolaemata, which includes the majority of fossil bryozoans and a few marine species; Gymnolaemata, which is predominantly marine and includes the majority of living species; and Phylactolaemata the freshwater species.

The group will be introduced by a brief description of Bugula, a gymnolaemate with a world-wide distribution.

Bugula

Bugula is common in shallow water up to the low tide mark, where it grows attached to rocks, seaweed or wooden piles, or even to other animals such as molluscs or sponges. Some species, such as B. neretina, invade deeper water.

Body form

The colony of Bugula is an erect, dichotomously branching structure a few centimetres long and composed of thousands of individual zooids. The zooecium is box-like and consists of cuticle with a layer of lightly calcified chitin. The sides of the box are thin, but with thickened edges, and with a variable number of spines (depending on species) on the upper edge. At the upper surface of the zooecium an opening, the orifice, allows the lophophore to be protruded. This has a kind of lid, the operculum, which can be closed when the zooid withdraws.

Zooids communicate with adjacent individuals through pores in the end walls of the zooecia. The majority are feeding structures, or autozooids, with a lophophore and gut. However on the sides of the colony there are avicularia (a name derived from their superficial resemblance to a bird's beak) which are highly adapted for defence and lack most of the normal body structures of zooids. Avicularia have an operculum modified as a moveable jaw, and may be sessile or have flexible stalks. They defend the colony, both from the settling of the larvae of other sessile organisms and from larger animals such as polychaetes which may crawl over and damage it. Thus their function is rather like that of the pedicellariae of some echinoderms. Vibracula are another kind of zooid modified for paddling and sweeping; the operculum is modified as a long bristle and when moved this assists in the creation of a water current around the entire colony which sweeps debris or settling larvae off the surface. (These structures are absent in Bugula). Ooecia are brood chambers for developing larvae. Each consists of a

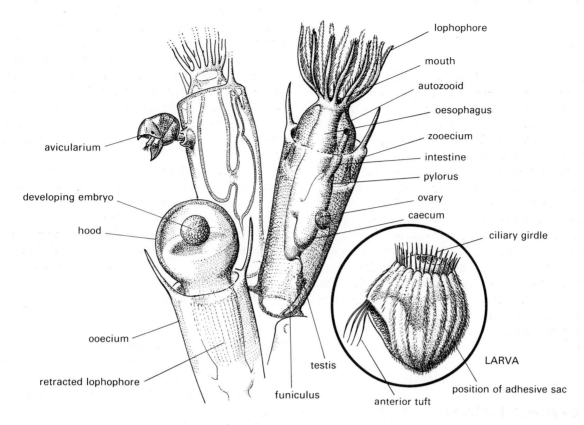

Anatomy of *Bugula*

pouch, formed by a large evagination of the zooid body wall producing a hood and a smaller evagination (into which the coelom extends) forming a base. The egg develops in the resultant space. Finally, some zooids are specialized for attachment, forming a stolon or rooting filaments.

Feeding
Bugula is a suspension feeder, and feeds on plankton which it captures by means of the lophophore. This is a large and elaborate structure which can be expanded by hydrostatic pressure to form a kind of funnel. The mesocoele extends into the tentacles of the lophophore and the increase in coelomic pressure necessary to extend them is achieved by contraction of transverse muscles attached to the walls of the zooecium. Retractor muscles between the lophophore and zooecium pull it inwards when the animal is not feeding. In some cases the tips of the tentacles may be turned inwards to enclose the prey. The tentacles have lateral tracts of cilia which beat to create a feeding current down into the funnel and out between the tentacles. Food particles are bounced down to the base of the lophophore and into the mouth, which opens into a pharynx. Ciliary action then transports the food to the main part of the gut which is formed by a large stomach, separated from the pharynx by a valve, in which both intracellular and extracellular digestion occur. Most of the stomach volume is provided by a large

caecum of the central stomach which is the principal site of intracellular digestion and food storage. Movement of food within the stomach is by peristaltic contractions of the body wall. The posterior region of the stomach is a small pylorus in which undigested matter is compacted before passing into the rectum and anus. The faeces are carried away by the feeding current.

Gaseous exchange, circulation and excretion
Gaseous exchange takes place across the entire body surface, and more particularly across the lophophore. There is no specialized circulatory system and the coelomic fluid serves for local internal transport. Coelomocytes within the coelomic fluid engulf waste products but there are no specialized excretory organs. Because *Bugula* is a colony with specialized zooids, some of which do not feed, some mechanism for transporting food between individuals is necessary. This is provided by the funiculus, a cord of mesenchyme associated with the stomach wall which transports digested food intracellularly in the form of lipids. Plugs of cells in the pores of the base of the zooecium convey the lipids to the non-feeding parts of the colony.

Nervous system and sense organs
A single bilobed cerebral ganglion lies dorsally at the base of the lophophore adjacent to the pharynx; it supplies

nerves to the gut, the lophophore muscles, and to a nerve ring at the base of the lophophore which supplies the tentacles. In addition it innervates a sub-epidermal nerve net in the body wall. There are no specialized sense organs. There is some evidence for a colonial nervous system linking the zooids in bryozoans which presumably allows them to co-ordinate their activities and responses.

Reproduction

Like the great majority of bryozoans, *Bugula* is hermaphrodite. The gonads develop in summer or autumn in the autozooids. The ovary lies on the body wall and the testis on the funiculus and both eggs and sperm are shed into the coelom by rupture of the gonads. There are no special ducts for the sperm and they pass into the surrounding water through terminal pores of specialized tentacles. They are both swept away and collected by the feeding currents and may enter the coelom of another individual through a narrow opening, the coelomopore, in the region of the lophophore. Fertilization occurs in the coelom and the eggs, which are large and yolky, begin their development there. In due course they are squeezed through the coelomopore in an elongated stream and carried to the ooecia where they become round again.

The developing animal is fed first by the yolk, and then by a placenta-like connection to the maternal zooid. Eventually a motile non-feeding larval stage leaves the ooecium. This has a posterior adhesive sac, a girdle of cilia for locomotion, and an anterior tuft of long cilia. At the time of leaving the ooecium the larva is positively phototactic and therefore moves towards the light. Later it becomes negatively phototactic and settles on a substrate, everting its adhesive sac and forming an ancestrula. This buds off other zooids to give the characteristic branching colony.

Bryozoan diversity

Bryozoans show two principal areas of diversity: the shape of the lophophore and the type of larva. In Gymnolaemata and Stenolaemata the lophophore is circular, whereas that of the freshwater Phylactolaemata is horseshoe shaped. Three main types of larva occur, of which the non-feeding larva, characteristic of bryozoans which brood their young, has already been described. In contrast non-brooding species, such as *Electra*, produce a larva which has a fully functional digestive tract, and survives as a planktonic feeding organism for several months. Freshwater bryozoans, for instance *Pectinatella*, are common in quiet, slow-moving waters where they build large colonies on reeds or logs. These consist only of feeding zooids, their diet being chiefly rotifers, diatoms or protozoans. They reproduce sexually in spring and summer, brooding the developing embryo in an embryo sac in the coelom. When released, it is already a young ciliated colony which has budded off several zooids. Many freshwater species can overwinter as highly resistant statoblasts with large food

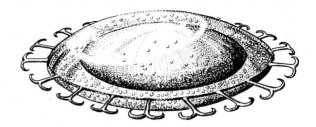

Pectinatella statoblast

reserves; these are formed asexually in late summer and in the autumn on the funiculus. In some species (for example *Pectinatella*) the statoblasts have an outer rim of air-filled horny cells and may float for a time at the surface thus promoting dispersal. Other kinds fall directly to the bottom.

Phylum Brachiopoda

The brachiopods or lampshells are an entirely marine group of lophophorates, which bear a superficial resemblance to bivalved molluscs since they have a bivalved shell made of horny material impregnated with calcium carbonate and secreted by a mantle. In fact these features led them to be regarded as bivalve molluscs until the mid-nineteenth century. However the brachiopod shell encloses the body dorsoventrally rather than laterally, and the ventral valve which is usually larger, is attached to the substrate. Because of their shell, brachiopods are well represented in the fossil record, and 30,000 fossil species have been described, extending back to the Cambrian. The 280 living species thus represent only a conservative remnant of a very old group which was formerly much more important.

In brachiopods the lophophore is usually modified as two long arms; the name of the phylum is derived from the Greek and means 'arm-feet', referring to these structures. The group is divided into two classes, Articulata, in which the valves are connected by a hinge as well as muscles, and Inarticulata, with an entirely muscular connection. *Lingula*, which has been selected for more detailed description, is a member of the Inarticulata.

Lingula

The genus *Lingula* is one of the oldest genera of living animals, and has persisted almost unchanged since the Ordovician. It occurs in shallow water in the tropics and subtropics and lives in vertical burrows in muddy sand.

Body form

Lingula has identical dorsal and ventral valves, which are secreted by the dorsal and ventral mantle lobes. They are blunt-tipped and greenish in colour, and formed from a mixture of chitin and calcium phosphate; the outermost layer of the shell is an organic periostracum. At the edges of the valves the mantle secretes a fringe of long chitinous setae thought to be both protective and sensory in function. The 'stalk' or pedicel of the animal is formed of a long block of longitudinal muscle fibres and connective tissue and has a thick, flexible and transparent cuticle. The posterior end has root-like extensions which anchor it firmly in the mud. Muscles allow the position of the body to be moved relative to the pedicel and also close (anterior and posterior adductors) and open (adjustors) the valves. If the animal is disturbed it contracts the pedicel and retreats into its burrow. When feeding the pedicel holds the anterior edges of the valves in an open position just above the surface of the substrate.

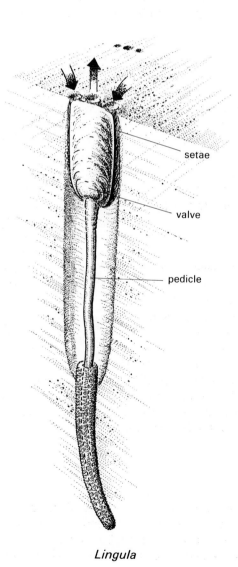

Lingula

The body wall is formed of an epithelial epidermis, with muscle fibres in the parts free of the shell, and a peritoneal lining. The mantle lobes are extensions of the body surface

and are therefore double structures with an extension of the metacoele between them. However, some fusion between the mantle walls divides the mantle coelom into a system of canals. The metacoele also extends into the pedicel.

Feeding

Like *Bugula*, *Lingula* is a suspension feeder, and feeds particularly on phytoplankton. The lophophore extends laterally in two elongated and spirally-coiled structures which are held in the vertical plane and provide a greatly increased surface area. During feeding the valves are held slightly apart through the combined action of a complex system of adjustor muscles. The cilia of the lophophore beat to create a feeding current which enters at the sides and leaves medially. Food particles caught by the lophophore pass down the tentacles and along a deep ciliated brachial groove to the mouth, by a similar process to that seen in the bryozoans. The mouth opens into a short oesophagus. This is surrounded by a large digestive gland which opens into it by paired dorsal and ventral ducts. Digestion is chiefly intracellular and takes place in the digestive gland. From the stomach, a short intestine and rectum extend to open at the right hand side of the mantle cavity.

Gaseous exchange and circulation

Gaseous exchange takes place over the surface of the lophophore and mantle lobes. Oxygen is transported by the coelomic fluid. This carries haemerythrin in coelomocytes and circulates through the mantle lobes. However, there is also a blood vascular system which is thought to be chiefly concerned with the transport of food. A contractile vesicle, the heart, lies dorsal to the stomach from which extend anterior and posterior channels to a sinus surrounding the gut. The blood, a colourless fluid containing coelomocytes, is transported in these vessels to the adjustor muscles, and to other regions of the body.

Excretion

Lingula has a pair of nephridia. The nephrostomes open into the metacoele and the nephridial ducts extend anteriorly to open into the mantle cavity at a nephridiopore at either side of the mouth. Particulate nitrogenous wastes are taken up by the coelomocytes and expelled through the nephridia.

Nervous system and sense organs

A nerve ring encircles the oesophagus, and nerves run from it to the mantle lobes, lophophore and muscles. A pair of statocysts have been described in one species of *Lingula*, and both mantle margin and setae are thought to have a sensory function.

Reproduction and growth

Most brachiopods, including *Lingula*, are dioecious and have four large gonads, lying within the metacoele in the mesenteries which also support the viscera. However there are some hermaphrodite genera. Typically gametes are released into the coelom and leave via the nephridia, though some brachiopods brood their eggs in the mantle cavity. A free-swimming larva develops which resembles the adult form. In *Lingula* there is scarcely any free-living phase; the larva attaches and takes up a sessile mode of life immediately.

Phylum Entoprocta

The phylum Entoprocta is a small group of about 60 species of animals most of which are sessile. In many ways they resemble the Bryozoa, and were originally included as a class within that phylum; their name Entoprocta refers to the fact that the anus lies inside rather than outside the food collecting organ (hence also Ectoprocta as an alternative name for the Bryozoa). Subsequent workers have assigned the entoprocts to a variety of places in the classification of invertebrates; for example an interpretation of their body cavity as a pseudocoelom led to their inclusion with the pseudocoelomate animals though their relationships to other forms at that level of organisation remained obscure. Their similarities with the bryozoans, notably the sessile mode of life, a food collecting organ consisting of a circlet of ciliated tentacles, a U-shaped gut, and a circum-oesophageal nerve ring, were interpreted as evolutionary convergence. However more recent work suggests that the features which separate them from the bryozoans (the position of the anus, a feeding current which enters the tentacle circle at its base and leaves along the longitudinal axis, and the possession of a pair of nephridia) are less fundamental than was previously thought, and that perhaps they should be reunited with the ectoprocts as bryozoans. For the present they are regarded as a separate phylum, but are discussed at this point to facilitate comparisons with the lophophorates, and particularly the bryozoans.

Entoprocts are marine, apart from one freshwater genus, and live attached to rocks, pilings, and other animals. All are very small, less than 5 mm long, and the majority are colonial. Many are cosmopolitan in distribution; a common example is *Pedicellina*, which is typical of the group as a whole.

Pedicellina

Pedicellina is commonly found between the tide marks, attached to almost any solid object. It is colonial, with individual zooids attached both to each other and the substrate by a horizontal creeping stolon.

Body form
Each zooid consists of a somewhat cup-shaped body (the calyx, which is attached to the stolon by a stalk) with a circle of ciliated tentacles which are extensions of the body wall. The tentacles enclose a space, the atrium, into which both

the mouth and anus open, their positions marking the anterior and posterior ends of the animal. The upper surface of the calyx is ventral while the lower, to which the stalk is attached, is dorsal.

The body wall, which forms the stalk and tentacles as well

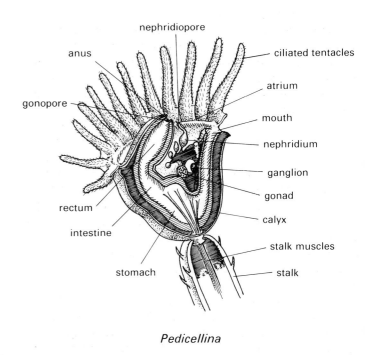

Pedicellina

as the calyx, is composed of an underlying epidermis and longitudinal muscle fibres. There is no body cavity; the region between the viscera and the body wall, and the hollows of the tentacles, are filled with a gelatinous mass of fixed and freely moving cells, the mesenchyme.

Feeding
Like bryozoans, *Pedicellina* is a filter-feeder on plankton and suspended organic matter. The beat of the lateral cilia of the tentacles creates a water current, which passes between the tentacles and medially out of the atrium. Suspended food particles are trapped by the lateral cilia and carried in mucus by frontal cilia on the inner tentacle face to the tentacle bases. Ciliated food grooves extend from the tentacle crown to the mouth. The U-shaped gut has a large stomach and food is moved through it by ciliary action.

Gaseous exchange and circulation
Pedicellina has no respiratory or circulatory systems.

Excretion
There are two nephridia which lie close to the oesophagus and open behind the mouth via a single median nephridiopore.

Nervous system and sense organs

A large ganglion lies within the mesenchyme ventral to the stomach; it gives rise to three pairs of nerves to the tentacles, and three pairs to the calyx and stalk. There are simple sensory cells of the bristle type scattered over the general body surface, with particular concentrations on the outer side of the tentacles and the calyx margin.

Reproduction and growth

All ectoprocts show asexual reproduction by budding, and the colonies are developed in this way in colonial forms. The genus *Pedicellina* includes both hermaphrodite and dioecious species. In dioecious forms a pair of rounded ovaries or testes lie ventral to the stomach, and paired gonoducts unite to open at a single median gonopore immediately posterior to the nephridiopore. In hermaphrodite species the paired testes are situated immediately posterior to the ovaries, and all four ducts unite before opening at the gonopore. Fertilization probably occurs in the ovaries, although the method is unknown. The gonoduct secretes a membraneous envelope around the egg which passes through the gonopore and attaches to the atrium wall. Embryonic development follows, with spiral cleavage, and a larva not unlike that of annelids and molluscs hatches and takes up a free living mode of life. The larva has an apical tuft of cilia at its anterior end, a ventral ciliated girdle, and a ciliated foot. There is also a frontal organ with which, in due course, it attaches to a substrate. After a short free-swimming period the larva sinks to the bottom and creeps about over the substrate using its foot. Eventually it attaches and undergoes metamorphosis to develop into the adult form.

Phylum Phoronida

The phoronids, or horseshoe worms, are one of the smallest phyla, with only about ten living species. They are all marine and typically live in chitinous tubes, buried in sand or attached to stones in mud or clay, in shallow water on the shore. Some species burrow into calcareous rocks or mollusc shells. Phoronids are simpler in structure than most other lophophorates and have a cylindrical worm-like body with no differentiation or appendages apart from a conspicuous lophophore in the shape of a horseshoe. Like the other lophophorates they are suspension feeders, and have a U-shaped gut with digestion taking place intracellularly in the stomach. Also like other lophophorates they have no respiratory organs, but a blood vascular system is present. A dorsal vessel supplies blood (which has corpuscles containing haemoglobin) to the lophophore, and a ventral vessel supplies the remainder of the body. Both vessels are contractile. A pair of nephridia are present in the metacoele.

The phoronid nervous system consists of a sub-epidermal nerve net, which is continuous with a nerve ring at the base of the lophophore. This supplies nerves to the tentacles, the longitudinal muscle and to a single giant fibre which passes down the left hand side of the body in the epidermis as a lateral nerve.

Phoronids are generally hermaphrodite with external fertilization. The egg typically gives rise to an elongate, ciliated larva, the actinotroch, which has a free-living planktonic phase before settling on the bottom and secreting a tube.

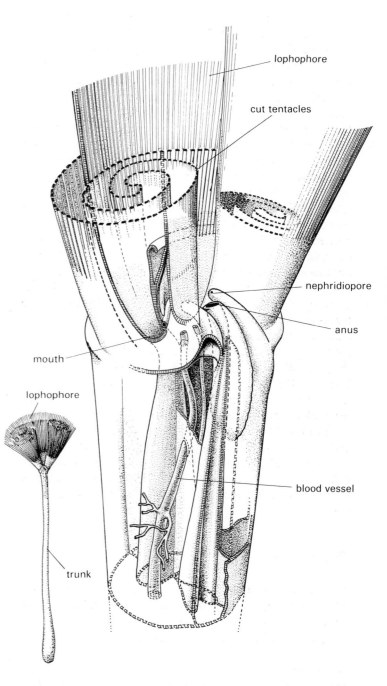

Phoronis

Relationships of the lophophorates

The lophophorates are a particularly interesting group since they exhibit both protostome and deuterostome characteristics. Thus they apparently develop a typical protostome anus, but cleavage is radial, a deuterostome characteristic. Also the typical deuterostome body form consists of three regions (protosome, mesosome and metasome) each with a coelomic compartment (protocoele, mesocoele, and metacoele) and this type of organization is reflected to a marked degree in the lophophorates. For the present the relationship of the lophophorate phyla, both to the other invertebrates and to each other, must remain uncertain.

The invertebrate deuterostomes

Apart from the major deuterostome phylum Chordata, which includes two invertebrate subphyla in addition to the subphylum Vertebrata, there are three deuterostome groups: Echinodermata, Chaetognatha, and Hemichordata. Of these three, the phylum Echinodermata is a large group with a long fossil record and a wide distribution. The others are small and will be discussed together as the lesser deuterostomes.

Phylum Echinodermata

The phylum Echinodermata includes approximately 6,000 species of marine animals, including such well known forms as starfish and sea urchins, which are widely distributed, brightly coloured and amongst the most beautiful of the sea animals. Their long fossil history extends back to the Cambrian, and about 20,000 fossil species have been described.

As a group, echinoderms are immediately recognizable by their pentaradiate symmetry: in other words, the body can be divided into five parts radiating from a central axis. This is most apparent in many of the starfish (such as *Asterias*) where the five parts are each represented by an arm, but is found throughout the group. The pentaradial symmetry of echinoderms must be regarded as a secondarily acquired characteristic, derived from an ancestral bilateral symmetrical form. The echinoderms are not related to other radially symmetrical animals, namely sponges, cnidarians, and ctenophores.

In talking about echinoderms it is convenient to recognize the body as divided into an oral region which is the part surrounding the mouth, and an aboral region on the opposite side to the mouth. Since the position of the mouth relative to the substrate is not the same throughout the group the terms 'upper' and 'lower' or 'dorsal' and 'ventral' are less useful. The body wall typically has an internal skeleton of calcareous plates. These are unusual among invertebrates in being formed by the deeper layers of the body wall and lying within it; however in many echinoderms the skeleton bears numerous spines or tubercles which project above the outer surface giving it a spiny or knobbly appearance. The name of the phylum is derived from this feature (it means 'spiny-skinned'). The skeletons or 'tests' of echinoderms are sometimes sold as ornaments; a large sea urchin skeleton can even be used as a lampshade.

Echinoderms have a body cavity which is a true coelom and is subdivided into three compartments; the main perivisceral cavity surrounding the gut, and two canalized coelomic structures, the water vascular and perihaemal systems. The water vascular system consists of a circum-oral ring joined by a series of fluid-filled tubes with a ciliated lining. The fluid is similar in chemical composition to sea water, but has a higher concentration of potassium ions and contains some protein. It also contains amoeboid cells, the coelomocytes, which are produced in greatly folded pouches arising from the circum-oral ring and called Tiedemann's bodies. Branches of the water vascular system extend into numerous podia (tube feet) which function hydraulically. The structures are highly characteristic of the phylum and may be used for feeding, locomotion, or for sensory purposes. The tube feet are arranged in rows or ambulacrae of which there are normally five (one per radius).

The perihaemal coelom also consists of a circum-oral ring connected to a series of intercommunicating canals. Like the water vascular system, the perihaemal system has canals which extend vertically from the circum-oral rings to the aboral pole. It contains a fluid which is essentially the same as that of the water vascular system.

Echinoderms are chiefly bottom-dwelling animals although the group does include a few free-swimming or pelagic representatives. They have bilaterally symmetrical barrel-shaped early larvae which are totally different from the trochophore larva of annelids and molluscs. In a simple, hypothetical, dipleurula form, the larva has cilia arranged in two lateral longitudinal bands, which connect anteriorly in front of the mouth, and posteriorly in front of the anus. These structures are used both for locomotion and for feeding. The early larval development is of particular interest for the light it casts on echinoderm relationships with other groups. Echinoderms are the only major group of invertebrate deuterostomes (animals in which the anus is derived from the blastopore) and similarites of larval form suggest they are more closely comparable to primitive chordates than to any other invertebrate group. Furthermore the bilateral symmetry of the larva is strong evidence that the pentaradiate symmetry of the group is secondarily acquired. Whilst echinoderm larvae show considerable superficial variation all can be derived quite simply from the dipleurula type described.

As the echinoderms are both common and easily collected, they have been studied in detail and were among the many animals investigated by Aristotle. However their economic importance to man is, with a few notable exceptions, rather limited.

Both sea cucumbers and sea urchins are eaten in many parts of the world. There are many poisonous sea urchins which may cause injury to humans; one form eaten in Japan

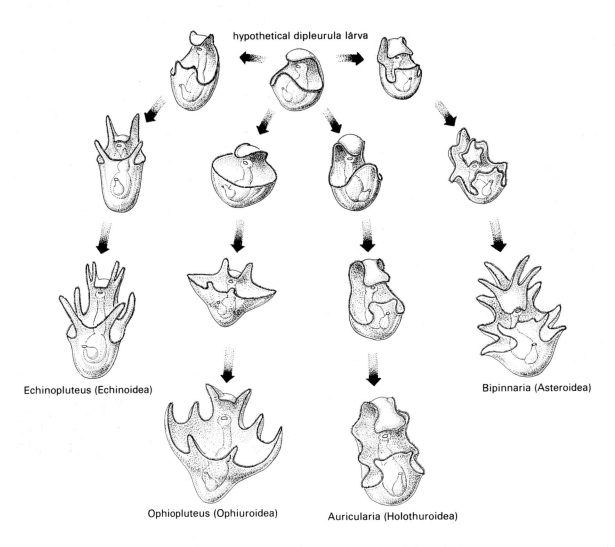

Derivation of echinoderm larval stages from the dipleurula form

has extremely poisonous ova and failure to extract them completely has fatal consequences. Of rather more importance is the destructive effect of starfish on commercial mussel and oyster beds where they are serious predators; one starfish may eat as many as 20 oysters in a day. The coral-eating crown-of-thorns starfish (*Acanthaster*) has recently become notorious for its devastating effect on large areas of the Great Barrier Reef. *Acanthaster* can occur in very large numbers, as many as 15 per square metre of coral reef, and an individual may eat an area equivalent to its central body disc in a day. The problems of predation are made worse because echinoderms are difficult to kill: they have considerable powers of regeneration and early attempts to destroy crown-of-thorns starfish by chopping them up, merely resulted in an increase in the population. Similar attempts to rid oyster beds of starfish have proved equally unsuccessful.

There are five present day echinoderm classes: Crinoidea (sea lilies and feather stars), Asteroidea (starfish), Ophiuroidea (brittle stars), Echinoidea (sea urchins) and Holothuroidea (sea cucumbers). (Some authorities unite the asteroids and ophiuroids as one class, Stellaroidea). Although superficially these groups appear very different from each other all can be derived from a basic sphere with pentaradiate symmetry of both the body wall and the internal organs. In fact the organ systems are remarkably stable and most of the modifications of the group concern its external structure.

The crinoids include the most primitive living echinoderms, and their fossil remains have been found in early deposits. However, the Asteroidea show many features of the group in their simplest form, and a detailed description of *Asterias rubens*, a readily available asteroid, is used as an introduction to the phylum.

Class Asteroidea: *Asterias rubens*, the starfish

The genus *Asterias* is cosmopolitan. *A. rubens* is a bottom dweller in rocky or stony areas from the lower shore down to about 200 m. It is often found associated with mussel or oyster beds on which it feeds, and is therefore a major economic pest.

Body form

Asterias has the typical starfish shape: it is strongly flattened along the oral-aboral axis and consists of a central disc from which project five blunt-tipped arms (regarded as radial in position). The central disc bears the mouth in the centre of the lower (oral) suface, and an inconspicuous anus on the upper (aboral) surface. Another opening, the madreporite, lies inter-radially near the centre of the aboral surface and opens into an extension of the water vascular system, the stone canal. The aboral surface is brownish yellow and covered with numerous irregularly arranged blunt spines; the oral surface is paler and has about 200 tube feet which project in two double rows in grooves (the ambulacral grooves) extending from the mouth along each arm. The grooves are protected by long stout ambulacral spines which are movable and can be held projecting downwards or across the groove.

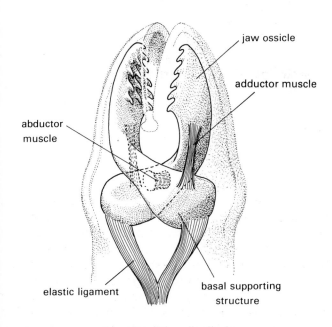

An asteroid pedicellaria

On both the oral and aboral surfaces there are pedicellariae, which are small pincer-like structures, easily studied with the aid of a hand lens. Each pedicellaria is formed from three ossicles: a basal supporting structure, and a pair of ossicles which form the pincers or jaws

themselves. Two pairs of muscles originate on the basal ossicle and extend to the jaws; the abductors swing them open, and the adductors close them. The pedicellarial jaws may cross each other, like a pair of scissors, or touch at the tips, like a pair of forceps. Pedicellariae may be fixed or stalked with a long reach and considerable flexibility. Stalked pedicellariae are particularly characteristic of the ambulacral grooves while those of the aboral suface are more usually fixed and are arranged in rings round the spines. Pedicellariae are used to clear detritus off the body surface, and in addition to kill and discard small animals which touch them. In this way they are used by the starfish to prevent sessile animals, such as barnacles, from settling on their surface. Without pedicellariae a slow moving animal with a firm surface, such as a starfish, would prove vulnerable to colonizers, as well as to small predators.

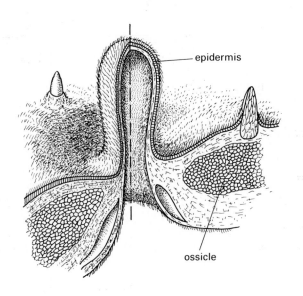

An asteroid papula

The body wall

The outermost layer of the body wall is a thin epidermis of ciliated cells, mucous gland cells, and sensory cells covered by cuticle. The mucous cells secrete a protective coating, and the epidermal cilia sweep detritus off the surface. A sub-epidermal nerve plexus lies immediately below the epidermis; and below this is a connective tissue dermis in which the ossicles are embedded. These are formed of a mixture of calcium carbonate (90%) and magnesium carbonate (10%) and are three-dimensional matrices, with spaces which are filled by soft tissue, and are joined to each other by a network of collagen fibres. The oral wall ossicles are closely but flexibly fitted and those of the ambulacral grooves are particularly large for muscle insertion; in contrast the aboral wall ossicles form an irregular lattice. The body wall is completed by inner longitudinal and outer circular smooth muscle fibres and by a layer of peritoneum.

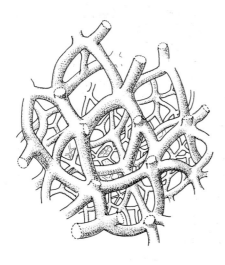

Asterias: structure of an ossicle

The water vascular system

The madreporite opens into a small space, the ampulla, which leads via the vertical stone canal (so called from calcareous deposits in its walls) to a ring canal. The lumen of the stone canal is subdivided so that fluid can pass orally and aborally. Five ciliated radial canals (one per arm) extend from the ring canal, and each terminates in a sensory structure, the optic cushion, which is a modified tube foot. Lateral canals extend from each side of the radial canal, each opening into an aboral bulb (ampulla) and oral tube foot (podium). The opening is guarded by a valve. These canals are alternately long and short on each side so that the tube feet on either side are arranged in two rows. The radial canals and tube feet lie outside the main part of the body in the ambulacral groove, but the ampullae project into the body cavity. A total of nine Tiedemann's bodies arise inter-radially from the ring canal; the position of the tenth (expected from pentaradial symmetry) is occupied by the stone canal.

The podia have broadened sucker-like tips and function hydraulically. Rings of smooth muscle in the wall of the ampulla contract, the lateral canal valve closes, and fluid is thus forced into the podium causing it to elongate. When the sucker of the podium touches the substrate its centre is withdrawn producing a slight vacuum and therefore causing it to stick by suction. The epidermis of the sucker region has numerous sensory cells, as well as gland cells which produce a copious adhesive secretion. The podium wall is not impermeable and as it is extended some fluid is squeezed out. It was thought that replacement of fluid lost in this manner occurred through the madreporite; however a more recent suggestion is that fluid is secreted from the coelom across the ampulla wall by a potassium pump and that there is no circulation of fluid between the radial canal and the ampulla and podium. In this theory the madreporite is thought to equilibrate internal and external pressure over the remainder of the water vascular system.

Locomotion

Asterias crawls slowly, typically with one arm leading, although it may extend two arms in front or even lead with an inter-radial region. Any arm can be used to lead and the same one is not used all the time. Propulsion is provided by the tube feet which move in a series of steps (3–10 per minute). Each foot in turn is extended, fixed to the substrate, and used as a lever to push the body forwards. Muscles in the podial walls are used to retract the foot, as well as to orientate it during use, and provide the necessary suction for attachment. Many of these originate on the ambulacral ossicles. In addition the tube has a layer of connective tissue which prevents the foot from bulging rather than becoming extended.

Feeding and digestion

The alimentary canal is aligned in the vertical axis, extending through the disc from the mouth on the lower side to the anus on the upper side. All the gut structures are surrounded by the perivisceral coelom. The mouth, which is closed by a sphincter muscle, opens into a short narrow oesophagus and then a large, pouched, cardiac stomach. The walls of this stomach are connected to the ambulacral ossicles of the arms by ten pairs of triangular gastric ligaments. The cardiac stomach opens into the smaller pyloric stomach; this is flattened and appears star-shaped because of 10 blind-ending pyloric caeca, or digestive glands, which extend from it, one pair down each arm. A short common duct joins each pair to the pyloric stomach. The digestive glands are hollow ciliated structures composed of masses of glandular cells which secrete enzymes and also function for food absorption and storage. From the pyloric stomach, a short intestine, from which arise the short, blind-ending rectal sacs, extends to the anus.

Asterias is a predaceous carnivore feeding principally on bivalve molluscs, especially *Mytilus*. It holds the bivalve so that the ventral (non-hinged) surface is held firmly pressed against the mouth, attaching its arms to the valves by the tube feet. Retraction of the tube feet then forces the mollusc

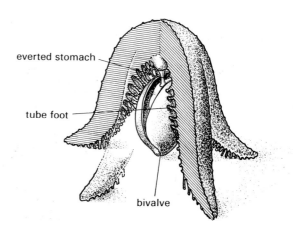

everted stomach

tube foot

bivalve

Asterias feeding

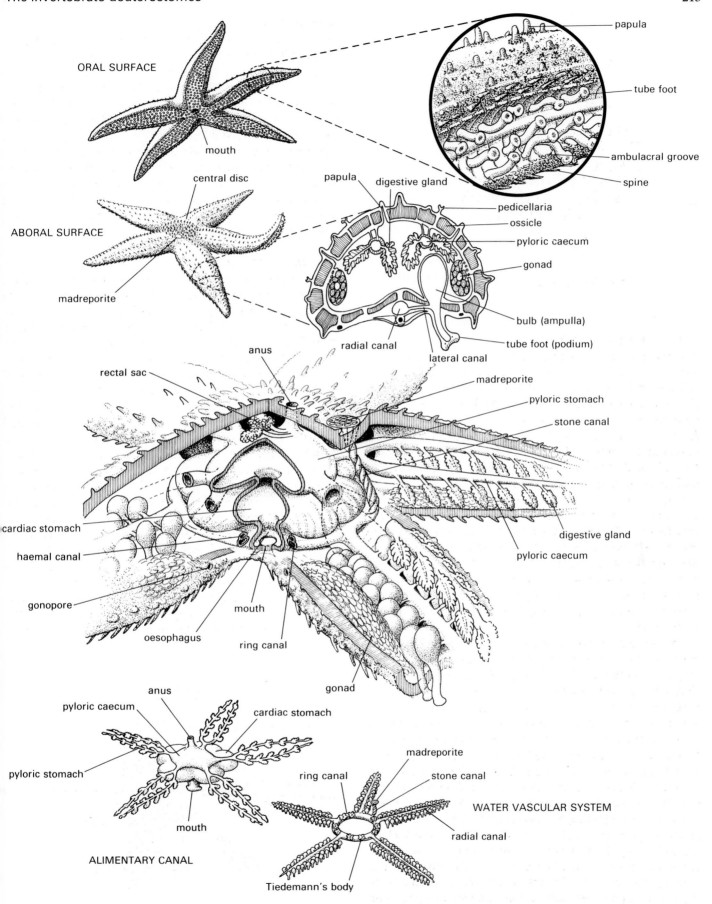

ORAL SURFACE

mouth

papula

tube foot

ambulacral groove

spine

central disc

ABORAL SURFACE

madreporite

papula   digestive gland

pedicellaria

ossicle

pyloric caecum

gonad

bulb (ampulla)

tube foot (podium)

radial canal

lateral canal

anus

rectal sac

madreporite

pyloric stomach

stone canal

cardiac stomach

haemal canal

digestive gland

pyloric caecum

gonopore

mouth

oesophagus   ring canal

gonad

anus

pyloric caecum

cardiac stomach

pyloric stomach

madreporite

ring canal

stone canal

WATER VASCULAR SYSTEM

radial canal

mouth

ALIMENTARY CANAL

Tiedemann's body

Anatomy of *Asterias*

slightly open. Once the shell is open (0.1 mm is enough) *Asterias* everts part of its cardiac stomach into the prey; this is brought about by contraction of the body wall muscles leading to increased pressure in the perivisceral coelom. Proteolytic enzymes with lipase, amylase, and invertase are poured into the prey and a partly digested soupy mixture is extracted from it. Fluid and particles are moved by ciliary action through the digestive tract and into the pyloric caeca where food is absorbed and stored.

Gaseous exchange, excretion, and circulation

Gaseous exchange takes place across the tube feet and small evaginations of the body wall called papulae. The tube feet and papulae are also the site for removal of nitrogenous waste in the form of ammonia; apart from these structures echinoderms have no specialized excretory system. Since the body fluids are isotonic with sea water, osmoregulation is unnecessary. The fluid-filled perivisceral coelom provides the principal means of internal transport; there is a continuous circulation of coelomic fluid brought about by beating of the peritoneal cilia.

Asterias has a poorly developed haemal system; this consists of oral and aboral rings, a radial sinus paralleling the water vascular canal in each arm and a canal which parallels the stone canal and is surrounded by a mass of spongy tissue, the axial gland. The axial gland is thought to destroy any micro-organisms which get into the body. The haemal system fluid contains coelomocytes which may play a role in food transport. A dorsal sac in the madreporite pulsates and functions as a heart.

Nervous system and sense organs

Apart from the sub-epidermal nerve net, which extends all over the body, *Asterias* shows some concentration of nerve fibres to form a circum-oral nerve ring which surrounds the mouth. Five radial nerve cords arise from the nerve ring, extending one along each arm; these are continuous with the sub-epidermal plexus and are involved in the co-ordination of the podia. In addition, thickenings of the sub-epidermal plexus form marginal nerve cords which extend along the sides of the ambulacral grooves. The sub-epidermal system is predominantly sensory and a deeper system of fibres extends from the radial nerves to supply the body wall muscles, podial muscles and ampullary muscles.

Asterias has large numbers of epidermal sensory cells (tactile, chemical and light sensitive receptors) scattered over the general body surface, but concentrated particularly along the margins of the ambulacral groove and on the podia. At the tip of each arm there is an optic cushion formed from a collection of about 150 pigment cup ocelli. *Asterias* is markedly phototactic.

Reproduction and growth

The sexes are separate. *Asterias* has five pairs of gonads (either ovaries or testes) which are suspended by mesenteric strands within the perivisceral coelom and open, via ciliated gonoducts, at a cluster of gonopores near the junction of the arms. The site and extent of the gonads varies seasonally: near spawning, which takes place in the spring, they may almost fill the arms.

The sperm and eggs are shed into the water, and fertilization is external. At a single spawning a female *Asterias* may release 2½ million eggs in a period of about two hours. The presence of eggs in the water acts as a stimulant for the shedding of sex cells by other individuals both male and female. Possibly this is due to a neurohormone, produced by the radial nerves and detected in the water by the axial glands of other individuals. The sperm stick to the ova.

After fertilization the larva develops to the dipleurula-like stage, with a ventral mouth, and with an anus derived from the blastopore (the deuterostome condition). At first cilia are widely distributed over the surface of the larva, but this general ciliation soon degenerates and is reduced to a definite ciliary band. This consists of two longitudinal bands connected in front of the mouth and behind the anus. The characteristic larval stage of *Asterias* is a bipinnaria larva arrived at by the development of two groups of arms along which the ciliated bands extend. It is planktonic for several weeks and feeds on pelagic organisms such as diatoms. It eventually sinks and becomes temporarily attached to the substrate, and develops into a branchiolaria larva with three additional anterior arms and an adhesive sucker. It then undergoes the final stage of metamorphosis, becoming sexually mature after about a year.

Echinoderm diversity

Asterias is typical of one type of echinoderm body form in which the pentaradiate symmetry is obvious from the five arms radiating from a central disc. However within the group to which it belongs (Asteroidea) there are variations. In some asteroids, for example *Porania*, the arms are short. In others the characteristic five-armed pattern is not adhered to; forms such as the sun stars, for example *Solaster* and *Crossaster* can have between 7 and 13 arms, and *Luidia ciliaris* is a very unusual form looking like a seven-armed *Asterias*.

Porania

Solaster

Class Ophiuroidea: the brittle stars.

The ophiuroids also have arms which radiate from a central disc. In this group the number five is more strictly adhered to although there are a few exceptions such as the six-armed *Ophiactis*. However the arms are more distinct from the central disc than those of *Asterias* and appear more fragile and delicate. They are greatly elongated and easily broken,

hence the common name for the group, brittle stars. Some ophiuroids can sever their arms deliberately as an escape response if seized by a predator, and regenerate them later.

The ophiuroids are a widely distributed group, with 2,000 species, but are probably less well known to the casual observer than the starfish. Although they are abundant in the littoral zone, they are easily overlooked because they hide amongst rocks or seaweed, and under stones and shells. Most are small animals, some are remarkably agile and all ophiuroids can move much more rapidly than starfish.

Arm structure and locomotion

The structure of the ophiuroid arm is very different from that of *Asterias*. Instead of having an open ambulacral groove it has an epineural canal formed by the body wall which has grown over the groove. The tube feet protrude through pores in the skeletal plates. The arms are formed of a longitudinal series of ossicles, hinged to each other and worked by muscles which give brittle stars a rather jointed appearance. Ophiuroids (such as *Ophiothrix*) can move swiftly by means of snake-like flexures of their arms. The tube feet are therefore not directly involved in locomotion though they may be thrust out to help the arms in their push against the substrate. It is not surprising that in agile animals like ophiuroids, pedicellariae, which in starfish prevent debris and sessile animals settling on the body surface, are absent.

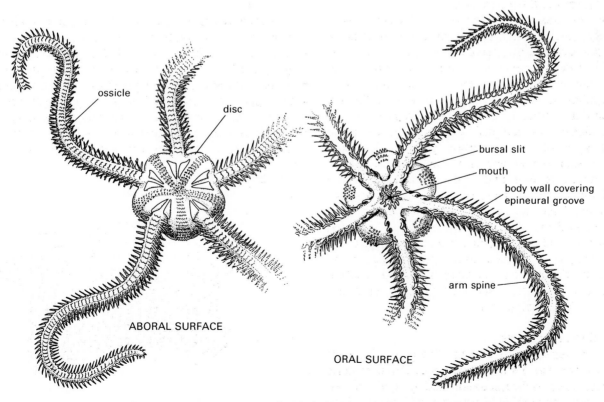

ossicle

disc

ABORAL SURFACE

bursal slit

mouth

body wall covering epineural groove

arm spine

ORAL SURFACE

Ophiothrix

Feeding

Ophiuroids are mainly scavengers, deposit feeders, and filter feeders. Some species may feed on animals which are small enough to be swallowed whole which they pick up with their tube feet and transfer to the mouth. Others trap particles of organic matter in webs of mucus secreted by the tube feet and use ciliary currents to transport the food and mucus to the mouth. *Ophiothrix* can hold itself in position by two of its arms and extend the others vertically, trapping particles of food by modified papillate tube feet. As in *Asterias* the oral surface is the lower surface. The gut structure is the simplest found in echinoderms for there is no intestine, anus, or pyloric caecum. A short oesophagus connects the mouth to a large ten-pouched stomach which is held into place by mesenteries and occupies most of the volume of the body disc. The stomach is the site for digestion and absorption, digestive enzymes being secreted mainly in the pouches. Ciliated cells (for mixing the stomach contents) and mucous gland cells are also present in the stomach. Undigested matter is discarded through the mouth.

Coelom

As a result of the narrowed arms and the increased stomach size, the perivisceral coelom is much reduced. The water vascular system is essentially similar to that of the asteroids, though ampullae are absent, and the lateral canals arise in opposite pairs rather than alternately. The perihaemal system also resembles that of *Asterias*.

Gaseous exchange and excretion

Ophiuroids have specialized regions for gaseous exchange called the bursae. These are invaginations of the oral disc between the stomach pouches which open to the exterior through slits in the oral margins. The bursae have ciliated linings and the ciliary beat creates the respiratory current.

Nervous system and sense organs

The nervous system is basically similar to that of *Asterias*. Although ophiuroids are active animals there appear to be no specialized sense organs and they rely on the general sensory structures of the epidermis. In association with their secretive habits they are negatively phototactic.

Reproduction and growth

In temperate waters *Ophiothrix* breeds in the spring. The gonads are associated with the bursae, and discharge into them, so that the eggs or sperm leave with the respiratory current. Fertilization is external and the early larva develops into an ophiopluteus which has four pairs of elongated arms supported by calcareous rods. The ophiopluteus is quite distinct from the bipinnaria in appearance but is also planktonic, feeding on other smaller planktonic organisms.

Many ophiuroids are hermaphrodite, and have internal fertilization. In these species the young are brooded in the bursae.

Class Echinoidea: the sea urchins, sand dollars and heart urchins.

Asteroids and ophiuroids are relatively active animals and can move with the aid of well-developed arms. In contrast the echinoids lack arms and are generally less active although they are capable of slow movement. The sea urchins are more or less spherical (regular echinoids) and retain the radial symmetry characteristic of echinoderms. Sand dollars and heart urchins are oval- or heart-shaped and are flattened to a greater or lesser degree; they have a secondarily acquired bilateral symmetry (irregular echinoids). These differences in form are reflected in different modes of life: while the sea urchins are generally adapted for life on hard surfaces such as rocks or gravel, the irregular echinoids are burrowers and have a definite anterior end.

Echinoids (their name means 'hedgehog-like') are characterized by spines of varying length, some very long, which are attached by ball and socket joints which allow movement in any direction. The spines arise from tubercles on the skeleton and are surrounded by two tissue layers. The outer layer is muscular and is used to move the spine, the inner layer is fibrous and can be used to hold the spine rigid as a defence mechanism.

The ossicles are flattened and sutured to form a solid structure, the test. In other respects the construction of the body wall is similar to that of *Asterias*, although of a less flexible nature. Pedicellariae are present which are both more varied and more numerous than in *Asterias*; sea urchins having even greater problems than starfish from debris and sedentary organisms. They are typically three-jawed structures with flexible necks. The tridactyl pedicellariae have elongated jaws and are modified for extremely rapid closure; the ophiocephalous pedicellariae have short jaws, a locking mechanism, and highly flexible necks. Some species (but not *Echinus*) have pedicellariae with glands associated with the jaws which secrete toxins extremely effective at driving away or paralysing small animals or even starfish.

Locomotion

Echinus can crawl slowly using its ventral spines as levers and can also climb using its tube feet, which are concentrated in five double rows in the ambulacral regions.

Feeding

Echinus grazes on seaweed, small algae, or encrusting animals, which it scrapes off the rocks using a specialized skeletal masticatory organ, originally described by Aristotle and called Aristotle's lantern. It is formed of five triangular calcareous plates, the pyramids, which are arranged inter-radially with their points directed towards the mouth. Each is joined to adjacent plates by transverse muscle fibres. A band of especially hardened calcareous material along the inner side of each pyramid projects above the pyramid tip forming a tooth. Each band originates in a dental sac, and is secreted continuously from the oral end as it is worn down in use. The lantern is attached by muscles to the plates of the test and can be partly protruded from the mouth which is surrounded by a flexible area, the peristome. Protractor muscles pull the lantern outwards and exposes the tooth ing; retractor muscles withdraw it.

The general arrangement of the gut of *Echinus* is comparable with that of *Asterias*. The intestine is long and coiled, and held suspended in the perivisceral coelom by mesenteries; digestive caeca are absent and the intestine is thought to produce digestive enzymes in its anterior region (sometimes referred to as the stomach) and also to absorb digested food.

The animal moves on top of the food and chews it into pieces with its teeth before pulling it into the buccal cavity, which lies within the lantern. Here it is formed into small mucus-covered pellets which are transported by the ciliated gut lining. The mucus is permeable to enzymes but separates the food from excess water taken in with it. A short oesophagus extends beneath the lantern to the intestine; at the junction a narrow tube, the siphon, separates from the main intestine to extend along the border of its anterior region. This is thought to act as a bypass for excess water past the anterior secretory region. The mucus which covers the pellets is tough, and the pellets retain their shape even after they have passed through the entire

alimentary tract to leave through the anus, which lies opposite the mouth on the aboral surface.

Both the compacting of the food into pellets and the bypass arrangement are thought to be adaptations to reduce the volume of the gut contents as far as possible, in response to the restriction of the body volume imposed by the test. The passage of food through the gut is generally slow and a high proportion of digestible matter is assimilated.

Gaseous exchange and circulation
The three main components of the coelom (the perivisceral coelom, the water vascular system, and the haemal system) are arranged in essentially the same way as in *Asterias*, the perivisceral coelom being important in the transfer of food and waste products.

In *Echinus* gaseous exchange takes place through the water vascular system and tube feet as in *Asterias*, as well as through highly branched extensions of the body wall, the peristomial gills. These have a coelom continuous with the peripharyngeal coelom, a subdivision of the perivisceral coelom which surrounds the lantern. The peripharyngeal coelom has a pumping mechanism, composed of specialized ossicles and their associated muscles, which pull on the partitioning membrane.

The water vascular system is similar to that of *Asterias*, with the circum-oral ring lying near the aboral end of Aristotle's lantern. A stone canal extends to the

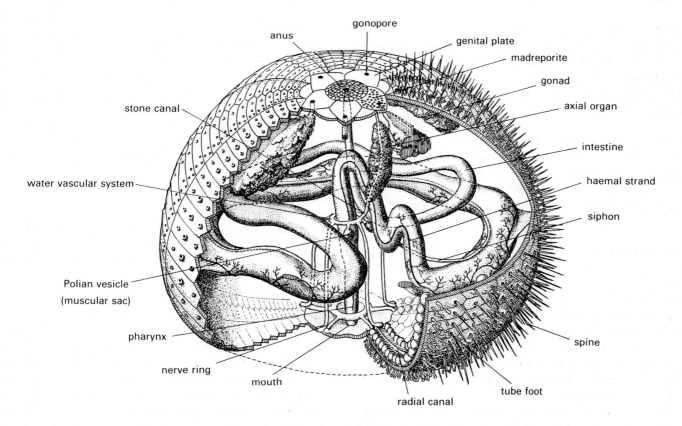

Anatomy of *Echinus*

madreporite. The haemal system is also essentially the same as that of *Asterias*, but the axial gland is a long, dark, blind-ending structure lying freely in the perivisceral coelom.

Nervous system and sense organs

The nervous system is similar to that of *Asterias*. Light sensitive cells are scattered over the body surface and, like the ophiuroids, most echinoids are negatively phototactic.

Reproduction and growth

The sexes are separate. Five gonads are present, suspended along the interambulacra. Short gonoducts lead to gonopores, each situated aborally on one of five specialized genital plates. Fertilization is external and an echinopluteus larva similar to the ophiopluteus with six pairs of arms, develops. This stage, during which the animal moves and feeds by means of ciliated bands, is relatively brief. Metamorphosis is rapid and a young sea urchin, at first only 1 mm in diameter, soon develops. Young urchins often settle near groups of adults, thus gaining some protection from their spines, but move away as they get older.

The irregular echinoids

Irregular echinoids have developed a secondary bilateral symmetry in association with a burrowing habit. Both the mouth and the anus are on the lower surface. In some species they are close together, but in others they occupy anterior and posterior positions respectively.

The heart urchin, *Echinocardium*, lives buried in the sand, and feeds on the films of inorganic matter on the sand grains which it picks up by special elongated tube feet near the mouth. It burrows, using modified digging spines on the inter-ambulacral area posterior to the mouth (the plastron) and on the lateral ambulacra. Sand is heaped up in front and along the sides of the animal, which sinks vertically down. During this process the sides of the burrow are plastered with mucus produced by specialized tube feet of the anterior ambulacrum and wiped onto the walls by a specialized tuft of spines. As the burrow gets deeper, the burrowing action is taken over by the tube feet. Forward movement through the sand is provided by the spines of the plastron and the anterior region of the test. Cilia of the body surface beat to bring a respiratory current through the burrow.

Sand dollars, for example *Echinocyamus*, are mainly small and very flattened, with a coat of thickly set short spines. Tube feet are restricted to the upper surface and are arranged in characteristic petalloid patterns. The sand dollars are less highly modified for burrowing than heart urchins and move about just beneath the surface of the sand using their spines and tube feet.

The minor echinoderm classes

Class Holothuroidea : the sea cucumbers

The holothuroids are one of the smaller echinoderm classes with about 900 species. Like the irregular echinoids, they have developed bilateral symmetry, though the plane of symmetry is different. Whereas the irregular echinoids are flattened in the oral-aboral plane and have come to move with one ambulacrum (the anterior ambulacrum) leading, holothuroids are greatly elongated in the oral-aboral plane and lie on their side, with an anterior mouth and a posterior anus. Their common name refers, rather fancifully, to this elongated shape. The underside (trivium) is formed from three ambulacral areas and two inter-ambulacral areas; the ambulacra have suckered tube feet placed in definite rows which are used for locomotion. The upper side (bivium) has two ambulacral areas with scattered tube feet which lack suckers and have a warty appearance.

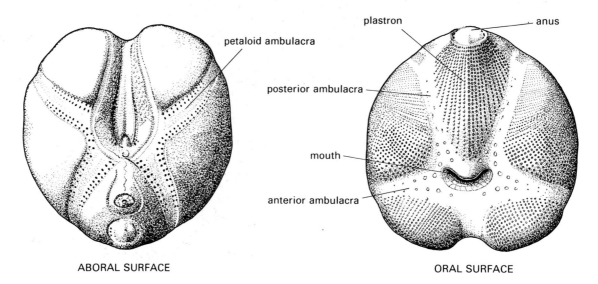

ABORAL SURFACE   ORAL SURFACE

Echinocardium test

The body wall of holothuroids is more closely similar to that of soft-bodied invertebrates than of other echinoderms; it has no spines or tubercles but is tough, smooth and muscular. It has a glandular epidermis with a cuticle, a sub-epidermis layer of collagen fibres (which function in a similar way to those of other soft-bodied forms), and layers of circular and longitudinal muscle. The skeleton is reduced to small calcareous spicules embedded in the body wall. Although holothuroids are relatively sluggish, their soft body wall makes them unattractive to sessile animals and pedicellariae are not present.

Holothuria is a common genus of sea cucumber, found on soft substrates on the lower shore, and to a depth of 70 m below low water. It creeps over the substrate by means of its ventral tube feet, whose suckers, which are richly supplied with mucus cells, provide the principal adhesive force. In addition contraction of the body wall muscles assists the process of locomotion. *Holothuria* feeds on particles of detritus which it shovels into its mouth using a circlet of 20 oral tentacles surrounding the mouth. These are highly modified tube feet and are branched and highly muscular, with mucous and sensory cells. When not in use the tentacles are contracted and the body wall folded over them. Like the echinoid gut the holothurian gut is long and coiled into three large loops. At the posterior end it enlarges to form a cloaca into which open ducts form a pair of branching respiratory organs, the respiratory trees. These are branched structures with a large surface area across which gaseous exchange can occur by diffusion directly in and out of the coelomic fluid. Nitrogenous waste probably leaves by the same route. *Holothuria* obtains about 60% of its total oxygen requirement via the respiratory trees and the remainder over the general body surface. Water is pumped in and out of the respiratory organs by slow

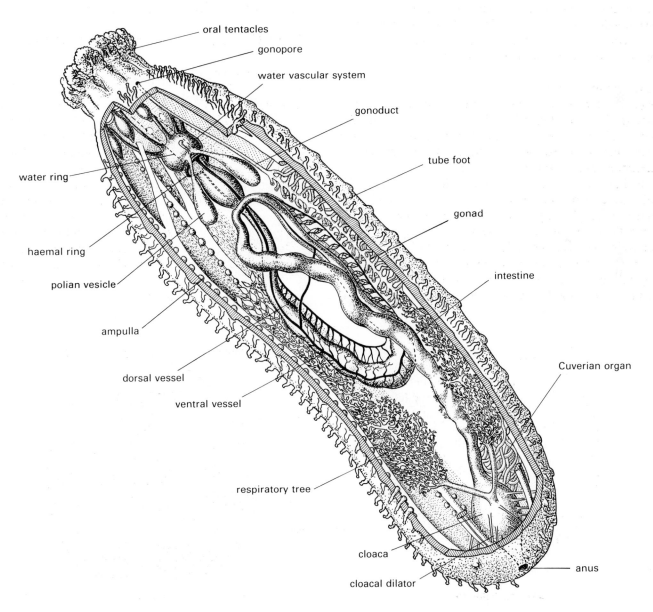

A holothurian

pumping movements of the cloaca; between 6–10 contractions are required to fill them and they are emptied by a single contraction.

Tufts of long, blind-ending tubules, the Cuvierian organs, are associated with the ducts of the respiratory trees and provide a remarkable form of self-defence. They have a narrow lumen, continuous with that of the respiratory tree ducts, a complex wall composed of peritoneum, a layer of gland cells which secrete polysaccharide, and a layer of spirally arranged collagen fibres held tightly coiled by muscles. If the animal is molested by a would-be predator, it points its anus at the predator and contracts its body wall muscles. This increases the internal pressure of the coelom ad forces water from the cloaca and respiratory trees into the lumens of the Cuvierian organs.. The Cuvierian organs elongate within the coelom and rupture the cloaca, shooting out of the anus in the form of long sticky white threads. They are an extremely effective defence mechanism, because they can entangle even such active predators as lobsters and prevent their movement whilst the holothurian creeps away. This response is the basis of the common name 'cotton spinner' sometimes applied to these animals. Occasionally *Holothuria* will also shoot out its entire alimentary tract and its gonad by the same process; all these structures are later regenerated.

The water vascular system is similar to that of other echinoderms, although the madreporite has no external opening but opens into the perivisceral coelom as a madreporic body. The haemal system is well developed and has basically the same organization as that of other echinoderms. Dorsal and ventral vessels extend along the intestine suggesting that the haemal system is involved in food absorption and transport.

The nervous system is similar to that of other echinoderms, with a circum-oral nerve ring which supplies nerves to the tentacles and pharynx, and five radial nerves which extend the length of the ambulacra beneath the dermis. The general epidermal sense organs show some concentration at the anterior and posterior ends.

Holothuroids are unique amongst the echinoderms in possessing a single gonad. This lies in the anterior part of the coelom and opens via a single gonopore. Fertilization is external, and a planktonic auricularia larva (similar to the bipinnaria stage of asteroids) develops. Further development gives barrel-shaped doliolaria larva, with 3–5 ciliated bands. Gradual metamorphosis takes place in the plankton, and eventually the young sea cucumber settles and takes up its benthic existence.

Class Crinoidea

There are only about 80 species of living crinoids (the sea lilies and feather stars) but the group has a long fossil history and crinoids are generally regarded as the most primitive echinoderms. Amongst the living forms, the sessile sea lilies, with a long basal stalk (up to 1 m length), live in deep water (100 m or more) and are poorly known. The more numerous free-swimming crinoids, the feather stars, are better known, forms such as *Antedon* are found in shallow water or on the lower shore of rocks, stones and among seaweed.

The orientation of the crinoids is different from that of any other echinoderm classes for the mouth is held uppermost (away from the substrate) and the anus is also on the oral surface. The aboral pole is either extended as a permanent stalk as in the sea lilies, or may form temporary attachments, by five-jointed cirri, in animals like *Antedon*.

The crinoid body consists of an inconspicuous cup-shaped body, the calyx, which has a basic pentamerous symmetry and bears five arms which fork at their bases to give a total of ten. Each bears an ambulacral groove and numerous jointed lateral extensions or pinnules which give the animals the feather-like appearance to which their common name refers. Five ambulacral grooves extend from the mouth to the arm bases. The body wall has numerous skeletal ossicles within the dermis and an almost solid construction. In the region of the ambulacral grooves the epidermis is ciliated, but otherwise cilia are poorly developed.

Like ophiuroids, crinoids do not rely on their tube feet for locomotion. Feather stars swim by sweeping their arms up and down in alternating sets of five. They can also crawl slowly on the tips of their arms, holding the body away from the substrate. Neither method of locomotion is particularly effective, and feather stars spend long periods of time attached to the substrate by their cirri. When not in use for feeding or locomotion, the arms are held curled over the body.

Crinoids are exclusively filter feeders, and catch their food by means of a food trap which is made by the tube feet. In *Antedon* the tube feet are arranged in groups of three, and extend along the arms and the lateral pinnules. Those of the arms are all the same length, but those of the pinnules are three different lengths, the outermost being the longest. The tube feet have numerous fine processes or papillae and secrete mucus. A ciliated food groove extends along the oral surface of each arm to the mouth and food caught by the extended tube feet is flicked into this; from there it is transferred by ciliary action to the mouth.

The mouth opens into a short oesophagus and thence into a coiled intestine which lies within the body disc. Several digestive caecae which secrete digestive enzymes extend from the intestine. The faeces are formed into pellets surrounded by mucus in the rectum; the anus lies at the end of a muscular papilla which is used to direct the faeces away from the mouth.

The coelom has been greatly invaded by connective tissue, and is reduced to a series of communicating spaces. The haemal system is similarly a series of spaces, with a branch of the haemal system extending along the two branches of each arm. The water vascular system forms a ring around the mouth, and from it a radial canal extends into each arm beneath the ambulacral groove. This branches into the two parts of each arm, and extends into all

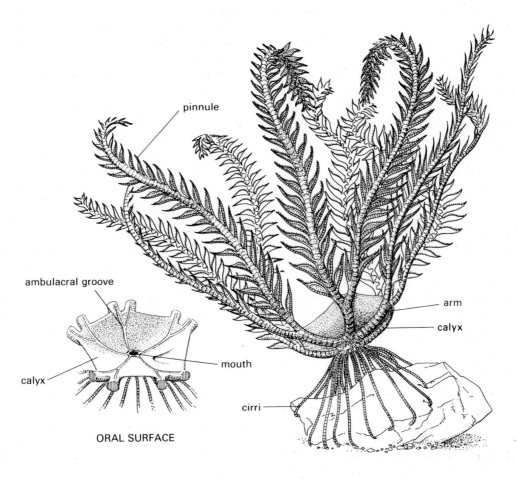

pinnule

ambulacral groove

arm

calyx

calyx

mouth

cirri

ORAL SURFACE

Antedon

the pinnules. Gaseous exchange takes place across the tube feet. No madreporite is present but the circum-oral ring gives off numerous (a total of about 50 in *Antedon*) short stone canals at each inter-radius. These connect with short ciliated funnels, extending between the oral surface and the underlying coelomic spaces, and probably allow the internal pressure to adjust to the external pressure.

Crinoids have elaborate nervous systems composed of three interconnecting divisions. An oral nerve ring surrounding the mouth gives off sub-epidermal radial nerves which extend below the ambulacral grooves and correspond to the nervous system of other echinoderms. This part of the system is sensory. There is also an aboral system in the aboral region of the calyx which supplies nerves to the cirri as well as to the muscles of arms and pinnules: the aboral system therefore controls locomotion. In addition, beneath the oral system a deeper hyponeural system supplies the tube feet and pinnules.

Crinoids have no distinct gonads. In *Antedon* the gametes develop within coelomic extensions into the pinnules; when they are mature the pinnules rupture and the eggs are cemented onto the outer walls of the pinnules where fertilization occurs. A barrel-shaped non-feeding larva, like

that of the holothuroids, develops. After a brief free-swimming existence it develops as a stalked sessile stage, the pentacrinoid larva which develops to the adult after several months. Like other echinoderms crinoids have considerable powers of regeneration.

Echinoderm relationships

In spite of the abundant fossil record, the relationships of the echinoderms are not fully understood. However they are generally accepted to have evolved from bilaterally symmetrical ancestors with a tripartite coelom and to have evolved their radial symmetry and skeleton in response to a sessile mode of life, a stage represented by crinoids. One of the oldest established theories of echinoderm origins postulates a pentactula ancestor, a bilateral form with tentacles round the mouth in an arrangement very like a lophophore. Such a form could well be the ancestor of both the lophophorates and the echinoderms; indeed a sipunculate has been postulated as the link between these groups.

Synopsis of the Echinodermata

Class Crinoidea (*Antedon*)
Class Asteroidea (*Asterias, Poronia, Solaster, Crossaster, Luidia*)
Class Ophiuroidea (*Ophiothrix, Ophiactis*)
Class Echinoidea (*Echinus, Echinocardium, Echinocyamus*)
Class Holothuroidea (*Holothuria*)

THE MINOR DEUTEROSTOME PHYLA

Phylum Chaetognatha

The phylum Chaetognatha is a group of marine animals which are extremely common in oceanic plankton. They are easily recognized from their elongated torpedo-shaped body with paired lateral fins and a tail fin, which gives them their common name of arrow worms. Their anterior mouth has strong grasping spines which are the basis of the phyletic name, Chaetognatha meaning 'bristle jaws'. Like many other planktonic animals, chaetognaths are transparent and therefore difficult for a predator to detect. There are about 50 species showing rather little structural diversity and belonging to a single class. One genus, *Spadella*, is bottom living and includes a European species, *Spadella cephaloptera*, which may be found attached to rocks or seaweeds in rock pools.

Many chaetognaths are associated with particular ocean currents and are therefore used as indicator species by oceanographers when they study the path of water currents or the origin of planktonic communities. For example, *Sagitta setosa* is characteristic of coastal waters of the North Sea, and the English Channel. The genus *Sagitta* was the earliest chaetognath type to be recognized (the name means 'an arrow') and most species are assigned to it.

Sagitta

Body form

Sagitta setosa is about 20 mm long, although other members of the genus are longer. The body is elongated and divided into a small head, a long trunk, and a post anal tail. Two pairs of lateral fins are present: an anterior pair midway along the trunk and a posterior pair which terminate just before the spatulate tail fin. The head is separated from the trunk by a slight constriction, the neck, and has an invagination of the ventral surface, the vestibule, into which the mouth opens. Curved chitinous spines hang down from each side of the head and are used for catching prey. When the spines are not in use a hood, formed from a fold in the body wall containing a coelomic space, is drawn over them

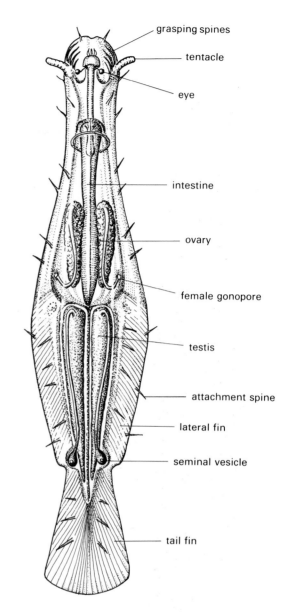

Spadella

and is thought to be used to protect them when the animal swims.

The body wall consists of the epidermis, which secretes a thin cuticle, a basement membrane, and (in the trunk and tail region) paired dorsolateral and ventrolateral muscle bands. Anteriorly the epidermis includes large vacuolated cells, thought to provide some buoyancy. The head musculature is intricate, with special muscles to operate the hood, spines, and anterior and posterior rows of teeth formed by the thickenings of the vestibule chitin. The fins function as stabilisers and flotation aids; they are not muscular but the basement membrane is thickened to provide support.

The coelom is divided into the head coelom which extends into the hood, paired trunk coeloms separated by a longitudinal septum, and subdivisions of these which form the tail coelom. It does not have a peritoneal lining and therefore resembles a pseudocoele; however a study of chaetognath embryology reveals its true coelomic nature.

Locomotion

Sagitta swims with a jerky, darting motion, caused by rapid contractions of the longitudinal muscles. Periods of activity alternate with intervals of rest, during which it slowly sinks.

Feeding

Chaetognaths are among the most voracious predators of the planktonic world. *Sagitta setosa* feeds particularly on young copepods, while *S. bipunctata* is recorded as feeding on young herring of the same size as itself. To catch its prey the animal darts forwards and siezes it with outspread spines which are then snapped rapidly together. The prey is dragged into the vestibule, where the thickened cuticle protects the arrow worm against possible damage, and is bitten with the chitinous teeth. In some species there is a glandular structure, the vestibular pit, which is thought to secrete venom to kill or immobilize the prey. The alimentary canal consists of a muscular pharynx and an intestine which extends through the trunk and has a pair of anterior lateral diverticula. The food is dragged into the pharynx, which secretes a moisturizing and adhesive substance, and moved to the posterior end of the intestine where digestion occurs.

Gaseous exchange, circulation and excretion

Sagitta has no special organs for either gaseous exchange or excretion, and these processes are thought to take place over the general body surface. The coelomic fluid provides transport for nutrients, metabolites and wastes.

Nervous system and sense organs

The nervous system is fairly complex, and shows a high degree of cephalization. A nerve ring with a large bilobed cerebral ganglion and a pair of ventral ganglia surrounds the pharynx. The eyes are supplied by nerves from the cerebral ganglion; and the vestibule, head muscles, and pharynx receive nerves from the ventral ganglia. Paired sub-enteric connectives connect the cerebral ganglion to a sub-enteric ganglion from which nerves supply a sub-epidermal plexus of the trunk and tail. There are fan-like groups of sensory hairs arranged in longitudinal tracts along the trunk which can detect water vibrations, and these are the principal means of detecting prey. In addition *Sagitta* has two eyes, each composed of five pigment spot ocelli. A U-shaped loop of ciliated epidermis over the head, neck, and anterior trunk, the corona, is supplied by nerves from the cerebral ganglia and may function as a chemoreceptor or to detect the direction of water currents (rheoreceptor).

Reproduction and growth

Sagitta is hermaphrodite and has a pair of elongate ovaries in the trunk coelom, and a pair of testes in the tail coelom. A sperm duct arises from each testis and terminates in a

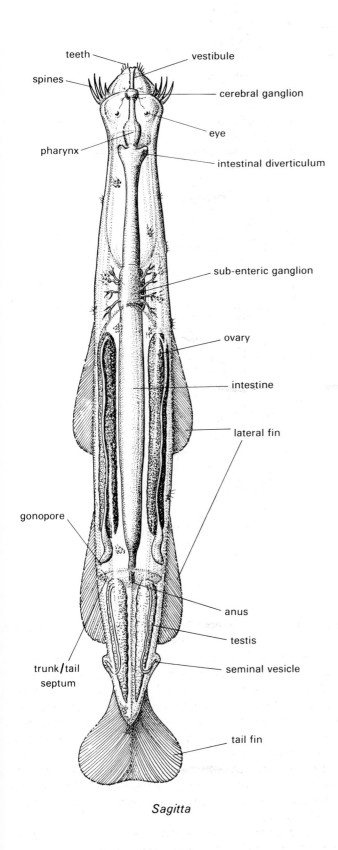

Sagitta

seminal vesicle in the lateral body wall. Sperm are shed into the tail coelom where they mature. They then enter the sperm duct through a ciliated funnel and are formed into a spermatophore in the seminal vesicle. The ovaries each have an oviduct which opens to the exterior at a gonopore immediately anterior to the trunk tail septum. The details of sperm transfer are not known, it is assumed that the spermatophore is transferred in some way to another individual and then ruptures to let the sperm escape. Whatever the mechanism, the eggs are fertilized in the oviduct, surrounded by a coat of albumen, and released through the gonopore. Some chaetognaths carry their eggs, others leave them to float in the plankton, and the benthic *Spadella* attaches them to algae or stones. The young arrow worm hatches as a miniature replica of the adult and takes up a similar mode of life.

Chaetognath relationships

The arrow worms are probably the most difficult to place of all the invertebrate phyla. Their radial cleavage suggests that they are deuterostomes, but the coelom shows a peculiar form of development, more closely comparable with that of a pseudocoele. Any evidence for the relationships of such highly-adapted and unusual animals is weak, and their grouping in this book with the deuterostomes is largely one of convenience.

Phylum Hemichordata

The phylum Hemichordata, with about 100 species, includes two classes of marine animals; these are the free-living Enteropneusta or acorn worms, and the sessile tube-dwelling Pterobranchia.

As their name suggests the hemichordates were once regarded as a subphylum of the phylum Chordata. The group has had a chequered taxonomic history; when the first adult enteropneust was described (in 1825), it was regarded as a highly aberrant holothurian. The characteristic swimming larval stage of the enteropneusts was interpreted as a starfish larva from its resemblance to the bipinnaria larva. In fact it was not until its development was observed in 1890 that its proper relationship was known. Pterobranchs on the other hand, were initially described as bryozoans; however the subsequent recognition of gill slits (one of the fundamental characters of chordates), in both some pterobranch genera and in the enteropneusts, led to their combination as hemichordates and inclusion within the phylum Chordata. More recently they were regarded as sufficiently different from the chordates to justify their establishment as a separate, though related phylum.

Hemichordates are readily distinguished from the other animals discussed so far by their body, which is clearly divided into three regions: protosome, mesosome, and metasome. The protosome is modified as a short, conical proboscis (from which the name acorn worms is derived) in the enteropneusts, and as a shield-shaped structure, used to secret the tube, in the pterobranchs. The mouth opens at the anterior edge of a collar, which in the pterobranchs is elaborate, bearing arms and tentacles which form a food gathering organ. The metasome forms a greatly elongated abdomen in the enteropneusts but is short and somewhat barrel-like in the pterobranchs. In common with many other sessile, tube-dwelling animals, pterobranchs have a U-shaped gut.

Class Enteropneusta

Enteropneusts occur in both shallow and deep waters. They live in U-shaped burrows which they dig in mud, sand or clay or may be found under stones or seaweed. Their body length usually ranges between 9 and 450 mm but at least one species, *Balanoglossus gigas*, is recorded as growing to 1.5 m long. The common British and Mediterranean species, *B. clavigerus*, grows to 300 mm long.

Balanoglossus

Body form
Balanoglossus has a short proboscis which is connected to the collar by a narrow proboscis stalk. The collar is a short cylinder which extends forwards to surround the proboscis stalk and to enclose the ventral mouth, which opens at the base of the proboscis. The long abdomen is flattened at the anterior end and extended laterally, forming genital flaps which contain the gonads. A row of about 50 pairs of gill pores, partially obscured by the genital flaps, connects the anterior region of the gut to the outside on either side of a shallow mid-dorsal groove. This groove marks the area of attachment to the body wall of a well-developed region of nervous tissue. The body cavity consists of a single coelomic sac in the proboscis, with paired bilateral sacs in the collar and abdomen. The coelom of both the collar and the proboscis open to the outside by minute dorsal pores. However much of the coelom is filled by connective tissue and muscle fibres, which replace the more usual body wall muscles of other animals. The body wall is covered by a ciliated epidermis which in the collar and proboscis regions is highly glandular and secretes mucus which is used to line the burrow.

Locomotion
Balanoglossus is a sluggish animal and seldom, if ever, leaves the burrow which it excavates by means of peristaltic movements of its proboscis. It uses body cilia to move within the burrow and retreats backwards to the surface to

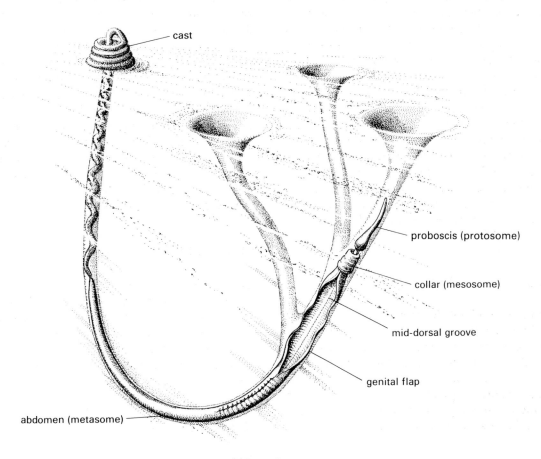

cast

proboscis (protosome)

collar (mesosome)

mid-dorsal groove

genital flap

abdomen (metasome)

Balanoglossus

defaecate, leaving characteristic casts rather like those of polychaetes.

Feeding and excretion

Balanoglossus clavigerus ingests large quantities of sand or mud from which it digests organic matter. However many enteropneusts are filter feeders which feed by extruding the proboscis from the burrow and slowly waving it about; particles of food from the substrate stick to the mucus and a ciliary beat carries them to the mouth. At the base of the proboscis the ciliary beat is lateroventral and the mucus-entangled food particles are directed towards the mouth. If the animal does not wish to feed it covers the mouth with its collar, and particles then pass over the collar. This provides a rejection mechanism for undesirable particles, such as those which are too large. The gut extends as a straight tube to the end of the abdomen. The mouth opens into a buccal cavity which lies within the collar and has a dorsal diverticulum extending forwards into the proboscis. This is historically interesting as it was originally interpreted as a skeletal structure comparable to the chordate notochord and was responsible for the inclusion of the hemichordates in the phylum Chordata. The diverticulum is associated with the central sinus of the circulatory system and a mass of peritoneal tubules called the glomerulus. Together these

three structures form the proboscis complex which is believed to be excretory in function. The pharynx extends posteriorly from the buccal cavity, and is perforated by the gill slits which are supported by thickenings of the basement membrane. The pharyngeal slits are thought to have evolved primarily as feeding structures, and certainly fulfil this function in primitive chordates. However in enteropneusts their role in feeding is limited and their principal function is thought to be respiratory.

The pharynx leads to an oesophagus in which the food is compacted into a cord; in some species there are external oesophageal pores which allow excess water to escape at this stage. The gut is completed by an intestine which is the principal region of digestion and absorption, and a terminal anus. Little is known of hemichordate digestive physiology.

Respiration and circulation

Gaseous exchange probably takes place over the entire body surface but the pharyngeal slits are particularly important in this process. The pharyngeal wall between adjacent slits (septum) and the tongue-like projection of the dorsal wall (tongue bar) which partially divides each cleft to give a U-shaped structure, are ciliated. The cilia beat to create a current which passes in through the mouth and out at the gill pores. The septa and tongue bars have a plexus of blood sinuses, and gaseous exchange occurs between the blood and the water passing through the slits.

The blood is a colourless fluid with no respiratory pigment and few amoebocytes. A dorsal longitudinal vessel extends in the mesentery which supports the gut, and transports the blood anteriorly. In the collar the vessel passes into a venous sinus and thence to the central sinus at the base of the proboscis, from which blood passes into the glomerulus. Blood from the glomerulus passes into a ventral longitudinal vessel which runs below the gut and supplies a system of sinuses in the pharyngeal slits, gut wall and body wall. These drain into the dorsal vessel. Blood is pumped through the system by pulsation of a closed, fluid-filled sac dorsal to the central sinus called the heart vesicle, and by contraction of the longitudinal vessels. It should be emphasized that the blood does not enter the heart vesicle; however the effect of the vesicle contractions is to create a current through the central sinus.

Nervous system and sense organs

The nervous system is formed from a sub-epidermal plexus of nerves which lies just external to the basement membrane. In various regions, notably the mid-dorsal regions at the proboscis and the mid-ventral region of the trunk, the nerves show local concentration and distinct thickening to form longitudinal dorsal and ventral nerve cords. The dorsal nerve cord continues through the collar (the collar cord), but is separated from the epidermis as an internal structure. In some species it is hollow in this region, providing another revealing point of similarity with the chordates. The collar cord contains giant fibre cells and probably provides for the rapid conduction of nerve impulses involved in peristaltic movements of the proboscis in burrowing. If the collar cord is cut these movements can continue, but less effectively, with conduction of impulse occurring, presumably, through the general sub-epidermal plexus.

There is a U-shaped pre-oral ciliary organ on the proboscis which is probably a chemoreceptor, and it appears to be the only special sense organ. General sensory cells are scattered through the epidermis.

Reproduction and development

The sexes are separate. Reproductive structures are simple and take the form of numerous sac-like gonads, each opening to the surface through a gonoduct beside a gill pore. Eggs are extruded from the burrow, in masses of 2,000 to 3,000 embedded in mucus, and fertilization occurs externally. The fertilized eggs are scattered and develop into tornarian larvae which are structurally similar to the early bipinnarian larvae of some echinoderms (asteroids). There is a convoluted band of cilia at the anterior end, which creates a feeding current, and a posterior ciliated girdle for locomotion. The larva has a free-living life of several weeks, during which it undergoes metamorphosis by developing a constriction at the proboscis-collar region and undergoing elongation to the adult form.

Class Pterobranchia

The pterobranchs are sessile, benthic animals of deep water and are far less well known than the enteropneusts. Three genera have been described, of which two, *Rhabdopleura* and *Cephalodiscus*, are reasonably well known. They are smaller than acorn worms, *Cephalodiscus* being about 5 mm in length, and *Rhabdopleura* only 1 mm in length.

The proboscis is modified as a cephalic shield, used by both *Rhabdopleura* and *Cephalodiscus* to secrete protective tubes (coenecia) constructed in a series of rings. *Rhabdopleura* forms colonies of individuals connected by a stolon whereas *Cephalodiscus* occurs in groups of independent stalked individuals.

The most striking feature of the pterobranchs is the collar which is modified into hollow arms containing extensions of

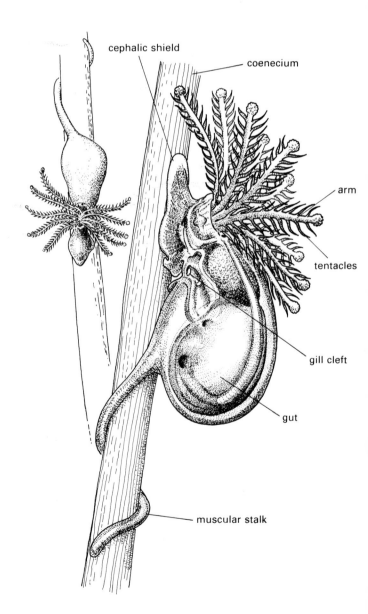

Cephalodiscus: feeding outside the coenecium

the mesocoele. The arms, one pair in *Rhabdopleura*, 5–9 pairs in *Cephalodiscus*, bear numerous ciliated tentacles. The collar can be compared both structurally and functionally with the lophophore of the lophophorate phyla. In addition to collecting food with the arms, *Cephalodiscus* collects food particles over its general body surface, transporting them by short body cilia to the mouth.

The pharynx has a single pair of gill clefts in *Cephalodiscus* but none are present in *Rhabdopleura*; the reduction of these structures, and their absence in some forms, is thought to be associated with the small size of these animals.

The gut is U-shaped and the anus opens at the dorsal surface of the collar. In other respects the organ systems appear to correspond with those of the enteropneusts.

Little is known of pterobranch reproduction or development; apparently a tornaria-like larva develops which passes through an enteropneust-like stage with a terminal anus. Later in development the intestine bends to give rise to the characteristic U-shape.

Phylum Chordata

Introduction to the phylum Chordata

The phylum Chordata, which includes the vertebrates, is the largest deuterostome group, and the most recently emerged in terms of geological time. While the invertebrate fossil record extends back over 1,600 million years, fossils of the earliest vertebrates are only 500 million years old. However, in spite of their relatively recent emergence, chordates have established themselves as one of the major animal phyla, with dominance of the water, land, and air. They cannot rival arthropods in terms of numbers of either species or individuals, but outstrip them in terms of total biomass and ecological dominance.

Chordates are coelomates, and as such are another example of the high level of diversification achieved at that level of organisation; however there are fundamental differences between chordates and all other members of the Animal Kingdom, three chordate characteristics being regarded as diagnostic. Firstly, all chordates have, at least at some stage in their life history, an internal axial skeletal structure in the form of a stiffened but flexible rod known as the notochord. This structure lies immediately ventral to the central nervous system and dorsal to the gut, and in more primitive chordates persists throughout the animal's life. However in advanced chordates (subphylum Vertebrata) the notochord is wholly or partly replaced by a backbone or vertebral column, composed of a series of separate bones, the vertebrae. The best known chordates (fishes, amphibians, reptiles, birds and mammals) are all vertebrates.

Secondly all chordates possess pharyngeal clefts or gill slits. These have already been mentioned in the context of phylum Hemichordata and must now be discussed in rather more detail. They develop in the embryonic chordate as a series of paired evaginations of the lateral pharyngeal walls. Their development is completed by corresponding invaginations of the body wall with a breakdown of intervening structures so that the clefts come to open to the exterior. Pharyngeal clefts appear during the embryonic development of all chordates and subsequently develop to serve different functions in different chordate groups. In the most primitive aquatic chordates they are modified as feeding structures and in more advanced aquatic forms as respiratory structures. In the majority of non-aquatic adults they close, and most of them disappear entirely.

The third diagnostic chordate character is the possession of a hollow dorsal nerve tube at some stage during life. This can normally be subdivided into an expanded anterior region, the brain, and a spinal cord which extends the full length of the animal's trunk. The brain is protected by a box-like skeletal structure, the cranium. In the least advanced chordates the brain and cranium are absent, and in some specialized lower chordates even the dorsal nerve tube degenerates when the adult develops. These lower chordates are sometimes grouped as Acrania or proto-chordates, in contrast with the vertebrates or Craniata in which a brain and cranium is always present.

In addition to their diagnostic features there are two other points of contrast between the majority of chordates and most other animals. Firstly, there is usually a post-anal segmented structure, the tail; this is flexible and muscular in most aquatic vertebrates and forms their chief means of propulsion. In terrestrial and aerial forms, where the tail is not used for propulsion, it is usually shortened. Secondly the chordate heart is a modification of the ventral blood vessel and lies ventral to the gut, whereas the non-chordate heart is usually a modification of the dorsal longitudinal blood vessel.

The phylum Chordata is subdivided into three subphyla: Urochordata, Cephalochordata, and Vertebrata. Whilst the Vertebrates, with approximately 38,000 species, are the dominant chordate group, both the Urochordata and Cephalochordata (together sometimes known as Acrania or protochordates) are of considerable biological interest. The Urochordata have a microscopic larval stage which is obviously chordate, whereas the adults bear little resemblance to other members of the phylum. Cephalochordates are also specialized and peculiar in some respects but demonstrate a simple adult chordate condition.

Subphylum Urochordata

Urochordates, often known as tunicates, are common marine animals which as adults show a closer, though entirely superficial, resemblance to invertebrate forms such as cnidarians than to other chordates. However the larval form displays all the diagnostic features of a chordate, including a notochord, pharyngeal clefts and a dorsal hollow nervous system, and also has a post-anal tail. The subphylum may be regarded as a group which branched off at an early stage from the main chordate evolutionary line; they show no evidence of the metameric segmentation characteristic of other chordates and the coelom is absent or reduced. The majority of the adult peculiarities are adaptations to a sessile mode of life.

Of the 1,300 species which make up the subphylum, the majority belong to the class Ascidiacea (the sea squirts), a group of sessile animals with a world-wide distribution in shallow sea waters. The other urotochordate classes, Thaliacea and Larvacea, are specialized for a marine planktonic mode of life. Since the majority of urochordates are ascideans, one of that class, *Ciona intestinalis*, has been selected for more detailed description.

Ciona intestinalis

Body form

Ciona is a translucent yellow-green sac-like animal around 100 mm in length, which is found growing in large numbers on rocks, piers, piles and on seaweed. It is also found growing on ship hulls, which may in part account for its wide distribution. Like many sessile animals it is highly specialized as a filter feeder. However its specializations are not immediately apparent since most of the body is enclosed in a thick protective outer layer secreted by the epidermis and called the tunic, from which the common name of the group (tunicates) is derived. The tunic is largely acellular, consisting of a fibrous matrix of tunicin (a polysaccharide similar to cellulose) with protein and only a few cells. One

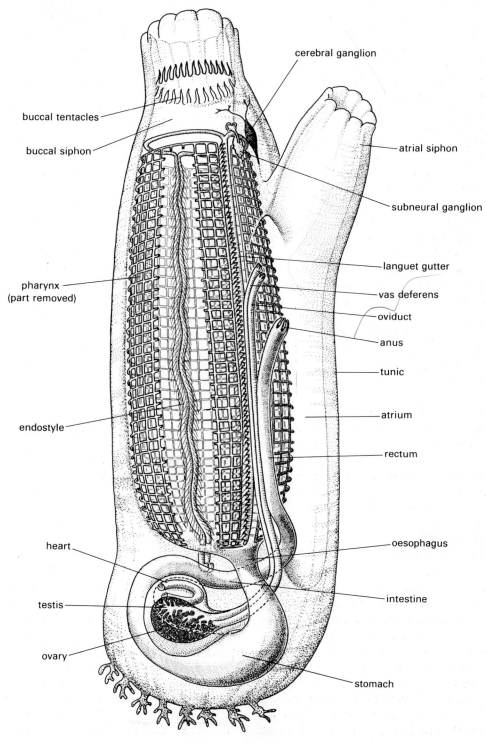

Ciona intestinalis

end is closely attached to the substrate by a series of projections, the holdfast, and the other has two openings at the end of short extensions, the buccal and atrial siphons. Primitive ascideans have a body divisible into three distinct regions: pharyngeal, abdominal, and post-abdominal. However in the majority of species the post-abdominal region is unclear, and in *Ciona* the distinction lies between a large barrel-like pharynx and the remainder of the body. The greater part of the internal volume of the tunic is occupied by the pharynx, into which the buccal siphon opens. The pharyngeal wall is perforated by numerous small slits (stigmata) which are the only obvious phylogenetic link between the adult ascidean and the chordates. These are formed by increase and subdivision of six pairs of initial protostigmata and allow a water current to pass through the pharynx into a surrounding space, the atrium, which opens at the atrial siphon and is criss-crossed by strands of tissue. The alimentary tract is completed by an oesophagus, stomach, intestine and rectum. Like that of many sessile animals it is U-shaped, opening into the atrium below the atrial sinus.

The body wall below the tunic (sometimes called the mantle) consists of a single-layered epidermis overlying a thicker fibrous connective tissue dermis, and bands of circular and longitudinal muscle. The circular muscle is particularly well developed around the siphons forming sphincters. General contraction of both longitudinal and circular muscle bands at intervals during feeding, as well as when the animal is disturbed or exposed, forces water out from the atrium and pharynx as jets, hence the common name sea squirts.

Tunicates do not have a true coelom, but are so obviously related to the coelomates that this loss must be secondary. Two body cavities are present: the pericardial cavity, which folds to form and surround the heart and is sometimes interpreted as a vestigial coelom; and the highly unusual epicardium which arises as a double evagination from the base of the pharynx. This evagination enlarges and unites to form a single tube which surrounds the viscera in the same way as a coelom. The epicardium is surrounded by mesenchyme, and the whole structure, including the alimentary tract and gonads, forms the visceral mass.

Feeding

Ciona feeds on diatoms and other unicellular algae as well as other organic matter of a suitable size which it filters out of the surrounding water. However the method of food collection is fundamentally different from non-chordate filter feeders, including hemichordates, since it depends entirely on filtration through the pharynx. Lateral cilia projecting across the stigmata from the pharyngeal bars beat with a metachronal rhythm to produce a water current which brings suitable food particles into the pharynx from the surrounding water. A circlet of buccal tentacles projecting within the buccal siphon excludes any overlarge particles. On the ventral side of the pharynx (opposite the atrial siphon) is a deep groove called the endostyle which

has mucus-secreting cells, basal flagellated cells, and lateral ciliated cells. Opposite the endostyle on the dorsal mid-line is a row of curved, finger-like dorsal structures, the languets, which form a ciliated gutter extending to the oesophagus.

Mucus secreted by the endostyle is swept over the walls of the pharynx in a thin sheet by the beating of frontal cilia which project into the pharyngeal chamber from the pharyngeal bars. As the feeding current passes out of the pharynx, food particles are trapped in the mucus which is moved towards the dorsal gutter. The pressure of the outflowing water current holds the mucus and food particles pressed against the cilia. The food-laden mucus collects in the gutter where it is rolled into a compact cord and propelled backwards towards the oesophagus. The oesophageal opening in *Ciona* has an expanded ciliated lip and the beating of these cilia draws the food cord into the oesophagus. The filter-feeding mechanism appears to be very effective; *Ciona* can extract particles as small as 1 to 2 microns, and produce a current of 3,000 ml per hour.

The remainder of the alimentary tract is fairly simple. At the base of the U-shaped tract the narrow oesophagus leads to an expanded ovoid stomach which has two spirally arranged bands of cilia, and mucus-secreting cells. More mucus is added to the food cord and the ciliary beat causes it to rotate and pulls it down through the oesophagus. Enzymes, chiefly amylase, saccharase, lipase, and peptidase, are secreted in the stomach, which is the principal site of extracellular digestion. From the stomach a straight intestine, a region of further digestion and absorption, ascends to the anus. Food is stored as glycogen in the intestinal wall.

Circulation, gaseous exchange and excretion

The blood vascular system is unusual. A fold of the pericardial sac forms a U-shaped tubular heart at the base of the gut loop. A ventral abdominal sinus, connected to the dorsal end at the heart, opens into an extensive system of sinuses which do not have walls and are simply spaces within the mesenchyme around the viscera. From the visceral sinuses a median dorsal sinus extends to the pharynx where there is a further system of sinuses known as the pharyngeal lacunar plexus. From the ventral side of the pharyngeal lacunar plexus a sub-endostylar sinus extends to the ventral end of the heart. The blood circulation is therefore essentially triangular and blood can, and does, travel through it in both directions. Peristaltic muscular constrictions pass along the heart, but after the blood has made about twenty circuits, the direction of the peristalsis and hence of blood flow, is reversed. This arrangement is unique to ascideans, and its functional significance is not understood. The blood of tunicates is also unusual since it is colourless, and the plasma, which is isotonic with sea water, contains several different types of cells: lymphocytes, amoebocytes, nephrocytes and morula cells. The last three cell types are all lymphocyte derivatives: amoebocytes are nutritive and phagocytic, nephrocytes store nitrogenous wastes (chiefly in the form of uric acid) and morula cells function in tunic formation. The necessary surface for gaseous exchange is provided by the wall of the pharynx and the pharyngeal lacunar plexus. The greater part of nitrogenous waste produced by protein metabolism is also shed by diffusion at the pharyngeal wall in the form of ammonia.

Nervous system and sense organs

Ciona has a simple nervous system. A hollow elongated cerebral ganglion lies in the mesenchyme between the siphons and extends posteriorly as a cord of tissue. Nerves arising from the anterior cerebral ganglion supply the buccal siphon and mantle muscles, those from the posterior part supply the remainder of the body. A glandular structure of unknown function, the neural gland, lies beneath the cerebral ganglion. There are no special sense organs, but sensory cells are scattered over the body surface and are particularly concentrated on the siphons, buccal tentacles, and in the atrium.

Reproduction and growth

Ciona intestinalis is essentially an annual animal with a high mortality rate in the winter. It has no definite breeding season, and the breeding period is related to the temperature of the water in which it lives. In colder waters, for example off the coast of Scotland, breeding is restricted to a short period in the summer, whereas in warmer waters it may be continuous. Like the majority of ascideans, *Ciona* is hermaphrodite. The reproductive system lies within the gut loop and consists of a single, compact, hollow ovary partly overlain by a branching testis. Eggs are shed into the ovarian cavity from which an oviduct, which is continuous with the cavity, extends to a gonopore opening into the atrium near the atrial siphon. The testis consists of a mass of thin, branching tubules overlying the ovary and part of the intestinal wall; a vas deferens extends to open near the female gonopore. Both eggs and sperm are shed through the atrial siphon.

Fertilization is external, and in *Ciona intestinalis* development is rapid. After total radial cleavage the egg develops to a larva which hatches. The larva is minute and tadpole-shaped, so that it is often referred to as the ascidean tadpole. It shows ascidean affinities with the chordates far

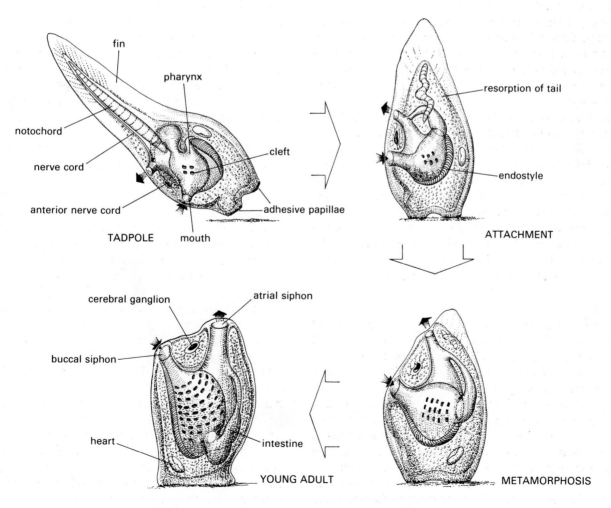

The ascidean tadpole and its metamorphosis

more clearly than the adult, having a notochord formed of twenty to forty vacuolated cells within a connective tissue sheath, a dorsal hollow nerve cord and a long tail. A tunic forms the outer body surface and extends on either side of the tail as a fin. The pharynx has a few pharyngeal clefts opening into the atrium which extends dorsally. An anterior mouth, which eventually forms the buccal siphon, is present, however ascidean tadpoles are short-lived larvae which do not feed and the mouth frequently does not open until metamorphosis. A dorsally directed twisted intestine completes the alimentary tract.

The larva swims by side to side movements of the tail, with bands of longitudinal muscle fibres on either side contracting alternately. This movement illustrates the value of a notochord: in the absence of an incompressible axial strut contraction of these fibres would, as in soft-bodied invertebrates, shorten the animal. As it is, muscles on either side of the notochord are antagonistic to each other.

The sense organs of the ascidean tadpole are more elaborate than those of the adult. The anterior end of the nerve cord is swollen and contains a unicellular statocyst and a pigmented cup which is a photoreceptor. The posterior end is formed of a cord of cells which contain no nervous tissue and have no obvious function. On hatching, ascidean tadpoles swim upwards towards the light, but after a short free-swimming period their behaviour changes and they swim away from the light and towards the bottom. The effect of the larval stage is to disperse the species, and to enable the young ascidean to select a suitable site for its adult life.

The tadpole adheres to the substrate by means of three adhesive papillae at the anterior end and undergoes metamorphosis. The tail is resorbed, the pharynx enlarges, the buccal siphon opens and the atrial siphon develops. The larval 'brain' forms the adult cerebral ganglion.

Colonial ascideans are capable of asexual reproduction, and have considerable powers of regeneration.

Urochordate diversity

Class Ascidiacea

As a class the ascideans show relatively little variation, and *Ciona* may be regarded as being fairly representative. Many ascideans, including *Clavelina* and *Aplidium*, are colonial; the individuals of *Clavelina* are joined together by a flat branching stolon, and *Aplidium proliferum* forms encrusting colonies. Some ascideans are specialized for attachment to soft substrates such as mud or sand; in these forms the larval stage tends to be reduced as the role it plays in site selection is less important.

The remaining tunicate classes, Thaliacea and Larvacea, show specializations for a pelagic mode of life.

Class Thaliacea

This is a small group of only six genera. Thaliaceans are transparent and look like small barrels of jelly, with their buccal and atrial siphons at opposite ends of the body. Because of this arrangement a thaliacean, such as the tropical *Doliolium*, can use the feeding current not only for gaseous exchange but also for locomotion. In *Doliolium* the feeding current is produced not by cilia but by rhythmic contractions of bands of circular muscle which drive water

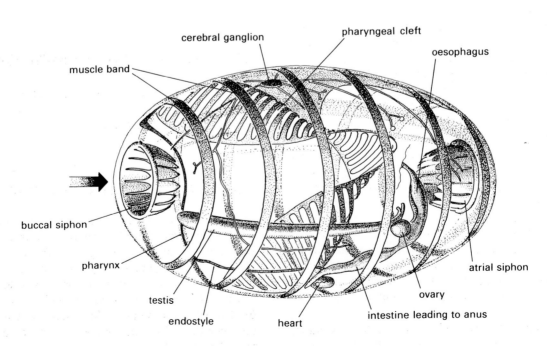

Doliolium

through the pharynx and atrium. The buccal siphon is closed when the body wall is contracted, and open in the relaxation phase; in this way the water current always enters the buccal siphon and leaves via the atrial siphon. Food particles are entrapped in a cone of mucus within the pharynx and drawn into the oesophagus by ciliary action.

Salps, for example *Salpa*, are generally similar to *Doliolium*, but the circular muscle bands are incomplete. *Salpa* has only two greatly enlarged gill clefts which reduce resistance to the water current. Unlike *Doliolium*, which is always solitary, some salps are colonial at certain stages of their life history. *S. democratica*, for example, may be found in long colonial groups of thirty or more asexually reproduced blastozooids which eventually separate to take up an independent existence.

The tropical and subtropical genus *Pyrosoma* is noted for its exceptionally brilliant bioluminescence. It forms large colonies (up to 3 m in length), which take the form of blunt-ended cones. The individual zooids are all orientated so that their buccal siphons open to the exterior of the cone and their atrial siphons into the central chamber. Their structure is the most closely similar to the ascideans.

Reproduction in thaliaceans is complex. Individuals may either reproduce by asexual budding, to produce blastozooids, or sexually, with direct development from the fertilized egg. A tadpole stage does not occur.

Class Larvacea

In contrast to other tunicates larvaceans have a tadpole-like form throughout their life. The tail is very long and arises at right-angles from the centre of the lower dorsal surface. They are tiny animals with a world-wide distribution in the marine plankton; *Oikopleura* is a common example.

Larvaceans do not have a cellulose tunic but have a delicate transparent 'house' secreted by the trunk epidermis. In *Oikopleura* the house is expanded by the water current, which the animal creates by beating its tail, and is larger than the animal; about the size of a walnut. It is replaced frequently, each house lasting for no more than four hours. The water current is used for feeding as well as slow propulsion entering through a fine grid which excludes all but the smallest particles, and leaving through a finer grid near the mouth which is capable of stopping particles of less than 1μm in diameter. As a result *Oikopleura* can feed on particles not normally available to other filter-feeding planktonic animals.

The pharynx has a single pair of clefts. The beat of cilia round the openings draws concentrated food suspension into the pharynx where it is filtered through a sheet of mucus. Water leaves through the pharyngeal clefts and passes out of the exhalent opening of the house. There is no atrium.

Larvaceans reproduce sexually to give a free-swimming larva which undergoes metamorphosis to the adult form.

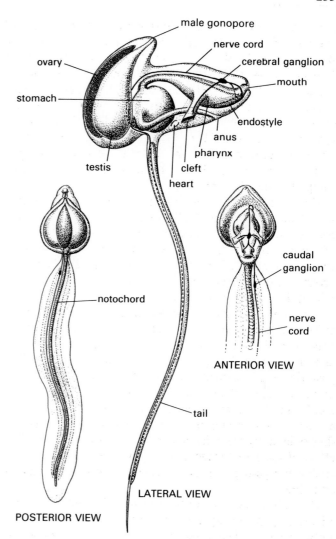

Oikopleura (house omitted)

Synopsis of subphylum Urochordata

Class Ascideacea (*Ciona, Clavelina, Aplidium*)
Class Thaliacea (*Pyrosoma, Doliolium, Salpa*)
Class Larvacea (*Oikopleura*)

Subphylum Cephalochordata

The subphylum Cephalochordata is extremely small with only two recognized genera, *Branchiostoma* and *Assymetron*, which are distinguished from each other by the position of their gonads. *Assymetron* derives its name from the fact that it has gonads on the right side of the body, whilst *Branchiostoma* has gonads on both sides. *Branchiostoma* was known for a long time as *Amphioxus*, however *Branchiostoma* is the correct name according to the rules of zoological nomenclature and is now in general scientific use; amphioxus has passed into English as a common name.

Cephalochordates are of significance in zoology in that, while having pecularities of their own, they also demonstrate a simple type of chordate organization. They share with the main chordate group, the vertebrates, the condition of metameric segmentation already described. The body wall musculature is arranged in segmental blocks, the myotomes, and both nervous and blood systems are modified in association with these.

Branchiostoma is frequently used as an introduction in courses in vertebrate zoology, in fact it is all too easy to forget that it is not a vertebrate at all.

Cephalochordates have a bottom-dwelling, filter-feeding mode of life, but the larval stage of *Branchiostoma*, the amphioxides larva, is pelagic. It shows an advanced level of larval differentiation to the extent that when first discovered it was assigned to a separate genus, *Amphioxides*.

Branchiostoma

The adult *Branchiostoma* is a bottom-dweller and is normally sedentary, living wholly or partially buried in the substrate, but is also capable of extremely rapid movement. Whilst the genus has a world-wide distribution in shallow water it is restricted to a great extent by the nature of the substrate, which is usually clean pure sand or shell gravel.

One of the best known species is the lanceolet, *B. lanceolatum*, which is found in both tropical and temperate waters within a wide range of temperatures of 3 to 27° C.

Body form
B. lanceolatum is a small (40 to 50 mm long), semi-transparent animal with a superficially fish-like appearance. It is strongly compressed laterally but in profile is roughly cigar-shaped. The body wall is extended as a fold of skin along the dorsal mid-line, round the tail, and anteriorly along the posterior third of the ventral surface to an opening, the atriopore. This skin fold is, in effect, a continuous dorsal caudal and ventral fin which is expanded in the caudal (tail) region. Anterior to the atriopore, a dorso-lateral skin fold extends forwards from each side as a pair of metapleural folds. These are continuous with an anterior integumentary fold which forms a hood, the oral hood, over the anterior part of the animal. The oral hood surrounds a cavity, the vestibule, which leads to the mouth. Above the mouth, about twenty stiff processes, the oral cirri, arise from the oral hood. The metapleural folds form the side walls of a cavity called the atrium or peribranchial cavity, which surrounds the greater part of the digestive tract and opens at the atriopore. Water from the pharyngeal slits runs in the atrium and leaves via the atriopore.

The body wall
The epidermis is a single layer of cubical epithelial cells,

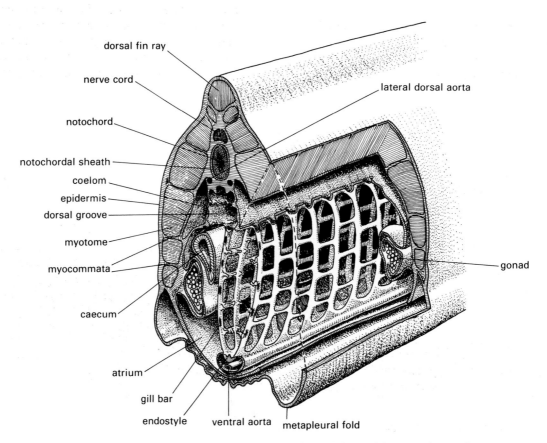

A diagrammatic section of the pharyngeal region of *Branchiostoma lanceolatum*

which secrete a thin, non-cellular cuticle perforated by minute pores. Beneath the epidermis the body wall is composed of a dermis of dense fibrous connective tissue and segmentally arranged V-shaped blocks of muscle fibres, known as myotomes and arranged with the fibres parallel to the longitudinal axis of the body. The innermost layer of the body wall surrounds the coelom and consists of peritoneum. The muscle blocks are enclosed in connective tissue envelopes, whose anterior and posterior walls form the myocommata on to which the muscle fibres insert. The myocommata are continuous with the dermis and with an inner connective tissue layer next to the peritoneum.

The notochord is embedded in the body wall muscle. It is cellular and formed of alternating soft gelatinous and fibrous cell discs. The gelatinous cells have a considerable turgor pressure and confer a degree of rigidity on the structure. The notochord, in combination with a thick outer layer of collagen fibres, the notochordal sheath, forms a substantial and effective axial skeleton.

Connective tissue provides a protective sheath for the nerve cord, which lies dorsal to the notochord. It also contributes to stiffening structures, in the form of rods of gelatinous matter enclosed in sheaths of connective tissue. Rods of this type support the oral cirri, pharyngeal bars, endostyle and oral hood. The oral hood is supported by a jointed hoop. The dorsal fin is stiffened by a single row of fin rays, and the ventral fin by a double row of fin rays.

Locomotion

Branchiostoma both swims and burrows by means of lateral sinusoidal movements of the body, rather like those of an eel. The myotomes on either side are contracted alternately, pulling the body into a series of curves.

Feeding

Branchiostoma is a ciliary filter-feeder, comparable in many respects with the ascideans. There is a well-developed pharynx and the feeding current is generated by the lateral cilia of the gill bars. Water is taken in through the mouth and passes through the pharyngeal clefts into the atrium to leave via the atriopore. During feeding the oral cirri are folded over each other forming a sieve which excludes over-large particles. On the inner surface of the oral hood there is a complicated series of ciliated bands called the wheel organ, this surrounds a mucus-secreting glandular structure known as Hatschek's pit. Particles of food which fall out of the incurrent stream are caught in the mucus and swept by the beat of the wheel organ cilia towards the mouth. The posterior margin of the oral hood is formed by a partition, the velum, which is perforated by an aperture whose margins form numerous protective tentacles. These velar tentacles are ciliated and project backwards into the pharynx to act as a further device for excluding large particles.

The pharynx is large and has about fifty double pairs of pharyngeal clefts. These are separated by primary gill bars and are subdivided by secondary gill bars, or tongue bars. They arise as downgrowths from the dorsal surface of the original cleft so that each cleft becomes divided into separate anterior and posterior sections. The bars have skeletal supporting rods, blood vessels and connective tissue, and the primary gill bars also have coelomic canals. They are linked to each other by cross connections, the synapticula, with a similar internal structure to the gill bars themselves. The pharyngeal wall therefore forms a network, reminiscent of the branchial basket of ascideans. A mucus-producing groove, the endostyle, lies in the

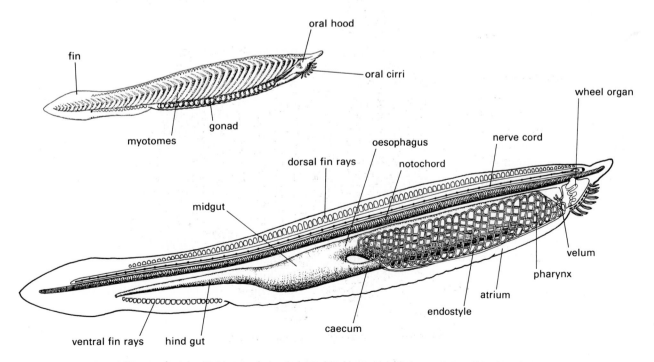

Branchiostoma lanceolatum: whole animal and dissection with gonads removed

ventral mid-line of the pharynx. The secretion is driven up the pharyngeal wall by the beating of the frontal cilia of the gill bars. Food and mucus collect in a cord along the dorsal mid-line of the pharynx in a ciliated dorsal groove and are carried backwards by ciliary action to the digestive region of the gut. This consists of a short ciliated oesophagus, a wide midgut from which a caecum extends along the right side of the pharynx, and a narrow hindgut, terminating at the anus on the left hand side of the ventral fin near the posterior end of the body. The midgut and hindgut are separated by a narrow and strongly ciliated region known as the ilio-colonic ring. At the junction of the midgut and the ilio-colonic ring the gut wall is thickened. As the food cord passes into the ilio-colonic ring it is rotated by the action of the ring cilia, which are specialized to fulfil this function. On the left hand side of the ring the cilia beat obliquely backwards and downwards, while on the right hand side they beat obliquely forwards and upwards.

As a result of its position between the lateral body wall and the pharynx, the midgut caecum is strongly laterally compressed and its roof and floor effectively form narrow ciliated grooves. The cilia beat strongly creating a current anteriorly along the roof of the caecum and posteriorly towards the midgut along its floor. The walls of the caecum have numerous absorptive cells, and gland cells which secrete proteolytic, lipoclastic and amyloclastic enzymes. The midgut secretions are driven into the midgut lumen below the rotating food cord and are swept up into it. The food cord thus becomes a mixture of food, mucus, and digestive enzymes. As the mass continues to rotate the surface fragments and particles are drawn into the midgut caecum by the beat of the roof cilia. Large particles sink to the floor of the caecum and eventually rejoin the main food mass, while smaller and lighter particles come to rest against the lateral walls of the caecum epithelium where they are ingested. Digestion is completed internally. At intervals food material breaks off from the rotating mass and passes into the hindgut. As it is transported along the hindgut further digestion and absorption of diffusible products takes place. Undigested matter is forced out of the anus by a strongly ciliated area of the posterior end of the hindgut.

Respiration and circulation

No special organs of respiration are present, and gaseous exchange takes place by diffusion through the body surface, and particularly at the gill bars. The blood circulatory system of Branchiostoma is unusual in comparison with that of the majority of chordates since there is no heart and the blood contains no respiratory pigment. Blood is transported to the primary gill bars by branches of a main ventral blood vessel, the ventral aorta. The branches (branchial vessels) drain into paired dorsal vessels to join the main dorsal vessel, the dorsal aorta. The secondary gill bars are supplied from the primary bars by vessels in the synapticula. The dorsal aorta distributes blood to the various organs. Peristaltic contractions of the ventral vessel and of small expansions, bulbilli, at the base of the branchial vessels,

cause blood to move forwards in the ventral vessel and posteriorly in the dorsal vessel.

In the region of the gut, the ventral vessel receives a plexus of short vessels from the gut wall; in the midgut region these unite forming a short wide vessel (hepatic portal vein) which supplies a capillary plexus in the wall of the midgut caecum. This plexus is drained anteriorly by a single vessel (hepatic vein) which is also joined by vessels which drain the body wall (right and left Cuvierian ducts). The whole extends anteriorly as the ventral aorta, beneath the branchial clefts. The general organization of the blood system is comparable with that of the vertebrates, and the names given to the various vessels are derived from comparisons with that group.

Excretion

Although in most respects Branchiostoma may be regarded as a simple chordate, its excretory organs are very different from those of other chordates and show a closer resemblance to those of invertebrates. About ninety pairs of nephridia are present, lying in the pharyngeal region and opening into the atrium. Each nephridium consists of a small bent tube, the excretory canal, with an upper limb which opens into the atrium by a pore at the dorsal edge of the gill slit. The lower limb extends ventrally and ends blindly. Numerous short branches arise from the excretory canal, which themselves have a number of fine tubules ending in a typical flame bulb or solenocyte. These project into the dorsal coelomic canal and are bathed in coelomic fluid from which they collect waste products. The nephridia are associated with blood vessels. In addition to the paired nephridia there is a single large nephridium, with essentially the same structure, which lies in the roof of the pharynx lateral to the left dorsal vessel and opens into the pharynx.

Nervous system and sense organs

Branchiostoma has a typical chordate nervous system although in a very simple form. A hollow dorsal nerve cord enclosed in a sheath of collagen fibres extends along the greater part of the notochord on its dorsal side. The central canal of the nerve cord is enlarged at the anterior end to form a cerebral vesicle; this region of the nerve cord may be compared with the brain of vertebrates. Sensory nerves extend to it from the oral cirri and anterior end. Other nerves arise from the nerve cord in a segmental series with an arrangement comparable with that of the vertebrates. The dorsal nerves are sensory while the ventral nerves are motor nerves. There are no segmental ganglia.

Branchiostoma responds to light, chemical and tactile stimulation but does not have elaborate sense organs. Photoreceptors, each composed of a photoreceptive cell and a cup-shaped pigment cell, lie in the nerve cord and the animal responds to intense light by rapid swimming movements. Tactoreceptor cells are present on the general body surface, with chemoreceptors being concentrated on the oral cirri and in a flagellated pit near the front of the nerve cord.

Reproduction and growth

Branchiostoma has 26 pairs of gonads, arranged in two latero-ventral rows in the pharyngeal and post-pharyngeal regions and bulging into the atrial cavity. The sexes are separate, although the testes and ovaries cannot be distinguished externally, and fertilization is external. The gonad and the atrial lining which surrounds it ruptures, and the gametes are shed into the atrium to leave via the atriopore. The newly-hatched larva is similar in many respects to the tornarian larva of echinoderms, and swims actively by means of its cilia. It develops into an asymmetrical pelagic amphioxides larva which undergoes metamorphosis to develop the adult form and assumes the adult mode of life. The amphioxides larva is highly differentiated and may even develop its gonads before the onset of metamorphosis. The details of *Branchiostoma* embryology will be discussed more fully in the chapter on chordate embryology.

Synopsis of the phylum Chordata

Subphylum Urochordata
    Class Ascidiacea (*Ciona*)
    Class Thaliacea (*Doliolum*)
    Class Larvacea (*Oikopleura*)
Subphylum Cephalochordata (*Branchiostoma*)
Subphylum Vertebrata
    Class Agnatha (lampreys and hagfish)
    Class Elasmobranchiomorphi (cartilaginous fishes)
    Class Teleostomi (Osteichthyes) (bony fishes)
    Class Amphibia (salamanders, frogs and toads)
    Class Reptilia (snakes, lizards, turtles and crocodiles)
    Class Aves (birds)
    Class Mammalia (mammals)

INTRODUCTION TO THE SUBPHYLUM VERTEBRATA

The majority of chordates belong to the subphylum Vertebrata, a group which, as its name suggests, is characterized by the possession of a metameric series of small skeletal structures, the vertebrae. These are associated with the notochord during embryonic development but in most cases replace it in adults. Like the notochord the vertebral column functions as an axial supporting rod, but it has the additional function of protecting the nerve cord. In vertebrates the anterior end of the nerve cord is greatly expanded to form a brain, and a skeletal structure, the neurocranium (or, more simply, the cranium) has developed to protect it.

In addition to the possession of a vertebral column, brain, and cranium all vertebrates have a series of visceral arches associated with the pharyngeal clefts. These are supported by skeletal elements. Throughout the vertebrates they show extensive modification to serve various functions; in particular the anterior visceral arches of most vertebrates have become greatly altered to form jaws. Such vertebrates are sometimes known as Gnathostomata (meaning 'jawed mouths') to distinguish them from the Agnatha (jawless forms). Gnathostomes are themselves sometimes subdivided into two groups: Pisces, fish; and Tetrapoda, which are characterized by having four limbs, although in some tetrapods these have undergone extensive reduction and in a few have been lost altogether.

The structural diversity of vertebrates is considerable. Although many different classifications have been suggested, the majority recognize seven major subdivisions: Agnatha (the jawless vertebrates), Elasmobranchiomorphi (the cartilaginous fishes), Teleostomi (the bony fishes), Amphibia (amphibians), Reptilia (reptiles), Aves (birds)

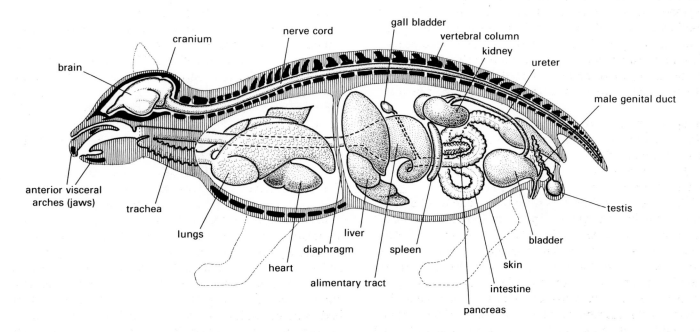

A representative mammal, summarising vertebrate characteristics and principal organs

and Mammalia (mammals). These may be grouped in a variety of ways, for example some classifications, as suggested above, place the greatest emphasis on the presence or absence of jaws, while others stress the importance of limbs. Another approach is to place together in one group (Anamniota) the Agnatha, Elasmobranchiomorphi, Teleostomi and Amphibia because they have fairly simple embryonic development with eggs and young developing in water. In contrast the Amniota (reptiles, birds and mammals) have a more complex embryonic development. Yet another approach recognizes seven vertebrate classes of equal rank with no attempt to group them.

In comparison with the groups already discussed, vertebrates are highly complex. They are essentially bilaterally symmetrical, but there are many exceptions to this general rule. Common adaptations involve displacement of organs which primitively lie in the mid-line (such as the heart) or the reduction or loss of one of a pair of organs. A loss of external symmetry is comparatively rare, although it does occur; for example the plaice is an animal which has come to lie on its side and in the process the normal symmetry has become distorted. Terms normally applied to bilaterally symmetrical animals, for example, dorsal, ventral, anterior and posterior, are often used in describing vertebrates, but sometimes cranial (head) and caudal (tail) are substituted for anterior and posterior respectively. Additional complications arise out of the different terminology used by human anatomists. For example in an upright bipedal form, the terms anterior and posterior as used for other vertebrates are obviously inappropriate, so superior and inferior are commonly used instead. However it is both unfortunate and confusing that in the case of man the terms anterior and posterior have come to be substituted for ventral and dorsal! Terms which were originally used in describing human anatomy, for example the superior and inferior vena cava, are applied more generally in vertebrate organ systems, sometimes with confusing results. Additional adjectives commonly found in anatomical description include 'medial', which is applied to structures close to the midline, as opposed to 'lateral' for those which are more distant from it, and 'proximal', which is applied to structures close to an important reference point such as a shoulder joint, in contrast to 'distal' for those further away. This terminology is particularly helpful when describing limbs.

Because vertebrates are much more complex than most other animals we must consider not only their gross functional anatomy, but also a certain amount of detail of the various organ systems. These may be discussed conveniently under eight headings: integumentary, skeletomuscular, digestive, respiratory, circulatory, urinogenital, nervous and endocrine. As an introduction to this it is essential to give some attention to the tissues of which the various organ systems are made.

Animal body tissues

Most of the animals discussed so far are small and fairly simply organized; they have therefore been discussed mainly from the point of view of their gross anatomy. For an adequate understanding of vertebrates we must consider the detailed structure of the various organs and organ systems. These are made up of tissues, which are themselves

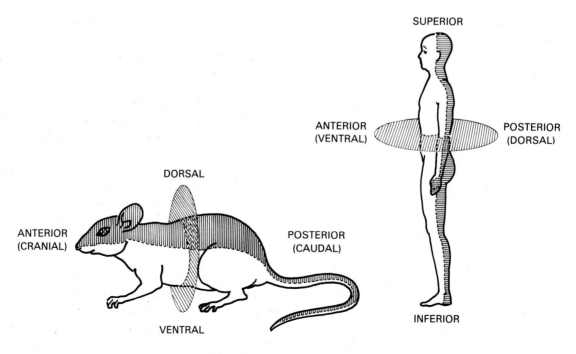

Vertebrate positional terms

composed of cells or cellular derivatives. Tissue may be defined loosely as a collection of similar cells with a characteristic organization and serving the same general function. However this definition can be misleading, since although many tissues consist mainly of one cell type, others may be present and essential for the functioning of the tissue as a whole. To give a simple example, muscle is not composed solely of muscle cells but has associated nerves, blood vessels and connective tissue, all of which are essential to its function.

The branch of biology dealing with the structure of tissues is histology; it could equally well be described as microscopic anatomy. Histological study has recently been revolutionized by the application of electron microscopical techniques, and in this way hitherto unrecognized and undescribed structures have been clearly demonstrated. As a consequence, our understanding of how particular tissues function has advanced enormously. In addition considerable progress has been made in the techniques of studying living tissue in the animal or in culture.

Tissue is formed by cell multiplication, followed by a process of differentiation in which the cell undergoes physical and chemical changes which fit it for its future function. As a broad classification, four tissue types are recognised in most multicellular animals: epithelial, connective, muscular and nervous. Animal tissues show considerable variation and the following description is based primarily on mammalian and particularly human tissue.

Epithelial tissue

The word epithelium means a covering: epithelial tissue forms a protective layer over the external surface of the body as well as forming a lining to its internal spaces (internal linings may also be called endothelia). Epithelia characteristically occur as sheets of closely connected cells joined together by a minimal amount of intercellular substance which is produced by the cells themselves. Almost invariably the cells rest on a delicate structureless or fibrous layer, the basement membrane. This is an extracellular substance, produced not by the epithelium itself but by underlying connective tissue and often contributing to the strength and resilience of the epithelium. The intercellular substance between individual epithelial cells was formerly regarded simply as a type of cement; electron microscopy reveals that it takes the form of a structureless matrix through which pass interdigitating prolongations, or bridges, between the cells.

Epithelia are important in animal metabolism since they are inevitably concerned in any processes occurring across surfaces. They are involved in respiration, secretion, absorption, assimilation and elimination of waste products. Their structure varies a great deal, but can be conveniently classified into three kinds: simple, transitional and stratified.

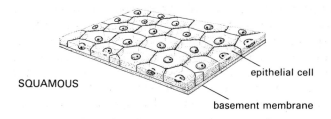

SQUAMOUS

epithelial cell

basement membrane

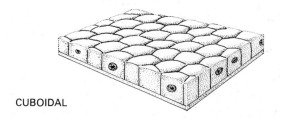

CUBOIDAL

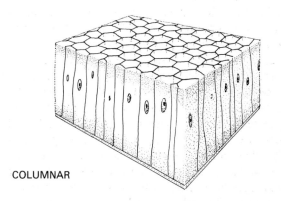

COLUMNAR

Simple epithelia

Simple epithelia consist of a single layer of cells on a basement membrane and are common on most surfaces where there is little wear and tear. They can further be described by words such as squamous, cuboidal and columnar depending on the shape of the cells. A squamous or pavement simple epithelium consists of a single layer of flattened, plate-like cells such as can be found lining the alveoli of the lungs, and the blood vessels. Cuboidal cells are common in the lining of many glands including the thyroid gland and sweat glands. (They are also common in the epidermis and gut of invertebrates). Columnar cells are elongated and frequently ciliated, and may be modified as mucus-producing goblet cells. Columnar epithelia are extremely common in invertebrates; in vertebrates they occur as linings to the alimentary, respiratory and reproductive tracts.

Transitional epithelium, which occurs as a lining in the urinary tract, for example in the bladder and urethra, is formed by two or three layers of cells, most of which are attached to the basement membrane. Part of its importance lies in the fact that it can be stretched, transforming it into a single or double cell layer.

As its name suggests, stratified epithelium consists of a number of layers of cells; the outer cell layers are continuously replaced and so the epithelium is highly suitable for regions of the body which are subject to friction.

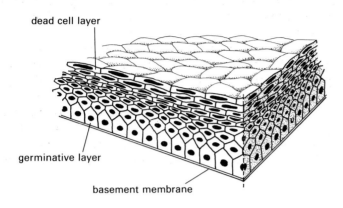

Stratified epithelium

The innermost cell layer is of columnar cells, which are always in a state of active cell division and for this reason this layer is sometimes called the germinative layer or stratum germinativum. The more superficial layers of cells are flattened and the outermost layer is of dead cells which are continually shed (desquamated) and replaced from below. Stratified epithelium forms the outer layer of the skin in terrestrial vertebrates, and lines both ends of the alimentary canal.

Connective tissues

The second of the four main tissue types has been given the general name of connective tissue, and includes all tissue characterized by considerable quantities of intercellular substance or matrix, (to such an extent that the cells make up only a small proportion of the tissue). Its principle functions are packing, binding and support, and it is highly variable in structure. However, all connective tissue consists of cells, with a matrix which is usually secreted by those cells. These are four main types of connective tissue: connective tissue (*sensu stricto*), blood and lymph, cartilage and bone, as well as a number of intermediate forms.

Connective tissue (*sensu stricto*)

The term connective tissue implies a tissue which connects more organized structures of the body together, and the various forms of tissue which fulfil this function are collectively known as connective tissue. These form a general matrix through the body (areolar tissue) as well as more specialized regions of fat deposition (adipose tissue), tough fibrous sheets (fascia), ligaments and tendons. Areolar tissue is the most widely distributed and the most easily studied. It forms a loose mesh and has a semi-fluid gelatinous matrix composed of protein and polysaccharide and in which are embedded fibres and cells. Three types of fibres may be present: they are white, yellow and reticular fibres. Inelastic white fibres of collagen are normally arranged in characteristically wavy bundles (except when under tension) and are relatively common. Yellowish fibres

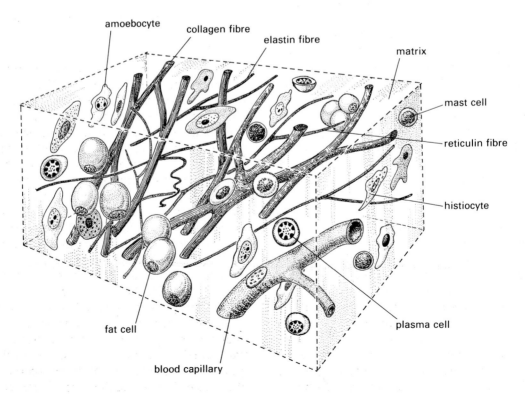

Areolar connective tissue

of elastin are less common, they run singly and branch and are capable of considerable stretching. Reticular fibres are somewhat similar to white fibres but have a different arrangement, forming a fine network throughout the tissue, and are formed of reticulin. All these fibres are produced by specialized, large, slender, stellate cells called the fibroblasts.

Apart from fibroblasts, several other kinds of cell may be present in areolar tissue. Macrophages (histiocytes) can ingest foreign material and also dispose of degenerating tissue prior to regeneration. Mast cells are round or polygonal in shape and are concerned with the production in the matrix of an anticoagulant (heparin) and of histamine (which increases blood flow through capillaries and is released when the tissue is injured). Plasma cells are oval in shape with a characteristic cartwheel-shaped nucleus. They are concerned with antibody production and, though generally present, are found in large quantities during chronic infections. Pigment cells, (chromatophores) are branched structures containing pigment (for example the black pigment melanin). They may be present in sub-epidermal or subcutaneous connective tissue, particularly in the skin, and the iris of the eye.

Fat cells, each consisting of a globule of fat surrounded by a cytoplasmic envelope, are common in areolar tissue. Adipose tissue is particularly abundant beneath the skin of mammals and in folds of mesentery among the abdominal organs. It is generally pale in colour. Darker fat, known as brown fat, is prominent in hibernating or new-born mammals and occurs in distinct patches. It is thought to be used specifically to produce heat. Fat cells are closely packed in adipose tissue where they may swell to a considerable size.

Tough fibrous connective tissue (fascia) forms a deeper layer beneath the skin in mammals and also forms a well defined tight-fitting sleeve around the muscles in vertebrates. Dense concentrations of white inelastic fibres are found in tendons, which connect muscles to the connective tissue sheath which surrounds the bone. In contrast elastic fibres are particularly abundant in ligaments (which principally join bones together) where some elasticity is required though some ligaments contain a higher proportion of collagen fibres and the distinction is not absolute.

Fluid connective tissues: blood and lymph

It may seem rather surprising to classify blood and lymph as connective tissues, however they fit the definition of connective tissues since they are composed of cells within a non-cellular matrix. They are fluid tissues in which the typical connective tissue matrix is represented by a fluid, plasma. Plasma is a complex mixture of inorganic constituents in true solution (chiefly sodium chloride and bicarbonate, potassium chloride and bicarbonate, sodium phosphate and sodium sulphate in mammals) with a concentration of about one per cent, and a colloidal solution of proteins including albumen, globulin and fibrinogen.

A variety of cells are present within this fluid matrix, making up slightly less than half the volume of blood. By far the most numerous in mammals are the red blood cells (erythrocytes) which are biconcave discs consisting of an envelope of protoplasm surrounding a solution of haemoglobin in a colloidal matrix. They develop in red bone marrow, a pulp-like tissue which is present in most bones at birth, but is restricted to a few in the adult. (These are the vertebrae, ribs, sternum, scapulae, pelvis and the proximal and distal ends of the femur and humerus.) Tissue of this type is referred to as haemopoietic (blood-making). Circulating erythrocytes have no nuclei and are destroyed by specialized white blood cells (phagocytes) or by the spleen after a few weeks. Their function is the transport of oxygen and carbon dioxide between the respiratory surface and other tissue cells, by means of the haemoglobin which they contain.

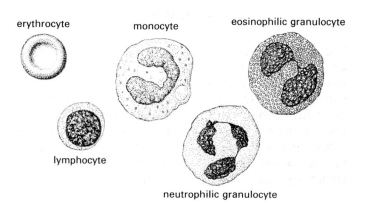

erythrocyte   monocyte   eosinophilic granulocyte

lymphocyte

neutrophilic granulocyte

Representative blood cells (human)

White blood corpuscles are larger, nucleolated, and fewer in number than the erythrocytes. They are produced in red marrow, as well as in the lymph organs and in specialized regions of the liver and spleen. They serve a variety of very important protective functions and are involved in processes such as the removal of unwanted cells or foreign particles. Their cytoplasm may be granular (neutrophils, eosinophils, basophils) or non-granular (lymphocytes, monocytes). The most numerous are neutrophils, which are amoeboid and phagocytic and have a characteristic lobulated nucleus and the granules stain with neutral dyes (hence their name). Basophils have granules which stain with basic dyes and are smaller with a kidney-shaped nucleus. Their function is obscure but the granules chiefly contain heparin and histamine. Eosinophils stain with acid dyes such as eosin, and are thought to be involved in allergic reactions.

Lymph consists mainly of lymphocytes in a fluid which is similar in composition to that of blood plasma, and is transported in lymph vessels which parallel the venous system. The protein content varies according to the region of the body from which the lymph is collected and in some regions the lymph may also contain emulsified fat, especially after a fatty meal. Lymphocytes have a variety of functions including ingesting foreign material and therefore providing an important defence against infection, and the elaboration and storage of enzymes. They are produced in specialized bean-shaped aggregations of tissue, the lymph nodes, as well as in other tissues.

Skeletal connective tissues

Cartilage

Cartilage ('gristle') is a type of supporting tissue which combines rigidity with flexibility and resilience. It is the only skeletal material in some lower vertebrates such as sharks, and is also an important part of the higher vertebrate skeleton, although in the adult it is replaced to a large extent by bone. Cartilage has a firm clear matrix composed chiefly of a sulphate mucopolysaccharide which is deposited intercellularly by the cartilage cells or chondroyctes. Also present in the matrix to a greater or lesser extent are the characteristic connective tissue fibres which contribute to its toughness. Cartilage does not have a blood supply and the chondrocytes dispose of wastes and receive metabolites by diffusion through the matrix.

Five types of cartilage are recognized, although only one, hyaline cartilage, is of widespread occurrence. This is translucent and bluish in colour and is especially prominent in vertebrate embryos as it forms the basis for the process of ossification by which the greater part of the vertebrate skeleton is formed. It is the most rigid variety and has large numbers of rounded chondrocytes, each surrounded by a small space (lacuna) and a few collagen fibres. It persists in the higher vertebrate adult on the articular surfaces of

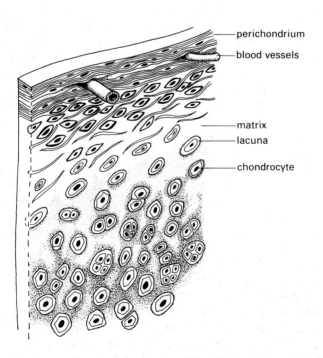

The histological structure of hyaline cartilage

bones, and forms the rib cartilages as well as the supporting cartilages of the respiratory tract. Except on articular surfaces of bones, hyaline cartilage is enclosed in a connective tissue sheath (perichondrium) to which tendons and ligaments are attached. In the early stages of the development of hyaline cartilage the cells lie close together, but as they secrete more matrix they gradually push further apart. Additional increase in the size of a cartilage is provided by proliferation of new cells from the inner layer of the perichondrium.

Both elastic cartilage and fibrocartilage are less abundant than hyaline cartilage. The matrix of elastic cartilage contains large numbers of elastin fibres which increase its resilience. It is restricted to regions where increased flexibility is important, such as the epiglottis and the external ear. In contrast, fibrocartilage contains bundles of dense white fibres but only a few cartilage cells. In vertebrates it is found in the intervertebral discs as well as in regions of attachment of ligaments or tendons.

Two types of cartilage, calcified and chondroid, are particularly common in lower vertebrates. Calcified cartilage has calcareous salts deposited in the matrix, forming a hard brittle substance which is superficially similar to bone and usually occurs in rather stouter skeletal structures. It is somewhat stronger than hyaline cartilage and is common in animals such as sharks in which the adult skeleton remains largely cartilaginous. By contrast, chondroid cartilage is soft and gelatinous and is restricted to parts of the skeleton of some larval agnathans and sharks.

Bone

Bone is the dominant skeletal tissue in the majority of vertebrates. Unlike cartilage it is hard and opaque, consisting mainly of calcium phosphate and carbonate which impregnate a matrix supported by a network of collagen fibres. Bone matrix is impervious to metabolites, and living cells are supplied by blood vessels which run through Haversian canals. A basic bone unit (Haversian system) consists of concentric layers of matrix around a Haversian canal, with bone cells, or osteocytes, lying in lacunae between the layers. In contrast with chondrocytes, osteocytes are irregular, elongated structures. In the early stages of bone development osteocytes give out numerous, fine, branching projections which extend outwards from the lacunae in fine, branching canals. Later the protoplasmic extensions are withdrawn and the canals (canaliculi) become filled with tissue fluid, providing the final means of transferring metabolites between the blood vessels and the osteocytes.

Bone tissue appears in two forms, compact and cancellous bone, which are not fundamentally different but represent different arrangements of the same tissue. Compact bone is very dense, and similar in texture to ivory. It consists of Haversian systems, with lacunae and interstitial systems which represent the remains of old Haversian systems which have been broken down and reformed during the process of bone growth. Cancellous bone, also referred to as

(labels on figure:)
perichondrium
blood vessels
matrix
lacuna
chondrocyte

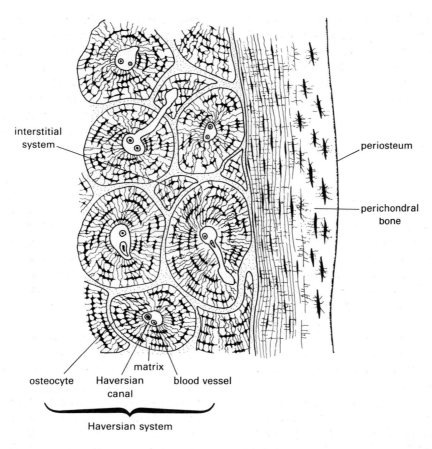

interstitial
system

periosteum

perichondral
bone

osteocyte   Haversian   matrix   blood vessel
            canal

Haversian system

The histological structure of compact bone (T.S.)

trabecular or spongy bone, consists of a mesh of thin sheets of bone (trabeculae) within which are relatively larger spaces (cancelli) filled with bone marrow, a vascular or fatty tissue. Red bone marrow is the site of blood cell formation in many vertebrates.

Most bony structures consist of an outer layer of compact bone, with an infilling of cancellous bone. The arrangement of the trabeculae and the relative proportions of compact and cancellous bone, is such as to give maximum strength with maximum economy of material and minimal weight. The trabeculae normally follow lines of stress. In addition the shape of individual bones is moulded by attached muscles or adjacent organs.

Bones (particularly in humans) may be classified according to their shape: long, short, flat, or irregular. Long bones, such as the limb bones of tetrapods, have a shaft consisting of a cylinder of compact bone filled with spongy bone and bone marrow. The ends of long bones and short bones such as those at the wrist consist of spongy bone with a thin outer layer of compact bone. Flat bones, such as the scapula and some of the bones of the skull are specialized for muscle attachment or protection and consist of two layers of compact bone with a thin layer of spongy bone between them. Irregular bones, such as vertebrae, have an outer layer of compact bone, and inside may contain red bone marrow, or air spaces (pneumatic bone).

During the course of development, bone may be formed either by ossification of a connective tissue matrix or by replacement of a pre-existing cartilage. The former is referred to as membrane or dermal bone, while the latter is endochondral or cartilage bone. The process of forming membrane bone is the simplest, with the osteoblasts laying down a dense matrix in which bone salts are deposited. In endochondral bone, first of all a cartilage is formed in the shape of the future endochondral bone, although of a far smaller size. The main part or shaft of a long cartilage bone is called the diaphysis. Bone formation begins with the appearance of a ring of bony tissue around the centre of the cartilage model, which becomes the diaphysis. Within the bony ring, the original cartilage is eroded, to be replaced by a network of cartilaginous bars. These become calcified, and serve as centres of ossification around which osteocytes (derived from bone marrow) congregate to form trabeculae. Increase in diameter is brought about by deposition of further compact bone by cells derived from the periosteum (the modified perichondrium). Since the process of ossification normally occurs in younger and growing animals the bones must elongate as they develop. Bones increase in length by further deposition of cartilage at the ends of the shaft and the centre of ossification in the diaphysis also gradually extends to the ends of the bone. In mammals, and to a lesser extent in reptiles, secondary

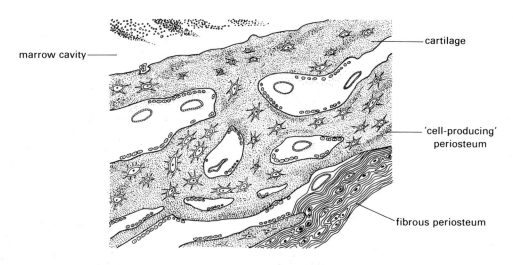

Section of developing endochondral bone showing ossification

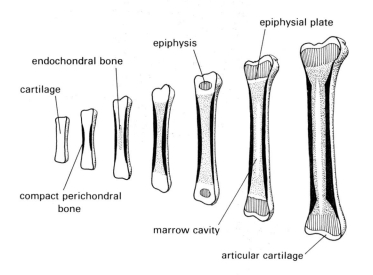

Ossification and growth of a mammalian long bone

surface covered by hyaline cartilage, and a fibrocartilage disc between them.

Freely movable joints may take a variety of forms according to their particular function, for example ball and socket (shoulder and hip), hinge (elbow and knee), pivotal (the atlas and axis of the neck) and gliding (wrist and ankle). Another name for this type of joint is synovial joint, because the articulating bones are separated by a fluid-filled synovial or joint cavity. The articulating surfaces of the bones are covered by articular cartilage, which is normally hyaline, and are surrounded by a fibrous capsule. A synovial membrane lines all the inner surfaces of the capsule and secretes synovial fluid which acts as a lubricant. Fluid-filled sacs known as bursae may also serve to protect muscles or tendons which extend over the joint.

centres of ossification, the epiphyses, appear at the ends of the bone. However, cartilaginous areas, the epiphysial plates, remain between the diaphysis and epiphyses and provide a region for further elongation of the bone until the full adult size is reached. At this stage the epiphysial plates are replaced by bone and no further elongation can occur.

Joints

Articulations between bones or cartilages are collectively referred to as joints. Two general types, immovable and freely movable, are recognized. Some immovable joints are completely fixed, and are formed by a direct union of bone with bone, as in the elements of the hip girdle. Others retain a degree of flexibility and are formed by a thin sheet of fibrous connective tissue such as in the majority of joints or sutures of the skull. Cartilaginous joints (for example the intervertebral discs) are slightly movable and have a bony

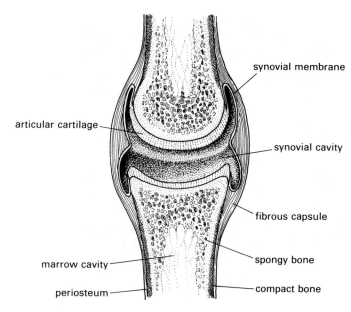

Longitudinal section through a synovial joint

Muscular tissue

Muscular tissue has developed contractility to a high degree and is responsible for all types of movement. Since individual muscle cells are very long they are normally referred to as muscle fibres. All are able to act in one way only, by contraction. Vertebrates have three well-defined types of muscle: striated, smooth and cardiac muscle. Striated muscle is specialized for rapid contraction and is associated mainly with the skeleton and diaphragm (in mammals) and therefore makes up most of the flesh of animals. Smooth or visceral muscle is capable of slow sustained contraction and occurs in the walls of the gut and most of the hollow organs of the body. Cardiac muscle is a special type of muscle found in the walls of the vertebrate heart. All muscle has a rich blood supply and is supplied by sensory and motor nerves.

Smooth muscle

Smooth muscle is structurally the simplest type of muscle. It consists of elongated spindle-shaped fibres less than a millimetre in length with a single oval nucleus surrounded by sarcoplasm which contains many simple contractile fibrilla. Smooth muscle fibres may be scattered, but generally individual fibres are bound together into interlacing bundles by delicate connective tissue. Smooth muscle is normally arranged in continuous sheets of either longitudinal or circular fibres and may be compared in several respects with the muscle of invertebrates such as worms. In vertebrates it is found where movement is required without the intervention of the conscious will (hence its alternative name 'involuntary muscle') and lines organs, such as the gut, which must contract rhythmically.

Striated muscle

Striated muscle is more complex than smooth muscle: the fibres can be extremely long and each contains numerous nuclei, sometimes as many as several hundred. The fibres are limited by a plasma membrane, the sarcolemma, and the nuclei lie immediately below this. Striated muscle fibres are generally unbranched but in specialized circumstances where great mobility is required (for example in the tongue of some animals) there is a certain degree of branching. Striated muscle is under the control of the will, hence its alternative name of voluntary muscle.

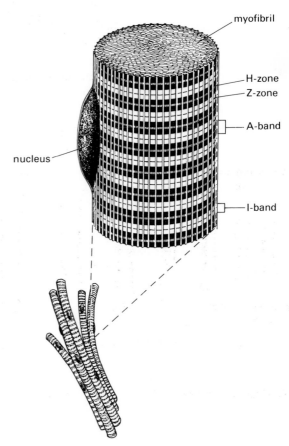

Striated muscle fibre

The individual fibrilla which lie within the sarcoplasm show a pattern of alternating light bands (isotropic or I-bands) and dark bands (anistropic or A-bands). These have a dark line (the Z-zone) crossing the I-band and a clear zone (Henson zone or H-zone) crossing the A-band. The fibrillae are arranged so that the light and dark bands coincide across the fibre and it is this that is responsible for the striped or striated appearance of the muscle from which its name derives. Fibrillae have been shown by electron microscopy to be built up of two distinct types of myofilaments: thick filaments consisting of a contractile protein, myosin, and thin filaments primarily of a contractile protein, actin. Cross bridges attach the two filament types to each other. A-bands consist of thick filaments, and I-bands consist of thin filaments, while the H-zone is a region of partial overlap.

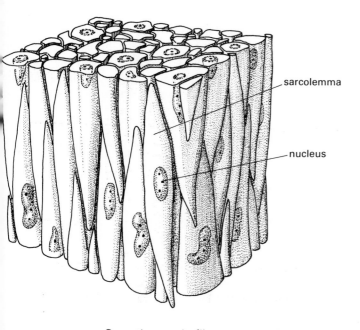

Smooth muscle fibres

The thin filaments attach to each other at the two lines. The portion of the myofibril between two consecutive lines is known as a sarcomere and represents a basic contractile unit of the muscle fibre. When striated muscle contracts the I-bands are seen to narrow and the H-zones to disappear as the Z-zones move closer together. Shortening of individual fibrilla is thought to be brought about by the cross bridges sliding the thin filaments in between the thick filaments.

At some point on the surface of each fibre, in a region without fibrilla, is a motor end plate (sometimes compared to an electrode) over which expands the tip of a terminal branch of a motor nerve fibre. Each nerve fibre normally divides to supply many muscle fibres through motor end plates so that the muscle fibres act in unison.

Striated muscle is fundamentally similar in vertebrates and arthropods. Bundles of fibres are normally arranged in blocks surrounded by a connective tissue sheath, the epimysium. Individual muscle fibres are surrounded and attached to neighbouring fibres by fibrous connective tissue (endomysium) and the bundles of fibres are also enclosed in connective tissue (perimysium). These blocks, referred to as muscles, are attached to skeletal elements which are moved as the muscles are contracted. Movement between parts is brought about by opposing sets of muscles. For example in the human arm the biceps and triceps work antagonistically to flex and extend it. When a vertebrate muscle contracts, one part (the origin) normally remains fixed whereas the other (the insertion) pulls upon the point of attachment and moves it towards the origin.

Cardiac muscle

Cardiac muscle, which as the name suggests is restricted to the vertebrate heart, is in some respects intermediate between striated and smooth muscle. The fibrilla are striated, but the muscle fibres have only a single central nucleus. The fibres are short, branched, and cylindrical and joined end to end at prominent intercalated discs to give a continuous network. Cardiac muscle is specialized for continuous rhythmic contraction which is essentially autonomous. The fibres are striated but involuntary and since they run together to form a continuous network, cardiac muscle both can and does contract *en masse*.

Nervous tissue

Nervous tissue is specialized for excitability and the conduction of nervous impulses. The basic functional unit is the neuron: this has an enlarged cell body which contains the nucleus, and two or more cytoplasmic extensions or nerve fibres along which impulses travel to the next neuron. Nerve fibres, which can vary in length from a fraction of a millimetre to several metres, are of two kinds: stout elongated axons carry impulses away from the cell body, while short branched dendrites carry impulses towards it. The axon of one cell forms a junction with a dendrite of the next cell at a synapse, but the axon and dendrite do not touch, there is always a gap between them. Impulses can only cross from axons to dendrites. The commonest are efferent or motor neurons, neurons which carry impulses away from the central nervous system to striated muscle. They have several dendrites but only a single axon, and may be termed monopolar. Afferent neurons carry sensory stimuli from the sense organs to the central nervous system and have two (bipolar) or sometimes more (multi-polar) long processes.

Various supporting cells are associated with the neurons. In the peripheral nervous system each fibre is surrounded by a sheath and may be either unmyelinated or myelinated. Both types of fibre are enclosed within cells called Schwann cells. In unmyelinated fibres, the fibre is simply invaginated into the Schwann cell. In myelinated fibres, the fibre is surrounded by many layers of Schwann cell membrane, called the myelin sheath. At intervals the myelin sheath of the myelinated fibre is deficient. These deficiencies are called the nodes of Ranvier. In the central nervous system, the supporting cells are called the neuroglia. They provide insulation between adjacent neurons and form the myelin sheath for the nerve fibres.

Neurons normally occur in groups. In particular, the cell bodies are concentrated in the brain, the central nerve cord, and in nerve ganglia. Nerves are bundles of nerve fibres bound together by connective tissue.

In vertebrates a distinction is made between the central nervous system, that is the brain and spinal cord, the peripheral nervous system composed of the cranial nerves and spinal nerves, and the autonomic nervous system which supplies the viscera. A further distinction is made in the autonomic nervous system between the parasympathetic system of short unmyelinated fibres which is responsible for

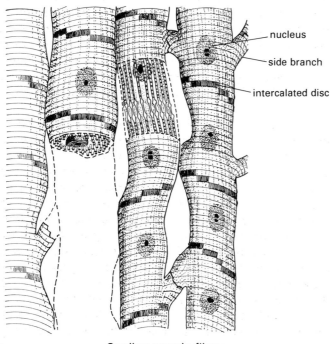

nucleus

side branch

intercalated disc

Cardiac muscle fibres

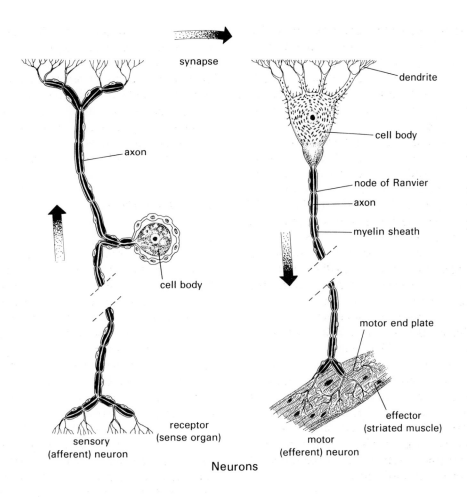

synapse

dendrite

cell body

node of Ranvier

axon

myelin sheath

axon

cell body

motor end plate

receptor
(sense organ)

sensory
(afferent) neuron

effector
(striated muscle)

motor
(efferent) neuron

Neurons

day to day activities (for example peristalsis in the gut), and the sympathetic nervous system which prepares the body for activities which are not routine (for example the flight or fight response).

Nerve impulses are often compared with electric impulses along a nerve although they are far less rapid and the physical processes are very different. A nerve impulse is an 'all or nothing' response in which either the fibre reacts fully or it does not react at all. The impulses are believed to be waves of changes in the polarity of a nerve membrane, caused by the migration of ions across the membrane.

A neuron may be envisaged as a long tube with a semi-permeable membrane separating two solutions. The inner side has a potassium organic ion solution which is normally positively charged while the outer side is negatively charged and is bathed in a sodium chloride ion solution. Stimulation of the fibre induces a local change in the membrane charge or depolarisation, which upsets the electro-chemical balance in adjacent parts of the nerve fibre and causes ions to migrate across the membrane. In this way the region of depolarisation passes along the fibre. As the wave passes, the surface is repolarised and the nerve prepared for another impulse. This repolarisation involves a local expenditure of energy and until it is completed the nerve remains insensitive. Thus unlike an electric wire there is a clear limit to the frequency with which a nerve can transmit impulses.

VERTEBRATE ORGAN SYSTEMS

The integumentary system

The integumentary system comprises the skin and a wide variety of skin derivatives such as nails, claws, feathers, beaks and a variety of glands. It is the largest and most visible organ system of the body. Most types of body tissue occur in the skin and its derivatives though they show a considerable degree of variation, both in different regions of the body and also between the various vertebrate classes and species.

At all levels of vertebrate organization the animal is dependent on the skin as a covering forming a barrier between vulnerable internal structures and the environment. This barrier is continuous over the entire body surface (even the transparent cornea which covers the eye is modified skin) and provides protection against such hazards as micro-organisms, harmful chemicals, excessive light and physical injury. In addition skin has other positive functions. It is usually tough and elastic, and is important in supporting internal structures and determining the shape of the animal. As the region of the body in closest contact with the animal's surroundings it is an important sense organ. In many vertebrate groups it has important physiological functions and may be involved in the control of body temperature, the regulation of internal water and salt content, and the elimination of nitrogenous waste. It also has a metabolic function, the formation of vitamin D from cholesterol in the presence of ultraviolet light.

Skin structure

Although the skin of most vertebrates appears superficially simple and uniform, it is in fact complex in structure and consists of two closely united parts: the epidermis (from the Greek *epi* meaning 'upon' and *derma* meaning 'skin'), and the dermis. In contrast to invertebrate epidermis these two layers are formed of several types of cell; generally vertebrate epidermis is thin and the dermis thicker. In larger vertebrates the whole structure is surprisingly thick, for example human skin is about 2 mm thick and altogether weighs about 3 kg. The epidermis is essentially a stratified epithelium, many cells thick. The cells are progressively delaminated from a basal layer of cuboidal cells. Dermis is chiefly composed of dense fibro-elastic connective tissue and has a fine network of capillaries. The outer layer is usually densely fibrous, deeper parts being more open in texture. The majority of skin derivatives, such as nails and hair, are epidermal in origin though some, such as fish scales, are derived from the dermis. Epidermis and dermis differ in origin, epidermis being derived from embryonic ectoderm, while dermis comes from mesoderm. The skin is attached to underlying muscle by a layer of loose connective tissue.

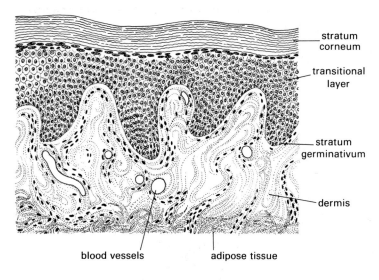

stratum corneum

transitional layer

stratum germinativum

dermis

blood vessels   adipose tissue

Tetrapod skin (semidiagrammatic)

Aquatic animals generally have a thin epidermis which consists entirely of living cells. With the assumption of a terrestrial mode of life the structure of the epidermis changes, for example the waterproof protein, keratin, is present in small amounts in the outer epidermal cells of aquatic vertebrates, but is present in larger amounts in the

dead cells which form the outermost layers of the epidermis of terrestrial vertebrates. The dermis shows less variation in structure, throughout the vertebrates it functions as insulation and protection against injury, its unyielding nature making it particularly suitable for this role. It is the dermis of certain mammals that is made into leather.

The skeleto-muscular system

The functions of the skeleto-muscular system may be summarized in three words: protection, support and movement.

The skeleton

The term skeleton (derived from the Greek word for 'dried bones') refers to any supporting framework of an animal's body. In vertebrates it is an internal structure (endoskeleton) composed of cartilage, bone or a combination of the two. Individual bones are joined together by ligaments (bands of connective tissue) or sutures. The whole structure forms a framework for the body, as well as protection for the most delicate organs, notably the brain and spinal cord. The skeleton also serves for the attachment of the body musculature which helps to maintain the animal's posture as well as to bring about movements essential for activities such as feeding, escape or reproduction. In addition to these functions the skeleton acts as a reservoir for certain minerals including calcium, phosphorus, magnesium, fluoride and chloride.

For comparative purposes the skeleton can be divided into the following three regions: the visceral skeleton, which in primitive vertebrates is associated with the pharyngeal wall and gills, but which in higher forms contributes to the skull; the axial skeleton, which comprises the vertebral column, ribs, sternum when present, and the greater part of the skull; and the appendicular skeleton, which is composed of the pectoral (shoulder) and pelvic (hip) limbs and girdles. Some authors combine the axial and appendicular skeleton as the somatic skeleton, others consider the visceral skeleton with the axial skeleton. As the latter practice is more convenient it will be followed here. Another important distinction is made between dermal or membrane skeletal structures, consisting of bone developed in the dermis, and the chondral skeleton (or endoskeleton) which is preformed in cartilage in the vertebrate embryo and in some cases remains cartilaginous throughout the animal's life. Some functional units of the skeleton, notably the skull and shoulder girdles of many vertebrates, have both dermal and chondral components.

The axial skeleton: the vertebral column and ribs
At an early stage in vertebrate development, the vertebral column and the base of the skull together replace the

notochord as the main longitudinal supporting and skeletal structure of the body. Although a vertebral column is less flexible than a notochord, it is considerably stronger, a factor of particular importance in land vertebrates. In addition to its supporting role, the vertebral column has a protective function since the individual vertebrae which make up the column extend dorsally to surround and protect the dorsal nerve cord. Vertebral structure varies considerably throughout the subphylum, and may also vary along the length of the vertebral column of the same animal. Fishes show a distinction between trunk and caudal vertebrae while in the higher vertebrates the vertebral column is divided into cervical, thoracic, lumbar, sacral and caudal regions. The cervical or neck vertebrae are freely movable to allow a high degree of movement of the head. The thoracic vertebrae bear the ribs which articulate ventrally with the sternum. Lumbar vertebrae provide a second freely movable region in many terrestrial forms while the sacral region has vertebrae of great strength for the support and attachment of the pelvic girdle. The caudal vertebrae are generally simple in structure and support the tail.

The simplest type of vertebra consists of a solid cylindrical structure, the centrum, which functionally replaces the notochord, and a neural arch which surrounds and protects the nerve cord. The centrum is typically spool-shaped and articulates at either end with adjacent centra by means of intervening cartilaginous discs. Various projections may extend from the walls of the neural arch. These include anterior and posterior articular processes (pre-and post-zygapophyses) which join the neural arches of adjacent vertebrae, and various lateral projections (the transverse processes) which provide surfaces for the attachment of ribs or muscles. Haemal arches extend ventrally from the centrum of the tail region of many vertebrates to surround and protect the major blood vessels of the tail. The vertebrae are bound together by ligaments to give a structure which combines a certain degree of flexibility with rigidity.

Vertebrae are segmental structures and the number present in the adult corresponds to the number of somites from which they are derived. However the vertebrae are so positioned that they alternate with the body wall muscles and each muscle is therefore associated with two vertebrae. This arrangement is particularly significant in the lower vertebrates where locomotion depends on alternate lateral flexures of the body. In fishes, muscular force is exerted on the myocommata between successive muscle blocks and on ribs which are formed within these structures and which articulate with the vertebrae. Thus, when the muscle blocks on one side contract, the body is flexed to that side, the vertebral column acting as a compression strut.

The head skeleton: the skull and lower jaw

The term 'skull' does not have a precise definition and can include any skeletal structure in the head, though it usually excludes the lower jaw. The main part of the skull is the chondrocranium, neurocranium, or brain case, which is a box-like structure representing the anterior end of the axial rod and forming a base which supports and protects the brain and surrounds the major sense organs. Other basic components of the skull include the anterior arches of the visceral skeleton with various dermal elements. These last roof and encase both the chondrocranium and the visceral elements. Thus the skull is a very complex structure.

The chondrocranium only appears as a distinct skeletal component in the agnathans and in cartilaginous fishes (where the dermal element is thought to have been lost during their evolution.) It is shaped like an old-fashioned sugar scoop, forming a floor, walls, and posterior roof to the brain. Four main regions can be distinguished from front to back: they are the ethmoid, orbital, otic, and occipital regions. Nasal capsules, lying on either side of the ethmoid region, enclose the olfactory organs; the orbital region is

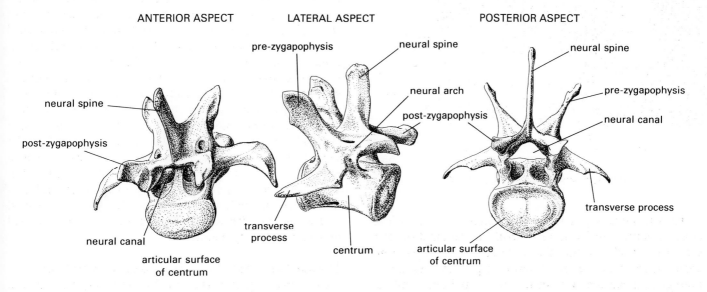

A lumbar vertebra (wallaby)

narrow, forming sockets for the eyeballs and their associated musculature; the otic region incorporates an otic capsule surrounding the internal ear on either side. The occipital region is the only part which is roofed by cartilage or cartilage bone.

The brain case has numerous small openings or foramina (from the latin *forare* meaning 'to bore') for the passage of cranial nerves and blood vessels. A large posterior opening in the occipital region, the foramen magnum provides for the passage of the spinal cord.

In primitive jawless vertebrates the visceral skeleton consists of seven pairs of V-shaped visceral arches which lie in the walls of the pharynx and primarily serve to support the gills. However in more advanced vertebrates these structures become so modified as to be almost unrecognisable. In gnathostomes the anterior two pairs of arches are modified as jaws. The first (mandibular) arch forms the upper and lower jaws, while the second (hyoid) arch helps in their support. In cartilaginous fishes, for example the dogfish, the entire jaw is derived from the mandibular arch, with the dorsal elements of the 'V' forming the upper jaw or palato-quadrate bar, and the ventral elements forming the lower jaw or Meckel's cartilage. In the majority of vertebrates a variety of dermal bones, principally the maxilla, premaxilla, and dentary become associated with or replace these cartilaginous elements. The remaining visceral arches (the third to the seventh) support the gills in lower vertebrates and are known as branchial arches. Higher forms show considerable modification of both the visceral skeleton and its enclosing dermal elements.

The anterior region of the skull is roofed by a shield of membrane bones forming the dermal skull roof. In most vertebrates this covers the top and sides of the head and extends to the jaws as the maxilla and premaxilla, which both bear teeth. The principal roofing elements are, from front to back, the nasals, frontals and parietals. The dermal skull roof has openings for the nostrils (external nares), eyes (orbits) and in many lower forms a median parietal foramen for the pineal eye. In addition, membrane bones have become closely applied to the ventral surface of the chondrocranium, forming the greater part of the palatal complex which lies within the roof of the mouth. The palatal complex also incorporates ossifications of the palato-quadrate cartilage. Its anterior region is a broad plate with lateral gaps for the internal nostrils (choanae). Posteriorly there are gaps (the sub-temporal fossae) between the margins of the palatal complex and the dermal skull roof, which allow the adductor muscles (which close the jaws) to pass from the skull to the lower jaw. In bony fishes and amphibians the palate forms the roof to the pharynx but in higher vertebrates there is a secondary structure which separates the nasal passages from the pharynx. In most reptiles and birds this is formed from a pair of longitudinal palatal folds which grow medially, while in crocodiles and mammals it consists of horizontal projections of the palatal bones and the maxillae and premaxillae. In mammals the secondary palate is extended posteriorly by a soft palate of connective tissue. As mentioned above, the lower jaw of cartilaginous fishes is formed by Meckel's cartilage which is part of the first visceral arch. However in vertebrates with a bony skeleton, this cartilage, or the chondral bone derived from it, persists only as a minor part of the lower jaw and is absent altogether in mammals. Instead the main element of the lower jaw is a tooth-bearing dermal structure, the dentary. This is the sole lower jaw element in mammals, though in other vertebrates it has a number of smaller dermal bones associated with it. In bony vertebrates, apart

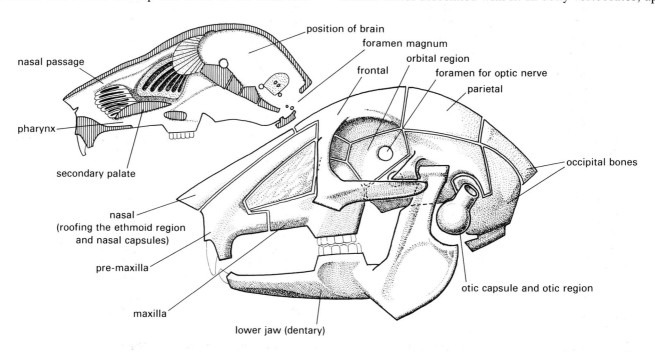

A vertebrate skull (mammal)

from mammals, the posterior part of the jaw is an ossification of Meckel's cartilage, forming the appropriately named articular bone by which the lower jaw articulates with the rest of the skull.

The appendicular skeleton

The appendicular skeleton is that part of the skeletal system associated with the appendages of which there are typically two pairs in vertebrates, taking the form of fins or limbs.

The anterior, or pectoral, appendages lie just behind the gills in fish and at the border of the neck and trunk in tetrapods. The posterior or pelvic appendages lie at the posterior part of the trunk.

Each limb skeleton includes not only the bones of the limb itself but a basal supporting structure known as a girdle, which lies within the trunk and provides a stable articulation point for the limb. In land vertebrates the girdle also functions to transfer the weight of the body from the vertebral column to the limb, and hence to the substrate, and provides a point of origin for the limb muscles. Generally the pelvic girdle is the simpler structure, consisting entirely of chondral bone. In most vertebrates it is composed of three elements: the ischium, ilium and pubis. The pelvic girdle of fishes is embedded in the ventral body

muscle and has no articulation with the vertebral column. In tetrapods it articulates with the vertebral column in the sacral region.

The pectoral girdle is more complex and includes structures of both chondral and dermal origin. The chondral elements form limb supports and the dermal elements (lacking in cartilaginous fishes) give added strength and attach the structure to the vertebral column. The chondral elements are the scapular and coracoid, while the principal dermal elements are the cleithrum and the clavicle (which may be absent).

The appendicular skeleton of vertebrates is very variable and some of the forms it takes are discussed in more detail in the relevant sections of the diversity chapter.

The muscular system

The muscular component of the skeleto-muscular system is chiefly muscle associated with voluntary activities and is derived either directly or indirectly from myotomes. It is composed of muscle blocks which consist of striated muscle

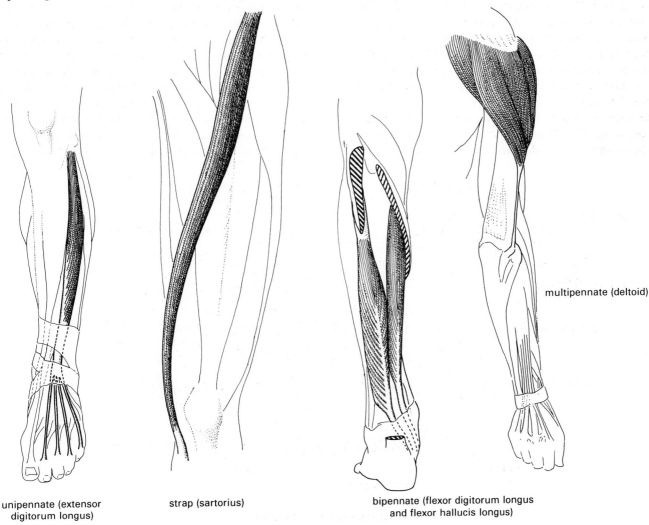

unipennate (extensor digitorum longus)

strap (sartorius)

bipennate (flexor digitorum longus and flexor hallucis longus)

multipennate (deltoid)

Arrangements of muscle fibres (based on human)

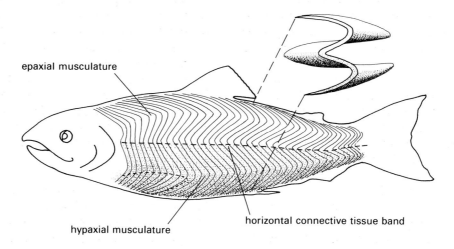

Axial musculature of a fish

fibres bound together by connective tissue (fasciae) and associated with nerves and blood vessels. The blocks may be attached to the bones either directly to their overlying connective tissue (periosteum) or more remotely by rope-like tendons. Like the skeleton, this part of the muscular system is normally sub-divided into the appendicular musculature (associated with limbs and girdles) and axial musculature (associated with the trunk, tail, and eyeball). Some of the muscle associated with the skeleton is anomalous in various ways. Although the branchiomeric musculature (associated with the visceral arch skeleton) is striated muscle, it is more closely similar both in its innervation and in its embryonic origin to the smooth muscle of the alimentary tract. It is therefore regarded as part of the visceral musculature, most of which is involuntary smooth muscle associated with the gut. In primitive vertebrates the branchiomeric musculature forms the muscles of the gill arches and in more advanced groups it is associated with visceral derivatives, forming, for example, jaw muscles.

The fleshy part of a muscle is usually referred to as the belly, while the end which moves least is the origin and the end which moves most the insertion. Another general rule is that the proximal end of a limb muscle is always referred to as the origin, and the distal end as the insertion irrespective of the relative mobility of the associated parts. Descriptive terms frequently applied to muscles include digastric for a muscle with two bellies and an intervening tendon, and pennate for a muscle consisting of numerous short fibres inserting onto a tendon lying along the muscle surface or even within the muscle. Unipennate, bipennate and multipennate are used to describe muscles with one, two or several groups of fibres within the muscle, while bicipital and multicipital describe muscles with two or several origins. Strap muscles have parallel fibres.

Many names applied to muscles refer to the action they produce, and it is important to know the significance of these names: muscles are often arranged in opposing pairs which work in an opposite way to each other, with flexors tending to bend a joint, while extensors straighten it. Adductors draw a segment towards the mid-line of the body, or towards a neighbouring part, while abductors move it away. Pronators turn a part downwards whereas supinators turn it upwards, and rotators turn it on its axis. A depressor lowers a part; in contrast an elevator raises it. Constrictors and sphincters surround orifices, and close them, for example around gills, similarly dilators open them.

The axial musculature forms the main muscle bulk of fishes and is arranged in segmental blocks or myomeres which are divided in all but the most primitive vertebrates by a horizontal band of connective tissue into dorsal (epaxial) and ventral (hypaxial) muscles. The epaxial muscles are effectively those of the back, and hypaxial those of the belly and sides. In tetrapods the bulk of the muscle is appendicular and although it is derived from the somatic musculature it is very different in its arrangement: most meat and poultry is appendicular muscle.

The digestive system

The function of the digestive system is to obtain food and prepare it for absorption into the body. The digestive processes are essentially similar to those of invertebrates and involve the enzymatic breakdown of complex chemicals into more simple molecules which can then be absorbed into the circulatory system through the gut wall. Only smaller molecules, such as water, salts and vitamins, can be absorbed unchanged. Complex carbohydrates must be reduced to simple sugars, proteins to amino acids, and fats to glycerol and fatty acids. Ingested matter which cannot be digested is passed along the digestive tract and eliminated.

Vertebrates have a very wide range of diets, but almost all of them are herbivores, predatory carnivores or omnivores.

Food gathering involves the mouth, oral cavity, and pharynx, with associated structures which in most vertebrates include the tongue, jaws and teeth. Digestion takes place in the remainder of the alimentary tract, often called the gut, which is adapted for the transport, physical and chemical breakdown, and subsequent absorption of food. Like the gut of most invertebrates it is basically a simple tube subdivided into three main regions: the oesophagus, stomach and intestine. However the detailed structure is highly variable, for example the stomach may be absent and the intestine frequently shows further subdivision.

The principal function of the pharynx is respiratory and its role in the digestive system is chiefly as a passage between the mouth and oesophagus. In fishes it is long and associated with the gill slits, but in terrestrial vertebrates it is rather short and is simply the region where the food and air passages cross. In all vertebrates, the swallowing muscles are in the wall of the pharynx.

The oesophagus connects the pharynx to the stomach and in most cases it is a simple conducting tube whose length varies with the length of the neck. It may be modified for food storage in some animals. In fish the oesophagus is very short and the junction with the stomach is almost imperceptible. As its wall is thrown into many longitudinal folds it is capable of considerable distension to cope with bulky food. The crop of birds is a modified pouch-like expansion of the oesophagus. An extreme modification of the oesophagus and stomach is seen in ruminants, for example the cow. The true stomach is the abomasum, whilst the remaining chambers, the rumen, reticulum and omasum, though popularly regarded as stomachs, are in fact sacculations of the oesophagus. The rumen is particularly large, constituting 80% of the total, and provides a chamber in which fermentation of herbage occurs as part of the digestive process. Grass is compacted into small balls of cud in the reticulum and returned to the mouth for further chewing. It then passes into the rumen and through the omasum which functions as a strainer to the true stomach.

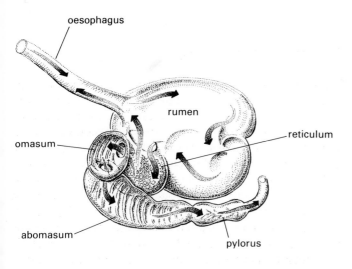

Modified oesophagus and stomach of ruminant

The oesophageal lining is of stratified epithelium, characteristic of areas subject to some friction, rather than the simple columnar type characteristic of the remainder of the gut. The oesophageal muscle fibres generally show a change from striated or voluntary muscle to the smooth muscle again characteristic of the alimentary canal. However ruminant mammals have striated muscle throughout the length of the oesophagus, since these animals must have an oesophagus whose contraction is under voluntary control, in order to return the cud to the mouth.

The stomach is a dilation of the gut, held in position by mesenteries within the large abdominal coelom. It is often a J-shaped pouch although its shape is to some extent related to the shape of the animal, for example in snakes it is greatly elongated. Since the primary function of the stomach is food storage it is thought to have been absent in the microphagous vertebrate ancestors. Obviously storage is of little adaptive value in jawless organisms which feed more or less continuously on small particles or liquid food. Thus it is hardly surprising that the lamprey has no stomach while that of the hagfish is poorly developed, both animals being agnathans. The development of a stomach is thought to be directly related to the development of jaws, and therefore of a diet which includes large pieces, rather than small particles of food. The digestive function of the stomach, involving mechanical breakdown and churning of food and the initiation of the chemical breakdown of proteins was probably a secondary development.

The stomach is divided into two regions, a cardiac region and a pyloric region, and is separated from adjacent parts of the alimentary tract by the cardiac and pyloric sphincters. After food enters the stomach, both sphincters close and muscular contractions of the stomach churn the food, breaking it up mechanically and, in mammals, mixing it with gastric juice. This is secreted by tubular glands in the stomach wall and contains hydrochloric acid and the proteolytic enzyme pepsin. The hydrochloric acid serves both to kill any micro-organisms which enter the stomach, and to provide a correct pH for the action of pepsin. Eventually the food is reduced to a creamy material (chyme) which is forced through the pyloric sphincter by peristaltic contractions of the pyloric region. In many birds the pyloric region of the stomach is highly modified for grinding as a gizzard, associated with their lack of teeth. The muscles are arranged to form two discs with tendinous centres and the cells lining the gizzard secrete a tough horny lining.

Like the stomach the intestine is suspended in the abdominal coelom mesentery. It is the principal site of digestion and absorption, and in most vertebrates has a large surface area obtained by increased length and by outgrowths (villi) and longitudinal or circular foldings of the intestinal lining. Primitive fish have a short straight intestine whose surface area is greatly increased by a helical fold of the lining known as the spiral valve. In more advanced fish and in terrestrial vertebrates the spiral valve is lost, and the

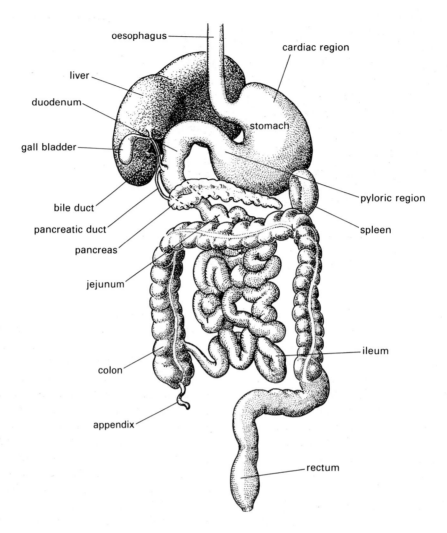

oesophagus

cardiac region

liver

duodenum

stomach

gall bladder

bile duct

pancreatic duct

pancreas

jejunum

pyloric region

spleen

ileum

colon

appendix

rectum

The mammalian alimentary tract

necessary absorptive and secretory surface is provided by a considerable increase in the length of the intestine which becomes more or less coiled. In tetrapods the intestine is differentiated into an anterior small intestine or ileum, and a posterior large intestine, the colon. The first part of the small intestine is the duodenum, and in mammals two subsequent divisions, jejunum and ileum, are recognised. This terminology is derived from the work of the early human anatomists: ileum is the Latin for intestine, duodenum is derived from the Latin *duodecimus* meaning 'twelve' because the early anatomists measured in finger-breadths and the average human duodenum is twelve average finger-breadths long. Jejunum is the Latin for 'fasting', referring to the fact that this part of the intestine is found to be empty soon after death.

In comparison with that of the other vertebrates the intestine of birds and mammals is very long and food moves along it rather rapidly. This is correlated with the higher body temperature and generally higher activity of these animals which therefore require a higher food intake and more rapid digestion. The intestine is thus a very active organ with rapid muscular movements. Herbivores, particularly mammals, generally have a longer intestine than carnivores and omnivores, reflecting the low digestibility of plant food: cows for example have 50 m of intestine and horses 29 m, compared with between 5 and 8 m in men and 3.4 and 7 m in women!

The wall of the alimentary tract shows a similar histological structure throughout. It is basically composed of three layers, muscular, submucous and mucous, with an additional serous layer in all regions except the oesophagus. The serous layer represents the visceral peritoneum and is therefore absent in the oesophageal region which lies outside the coelom. The muscular layer is composed of an

LUMEN

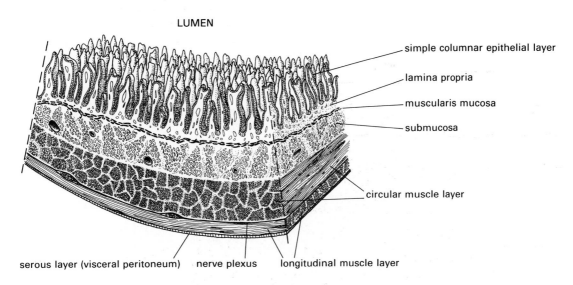

simple columnar epithelial layer

lamina propria

muscularis mucosa

submucosa

circular muscle layer

serous layer (visceral peritoneum)   nerve plexus   longitudinal muscle layer

The histological structure of the small intestine

outer layer of longitudinal muscle, and an inner layer of circular muscle, with a nerve plexus between them (the para-sympathetic myenteric plexus). The submucosa, which lies inside the muscular layer, is composed of dense connective tissue and elastin fibres surrounding intestinal glands, blood and lymph vessels. The mucosa is three-layered with an outer layer of smooth muscle fibres, again arranged in outer longitudinal and inner circular layers, a layer of connective tissue (lamina propria) and a columnar surface epithelium lining the gut cavity. Thus the alimentary canal is a complex structure well adapted for its functions. The muscles propel the gut contents, mix them, and reduce them to a pulp on which the gut enzymes can act. Some of the glands produce mucus which facilitates the passage of food along the gut and helps to bring it to the right consistency, while others secrete the necessary digestive enzymes. The thin lining is adapted for absorption and is supplied with blood and, in mammals, lymph, which function both to transport digested food away from the gut itself and also to support and renew the gut tissue. Digested food, chiefly sugars and amino acids, are transported in the blood via the hepatic portal vein to the liver. Digested fats are reconstituted once they have crossed the intestinal lining into the lymph system (in mammals) and carried to the general circulation via the lymph vessels.

Various glands are found associated with the vertebrate digestive tract, apart from those which are part of the gut lining. The most noteworthy are the salivary glands, which are connected to the oral cavity, the liver, whose duct marks the division between the fore- and hindgut, and the pancreas. Salivary glands are a feature of terrestrial vertebrates and their primary function is to moisten food and make it slippery for swallowing. Fish, and amphibians spending their entire life in water, have no glands other than simple mucous glands opening into the oral cavity. Aquatic

mammals such as whales and sea cows have secondarily lost their salivary glands. In certain vertebrates, for instance frogs, some birds and some mammals including man, an enzyme, ptyalin, is present in the saliva and initiates starch digestion. The salivary secretions of swifts include an adhesive substance which is used in nest building to stick mud together, while in some snakes and certain shrews the glands are modified to secrete venom.

Both the liver and the pancreas develop as outgrowths from the anterior part of the intestine. The liver is the largest gland in the body and is generally lobed but is very variable in shape even between individuals of the same species. The variability in mammals is such that in some ancient religions the shape of the liver in sacrificed animals was used as a means of telling the future. The liver is the major chemical processor for the body as a whole. It may synthesise protein, or convert proteins and fats into carbohydrates and transform nitrogenous wastes in the form of ammonia into urea or uric acids. Carbohydrate from the intestine is stored in the form of glycogen to be released again into the blood stream where needed. In some animals, notably the cod (hence cod-liver oil) and the shark, the liver serves as a region for fat storage.

The liver is made up of plates of polyhedral liver cells, supported by connective tissue. Blood leaves the liver by the hepatic vein and is supplied to it by the hepatic portal vein, which drains the digestive tract and spleen, and by the hepatic artery. The liver cells secrete bile which passes through the hepatic ducts into a common bile duct which enters the intestine, either directly or through a storage organ known as the gall bladder. The majority of vertebrates have a gall bladder (exceptions include most fish and some birds and mammals) and the bile which finally enters the intestine is highly concentrated as considerable amounts of water are resorbed there. Much of the bile is

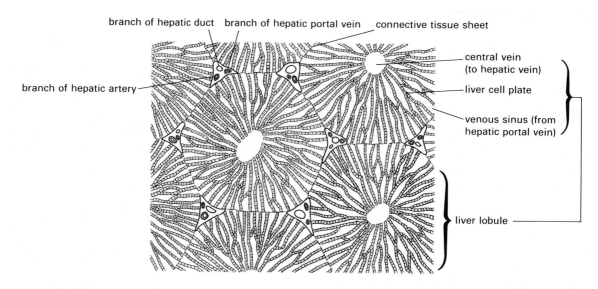

branch of hepatic duct branch of hepatic portal vein connective tissue sheet

central vein (to hepatic vein)

branch of hepatic artery

liver cell plate

venous sinus (from hepatic portal vein)

liver lobule

The histological structure of mammalian liver

excretory and consists of waste products from liver metabolism. These are the products of protein breakdown, and the bile pigments, which may be green, yellow, orange or red depending on the species, are derived from the haemoglobin of red blood cells which have been broken down. However bile also has a two-fold digestive function. As an alkaline fluid it neutralises the acid chyme which enters the intestine from the stomach, creating a natural pH which is optimal for the action of enzymes (amylase, maltase, sucrase, lactase) produced by the intestinal glands and the pancreas. In addition the bile salts emulsify fat, breaking it down into smaller globules, and making possible the absorption of both fats and the fat soluble vitamins (A,D,E,K). The salts themselves are resorbed in the intestine and recycled to the liver. A common malfunction of the gall bladder in humans is the formation of 'gall stones' of solidified cholesterol which may block the bile duct. One of the side effects is that the bile pigments are resorbed by the liver and gall bladder and deposited in the skin giving it a yellowish tinge.

The pancreas is an important digestive gland, of irregular shape and often lobed. It generally lies in the loop between the stomach and duodenum and produces large quantities of digestive enzymes, including amylase which digests starch, trypsin which digests protein, and lipase which digests fat. These enter the intestine, either through a separate duct or more usually via the bile duct. In addition pancreatic tissue includes patches of hormone-producing tissue, the islets of Langerhans, which produce the hormone insulin.

Most of the water ingested with the food, along with that added with the digestive secretions, is resorbed as the indigestible residue passes along the colon. Many bacteria are found in the colon and may bring about some further digestion of the more indigestible food materials. A valve between the ileum and colon, the ileocolonic valve,

prevents the bacteria from passing further back up the intestine. In some mammalian herbivores a long caecum containing colonies of cellulose-digesting bacteria opens out of the intestine at this point.

In most vertebrates the colon opens into a posterior cloaca which also receives the products of the urinogenital system. However in mammals the cloaca is subdivided into a ventral part (which receives the urinogenital products) and a dorsal rectum which opens at the anus.

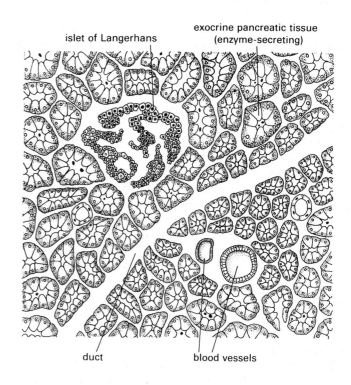

islet of Langerhans

exocrine pancreatic tissue (enzyme-secreting)

duct

blood vessels

The histological structure of mammalian pancreatic tissue

The respiratory system

The primary function of any respiratory system is the exchange of gases between the organism and its environment, specifically the intake of oxygen and expulsion of carbon dioxide. These processes have already been discussed for the invertebrates, and as we saw these animals differ greatly in their activity and hence their oxygen needs. In general, vertebrates need more oxygen than invertebrates, and birds and mammals in particular have very high oxygen requirements. The process of gas exchange has two phases: the first, ventilation, involves the exchange of gases between the animal's blood and its environment; the second is the transport of gases and their exchange between the blood and tissues or cells. The respiratory system has ventilation as its function.

The structure of respiratory systems can be closely related to both the energy requirements of an animal, and to the environment in which it lives. Gaseous exchange always occurs in solution by diffusion across a moist membrane well supplied with blood vessels. In a few vertebrates, for example eels and to some extent frogs, as in many invertebrates, the external surface of the body fulfils this role but all vertebrates also have specialized respiratory organs. These are the gills in most aquatic vertebrates and the lungs in terrestrial vertebrates. Both kinds of structures are derived from, and closely associated with, the pharynx. Like other aquatic animals aquatic vertebrates derive their oxygen from that dissolved in the water. However, since the solubility of oxygen in water is low there is a problem of obtaining a sufficient amount. This problem is not encountered by animals breathing air (21% oxygen by volume). The respiratory organs of aquatic animals reflect this problem; their surface area is increased by complicated plate-like outfoldings (lamellae) to give the maximum surface for gaseous exchange. In contrast terrestrial vertebrates have considerable problems of desiccation. Their lungs are structurally distinct from gills and are sac-like organs extending into the body with their moist respiratory surface usually increased by numerous pocket-like folds, the alveoli. The energy requirements of the animal are reflected by the size of the area for gaseous exchange, thus active vertebrates with large amounts of muscle usually have a large respiratory surface to keep those muscles supplied. The respiratory surface itself, whether in gill or lung, is a single-layer epithelium in order to reduce to a minimum the distance that gas must travel by diffusion. The epithelium always has a rich blood supply which moves gases between the surface and the body cells. In addition to supporting and protecting this delicate respiratory surface the respiratory organ of an active animal such as a vertebrate must provide some means of moving the external medium across it, be the medium water or air. This is the process of ventilation or breathing.

In fishes a pumping action of the mouth and pharynx causes water to flow across the gills, while in terrestrial vertebrates air is moved out of the lungs by pumping

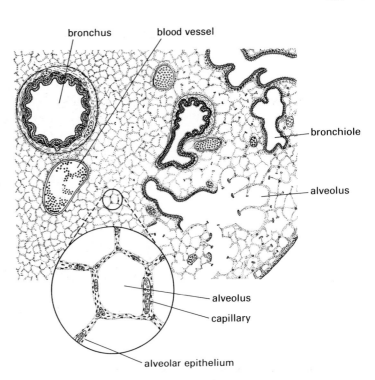

A section through mammalian lung tissue

movements of adjacent parts of the body. The pumping mechanism in amphibians is relatively inefficient and animals such as the frog supplement gaseous exchange through the lungs by cutaneous respiration, a factor that has undoubtedly restricted both their environmental distribution and their size. Higher vertebrates have more efficient ventilation mechanisms and a greatly expanded respiratory surface which permits them much higher oxygen requirements, and enables them to dispense with cutaneous respiration with its inherent problems of desiccation. Mammals maintain a generally high level of metabolic activity at all times and therefore need to have a high and constant supply of oxygen to the tissues. For example, the basal metabolic activity of *Homo sapiens* is 0.25 ml of oxygen consumed per gram of body weight per hour, compared with 0.05 ml for a toad. In addition to their efficient ventilation system based on suction the mammals have developed a secondary palate, a shelf of bone in the roof of the mouth, which separates the nasal and food passages and enables the animal to feed and breathe at the same time. Ectothermic animals with low oxygen requirements can interrupt their breathing to chew and no harm is done, but a mammal would die from lack of oxygen if it did this. As might be expected, bird lungs are the most highly specialized of all vertebrates to meet the very heavy oxygen demands of flight.

In addition to its primary function, in many vertebrates the respiratory system has the secondary function of producing sound. In mammals there is a special organ, the larynx, which is a differentiation of the entrance to the lungs above the trachea with a pair of ridges of elastic tissue, the vocal cords, stretched across it. Birds have a special organ for song production, the syrinx, which is similar to the larynx but lies further down the trachea.

In all vertebrates the blood vascular system is responsible for the transport of gases between the respiratory surfaces and the body cells. The respiratory pigment haemoglobin, a complex iron-containing substance linked with a protein, is present in the blood, enclosed within the red blood corpuscles. As oxygen diffuses through the respiratory surface into the blood it combines reversibly with the haemoglobin to form oxyhaemoglobin. In regions of the body where oxygen concentrations are low the oxyhaemoglobin gives up the oxygen. Haemoglobin, which is also present in solution in the blood of certain invertebrates, is a highly effective carrier and enormously increases the capacity of blood to transport oxygen. If oxygen were only carried in physical solution, blood could only carry about 0.2 ml per 100 ml, but the presence of haemoglobin increases this one hundred fold to 20 ml per 100 ml. Carbon dioxide reacts with water to form carbonic acid in the red corpuscles (about 30% of vertebrate carbon dioxide is transported in this way) and also combines reversibly with haemoglobin to form carbaminohaemoglobin. Jointly these increase the carbon dioxide transported by the blood from a possible 0.3 ml per 100 ml in simple physical solution to about 30 to 60 ml per 100 ml.

Circulatory system

The function of a circulatory system is to transport materials to and from cells. Materials transported to cells include oxygen, nutrients, and water, those transported from cells include carbon dioxide and other waste products of metabolism. In addition hormones are transported from the cells where they are formed, and nutrients are transported from the intestine where they are absorbed. The considerable modifications of the respiratory system associated with the change from an aquatic to a terrestrial mode of life are reflected in major changes in the circulatory system. Another important influence is the increased metabolic rate of higher vertebrates so that the greatest contrast is seen between ectothermic aquatic and endothermic terrestrial vertebrates.

The vertebrate circulatory system is always closed and consists of a pumping device, the heart, and a series of vessels. Arteries and their smaller branches, arterioles, transport blood away from the heart while veins and venules return it to the heart. Within the various organs, arterioles and venules are connected by networks of very fine capillaries which are typically only just large enough to allow an erythrocyte to pass.

The walls of blood vessels are usually composed of three layers, tunica intima, tunica media, and tunica externa. In the smallest and simplest vessels these are restricted to simple layers of endothelium, muscle fibres and connective tissue respectively, but in larger vessels they are thicker and more complicated. In particular, arteries transporting blood directly from the heart at high pressure have well-developed muscle layers and an additional sheath of elastic tissue. Veins have a simpler structure, with less muscle and elastic tissue and more connective tissue.

Vertebrate hearts show a great variety of structure though they all have some features in common. The heart always lies ventral (anterior in man) to the gut in a special anterior region of the coelom. In origin it represents a series of expansions of the main vascular trunk, and has a histological structure which is essentially the same as that of the other blood vessels. Three layers are present, referred to as the endocardium, myocardium, and epicardium, which correspond to the three layers of the blood vessels. The

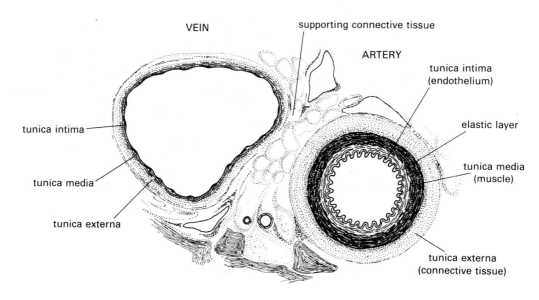

A section through an artery and a vein

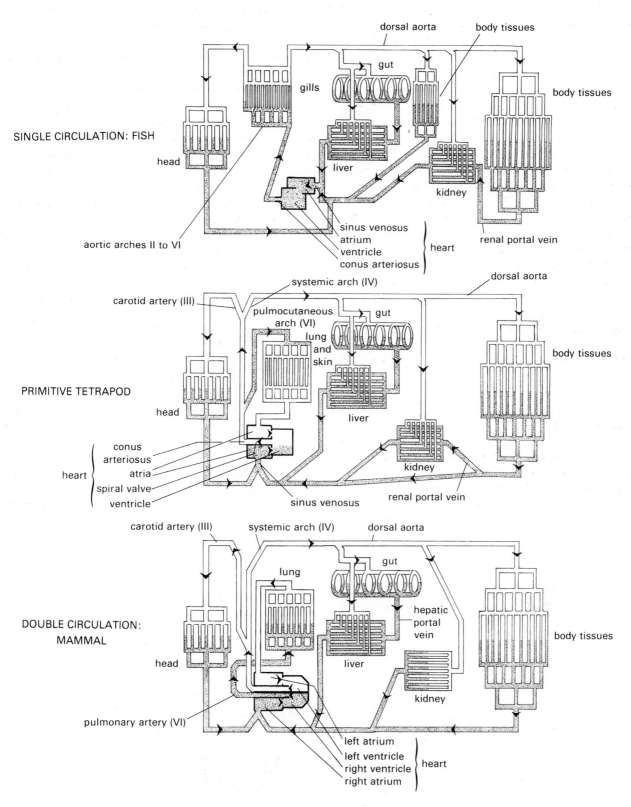

SINGLE CIRCULATION: FISH

dorsal aorta

body tissues

gut

gills

body tissues

head

liver

kidney

aortic arches II to VI

sinus venosus
atrium
ventricle
conus arteriosus

} heart

renal portal vein

PRIMITIVE TETRAPOD

systemic arch (IV)

dorsal aorta

carotid artery (III)

pulmocutaneous
arch (VI)

gut

lung
and
skin

body tissues

head

liver

heart {

conus
arteriosus
atria
spiral valve
ventricle

sinus venosus

kidney

renal portal vein

DOUBLE CIRCULATION:
MAMMAL

carotid artery (III)

systemic arch (IV)

dorsal aorta

lung

gut

head

hepatic
portal
vein

body tissues

liver

kidney

pulmonary artery (VI)

left atrium
left ventricle
right ventricle
right atrium

} heart

Vertebrate circulatory systems

main bulk of the heart is provided by the myocardium which, as its name suggests, consists mainly of cardiac muscle, with some connective tissue. (Cardiac muscle is discussed more fully earlier in the text.) Fishes have a single circulation with a heart consisting of four chambers, the sinus venosus, atrium, ventricle and conus arteriosus, arranged in a linear sequence. All the blood entering the heart comes from the tissues via the veins and therefore has a low oxygen and a high carbon dioxide content. The ventricle is the main contractile portion of the heart. Blood enters it at low pressure from the sinus venosus and atrium, and is driven from it through the conus arteriosus at high pressure. It passes through a median artery, the ventral aorta, which gives off five or six pairs of aortic arches, each of which forms a capillary network within the gills. There gas exchange occurs, so that carbon dioxide is removed and oxygen is added. The aortic arches extend from the gill capillaries to join a median dorsal aorta which gives off various branches to the rest of the body. As the blood flows away from the heart its pressure drops because of friction between it and the blood vessel linings, particularly in the capillary networks of the gills and the tissues. The capillaries of the tissues are drained by veins. Some of the blood is returned directly to the heart, but some passes through the kidneys first via the renal portal system (in some vertebrates) or from the intestine to the liver via the hepatic portal system. The function of the hepatic portal system is easy to understand as the liver plays an important role in food storage and metabolism. The need for a specialized renal portal system, which appears in the bony fish and persists up to the birds, is less clear.

Many changes have occurred in the arrangement of the aortic arches associated with the development of lungs. In amphibians the first two and the fifth aortic arches have been lost altogether, and the third is modified as a pair of internal carotid arteries which supply blood to the head. Arches four and six no longer supply gills and are not interrupted by a capillary network. The fourth remains as the systemic arch, which supplies blood to the body through the dorsal aorta, and the sixth as the pulmocutaneous arch, supplying blood to respiratory surfaces in the lungs and skin. New structures, the pulmonary veins, supply blood from the lungs to the heart, which therefore receives both oxygenated blood from the lungs and deoxygenated blood from the rest of the body. Blood from the two sources is separated in the atrium which is subdivided by a septum, but converges in the ventricle which remains a single structure. However, although some mixing inevitably occurs in the ventricle a functional separation is achieved to a considerable extent. The conus arteriosus has a spiral valve and the blood streams are deflected with the pulmocutaneous arch receiving primarily blood which has a relatively low oxygen content from the right atrium, the systemic arch receives mixed blood, and the internal carotids receive primarily oxygenated blood from the left atrium.

In birds and mammals the blood system has a double circulation with complete separation of oxygenated and deoxygenated blood. Both the atrium and the ventricle are completely subdivided. The sinus venosus is incorporated into the right atrium. The conus arteriosus is also subdivided, with one part contributing to the pulmonary artery which supplies blood to the lungs, while the other forms part of the aortic arch which leads to the body tissues. The renal portal system is completely lost, although the hepatic portal system remains. These changes result in a highly efficient blood system with no mixing of arterial and venous blood, a greater blood volume, and a higher blood pressure.

One consequence of the higher blood pressures found in more advanced vertebrates, is that blood fluid and plasma proteins tend to escape from the capillaries into the interstitial fluid bathing the cells. In higher vertebrates, and particularly in the mammals, a lymphatic system has evolved which parallels the venous system and functions as a route for returning both blood fluid and plasma protein. The lymph capillaries have thin, permeable walls and are not connected with the arteries but arise blindly in the tissues. The fluid they contain, lymph, is similar to that bathing the tissues and moves sluggishly through the lymph vessels. Lymph nodes lie at the points where many lymph vessels converge: notably in the neck, the axillae and in the groin. They play an important part in the body's defences against disease, and are major sites for the production of lymphocytes (which are a type of white blood cell). In addition the cells of the lymph nodes respond to invading bacteria, either by initiating antibody production or by phagocytosis.

The spleen is a large mass of reticular tissue, associated with the blood system and usually found suspended in mesentery near to the stomach. It is a discrete structure enclosed within a capsule of connective tissue and has a reddish colour. The spleen consists of two types of tissue: red pulp, which is a mass of all the elements of blood including a large proportion of red blood cells, and white pulp composed of clusters of white corpuscles. The lymph nodes have similar patches of reticular tissue, but their cellular contents are almost entirely lymphocytes. The spleen is important in the formation, storage, and destruction of blood corpuscles as well as a defence against disease.

The urinogenital system

The vertebrate urinogenital system combines the functions of excretion, osmoregulation, and reproduction. Such a combination may seem curious, but the various structures involved are closely associated in their development. Also the urinary system (which combines the functions of osmoregulation and excretion) and the reproductive system use common ducts, thus the two systems are morphologically inseparable. The urinary system of vertebrates consists of the kidneys and their associated ducts, but the gills, lungs, skin and intestines may also be involved in the processes of excretion and osmoregulation. The major components of the reproductive system are the gonads and the various ducts and glands associated with the transfer of eggs or sperm.

The vertebrate urinary system

Vertebrate kidneys are discrete, paired structures which lie dorsal to the coelom on either side of the dorsal aorta. They are composed of large numbers of kidney tubules or nephrons, which are blind-ending tubes that receive a filtrate from the blood. The number, arrangement and detailed structure of the kidney tubule vary according to the environment of the vertebrate concerned, but, all are built on the same general plan. The proximal part of the tubule consists of a cup-shaped capsule with a double wall of squamous epithelial cells (Bowman's capsule) which surrounds a knot of small blood vessels (the glomerulus).

The inner layer of the capsule is closely associated with the walls of the vessels so that only a thin membrane separates the blood from the cavity between the two walls of the capsule. This cavity is continuous with that of a convoluted tubule of cuboidal epithelial cells which may be of a considerable length and eventually connects with a system of ducts leading to the exterior. Throughout its length the convoluted tubule is closely associated with a capillary network. In the majority of vertebrates the tubule is separable into distinct proximal and distal regions, in the mammals a long straight loop, the loop of Henlé, lies between these two regions.

The kidney functions in two separate operations, one involving the capsule and one the convoluted tubule. The capsule functions as a very fine filtering device which produces a filtrate of blood plasma as a result of blood at high pressure passing through the small vessels in the capsule. Blood corpuscles and large molecules, such as the plasma proteins, are excluded by the membranes but smaller molecules are not, and the filtrate includes not only nitrogenous waste but valuable food substances

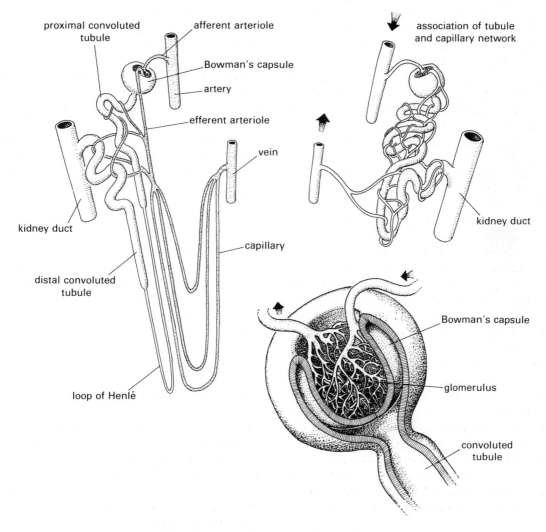

The structure of a mammalian nephron

(particularly sucrose) and a large quantity of water. As the filtrate passes through the tubule, food materials and water (particularly in terrestrial vertebrates) are reabsorbed. At this stage there is further active excretion of waste substances into the tubule. In this way the kidneys produce very close control of the chemical composition and water content of the blood.

The result of these activities is production of urine, an excretory fluid composed mainly of water but containing simple nitrogenous compounds (ammonia, urea, or uric acid) and salt ions such as sodium, potassium and calcium. As in invertebrates, the final product of nitrogenous excretion varies according to the animal's environment. Generally vertebrates which live in freshwater excrete most of their nitrogenous waste in the form of ammonia, which is the form in which surplus nitrogen leaves the cell. In both salt-water and terrestrial vertebrates this is transformed in the liver into one of the less toxic nitrogenous compounds, urea and uric acid.

The volume of urine excreted also varies according to the vertebrate's environment and the form of its nitrogenous waste. Freshwater fish excrete 100 to 300 ml per kg of body weight per day whereas in salt-water fishes the volume is only 5 to 55 ml per kg of body weight per day. The volume output of terrestrial vertebrates is generally lower but varies a great deal according to habitat.

It seems likely from comparative anatomical and embryonic studies that ancestral vertebrates had one nephron for each body segment. Their kidney (holonephros) extended the whole length of the coelom and the tubules drained into an archinephric duct which ran posteriorly to the cloaca. This type of kidney does occur in the larvae of certain agnathans, but not in any adult vertebrate. Adult fish and amphibians have a posterior kidney, or opisthonephros, in which the most anterior tubules are lost, and there is a posterior concentration and duplication of tubules. In the male some of the middle tubules are associated with the testis. In higher tetrapods all the middle tubules are absent apart from those associated with the testis with further posterior tubule concentration and duplication (metanephros). A higher metabolic rate leads to an increased production of nitrogenous waste and the number of tubules is particularly high in mammals and birds. For example whereas many amphibians have fewer than 100 tubules in each kidney, the number in the human kidney is about 1,000,000! In the higher tetrapods the archinephric duct becomes part of the male reproductive system and is replaced functionally by a ureter which evolved as an outgrowth from it.

Most vertebrates have some form of bladder for the storage of urine. In fishes this is generally fairly small and develops as a dilatation of, or near, the excretory ducts. Terrestrial vertebrates obviously have a greater need for such a structure and in tetrapods a bladder develops as an outgrowth of the cloaca. In those forms in which the excretory ducts open into the cloaca, urine flows across it to enter the bladder. In mammals the ureters open directly into the bladder which opens to the external body surface through a separate tube, the urethra. In the majority of mammals the cloaca is subdivided with the ventral portion contributing to the urethra and the dorsal part of the rectum. The tetrapod bladder has stout walls and is capable of great distension. The inner lining is transitional epithelium which can readily be stretched from a thick, apparently stratified, layer to a thin single or double layer of squamous cells. The bladder wall is completed by layers of smooth muscle. It is closed by a smooth muscle sphincter which is under voluntary control and can be relaxed to allow urine to escape.

A bladder is absent in the majority of birds (the ostrich is an exception) and in many reptiles including snakes, crocodiles and some lizards. In these animals the urine is mixed with the faeces. The loss of a bladder is undoubtedly associated with the use of uric acid, a precipitate, as an excretory product.

The reproductive system

In the great majority of vertebrates, reproduction is sexual and the sexes are separate. There is usually a pair of gonads, which in males are testes associated with efferent ducts, and in females are ovaries associated with oviducts.

In aquatic vertebrates fertilization is usually external, but in all terrestrial forms except the frogs and toads internal fertilization is the rule. In its simplest form internal fertilization may be brought about by the partners bringing their cloacae together, though a variety of accessory structures have developed to facilitate sperm transfer during copulation. In live-bearing forms additional structures have evolved to provide a place in the female for the fertilized egg to develop.

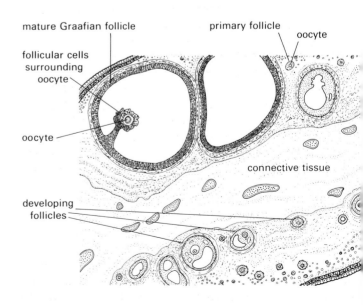

A section of a mammalian ovary

The gonads are held attached to the dorsal body wall by mesentery. Eggs are typically surrounded and nourished by follicle cells as they develop. The follicle cells also produce female hormones (oestrogen and, in mammals the pregnancy hormone progesterone) at certain stages in the breeding cycle. The ovaries are generally solid structures and eggs are shed into the abdominal cavity (ovulation) and pass into the funnel-shaped distal end of the oviduct. A few advanced bony fish have hollow ovaries and their eggs are shed into the lumen which is connected to the oviduct. Eggs move along the oviduct, either by peristaltic contractions of its muscular wall or by the beating of cilia which line the lumen. In many vertebrates (and particularly birds and reptiles) the egg has a supply of yolk to nourish the developing embryo. This is formed in the ovary, but additional protective layers of albumen or egg-white, shell membranes and shell are added, after fertilization, by glands in the oviduct wall. In the majority of vertebrates the oviducts open directly to the exterior but in therian mammals they unite and expand to form a muscular pouch, the uterus, in which the young develop. A short tube, the vagina, adapted to receive the male copulatory organ, leads from the uterus to the exterior.

In the breeding period the vertebrate ovary becomes distended and irregular in outline. In many vertebrates, particularly those in which there is parental care of the young, only a small number of ripe eggs are present at any one time, derived from a stock of immature eggs produced at an early stage in the ovary's development. However those species with external fertilization and little parental care may produce large numbers of eggs; the majority of frogs and toads, for example, produce hundreds or even thousands of eggs in the breeding season, and some bony fishes may shed millions of eggs. In these cases further germ cells, which develop to ova by a process known as oogenesis, must be proliferated by the germinal epithelium in the adult ovary.

The typical vertebrate testis is a compact organ made up of numerous seminiferous tubules in which sperm are produced and which communicate to the outside by means of ducts. The tubule walls are made up of cells known as spermatogonia which divide and give rise to the sperm, and also Sertoli cells which are nutritive in function. The testis also has a considerable number of interstitial cells which have an endocrine function, producing the male hormone, testosterone. In most vertebrates the testes are located in the dorsal region of the body cavity, in the equivalent position to the ovaries of the female and suspended in mesentery. Some mammals, which have a body temperature which is too high for spermatogenesis, have their testes lying outside the body cavity in scrotal sacs.

The efferent ducts, which conduct the sperm away from the testes, show a variety of arrangements, but there is always some type of connection between the seminiferous tubule and the archinephric duct. In some vertebrates (frogs for example) the efferent ductules run through the mesentery (suspensory ligament) to connect with the most

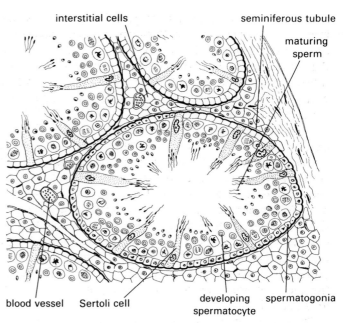

A small area of mammalian testis showing developing sperm

anterior kidney tubules, and sperm is thus conveyed from the testis, through the efferent ductules and kidney tubules to the archinephric ducts, which also have a urinary function. Most bony fish have developed a separate sperm duct. In other vertebrates the archinephric duct has entirely lost its urinary function and is modified for sperm transport as a ductus (vas) deferens. For most of its length the ductus deferens is a simple structure though parts of it may be modified for sperm storage and glandular secretion.

Fertilization always takes place in a fluid medium: in aquatic animals which use external fertilization this is provided by the water, but in other forms the necessary fluids are provided by the glandular areas of the male, and to a lesser extent the female. The seminal fluid produced by the male not only serves as a vehicle for the sperm, but may also protect, nourish, and activate them. A variety of copulatory organs are found in the many vertebrates which use internal fertilization. In the male mammal the sperm ducts empty into the urethra, which also transports urine from the bladder. Towards its distal end this is surrounded by a copulatory organ, the penis, which is used to transfer sperm to the female.

The majority of vertebrates only reproduce during a brief breeding season. Since the release of eggs and sperm must be synchronous for fertilization to occur the majority of vertebrates show some sort of courtship behaviour. This is usually initiated by the male, and may be a brief ceremony or may last for several days. Apart from bringing about the necessary physiological synchronisation, courtship enables the partners to identify correctly a member of the same species but of the opposite sex. Many vertebrates use specific mating calls to attract partners of the opposite sex. Courtship also tends to reduce aggressive tendencies, and is therefore of particular importance in animals which are predatory or strongly territorial.

The nervous system

The nervous system is responsible for regulation and co-ordination of body activities and the structural and functional unit of nervous systems is the neuron. It is fast acting, for example the conduction rate of the human sciatic nerve is 90 m per second, but the responses are of short duration. Each nerve impulse is a single rapid wave of depolarisation in the axon of individual nerve cells; if the effect is to be maintained the impulse has to be repeated.

The vertebrate system is made up of three major components: the central nervous system consisting of the brain and spinal cord, the peripheral nervous system and various sense organs. The peripheral nervous system is composed of cranial and spinal nerves which leave the protection of the cranium or vertebral column to extend either to the sense organs (receptors) or to the voluntary or involuntary muscle (effectors).

The central nervous system

A unique feature of the chordate nervous system is that the spinal cord and the brain are hollow.

The spinal cord
The spinal cord extends posteriorly from the anterior region of the neural tube which is differentiated as the brain, as a more or less cylindrical tube which tapers down to a thread posteriorly. It lies within, and is protected by, the neural arches of the vertebrae, though in many vertebrates its growth fails to keep pace with the growth of the backbone and it is shorter. Immediately surrounding the spinal cord are various protective layers known as the meninges, these protect the spinal cord against possible damage by the vertebrae. In mammals there are three layers of connective tissue known as the dura mater, which is closely applied to the vertebrae, the arachnoid mater (named from its spider's web appearance) and the pia mater. There is a space between the arachnoid and pia mater which is filled with cerebro-spinal fluid. This fluid is similar in composition to plasma but with less protein. The brain is also surrounded by meninges. Amphibians have a two-layered structure (a dura mater and an inner pia-arachnoid layer) while in agnathans and fish there is a single meninx, the meninx primitiva.

The spinal cord has two important functions: to transmit impulses to and from the brain, and to act as a reflex centre. Many vertebrate responses are in effect stereotyped, that is to say they are immediate and automatic and depend only on the anatomical relationship of the neurons involved. For example the response to handling something hot is of this kind. The heat stimulates receptors in the skin which transmit impulses to the spinal cord through a spinal nerve. Here they synapse with interneurons which transmit impulses to effector neurons. The effector neurons transmit impulses back through the spinal nerve to the muscles of the hand which contract and move the fingers away from the hot object. This stimulus and response is known as a spinal reflex, the response is immediate, and does not wait for the brain to 'feel' the pain and decide what to do. Many vertebrate activities, for example walking, are regulated to a large extent by reflexes, the majority of which have an obvious survival value for the animal concerned.

In cross-section the spinal cord can be seen to consist of two regions: an inner mass of grey matter and an outer mass of white matter. The grey matter consists of the nerve cell bodies and dendrites while the white matter consists of bundles of axons and derives its colour from their myelin sheaths. The grey matter is divided into paired dorsal and ventral 'horns' and in amniotes has the shape in cross-section of a pair of butterfly's wings. The dorsal horns are associated with the dorsal sensory nerve roots and contain cell bodies through which impulses from the sense organs are relayed. The ventral horns contain the cell bodies of motor neurons whose axons pass out of the spinal cord via the spinal nerves to the muscles. Cell bodies of interneurons are also present. The axons in the white matter are arranged in bundles. Those of like function are grouped together and form ascending tracts which carry impulses to the brain and descending tracts which control the spinal neurons which supply the effectors. In the centre of the spinal cord is a narrow canal filled with cerebro-spinal fluid.

The brain
The vertebrate brain is the enlarged anterior end of the neural tube and is therefore hollow with a continuous series of cavities or ventricles filled with cerebro-spinal fluid (the ventricular system). In the early embryo the region which will eventually form the brain enlarges and two constrictions appear, marking off three primary brain divisions: the forebrain (prosencephalon), midbrain (mesencephalon) and hindbrain (rhombencephalon). These may well be related to the development in an early stage in vertebrate evolution of the three major receptor organs: the nose, eye, and ear (and the lateral line in lower aquatic vertebrates). Each region develops a dorsal outgrowth of 'grey matter' associated with one of these sense organs: the cerebral hemispheres of the forebrain are primitively associated with smell, while the roof of the midbrain (tectum) is associated with sight, and the cerebellum of the hindbrain with hearing and balance. Both the fore- and hindbrain undergo further subdivision to give the five major divisions usually recognised in the adult brain. The forebrain divides to give the telencephalon and diencephalon, the hindbrain to give the metencephalon and myelencephalon. The mesencephalon retains its undivided state.

The forebrain
The telencephalon gives rise to a pair of olfactory bulbs which grow towards the nasal region. Posteriorly the roof and sides of the telencephalon bulge outwards as the cerebrum forming a pair of cerebral hemispheres. In fishes and amphibians the cerebral hemispheres are small and

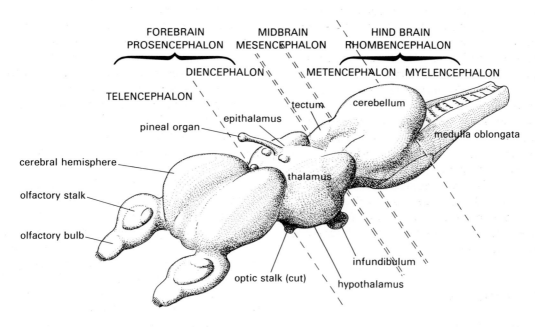

FOREBRAIN · MIDBRAIN · HIND BRAIN
PROSENCEPHALON · MESENCEPHALON · RHOMBENCEPHALON

DIENCEPHALON · METENCEPHALON · MYELENCEPHALON

TELENCEPHALON

tectum · cerebellum

epithalamus

pineal organ

medulla oblongata

cerebral hemisphere

thalamus

olfactory stalk

olfactory bulb

infundibulum

optic stalk (cut) · hypothalamus

A generalized vertebrate brain showing principal brain divisions and structures

chiefly concerned with integration of olfactory impulses but in amniotes they become of increasing importance and are greatly enlarged. In mammals they take over most of the function of the optic lobes of lower vertebrates and have a greatly convoluted surface to incorporate increased numbers of cell bodies. The diencephalon retains its unpaired structure; its walls become the thalamus, its roof the epithalamus, and floor the hypothalamus. The optic vesicles, which contribute to the developing eye, push out from the floor of the diencephalon with, more posteriorly, a downward median projection, the infundibulum. This becomes fused to an upgrowth from the roof of the mouth (Rathke's pouch) and together these structures become modified as the pituitary gland. An outgrowth (sometimes paired) from the roof of the diencephalon gives rise to the pineal organ. This usually remains as a knob of tissue of unknown function lying beneath the cranium but in a few vertebrates, notably some reptiles and agnathans, it forms a median pineal eye. The thalamus is an important relay centre for impulses passing to and from the cerebral hemispheres. The nerve centres of the hypothalamus integrate the functions of the peripheral nervous system with those of other sense organs. They control a variety of physiological processes including temperature regulation, water and carbohydrate metabolism, sleep rhythms and reproduction. Impulses from the eyes via the optic nerves enter the brain ventro-laterally to the hypothalamus and cross over, forming the optic chiasma, on their way to the optic lobes of the midbrain.

The midbrain

The roof of the midbrain develops as a pair of dorsal swellings which form the tectum. This is primarily a visual centre which receives the fibres of the optic nerves (which enter the brain at the hypothalamus), and integrates the information they carry. The tectum varies in size according to the importance of the eyes, and in many fish and birds is particularly well developed so that the dorsal swellings form a pair of distinct optic lobes. In fishes and amphibians the tectum also receives fibre paths from other sensory systems and is the principal area of the brain for controlling body activity. In reptiles and birds it is still of great importance although its role is diminished to some extent by the development and increasing importance of the cerebral hemispheres. In mammals the tectum is represented only by a group of four small swellings, the corpora quadrigemina, which function for visual reflexes and as a relay station for auditory stimuli on their way to the thalamus and cerebral hemispheres. However in some nocturnal mammals the posterior swellings (inferior colliculi), which are responsible for auditory stimuli, can achieve considerable size.

The hindbrain

The regions of the hindbrain are the metencephalon and myelencephalon. The posterior myelencephalon develops as the medulla oblongata, which, structurally, is the simplest part of the brain and the most similar in form, particularly in lower vertebrates, to the spinal cord. The grey matter of the spinal cord extends into the medulla oblongata in higher vertebrates and becomes discontinuous, breaking up into discrete groups of cell bodies usually referred to as nuclei. The central canal is greatly enlarged to form a ventricle so that the columns of grey matter become widely separated, with the dorsal, sensory, structures above, and the ventral, motor, structures below. The columns (or nuclei) form reflex circuits between receptor and effector organs of the head and brachial region. Further forward in the brain the grey and white matter lose their original relationship altogether and become more or less intermingled. In the higher vertebrates their position is reversed in the cerebellum and cerebral hemispheres so that the grey matter forms a distinct layer over the white matter.

In mammals, reflexes that regulate the respiratory movements, heart beat, salivary secretion, and swallowing among others are controlled by the nuclei of the medulla oblongata. However in lower vertebrates only a small proportion of spinal cord fibres reach the anterior parts of the brain, and for many activities the trunk and tail are largely independent of it. Fish and tailed amphibians, which move by rhythmic undulations of the trunk and tail, have a pair of giant cells of Mauthner in the medulla whose axons extend the length of the nerve cord. Although movement is regulated partly by local cord reflexes in these animals, the giant cells of Mauthner exercise an overall control.

The cerebellum, which is present in all vertebrates, arises dorsally from the metencephalon at the anterior end of the medulla oblongata. Its action is passive and essentially reflex, and it is of major importance in the co-ordination and regulation of motor activities and the maintenance of posture. In many of the lower vertebrates, whose movements are not complex, it is small and inconspicuous, but it is large in active animals such as birds and mammals. The cerebellum receives impulses from most of the sense organs, from the proprioreceptors in the muscles and tendons, and from the part of the ear concerned with balance. In active animals the surface of the cerebellum is highly convoluted, especially in mammals where much of the grey matter lies on the surface, allowing room for an increased number of cell bodies. Removal of the cerebellum in, for example, birds causes impaired muscle co-ordination: the animal can still move but thrashes around jerkily.

In the majority of vertebrates the floor of the metencephalon remains unspecialized and contributes to the medulla oblongata. However in mammals this region is swollen and is developed as the pons, which is a great mass of fibres extending crosswise into the two halves of the cerebellum; it relays impulses from the cerebral hemispheres to the cerebellum.

The ventricles

Although the cavity of the neural tube is reduced as a result of various thickenings, it persists in the form of a series of fluid-filled cavities. These form the lateral ventricles in the cerebral hemispheres which communicate through small foramina, the foramina of Monro, with the median third

ventricle of the diencephalon and the fourth ventricle of the medulla oblongata. In anamniotes there is a well developed ventricle in the midbrain, but in amniotes this is reduced to a narrow canal, the aqueduct of Sylvius, connecting the third and fourth ventricles. Areas of thin and highly vascular tissue, the choroid plexuses, frequently develop in the ventricles and provide a surface for the exchange of materials between the blood and cerebro-spinal fluid. In animals with pronounced olfactory lobes, additional cavities (rhinocoeles) are present in the lobes, connecting by small foramina with the lateral ventricles.

The peripheral nervous system

The peripheral nervous system is simply constructed and consists essentially of the nerves which emerge from the brain and spinal cord to extend to almost every region of the body. It is made up of paired spinal nerves and paired cranial nerves, all of which are composed of bundles of nerve fibres (axons).

All nervous functions of the body can be allocated to two main categories, somatic and visceral: somatic functions are those of the skin and its derivatives, the striated muscles and skeletal structures, while visceral functions are those of the digestive, respiratory, and similar major organ systems.

The spinal nerves

The spinal nerves of all vertebrates are essentially alike. For most of their length they contain both afferent (sensory) and efferent (effector) fibres. In the majority of vertebrates the nerves are formed from dorsal and ventral roots from the spinal cord. The dorsal root contains the afferent fibres and a prominent ganglion, (the dorsal root ganglion) which

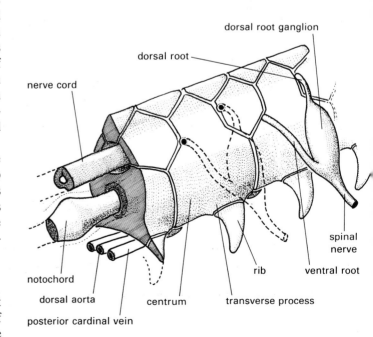

Vertebral column of a cartilaginous fish showing notochord, spinal cord and spinal nerves

contains their cell bodies. It enters the spinal cord dorsally in its side wall. The ventral root runs straight outward from the ventral margin of the cord, but the cell bodies of the efferent fibres lie in the grey matter of the spinal cord itself. In most vertebrates, apart from the most primitive, the roots join outside the vertebrae to form a main trunk, the spinal nerve, from which various branches (rami) diverge. In most segments of the trunk each spinal nerve is a discrete structure innervating the axial muscles, and the skin appertaining to its segment. Usually a dorsal ramus supplies the skin and muscles of the dorsal part of the segment, and a ventral ramus supplies the ventral and lateral parts; in addition there may be one or more branches from the rami supplying the visceral organs. However in certain regions, notably opposite the paired limbs, the rami of the several spinal nerves interweave to form an anterior brachial and posterior lumbo-sacral plexus. The number of spinal nerves varies; fish have a large number of trunk and tail segments and accordingly have a large number of spinal nerves. In contrast, frogs, whose bodies show reduction and telescoping of the terminal segments, have only 10. In humans there are 31 pairs of spinal nerves.

The cranial nerves

The brain gives rise to a series of cranial nerves, which, since they were first studied in man, have been given names and numbers based on their human position and function. Amniotes have 12 pairs of cranial nerves; fishes and amphibians only 10. Three pairs, those of the nose, eye, and ear, have evolved with the sense organs and are composed almost entirely of afferent fibres together with a few efferent fibres which modulate the organ's activity. In this respect they may be regarded as specialized somatic sensory nerves. The remaining cranial nerves contain both afferent and efferent fibres and can be considered as homologous with the roots of the spinal nerves in primitive vertebrates. However they are usually much modified, and some are the counterparts of the dorsal roots, while others represent the ventral roots. A complete list of the cranial nerves is given in the following table:

Cranial nerves

Number	Name	Function
I	Olfactory	Sensory (olfactory epithelium)
II	Optic	Sensory (eye)
III	Oculomotor	Innervates four of the eye muscles
IV	Trochlear	Innervates the superior oblique muscles of the eye
V	Trigeminal	A three-branched nerve. Supplies motor fibres to the jaw muscles and receives impulses from the teeth and skin receptors of the head.
VI	Abducens	Innervates the posterior rectus muscle of the eye.
VII	Facial	Innervates the muscles of the scalp and face, salivary glands, tear glands. Sensory fibres from the tastebuds of the anterior two-thirds of the tongue.
VIII	Vestibulocochlear (auditory)	Sensory, from the inner ear.
IX	Glossopharyngeal	Innervates muscles derived from the third visceral arch; salivary glands. Sensory fibres from the posterior third of the tongue, and the pharynx.
X	Vagus	Receptors in many internal organs such as heart, lungs, stomach. Supplies muscles derived from the remaining branchial arches eg. swallowing, speech. Also supplies muscles of the gut, heart, lungs. Supplies the gastric glands.
XI	Spinal Accessory	Motor nerve accessory to the vagus.
XII	Hypoglossal	Muscles of the tongue.

The spinal accessory and hypoglossal nerves are absent in anamniotes.

The autonomic nervous system

The internal organs, together with the ciliary and iris muscles of the eye, the small muscles associated with hair in mammals, and many glands, are innervated by a special set of nerves and ganglia which comprise the autonomic nervous system. This is strictly a part of the peripheral nervous system, but it differs from other parts of the system in several important respects. It functions automatically, with no voluntary control via the cerebrum, and motor impulses reach the effector organ from the brain or spinal cord by a relay of two neurons. The first (pre-ganglionic) neuron is in the brain or spinal cord, and the second (post-ganglionic) neuron is in a ganglion outside the central nervous system. Furthermore, in most vertebrates the autonomic system is subdivided into two parts, referred to as the sympathetic and parasympathetic systems. Most organs receive nerves from both parts, and the impulses they carry tend to have opposite effects. Thus stimulation of the sympathetic nervous system tends to increase activity, speed up the rate and force of the heart beat, and slow down digestive processes, preparing the animal for fight, flight, or frolic! The parasympathetic system on the other hand tends to reduce activity, slow down the heart rate, and promote digestion by speeding up peristalsis of the digestive tract. The significance of these twin systems is to permit normal metabolic activity to function efficiently and yet allow rapid and effective response of the animal to changed or dangerous circumstances.

The autonomic nervous system is best developed in mammals; in contrast cartilaginous and lower bony fish have no regional differentiation between sympathetic and parasympathetic systems and no double innervation of organs. In higher vertebrates the post-ganglionic neurons of the sympathetic system are within a chain of ganglia lying on either side of the spinal cord. Fibres connect these both to the effector organs and to each other; they also receive fibres via the intercommunicating rami. In contrast those of the parasympathetic system are close to, or within, the organs they innervate. The pre-ganglionic fibres originate in the brain, emerging via cranial nerves III, VII, IX and especially X (vagus) and also from sacral nerves 2, 3, and 4 which join to form the pelvic nerves.

Vertebrate sense organs

By tradition, we talk of the five senses: touch (general sensation), taste, sight, smell and hearing, and it is often assumed that these are the only senses that vertebrates have. However general sensation can be divided into several distinct senses. It includes sensitivity to pressure and temperature, and if the stimuli are excessive then there is a rather different sensation, namely pain. The senses of sight, smell and hearing are associated with conspicuous and specialized organs, the eyes, nose, and ears. However in addition to detecting sound waves the ear also contains an organ for detecting orientation and hence maintaining equilibrium. Apart from these well known and familiar senses there is increasing evidence that many vertebrates (for example birds and turtles) can detect the earth's magnetic field and so use it, literally like a compass, in navigation.

The receptor organs which receive stimuli from the outside world are sometimes described as exteroceptors but vertebrates, like invertebrates, also, have receptor organs which receive stimuli concerning conditions within the organism. These include proprioreceptors located in striated muscles and tendons and interoceptors in the internal organs.

All vertebrate sensory cells are associated with sensory nerves which relay impulses to the spinal cord or to the brain. Impulses from the outside world are received by the skin (the site of general sense receptors in vertebrates) and by the special sense organs. They are carried by somatic sensory nerves which also carry impulses from the proprioreceptors. The impulses from interoceptors, which do not normally reach the level of consciousness, are carried by visceral sensory nerves.

General sensation

The associated stimuli of warmth, cold, and pressure have the least specialized receptor organs. In lower vertebrates in particular these sensations are received by direct stimulation of free nerve endings. In higher vertebrates a variety of sensations are received by sensory structures which are generally microscopic and are widely distributed over the body surface. Such structures are particularly numerous in birds and mammals. Humans, for example, have especially high concentrations of receptors in various sensitive regions such as the finger tips and lips. The free nerve endings in the epidermis detect pain, while specialized nerve endings in the deeper epidermal layers and in the dermis detect pressure, temperature, and touch. The deep pressure receptors (Pacinian corpuscles) consist of a bare axon surrounded by a corpuscle consisting of lamellae interspersed with fluid. If the corpuscle is subjected to pressure the lamellae are deformed and stimulate the axon. Other receptors in the skin include Meissner's corpuscles and Merkel's discs (touch) and Ruffini's organs and the end bulbs of Krause (temperature).

While all vertebrates have a more or less sensitive body surface, some have developed particular senses to a very high degree. Pit vipers, for example, have a specialized temperature detector called the pit organ, which lies between the nose and eye and consists of a pit filled with vascular tissue and nerve endings. It is highly sensitive to the movement of a warm body, and is used by the snake to detect its mammalian prey. However one of the dangers of pit vipers to humans is their inability to distinguish between passing ankles and passing rodents!

Proprioreceptors are of considerable importance in the vertebrates, enabling them to perform complex actions involving co-ordination of several muscles, as well as to maintain balance. For example, in humans, every day activities such as writing, dressing, and tying a knot would be impossible in the absence of proprioreceptors. These structures both register the state of muscle contraction and give information as to the position of various parts of the body, enabling the animal to perform many complex actions. These structures are best developed in the mammals. Lower vertebrates have simple proprioreceptors in the form of nerve fibrils twisted round individual muscle fibres or spread with the tendons. Mammals have specialised muscle spindles which consist of bundles of modified muscle fibres called intrafusal fibres (in contrast to the normal extrafusal fibres of striated muscle), bound together in a sheath and associated with a mass of sensory nerve endings. In addition tendon organs or Golgi organs are found in regions where muscle fibres join a tendon.

The lateral line (acoustico-lateralis) system

The lateral line system is highly developed in fishes and larval or aquatic amphibians and is sensitive to changes in pressure, and to vibrations of low frequency. As a result it can detect water currents and small movements in water. The receptors are groups of hair cells (the neuromasts) similar to those found in the ear, and consist of an elongated sensory cell with a hair-like projection. The hairs are surrounded by a flexible mass of gelatinous material called the cupula which responds to pressure, vibration or water currents by bending the hairs. Stimuli from the neuromasts enter an acoustico-lateralis area of the brain which also receives stimuli from the ear suggesting a close functional relationship between the ear and lateral line system.

In fishes the neuromasts are arranged in a series of canals which are embedded in the skin and normally open to the surface through pores. A major canal, the lateral line, extends posteriorly along the length of the animal and also anteriorly onto the head where canals typically extend above and below the eyes and along the lower jaw. In many fish the position of the lateral line canal can be recognised externally as a distinct stripe. In addition isolated groups of neuromasts occur on the head. In the larvae of modern amphibians the lateral line system of the head has no linear arrangement. The system is lost during metamorphosis to the adult form and is entirely absent in the amniotes, even those which have returned to an entirely aquatic environment.

The ear: balance

Although we tend to think of ears as sound detectors, their most consistent function, maintained throughout the vertebrates, is to detect changes in the orientation of the body with respect to its surroundings and thus to help it to maintain its equilibrium. Clearly both vision and proprioreception play a part in this process, but it is primarily a function of the inner ear. The majority of inner

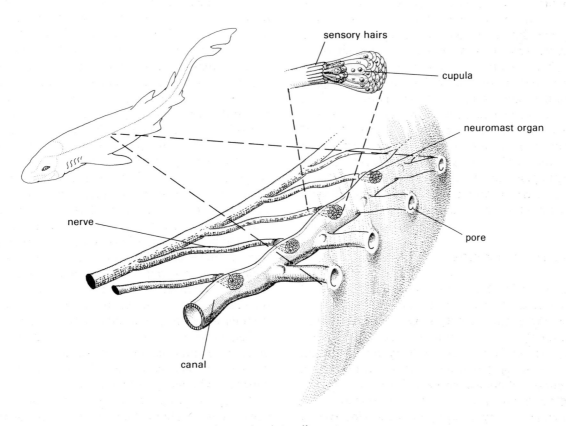

Acoustico-lateralis system

ear structures are related to equilibration and are common to all vertebrates. They consist of a complex of closed sacs and canals, the membranous labyrinth, filled with fluid (endolymph) and surrounded by a protective liquid cushion, the perilymph. The equilibrium receptors are two sacs, the sacculus and utriculus, which have patches of sensory hair cells comparable with the neuromasts of the lateral line system (the saccular and utricular maculae). The two sacs are filled with a gelatinous material containing calcium carbonate concentrations known as ear stones or otoliths which lie under gravity against particular sensory cells of the maculae, sending impulses via branches of the auditory nerve to the brain. Movement of the head causes the otoliths to move to stimulate the hairs of different sensory cells and these impulses are interpreted by the brain as a change in orientation.

Information concerning turning movements is given by the semicircular canals of which there are one or two in agnathans and three in gnathostomes. The semicircular canals are tiny fluid-filled tubes with a spherical expansion, the ampulla, at either end. Each canal lies in a plane at right angles to the others and is connected at either end to the utriculus. The ampullae have patches of sensory epithelium, each consisting of the familiar hair cells embedded in the gelatinous capsule (cupula). Movement of the head in any direction will move the fluid in at least one of the semicircular canals, thus moving the cupula and stimulating the sensory hairs.

The inner ear is surrounded and protected by the otic capsule of the skull.

The ear: hearing

The hearing (phonoreceptive) portion of the inner ear is represented in fishes, amphibians, and the majority of reptiles by an evagination of the sacculus, the lagena, surrounded by perilymph and lying within a large perilymphatic cistern. In mammals and birds and to a lesser extent crocodiles, the lagena and the perilymphatic cistern have evolved as a highly complex structure which is essentially a fluid-filled closed tube, coiled into a spiral. The resemblance of this structure to a snail shell, at least in mammals, explains its name, cochlea, from the Greek word meaning 'a shell'. The lagena has a patch of sensory epithelium with hair cells, the basilar papilla, which in the cochlea of higher vertebrates is greatly elongated and extends the full length of the tube as the organ of Corti.

The lagena of fishes has a rudimentary structure and it was assumed for a long time that fish were incapable of hearing. One reason for this assumption was that animal body tissues are composed mainly of water and vibrations generated in water will tend to pass through the fish without interruption. For fish to respond to these vibrations it is necessary for part of the body to respond differently to the passage of sound. In the majority of modern bony fish the saccular otolith, which is particularly massive, fulfils this role. In others a lung-like air-filled evagination of the pharynx, the swim bladder, acts as a resonating chamber.

From here vibrations are transferred to the sacculus through a chain of small bones, the Weberian ossicles, which are derived from the anterior ribs and vertebrae.

For the tetrapods, hearing is an important sense. To improve the detection of faint sounds which produce only small vibrations, various devices have evolved to amplify vibrations and to transfer them to the inner ear. These include additional ear structures, derived from the spiracular (second) gill cleft and the hyomandibular bone (part of the hyoid arch). The embryonic pouch representing the spiracle develops as an air-filled middle ear cavity with a tightly stretched external membrane, the tympanic membrane, which moves in response to vibrations of the outside air. (The name, from the Latin *tympanum* meaning 'a drum', refers to its resemblance to the skin stretched over a drum.) The simplest condition is seen in certain amphibians, such as the frog, where the tympanic membrane lies in a shallow depression on the side of the head. The inner surface of the membrane is attached to a rod-like bone, the stapes (a modified hyomandibula). This transfers the vibrations across the middle ear cavity to the sensory structures of the inner ear through an opening in the otic capsule of the skull (the fenestra ovalis). Reptile and bird middle ears are similar to those of the amphibia but the external depression is deeper forming the external meatus (outer ear) and the stapes may consist of several pieces. A Eustachian tube connects the middle ear to the throat and ensures equal pressure on either side of the tympanic membrane.

The hearing apparatus of the mammals is similar but more elaborate. The external meatus is a deep tubular structure and the tympanic membrane or eardrum occupies a protected position at its internal end. The outer ear has acquired a skin-covered cartilaginous projecting pinna or auricle which is the external structure commonly referred to as the ear. In many mammals, for example cats, the pinna is large and highly flexible and serves to collect the sound waves and funnel them into the ear. It also acts as an ear trumpet concentrating and increasing the pressure of the sound waves. A chain of three small bones, (the auditory ossicles: malleus, incus, and stapes) crosses the middle ear to the fenestra ovalis and transmits sound waves to the inner ear. The malleus and incus are examples of the remarkable transformations that have occurred during the evolution of the vertebrate skull, since they evolved from the posterior part of the ancestral lower jaw when a new articulation evolved anterior to the previous one. Together they form a lever system that reduces the lateral displacement amplitude and increases the pressure amplitude of the sound waves. As the fenestra ovalis is only about a twentieth of the area of the tympanic membrane, the pressure on it is greatly amplified. This is essential for the effective transfer of vibration from air to the much denser, incompressible fluid of the inner ear.

The cochlea is an elaborate structure consisting of three canals separated from each other by thin membranes. Two perilymph filled scalae (the scala vestibuli and scala

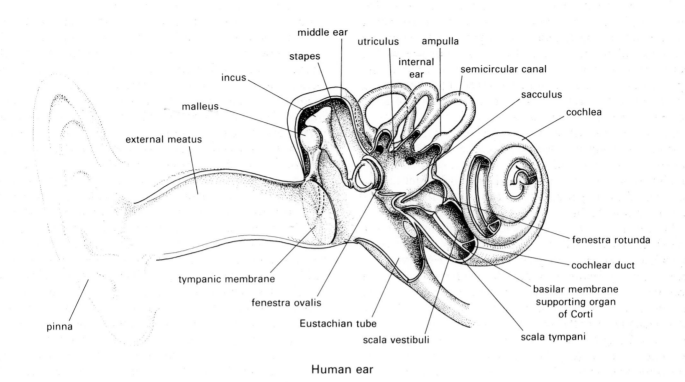

Human ear

tympani) are derived from the perilymph cistern. These lie on either side of the endolymph filled cochlea duct (scala media) and are connected at its apex. The cochlea duct represents an expansion of the lagena and contains the organ of Corti (an elongated basilar papilla) which is the organ of hearing. It consists of five rows of sensory cells, covered by a membraneous flap, the tectorial membrane, and resting on a basilar membrane which separates the cochlea duct from the scala tympani. The base of the scala vestibuli is connected to the fenestra ovalis, while the base of the scala tympani opens at the fenestra rotunda which is covered by a membrane which also leads to the tympanic membrane. Pressure waves formed by the stapes pass through the scala vestibuli, across the cochlear duct, and leave via the scala tympani and the fenestra rotunda. As the vibrations cross the cochlea duct they set the basilar membrane in vibration. Since the basilar and tectorial membranes are hinged in slightly different positions a shearing movement develops which stimulates the hair cells lying between them. The basilar membrane varies in width along its length and different parts respond to particular wavelengths; as a result the animal can discriminate between sounds of different pitch.

Chemoreception

Both taste and smell are examples of vertebrate perceptions of chemical stimuli. Taste buds are found in the mouth with a general distribution over the oral cavity in lower vertebrates and concentrated on the tongue and soft palate in mammals. In fishes and amphibians they may also occur on the skin. Individual tastebuds give particular sensations which can broadly be classified as salt, sour, bitter and sweet. Some can also detect water. They consist of bud-like groups of elongated taste cells associated with supporting cells which are sunk into an epithelium. The taste cells have a border of microvilli (previously described as taste hairs) which project through a tiny pore opening at the surface of the epithelium.

In tetrapods the nose is associated with breathing but its primary function is smelling, that is the detection and discrimination of the chemical characteristics of distant objects from molecules in the air. Smells, unlike tastes, cannot be assigned to classes with each chemical having a distinctive smell. In many vertebrates, smell is the most important sense (it will be remembered that the most strongly developed part of the brain in higher vertebrates, the cerebral hemispheres, developed from the part of the brain primitively associated with olfaction). However it is poorly developed in higher primates, birds, marine mammals and most teleosts.

Tetrapods detect smell through the nasal epithelium. In addition a large part of what is described as taste is in fact smelling of the mouth contents, with odours passing from the food in the mouth to the nasal chambers through the choanae or internal nostrils. The olfactory cells in the nasal epithelium are neurons which have a group of projecting cilia and are associated with supporting cells.

In most fishes the 'nose' is represented by a pair of nasal sacs (agnathans have a single olfactory pouch) which are placed anteriorly on the head and have two openings to the exterior. These have no internal openings in the mouth and fish are therefore incapable of 'taste' in the tetrapod sense. A flow of water passes through the sacs over a series of lamellae in which lie the olfactory cells.

The eye: photoreception

The vertebrate eye, while varying in its adaptation for seeing in air, under water, or at different light intensities, is essentially similar in structure throughout the group. It consists of a roughly spherical eyeball, situated in and protected by the orbits of the brain case and connected to the brain by an optic nerve. In its essential organization it is the same as a simple camera. It has a lens which can be focused for different distances, a diaphragm or iris which regulates the size of the pupil or light opening, and a light sensitive retina which corresponds to the film in a camera.

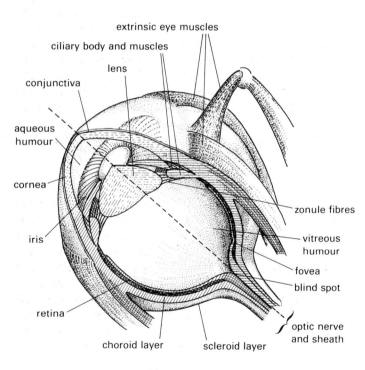

Human eye

At the back of the retina a layer of black pigment cells absorbs extraneous light and prevents internally reflected light from blurring the image, in exactly the same way as the black-painted inside of the camera.

The walls of the eyeball are three layered: scleroid, choroid and retina. The outer scleroid layer is a stiff tough sheet of connective tissue reinforced in most vertebrates by

bone or cartilage, which helps to protect the inner layers of the eyeball and to maintain its shape. It forms a complete layer both externally and internally. Some of the outer (visible) part is modified as a thin translucent cornea through which light enters. The overlying epithelium of the cornea is modified epidermis. The choroid layer contains blood vessels and supplies metabolites to the pigment layer of the retina. The retina is a double structure and is attached to the choroid layer. Both choroid and retina are highly modified towards the external surface of the eye. The lens is a transparent structure composed of elongated cells in a complicated series of layers and lying immediately behind the iris. It is surrounded by a ring-shaped ciliary body from which zonule fibres extend to hold it in place. Muscles within the ciliary body act on the zonule fibres either to move the position of the lens or alter its shape thus focusing the light rays onto the retina to form a sharp image. In tetrapods the light is partly focused by the cornea with the lens providing the fine adjustment but in fish the lens is the only agent for focusing. The lens of lower vertebrates is resistant to distortion and the eye is focused by moving the lens towards or away from the retina. However, in amniotes it is more elastic and focusing is achieved by altering its shape: the lens assumes a narrowed form if the object is distant (this is the resting position of the ciliary muscle) and a rounded form for a nearer object.

The principal cavity within the eyeball lies between the retina and lens and is filled with a thick jelly-like material known as the vitreous humour. There is also a smaller cavity, containing watery aqueous humour, lying between the cornea and the lens. The humours are important in maintaining the shape of the eyeball. In addition, since the cornea and lens lack blood vessels, they are nourished by means of the aqueous humour.

The iris is formed from the anterior part of the choroid and retina. In some fish the iris has fixed dimensions, but in the majority of vertebrates it includes muscle cells, arranged in circular or radial patterns, which expand or contract the pupil. The iris responds to changes in light intensity and can thus control the amount of light falling on the retina. Slit-shaped pupils which close readily to exclude the light but which can also open very wide, are characteristic of nocturnal forms. The iris is visible as a ring of blue, green, yellow or brown. Although it is derived from the retina layer it contains cells with all the structural and functional attributes of muscle fibres.

The retina is the light-sensitive part of the eye and consists of a double-layered hemisphere. The outer layer is composed of pigment cells, while the inner layer consists of sensory and nervous tissue. The light-sensitive cells are situated at the back of the retina and directed away from the light which passes through several layers of neurons to reach them. This curious arrangement is a consequence of the development of the retina as an outgrowth and subsequent invagination of the brain, the cells retaining their original orientation. The receptor cells are of two kinds called rods and cones according to their shape. Cones are responsible for vision in bright light, colour vision, and for perception of detail; they are normally concentrated in a central region of the retina especially in the centralis or fovea directly opposite the lens at the back of the eye. The rods are insensitive to colour but are more responsive to low light intensities. In general the retinas of deep water fishes and nocturnal vertebrates are chiefly composed of rods; cones are comparatively rare. A layer of bipolar cells lies internal to the rods and cones and transmit impulses to a layer of ganglion cells which forms the optic nerve. The retina therefore really has four layers: a pigmented layer adjoining the choroid, a layer of photoreceptive cells, a bipolar cell layer and a ganglion cell layer.

Each eye has six muscles which stretch between the eyeball and orbit and enable the eye to be rotated in different directions. Terrestrial vertebrates have a pair of movable eyelids, and lacrimal (tear) glands whose secretions clean and moisten the cornea. In all tetrapods except turtles a tear duct drains surplus fluid from the cornea of the eye to the nasal cavity. In reptiles, birds, and many mammals a nictitating membrane (effectively a third eyelid) lies deeper than the other eyelids.

Ancestral vertebrates had a median or pineal eye on the top of the head which probably functioned simply as a light detector. This is present in modern lampreys (agnatha) and a few reptiles, and develops as a dorsal out-pocketing of the brain, persisting in higher vertebrates as a glandular structure.

The endocrine system

The nervous system, discussed in the previous section, enables an animal to adapt quickly to environmental changes by means of nervous impulses. By contrast the endocrine system secretes 'chemical messengers' or hormones which are transported to other parts of the body in order to regulate or integrate its activities. The responses under endocrine control are slower than those under nervous control and are measured in minutes, hours, days or even weeks rather than milliseconds, but they are also longer lasting. They include long term adjustments of metabolism, reproduction and growth.

All glands can be described as exocrine or endocrine. Those present in, for example, the skin or gut, are exocrine and as their name suggests these have ducts through which their secretions are discharged directly to the sites where they are used. In contrast endocrine glands have no ducts and their products are discharged into the circulatory system and transported by the blood. Thus a hormone is classically defined as 'any substance normally produced by the cells in any one part of the body and carried by the bloodstream to distant parts which it affects.' This definition applies to the majority of hormones but excludes neurohormones which may pass along an axon, and prostaglandins which are transferred in the seminal fluid.

Hormonal systems are known in many invertebrates (including annelids, molluscs and arthropods) but reach their highest level of development in vertebrates. In primitive vertebrates most hormone-secreting cells are scattered through various tissues but in more advanced forms they are aggregated to form separate glands or discrete patches of glandular tissue. Most hormones are effective in extremely small quantities but for normal body metabolism they must all be present at their individual optimum concentrations. The chemical diversity of hormones is considerable and they include amines, amino acids, peptides, proteins, fatty acids, steroids, proteins and gibberellins.

The hormonal system comprises three parts: the secreting cell, the means of transport, and the target cell. Generally a specific hormone is synthesised and secreted by a specific type of cell to bring about a specific reaction. In higher vertebrates most secreting cells occur in glands of which the thyroid, thymus, parathyroid, adrenal, pineal and pituitary glands are the most important. Other hormone secreting regions include the islets of Langerhans (in the pancreas), the gonads, and the placenta. Neurohormones are secreted by the tips of axons (neurosecretion). The majority of hormones are transported bound to protein compounds in the blood serum, although some are transported in solution.

Although nervous and hormonal systems are very distinct in their modes of action, they are far from independent of each other. Some of the endocrine organs, such as the pituitary, consist partly of modified nerve cells and endocrine production may be strongly influenced by the brain. For example the activities of the pituitary gland are regulated by the adjacent hypothalamus. In the other direction the nervous system may be strongly affected by hormones.

The thyroid gland

The thyroid gland occurs in all vertebrates and is probably the most familiar gland in the endocrine system. It is generally a bilobed structure and develops as a ventral outgrowth of the pharynx: in adult fishes it lies below the gill chamber and in tetrapods occurs in the neck region. The thyroid gland has a rich blood supply, reflecting its endocrine function, and consists of numerous spherical follicles which are lined with a cuboidal epithelium and are separated from each other by connective tissue. The follicles contain a gelatinous colloidal substance.

The thyroid has a long evolutionary history since it is homologous with the endostyle of primitive chordates and larval lampreys. Thus it originated as an exocrine gland producing mucus and iodine compounds which were taken into the digestive tract, though in the vertebrate adult it is entirely an endocrine organ. In view of its origins it is interesting that thyroid hormones are the only hormones which remain effective when taken by mouth.

Thyroxine is produced by the thyroid gland and exercises a general control over the rate of metabolism of the whole body. In humans thyroxine deficiency produces a condition known as myxoedema in which the patient has a low metabolic rate. Over production of thyroxine produces Graves' disease in which the patient's basal metabolic rate increases to as much as twice the normal amount.

Parathyroid glands

The parathyroid glands are small masses of glandular tissue which occur in tetrapods and are derived from the gill pouch region of the embryo. They occupy varying positions in the neck but in man occur close to the thyroid tissue which is how they derive their name. However, both in structure and function they are very different from the thyroid. Their tissue is arranged in solid masses and cords rather than in follicles and they secrete parathyroid hormone (PTH or parathormone), a peptide chain of 84 amino acids which is involved in calcium, and to a lesser extent phosphorus, metabolism. Calcium is a major element essential in large quantities for the maintenance of the body skeleton and also required for the functioning of neuromuscular junctions. Parathormone promotes absorption of calcium from the intestine and reabsorption of calcium in the kidneys. It also controls the release of calcium from the bones. The absence of this hormone produces decreased concentrations of calcium in the body fluids and increased levels of phosphorus resulting from reduced phosphorus excretion. The effects of the calcium deficiency include muscular tremor, cramp and convulsions. An excess of parathormone is associated with high calcium and low phosphorus levels in the blood with increased excretion of both. As calcium is removed from the bones these become soft and brittle, but calcium may be deposited in undesirable regions such as the heart, lungs, kidneys and intestinal wall.

Although fish lack parathyroid glands, they possess ultimo-branchial bodies developed from the posterior embryonic gill pouch which are thought to be their equivalent. These bodies produce a hormone called calcitonin which lowers the level of calcium in the blood.

Adrenal glands

The adrenal glands are small paired structures found adjacent to or capping the kidneys in most tetrapods. They are formed of two distinct types of tissue which are intermingled in lower tetrapods but form distinct layers, the cortex and medulla, in mammals. In fishes they do not usually occur as discrete organs but there are masses of tissue representing either the cortex or the medulla between and around the kidneys and along the courses of the major blood vessels. The tissue of the medulla (reddish brown in mammals) is modified nervous tissue and is arranged irregularly around blood vessels into which it secretes two closely-related hormones, adrenalin and noradrenalin. Both hormones are comparatively simple amines, derived from the amino acid tyrosine, and their function is to prepare the body for sudden emergencies involving rapid action: 'fight, flight, or frolic'.

Adrenalin secretion is increased by stress, which may be induced by cold, pain, injury or emotion, and produces a wide range of metabolic effects. It increases the heart rate and blood pressure and dilates the major blood vessels; at the same time the smaller blood vessels of the skin become constricted (vasoconstriction) so that, in humans it becomes pale and goose-pimpled. The respiratory rate increases and the respiratory passages become dilated, thus increasing the oxygen available for internal respiration. More glycogen is released from the liver and muscle, thereby increasing the available energy. Adrenalin also enhances the rate of blood coagulation in wounds and causes the eye pupils to dilate. It stimulates the release of another hormone (ACTH, adrenocorticotrophic hormone) which is produced by the pituitary. Noradrenalin has similar but generally weaker effects to adrenalin, though it is a more powerful vaso-constrictor.

The cortex is pale yellow or pink tissue derived from mesoderm and consisting of three distinct cell layers: an outer glomerosa, a middle fasciculata and an inner reticularis. Of these the cells of the fasciculata layer are believed to be the most active hormone secretors. The cortex secretes various steroid hormones which aid the body to meet environmental stress of long duration. These hormones are involved in regulating the salt and water balance of cells and body fluids, and in metabolism. Glucocorticoids promote conversion of proteins to carbohydrates while mineral ocorticoids regulate sodium and potassium, and androgens have male sex hormone activity. In humans a suboptimal level of adrenocortical hormones results in Addison's disease, characterized by a wide range of symptoms including muscle weakness, increased sodium and chloride excretion, loss of body fluids and low blood pressure. The digestive tract functions less effectively and the patient loses weight in spite of an increased appetite for both food and water. The basal metabolic rate decreases and the ability to survive exposure to environmental stress such as cold is reduced. The development and function of the adrenal cortex is regulated by ACTH.

The pituitary gland

The pituitary gland is a major endocrine organ and among its many complex functions it exercises a control, through the hormones it produces, over various other endocrine glands. It lies beneath the hypothalamus, to which it is attached by a short stalk. It has a double origin, with one part composed of modified nervous tissue growing downward from the developing brain (the infundibulum), the other developing from a dorsal outgrowth from the roof of the mouth (Rathke's pouch). Like the adrenal, the pituitary is therefore a double structure with two quite different functions: one part, the adenohypophysis, derived from Rathke's pouch, and the other part, the neurohypophysis, derived from the infundibulum. In higher vertebrates the pituitary is three-lobed: the anterior and intermediate lobes form the adenohypophysis and the posterior or neural lobe the neurohypophysis. The anterior lobe has no nerve fibres and its activities are controlled by hormones transported to it in the blood supply, notably the blood coming directly from the hypothalamus. The anterior lobe has five different types of cell and it is inferred that each secretes a different hormone. In all six hormones are known to be secreted by the mammalian pituitary: growth hormone (somatotrophin), thyroid stimulating hormone (TSH), adrenocorticotrophic hormone (ACTH), follicle stimulating hormone (FSH), luteinizing hormone (LH) and prolactin or luteotrophic hormone (LTH).

Pituitary structure varies considerably, for instance the lamprey has no neural lobe, and in fishes the adeno-hypophysis is not subdivided.

Growth hormone

As its name suggests, this hormone controls general body and bone growth and its over-production leads to excessively tall growth while under-production produces midgets. Human growth hormone is a single peptide chain of 191 amino acids.

Thyroid stimulating hormone

This is a basic glycoprotein and stimulates growth and hormone production by the thyroid.

Adrenocorticotrophic hormone

ACTH is another single peptide chain. It is synthesized rapidly and stimulates the production of hormones by the adrenal cortex.

Follicle stimulating hormone, luteinizing hormone and prolactin

Like TSH, FSH and LH are glycoproteins. They control the development and functioning of the testis, FSH increasing the size of the seminiferous tubules and LH stimulating the interstitial cells of the testis to produce male sex hormones. In females both are necessary for the achievement of sexual maturity and regulation of the reproductive cycles. Prolactin is involved in the secretion of milk by female mammals after giving birth and also induces behaviour patterns leading to the care of the young.

The intermediate lobe of the pituitary secretes melanocyte stimulating hormone (MSH) which in lower vertebrates dispenses pigment in the chromatophores and causes the skin to darken. In mammals and birds the pituitary can be shown to be rich in MSH, but it appears to have little effect on pigmentation.

The posterior lobe of the pituitary contains two hormones, oxytocin and vasopressin which are produced by neurons with cell bodies in the brain but stored and released by the pituitary. Each is a peptide chain and their functions are highly specific. Vasopressin stimulates smooth muscle to contract and to act on the muscles in the walls of arterioles raising the blood pressure; it also regulates reabsorption of water in the kidney tubules. Oxytocin helps to stimulate contraction of the uterine muscles in the female mammal at the time of giving birth.

Control of pituitary function is itself hormonal and is provided by neurosecretion. These hormones are:

i) corticotropin releasing hormone (CRH) which controls the release of ACTH;

ii) thyrotropin releasing hormone (TRH) which controls the release of both TSH and prolactin;

iii) gonadotropin releasing hormone (GnRH) which controls the release of FSH and LH;

iv) growth hormone releasing hormone which controls the release of growth hormone.

Hence the pituitary gland, the master gland of the endocrine system, is in turn controlled by the hypothalamus.

The pineal gland

The pineal gland lies on the upper surface of the thalamus and is derived embryologically as an outgrowth of the pineal stalk of the brain. In mammals the pineal gland secretes melatonin which inhibits gonadal function either directly or indirectly by way of the pituitary. The secretion of melatonin may itself be controlled by the amount of light falling upon the retina since the pineal gland receives impulses from the retina via a small nerve, the inferior accessory optic tract. In jawless vertebrates, for example the lamprey, the pineal gland functions as a light-detecting body.

The islets of Langerhans

In the majority of vertebrates the islets of Langerhans consist of small groups of glandular cells with a rich blood supply dispersed throughout the pancreatic tissue. However in teleosts the insular tissue is spread in the general region of the gut and in a few may form a separate organ. The secretory cells produce a specific protein hormone, insulin, composed of two peptide chains joined by disulphide bonds which has an important regulatory effect on metabolism, particularly of carbohydrates. Interruption of its supply causes diabetes mellitus in which the animal excretes large quantities of sugar. To make up for this deficiency of blood sugar, protein is converted to carbohydrate which is also excreted so that there is steady weight loss. The condition can be controlled simply by regular injections of insulin. Glucogen (also a peptide chain) is produced by the islets, particularly in reptiles and birds, and is also involved in carbohydrate metabolism. It increases the blood sugar level by stimulating conversion of liver glycogen to blood glucose.

Sex hormones

The principal female sex hormones are oestradiol, oestrogen and progesterone (and in some mammals, relaxin) which are produced chiefly by the follicle cells of the ovary. Male hormones, produced by the interstitial cells of the testis, are testosterone and androsterone. Male and female gonads are not completely differentiated in their hormonal products and male gonads secrete a certain amount of female hormone and vice versa. In addition the placenta of mammals secretes a gonadotropin which is important in maintaining pregnancy. The role of these hormones will be discussed in more detail in the section on mammalian reproduction.

The early development of chordates

Chordates reproduce sexually and in consequence all begin their life as a single cell. Four major processes are necessary to change that cell into the adult form: cell division, growth, differentiation and changes in spatial distribution of cells associated with the change in shape from a single-layered sphere to an elongated, bilaterally symmetrical, multi-layered form. These processes represent a complicated, prolonged, and generally hazardous phase in the life of any animal. Some aspects of animal development have already been discussed in earlier chapters, and this section deals with chordate development in a rather more detailed way.

During the complex sequence of changes necessary for chordate development, the embryo has the same basic requirements of all organisms, for example nutrition, respiration and excretion. Of these the arrangements for nutrition have the greatest effect on the appearance of the egg itself, and also upon the early development of the embryo. At least in the early stages of development, food is normally provided in the form of yolk within the egg cell itself. Primitive aquatic chordates produce small eggs with little yolk and the period of embryonic growth within the egg is inevitably brief. The young chordate hatches and then follows a relatively long free-living larval stage. The tiny animal is liable to dangers which would not effect the adult, and therefore cannot share the adult habitat or mode of life straightaway. Higher vertebrates have a longer embryonic period, and development proceeds more directly to the adult form. At hatching or birth the young is a replica of the adult, albeit smaller and immature, but able to live in a similar habitat and to follow the same mode of life with a reasonable chance of survival. The egg in such forms becomes loaded with yolk to a point where it is grotesquely large for a single cell. Such eggs are an extremely attractive food source for other animals, and since each egg represents a considerable drain on the resources of the female, defence of the egg by the parents is common. In addition, the egg must be protected from water loss, and since the newly hatched animal has to be capable of an aerial life it has to have a mechanical strength not demanded of aquatic hatchlings. An effective solution to many of these problems is for the embryo to remain within the female, drawing metabolites from her and returning excretory products to her. This process, known as viviparity, is characteristic of the mammals although many other vertebrates also give birth to live young; in viviparous species there is again a small egg with little or no yolk.

The typical chordate egg is a spherical cell with a nucleus, cytoplasm and varying amounts of yolk. It is always surrounded by at least one membrane, the vitelline or primary egg membrane secreted by the egg cell or oocyte. In many cases there are also secondary membranes secreted by follicle cells which surround the egg as it develops in the ovary, and tertiary membranes may be deposited around the egg after fertilization as it passes down the oviduct. Examples of tertiary membranes include albumen (egg white), shell membrane, and shell, and are normally found in larger-yolked eggs.

The quantity of yolk present is immensely variable. For example the eggs of amphioxus and mammals have a small amount of yolk and are termed microlecithal. Mesolecithal eggs are slightly larger, with a moderate amount of yolk; the frog is a well known example of this type, and in macrolecithal eggs yolk makes up most of the egg volume. These eggs, characteristic of birds, reptiles and the sharks and skates, are large. Whereas microlecithal eggs generally have their yolk fairly evenly distributed throughout the cytoplasm of the cell, a condition termed oligolecithal, in macrolecithal eggs the heavier yolk becomes concentrated in the lower hemisphere and the egg is said to be telolecithal.

Zoologists have been interested to know the fate of the various parts of the egg in subsequent embryological development. The uneven concentration of yolk is one piece of evidence that there is organization within the egg. Even prior to its development it is possible to distinguish regions in the egg which give it a clear orientation. A region of relatively clear cytoplasm surrounds the animal pole, the yolk region is at the vegetal pole, and the axis of the egg passes through both poles. The equator separates the animal and vegetal hemispheres. It is important to realize that in chordates this orientation does not represent the orientation of the embryo. In fact the embryo axis (antero-posterior) in chordates lies at an angle of about 45° to the egg axis: roughly the animal pole corresponds to a position above its head and the vegetal pole is on the ventral surface of the developing embryo. A medial plane can be recognized separating the future right and left halves of the body. Polar bodies represent the remaining products of meiotic division.

To illustrate the various patterns seen in vertebrate development we will discuss four representative types: amphioxus, the frog, a chick and a mammal. Each will be followed through three successive stages: cleavage, ending in the formation of a hollow spherical blastula; gastrulation, in which the blastula cells are rearranged into the main body layers, and the early stages of organogenesis, in which the organ systems of the body are laid down. Growth follows to the adult stage.

The development of amphioxus

Cleavage

The process of cleavage has already been discussed but will be summarized briefly here. The first cleavage is vertical, and extends along a meridian between the animal and vegetal poles, resulting in two cells, or blastomeres, of equal size. These adhere to each other, but tend to round up as separate spheres, a process which follows all subsequent divisions. The second cleavage is also vertical, but at right angles to the first, to give four cells rather like four quarters of an orange. The third cleavage, producing eight cells, is at

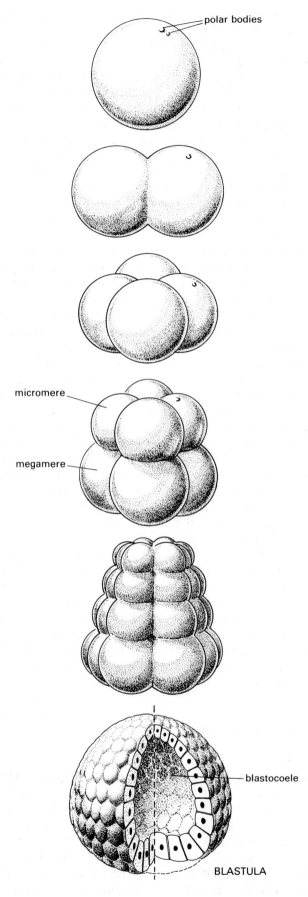

Amphioxus: cleavage

right angles to the first two and is essentially a horizontal division across the equator of the egg. Since a cell tends to divide not through the centre of its total mass, but through the centre of its cytoplasm without regard for biologically inert materials such as yolk, this division is slightly above the equatorial plane. The upper quartet of cells are slightly smaller, and may be termed micromeres, while the more yolky cells of the lower quartet are the megameres. At this stage it becomes impossible for the cells to form a spherical shape and still maintain contact with all of their former neighbours. The developing blastula therefore has a cavity, the blastocoel, within a sphere of cells. Following the eight-celled stage each cell cleaves horizontally (16 cells) and then vertically (32 cells). The seventh division gives a total of 64 cells, and beyond this stage increase in cell number is rapid, but the divisions become less regular and less synchronous. Eventually a simple hollow spherical blastula of several hundred cells is formed with smaller cells towards the original animal pole and larger, somewhat more yolky cells towards the original vegetal pole. Even at this early stage the blastula therefore shows differentiation of cell types, and the various cells have particular destinies in the embryo. As a rough generalization the cells of the original animal hemisphere will form the ectoderm or outer germ layer of the later embryo and adult. The cells of the original vegetal hemisphere form both the endoderm or inner germ layer, which eventually develops as the gut, and the third major germ layer, the mesoderm, which forms the greater part of the chordate body. The ectodermal area consists of columnar cells and occupies the bulk of the ventral half of the blastula. The endodermal layer is a plate of larger yolky cells across the dorsal part of the blastula, which are bordered both laterally and posteriorly by a crescent of smaller mesodermal cells. Two further specialized regions are recognized in the blastula. The chordamesoderm cells, which will eventually give rise to the notochord, lie between the endodermal plate and the ectodermal region, the neurectoderm. This is a region of specialized ectodermal cells lying immediately in front of the chordamesoderm, destined to form the central nervous system.

Gastrulation

The process of gastrulation involves a series of movements of the blastula cells towards the approximate positions which they will occupy in the adult. To understand the significance of the events at this stage of embryonic development we must be able to trace the movements of individual groups of cells. As we have seen, the cells can be distinguished to some extent by their appearance, but detailed study was only made possible by the development of dyes (in the 1920s and 1930s) which were both harmless to the cells, and were retained by their descendants (cell bound). The use of such dyes made it possible to construct maps commonly known as 'fate maps' in which the final destination of the cells in every region of the blastula is known. In many invertebrates the transition from the spherical blastula to a double-layered hemispherical

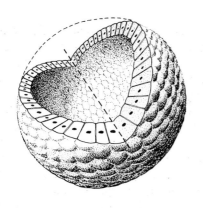

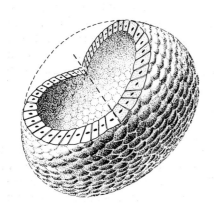

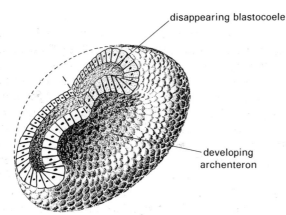

disappearing blastocoele

developing
archenteron

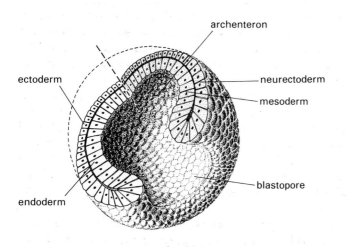

archenteron

ectoderm

neurectoderm

mesoderm

endoderm

blastopore

Amphioxus: gastrulation

gastrula takes place by a simple process of **invagination**. The process of gastrulation in amphioxus appears very similar, although it is in fact more complex. It proceeds by invagination of the endoderm and chordamesoderm with obliteration of the blastocoele. The large yolky cells which form the ectodermal plate first become flattened and then sink into the blastocoele which becomes smaller and finally disappears. As a result the embryo becomes cup shaped; the newly formed cavity of the cup is the archenteron, and the opening is the blastopore. The chordamesoderm rolls over the dorsal lip of the blastopore and pushes forwards internally, tending to lengthen the gastrula. At this stage the lateral and ventral lips of the blastopore are formed by prospective mesoderm cells, these follow the inward movement of the endoderm and move forwards and upwards to align themselves in a strip on either side of the prospective notochord. The future neural tissue comes to occupy a large area on the mid-dorsal outer surface, anterior to the blastopore, and is referred to as the neural plate. As gastrulation continues the embryo lengthens and the blastopore becomes smaller and is shifted to a posterior position by backward growth of the dorsal lip. The completed gastrula is a two layered structure: the outer layer is formed of ectoderm and neurectoderm, and the archenteron is lined with endoderm and roofed by chordamesoderm and mesoderm.

Organogenesis

The final phase of embryonic development is that of organ formation, beginning with the development of the neural tube, which brings the embryo to a stage known as a neurula.

Neural tube formation

The neural plate flattens and sinks beneath the surface ectoderm which rises up on either side as the neural folds. Starting posteriorly, these grow towards each other over the neural plate and finally meet in the dorsal mid-line, though for some time the anterior end of the neural tube remains open at the surface as a neuropore. At the posterior end the folds cover over the blastopore so that there is a connection (the neurenteric canal) between the primitive gut (archenteron) and the neural tube. In later development this closes and the neural tube and gut become separate structures. This method of neural tube formation is peculiar to amphioxus, and the central nervous system of higher chordates develops in a slightly different way.

The further development of chordamesoderm and lateral mesoderm

After gastrulation the chordamesoderm cells form a mid-dorsal strip in the wall of the archenteron where they undergo rounding and rearrangement to form a single row of disc-shaped cells. These later become vacuolated and surrounded by a sheath to form the notochord. At each side of the chordamesoderm the lateral mesoderm folds outwards from the archenteron as longitudinal mesodermal

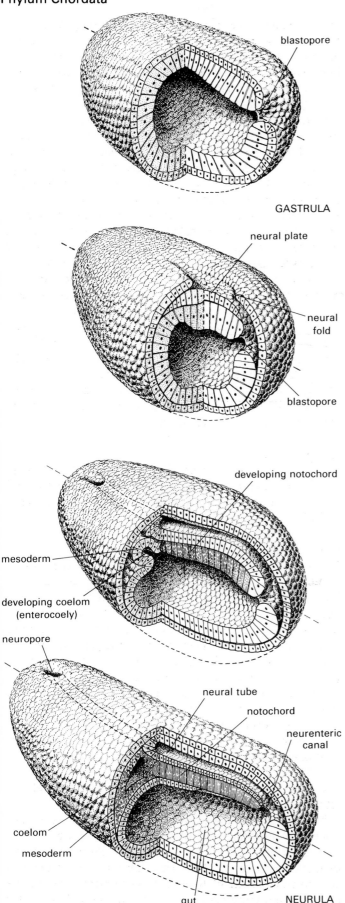

blastopore

GASTRULA

neural plate

neural fold

blastopore

developing notochord

mesoderm

developing coelom (enterocoely)

neuropore

neural tube

notochord

neurenteric canal

coelom

mesoderm

gut   NEURULA

Amphioxus: early organogenesis

grooves. These gradually become metamerically segmented into a series of paired mesodermal somites by a process which involves pinching off of successive pouches, each containing a coelomic cavity at the anterior end. The cavities enlarge and the ventral part of each pouch grows downwards between the endoderm and ectoderm towards the mid-ventral surface, where it unites with its partner in the mid-ventral line. The walls separating right from left pouches break down and the transverse partitions are also lost, establishing a single cavity, the splanchnocoele, which is continuous throughout the length of the embryo and extends into the gut. The splanchnocoele forms the greater part of the coelom and is bounded laterally and ventrally by layers of mesoderm. The more solid dorsal part of the mesoderm retains its metamerism forming a series of somites. The coelom and mesoderm development described here has important similarities to that seen in the echinoderms and acorn worms, and is a prime basis for associating the origin of chordates with that of the echinoderms.

Formation of the gut

After separation of the notochord and development of the mesodermal pouches the upper edges of the endoderm grow dorsally and towards each other to fuse below the notochord, completing the endodermal part of the gut.

Hatching

At hatching amphioxus is a small ciliated larva which is incapable of feeding because neither the mouth nor the anus are formed. It elongates rapidly and begins to swim actively using its myotomes. The mouth is formed on the left hand ventral surface of the head where a patch of ectoderm fuses with the endoderm and then perforates. In a similar fashion the anus develops on the left hand side of the ventral surface towards the posterior end. Visceral clefts develop in the anterior region of the gut, though at first they do not appear symmetrically nor in a regular sequence. The first cleft arises, before the mouth opens, as a diverticulum from the right hand side of the pharynx near the anterior end to form a coiled cigar-shaped gland which disappears at metamorphosis. Fourteen other clefts, which will eventually form the gill clefts of the left-hand side, perforate the ventral mid-line and move to lie just to its right. Eight clefts which will eventually form the right-hand gill clefts develop on the right-hand side of the pharynx above the fourteen clefts originally formed. Thus at this stage the larva is asymmetrical with both sets of gill clefts lying on the right-hand side. As larval development proceeds the left-hand clefts move across the ventral mid-line to take up their final position, and six of them close, so that the larva is left with eight pairs of clefts. These are subdivided by tongue bars, with fuller clefts being added to the posterior end of the series.

The endostyle also develops asymmetrically as a series of ciliated cells arranged in a V-shape on the right-hand side of the gut. This moves to the mid-line, where the arms of the

'V' fuse and the structure extends posteriorly in the pharyngeal mid-line.

The atrium develops as two downward folds, one on either side of the body, called the right and left metapleural folds, which eventually meet and join in the **ventral** mid-line. At its posterior end the atrium is open at **the atriopore** and the metapleural folds also remain separate anteriorly. At this stage the larva lives and feeds for some time in the surface plankton, before sinking to the sea bottom where it undergoes metamorphosis to the adult form.

Vertebrate development

The development of the frog

endostyle

cigar-shaped gland (first cleft)

visceral clefts (left side)

future Hatschek's pit

mouth

visceral clefts (right side)

tongue bar

Amphioxus: development after hatching

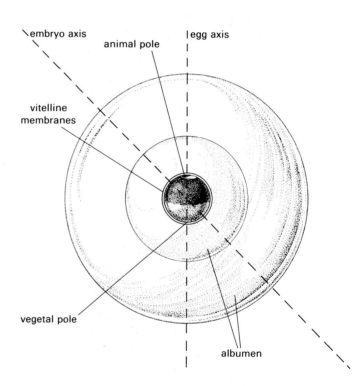

A representative chordate egg (frog)

The amphibian egg has rather more yolk than that of amphioxus and is much larger, about 1.6 mm diameter instead of 0.12 mm. As it passes along the oviduct it receives a thick coat of albumen which absorbs **water and** swells. When the eggs are laid the albumen protects them from physical damage and desiccation, and also causes them to stick together, forming the familiar frogspawn. The increased yolk content of the egg allows the frog to hatch at a more advanced stage than the larva of amphioxus.

The early stages of development in the frog, and indeed of the other vertebrates, follow the same general lines as those described for amphioxus. Again the egg divides totally (holoblastic cleavage) and the first two cleavages are longitudinal and at right angles to each other. However the presence of substantial amounts of yolk, concentrated on the vegetal hemisphere, has the effect of slowing the rate of cleavage in that region so that the second cleavage may begin in the animal hemisphere before the first is completed in the vegetal hemisphere. The third horizontal cleavage is well above the equator and there is therefore a very marked difference between the size of the cells in the two hemispheres. Furthermore the blastocoele is partly filled with large yolky cells in the vegetal hemisphere and to a lesser extent in the animal hemisphere, so the blastula wall is therefore several cells thick in places. By comparison with that of amphioxus, the blastocoele is relatively much smaller. As in amphioxus, the developing embryo shows no increase in size during cleavage.

Gastrulation

The process of gastrulation as described for amphioxus is basically similar in the frog though the reduced size of the blastocoele and the presence of a mass of large yolky cells influence the course of events. It is a physical impossibility for the potential endoderm to fold itself within the potential ectoderm cells formed from the animal hemisphere. Instead, in a region of the egg known as the grey crescent area (which develops opposite the point at which the sperm entered the egg during fertilization), a furrow forms on the surface of the blastula. This furrow is the beginning of the archenteron and may be regarded as corresponding to the dorsal blastopore lip of amphioxus. Endodermal cells from the region below the furrow stream into it. As this migration of cells continues the furrow becomes progressively deeper and the blastopore becomes correspondingly smaller. The shape of the blastopore changes as gastrulation continues. On either side of the blastopore mesodermal cells (lateral to the chordamesoderm) join the inturning movement and the blastopore is finally completed by the inturning of mid-ventral mesodermal material. After this the blastopore becomes gradually smaller by contraction of its margins. This is brought about by a spreading of the cells of the animal hemisphere over the yolky cells, known as epiboly. Downward growth of the dorsal lip carries it towards the original vegetal pole. Finally a small plug of yolky cells is left protruding through the much reduced blastopore.

Organogenesis: the formation of the neural tube

The development of the neural tube in amphioxus is slightly different from that of the vertebrates although the end result is the same. In amphioxus the potential ectoderm separates from the neurectoderm and grows over it as the neural tube forms. In the vertebrates the ectodermal tissues do not separate but fold upwards as neural folds on either side of a neural groove. The folds from each side meet mid-dorsally. Only at this stage do the ectoderm and neurectoderm separate and fuse with their opposite numbers and the ectodermal surface and neural tube are

blastocoele

EARLY BLASTULA

LATE BLASTULA

yolky cells

The frog: cleavage

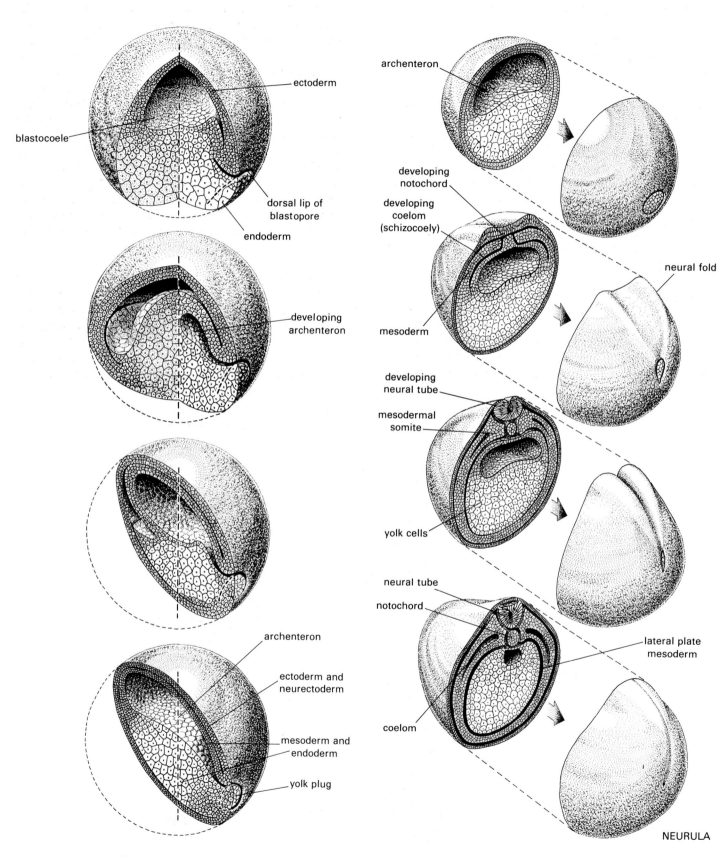

The frog: gastrulation

The frog: early organogenesis

completed. The anterior end of the neural tube remains open as a neuropore, and withdrawal of the yolk plug posteriorly produces a neurenteric canal.

The development of the notochord
This forms by a similar process to that described for amphioxus.

Mesoderm development and the development of the coelom
In vertebrates the process of mesoderm development is different from that described for amphioxus. Initially the mesoderm is unsegmented, and without cavities. It extends to form a hemicylindrical sheet of tissue, several cells thick, from either side of the notochord. As development proceeds a split appears within the mesoderm to produce a cavity which enlarges as the coelom. The split first appears dorso-laterally and extends ventrally on either side until eventually the wall between the two halves of the mesoderm breaks down and the coelom is continuous around the developing gut. This process of coelom formation is known as schizocoely, as opposed to the process of enterocoely described for amphioxus. In all vertebrates three divisions of mesoderm can be distinguished, extending the length of the trunk and arranged from the dorsal mid-line outwards. On either side of the notochord the mesoderm thickens and subdivides as a metameric series of cuboidal structures, the mesodermal somites. Each somite makes three contributions to future structures: the dermatome is the mesodermal contribution to the skin, the myotome gives rise to the body muscle, and the sclerotome produces the axial skeleton. Ventral to the somites an intermediate region sometimes becomes segmented and forms the nephrotome from which the kidneys and the deeper parts of the gonads develop. In the majority of vertebrates the remainder of the mesoderm remains unsegmented as the lateral plate mesoderm (modern agnathans are an exception). It is in this region that the general body cavity (perivisceral cavity) develops. Anteriorly, the structure of the somites becomes modified by the development of the sense organs and visceral clefts. Here the coelom is restricted to a ventral portion which becomes separated from the perivisceral cavity and eventually develops as the perivisceral cavity surrounding the heart.

Hatching and larval development
The time taken from fertilization to hatching varies with environmental conditions, but, on average, in temperate climates the tadpole is ready to hatch two weeks after the egg is fertilized. On hatching the tadpole is a small fish-like creature about seven millimetres in length with rudimentary sense organs, a rudimentary kidney, and four pairs of

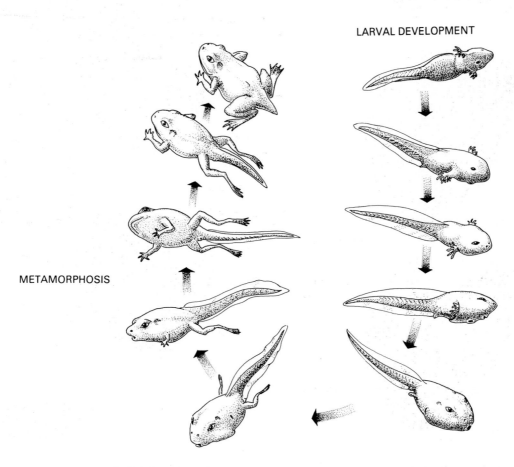

LARVAL DEVELOPMENT

METAMORPHOSIS

The frog: larval development and metamorphosis

visceral clefts which do not open to the exterior. The anus is formed, but the mouth has not yet opened. It is however at a more advanced stage than the newly emerged larva of amphioxus. After the tadpole has left its surrounding jelly it attaches itself to weeds or similar objects by means of a sticky secretion produced by a mucus gland at the base of the gill arches. For the first week after hatching it continues to feed on yolk, but at the end of this period a mouth develops, fringed by a pair of horny 'jaws' which it uses to rasp off small bits of vegetable matter. The gut becomes elongated and coiled and begins to function. The gill clefts also develop as functional structures: at first they open directly to the exterior but later they become covered by folds of skin. These opercular folds join ventrally so that the gills open into a common gill chamber with an opening at a spout-like spiracle on the right-hand side. At first the respiratory surface is external, but later true internal gills develop. By this stage the tadpole has the familiar tadpole body form with a globular body and a long propulsive tail. It feeds actively and grows rapidly.

By the time it is about 15 mm long the tadpole has a typically aquatic lateral line system and essentially fish-like respiratory and circulatory systems. However after about three months of larval life it develops lungs and begins to gulp air. The hind limbs can be seen to develop as buds from the body wall but the forelimbs develop beneath the operculum through which they emerge immediately prior to metamorphosis.

Metamorphosis

Metamorphosis begins towards the end of a month and is a rapid yet radical series of changes associated with a major change both in habitat and way of life. The tadpole ceases to feed and sheds its larval skin and jaws. The tail is absorbed and shortened and the legs continue to grow. The lungs become of increased importance and eventually the gill slits are closed. Associated with this development, the blood system undergoes modification to a typical terrestrial amphibian type. The gut and mouth undergo modifications associated with a change from a vegetarian to an insectivorous diet. The mouth widens, the true jaws and their associated muscles begin to function and the tongue enlarges and elongates. In relative terms the intestine becomes shorter but the stomach enlarges. The eyes take on the bulging form of the adult frog. At this stage the young frog leaves the water for the land where it feeds on insects and continues to grow. At first it is entirely restricted to damp habitats, but finally can range further and follow the adult mode of life.

Amniote development

The amniote egg, with its shell, evolved in the reptiles as a structure in which development can take place on land rather than in the water. The egg and the developing embryo of any animal are extremely vulnerable but this vulnerability is much greater in a terrestrial environment. Many anamniotes, such as the frog, have a layer of albumen which surrounds the egg, protecting it against mechanical shock and reducing the risk of desiccation. The amniote egg has additional membrane layers including a tough leathery or calcareous shell which protects it from drying. In addition the embryo itself becomes surrounded by a series of extra-embryonic membranes which give it further protection and aid its metabolic activities. These membranes, the amnion, yolk sac, chorion, and allantois, are sheets of tissue that grow out of the embryo itself.

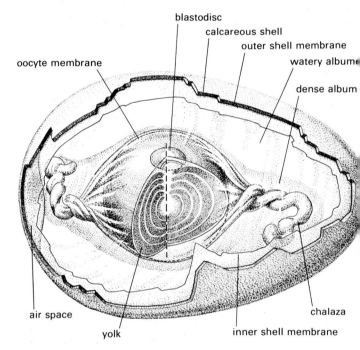

A representative amniote egg (chicken)

Birds provide examples both of macrolecithal eggs and of amniote development. Eggs of this type are also found in the reptiles but the egg of the domestic hen provides an example familiar to everyone. The oocyte is an enormous cell with a diameter of over 20 mm and containing a vast amount of yolk. The cytoplasm is confined to a small blastodisc which forms a cap at the animal pole and has the nucleus in its centre. The oocyte is surrounded by a membrane which separates it from the albumen (egg white). The albumen is deposited in successive layers as the fertilized egg passes down the oviduct and provides an extra store of protein and water for the developing chick. The innermost layer of albumen is thin and watery and this is covered by a layer of dense viscous albumen and then by a second layer of thin albumen. Part of the dense albumen forms twisted spring-like cords, the chalazae, which keep the blastodisc in an upright position. During its passage along the lower part of the oviduct the albumen becomes surrounded by two tough shell membranes formed of keratin fibres; these are in close contact with each other apart from a small air space at the blunt end of the egg. The water which makes up the bulk of the albumen layer is absorbed by osmosis in the lower part of the oviduct and causes the albumen to swell. Finally the egg is enclosed in a calcareous shell. This has inner and outer protein layers with pores between them which serve for gaseous exchange and are particularly numerous above the air space.

Cleavage

As the greater part of the egg mass is an inert yolk, it is impossible for it to divide completely, and cleavage and blastula formation are confined to the protoplasm. The first and second cleavages pass through the animal pole at right angles to each other. A horizontal division is then followed by rather irregular subdivisions resulting in a plate-like blastoderm representing, in effect, a flattened sphere. This type of cleavage is termed meroblastic, in contrast to the holoblastic cleavage of amphioxus. It begins soon after fertilization and continues as the egg passes down the oviduct, so that by the time the egg is laid, blastula formation is complete. Viewed from above at this stage the blastoderm can be distinguished as a clear central region, the area pellucida, overlying the blastocoele and a narrow darker region, the area opaca, which spreads over part of the yolk. The surface of the blastoderm is laid out in a pattern of germ layers which are essentially those of the amphibian egg, but spread out over the surface of the yolk rather than surrounding it.

Gastrulation and early organogenesis

Inevitably the geometry of gastrulation of the macrolecithal egg is different from that already described. It begins by a rapid expansion of the blastoderm. The prospective endoderm cells migrate downwards into the blastocoele (at this stage described as the subgerminal cavity) and spread out to form a layer of embryonic endoderm across the floor of the cavity. A region known as the primitive streak corresponds to the blastopore and appears on the posterior part of the zona pellucida and grows forwards as a longitudinal groove with raised parallel edges. The primitive streak is a region of intense cellular activity with continued movement of mesodermal cells which converge on the streak, migrate downwards through it and fan out laterally to interpose themselves between the outer ectoderm and inner endodermal layers. As a result of these processes there develops a raised area of tissue, Hensen's node, formed of prospective notochordal and neural cells at the anterior end of the primitive streak. The chordamesoderm turns downwards at Hensen's node and pushes forwards in the mid-line beneath the neurectoderm.

Neural tube formation, gut formation, and the further development of the chordamesoderm and mesoderm are essentially the same as those described for the frog.

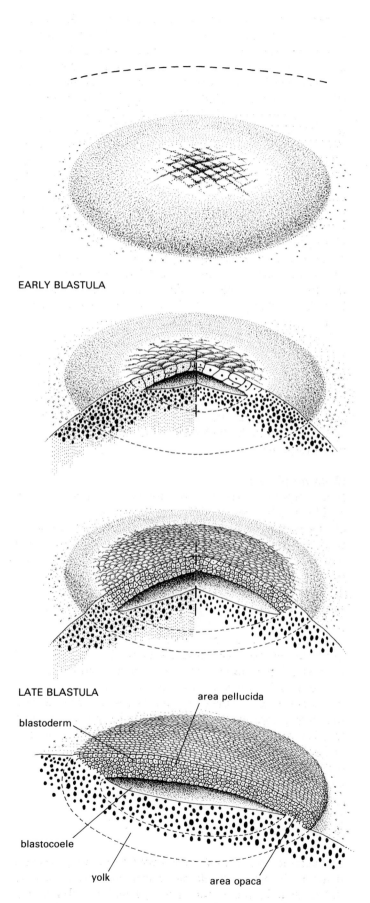

EARLY BLASTULA

LATE BLASTULA

area pellucida

blastoderm

blastocoele

yolk

area opaca

The chick: early blastula

CLEAVAGE

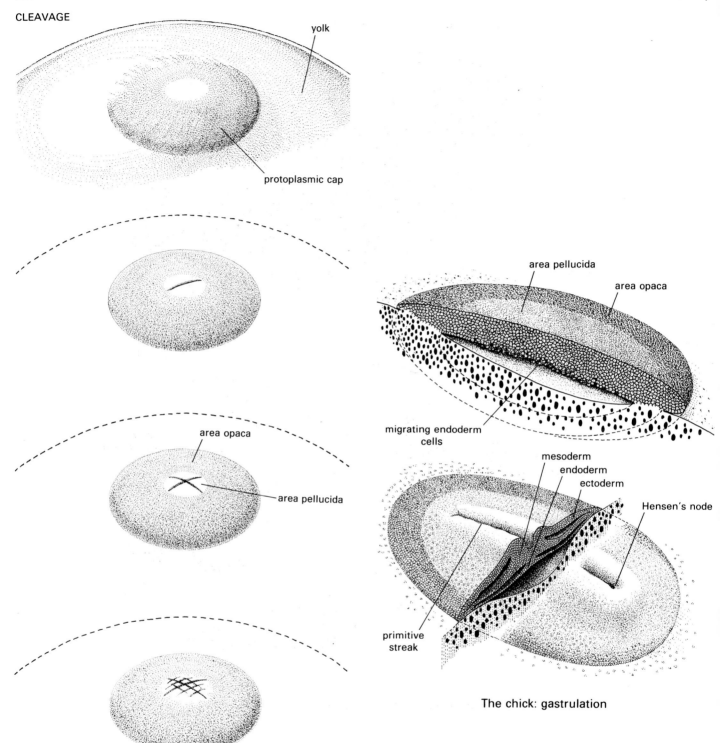

yolk

protoplasmic cap

area opaca

area pellucida

The chick: cleavage

area pellucida

area opaca

migrating endoderm
cells

mesoderm
endoderm
ectoderm

Hensen's node

primitive
streak

The chick: gastrulation

The extra-embryonic membranes

As mentioned earlier, amniotes have four extra embryonic
membranes: the yolk sac, chorion, amnion, and allantois.
Each of these is composed of two germ layers (endoderm,
ectoderm or mesoderm in different paired combinations)
which grow in close apposition to each other. In reptiles and
birds the yolk sac represents a down growth of endoderm
followed by a development of mesoderm, which eventually
encloses the yolk. Blood vessels develop in the mesoderm

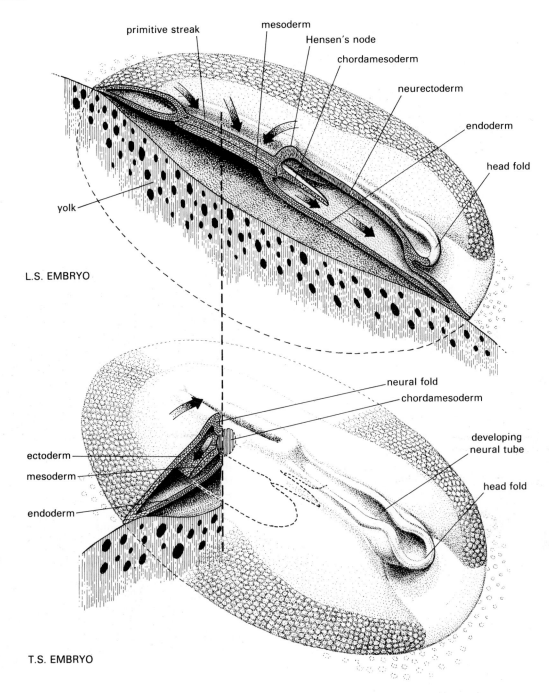

primitive streak
mesoderm
Hensen's node
chordamesoderm
neurectoderm
endoderm
head fold
yolk

L.S. EMBRYO

neural fold
chordamesoderm
developing neural tube
head fold
ectoderm
mesoderm
endoderm

T.S. EMBRYO

The chick: completion of gastrulation and beginning of differentiation

portion, and the connection between yolk sac and embryo eventually becomes constricted to a stalk. The yolk sac develops in macrolecithal anamniotes, as well as in the amniotes, and functions primarily for nutrition, by producing enzymes which digest the yolk and transporting products of digestion to the embryo. It may also play a part both in excretion and respiration. The other three membranes are found only in amniotes.

Both the amnion and chorion develop as folds of the body wall which come to surround the developing embryo. The amnion is composed of extra-embryonic ectoderm with accompanying mesoderm and develops as folds of tissue which join to form a fluid-filled sac. Thus the embryo develops bathed in amniotic fluid which can be regarded as a minute replica of the aquatic environment in which aquatic vertebrates develop. The chorion is also composed of extra-embryonic ectoderm and mesoderm and develops to enclose all the embryonic structures in the protective outer layer.

The allantois, composed of extra-embryonic endoderm and mesoderm, develops as an outgrowth from the posterior region of the embryonic gut, expanding to form a large sac underlying, and attached to, the chorion. Since embryonic development proceeds to an advanced stage in amniotes the kidneys begin to function early in development. The allantois functions as a storage bladder for excretory products and in both reptiles and birds it is equally important as a breathing organ with a respiratory surface formed by the combined chorion-allantois. Here oxygen, which enters through the porous shell, is absorbed and carbon dioxide, produced by the developing chick, is removed. Gases are transported to and from the respiratory surface by blood vessels which develop in the allantois. In the region near the embryo, the allantois shrinks to a stalk which, together with the stalk of the yolk sac, becomes enclosed in a tube of tissue developing from the ventral wall of the embryo to form an umbilical cord.

The speed and development of the embryo varies according to the temperature. The amnion and chorion are the first membranes to arise and the allantois develops after the hindgut is formed. At around 40° C the amnion and chorion begin to develop at about 46 hours after laying while the allantois develops after 72 hours.

The development of mammals

Mammals have a highly specialized form of development: unlike that of other vertebrates, with the exception of the primitive egg-laying mammals (monotremes), the embryo develops inside the mother and derives nutritive materials and oxygen from her blood stream by means of a specialized structure, the placenta. The mammalian egg is minute (about 0.1 mm in diameter) and practically devoid of yolk. It is therefore essential that the placenta, which is derived partly from maternal and partly from embryonic tissue, should develop early.

As amniotes, mammals have the same extra-embryonic membranes already described for the chick, though their development is rather different. Mammalian embryology is very variable and the following description is based on human development.

Cleavage and the development of the blastocyst
We might reasonably expect that the earliest stages of mammalian embryology would be identical with those of amphioxus since both have a microlecithal egg. Indeed there is a superficial similarity but the final results are very different.

The human egg is a small spherical cell surrounded by a thin, striated membrane, the zona pellucida. After fertilization, during the three to four days that it is moving along the Fallopian tube to the uterus by ciliary action and peristalsis, the egg undergoes a series of cleavage divisions, similar to those described for amphioxus. At the end of cleavage, the egg consists of a mass of blastomeres surrounded by the zona pellucida and is known as the morula. In the fully-formed morula two types of cell are recognizable: these are an expanded, hollow sphere of epithelial cells, the trophoblast, which is destined to become part of the placenta, and a compact inner cell mass the equivalent of the blastula, lying above the trophoblast cavity. The embryo reaches the uterine cavity in this condition. Here it comes into contact with the mucus membrane lining the uterus and absorbs the liquid which it secretes. The embryo swells rapidly, accumulates fluid in the cavity of the trophoblast and is then referred to as a blastocyst. Cell division continues, and the blastocyst increases in size. Development up until the blastocyst stage takes place with the embryo lying freely in the uterine cavity.

Gastrulation and the development of the extra-embryonic membranes.
The process of gastrulation in mammals is different from that of other vertebrates and is a result of the unusual cleavage, and as in cleavage, there are differences between various mammalian groups. Frequently the zona pellucida disappears about seven days after fertilization and the trophoblast becomes attached to the uterine lining, which is greatly thickened, by secreting enzymes that erode the uterine cells. Two cavities, which appear by the splitting of the trophoblast in the inner cell mass, expand and leave a flat bilaminar (two-layered) plate of cells between them which is comparable to the blastodisc of birds and reptiles. The upper cavity is surrounded by ectoderm and forms the amnion, the lower is surrounded by endoderm and forms the yolk sac. The intervening plate is the blastoderm or embryonic disc. The upper surface of the blastoderm has the same pattern of potential germ layers as that found in the blastodisc of other amniotes. The lower surface is the endoderm, which has acquired its position through the splitting process rather than through cell movement. As in other amniotes a primitive streak arises and the chordamesoderm and mesoderm cells roll inwards and downwards to fan out between the endoderm and ectoderm.

This process is highly variable. Sometimes (for example in man) the amniotic cavity develops by splitting, and the definitive yolk sac forms by downward migration of cells from the disc. In rodents the amnion develops as paired folds from an embryonic knob.

The embryonic disc is connected to the trophoblast by a stalk of mesodermal cells. As in the chick the allantois develops as an outgrowth of the gut which pushes along the stalk and expands beneath the trophoblast wall. The endoderm of the yolk sac and the ectoderm of the amnion become reinforced with mesodermal cells. These also sheath the trophoblast internally to form a chorion. The allantois, and the yolk sac nearest the embryo, shrink to form an umbilical cord.

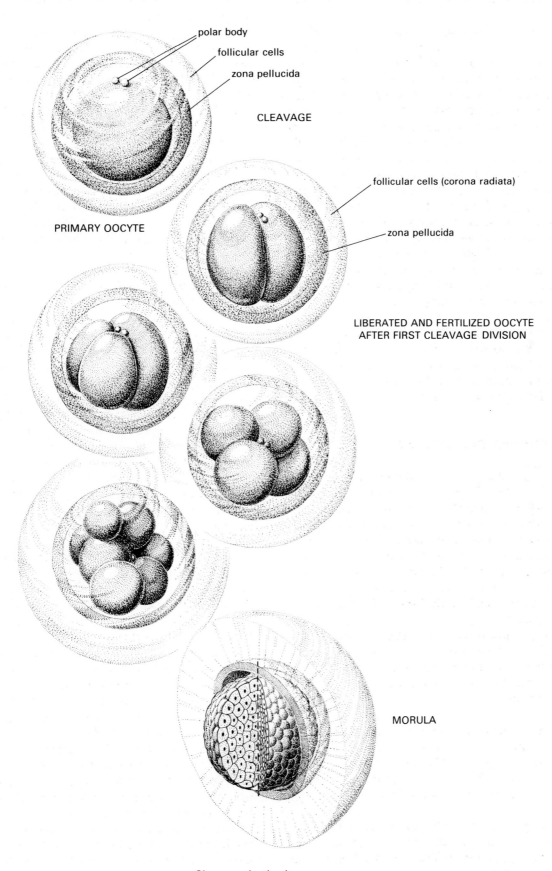

polar body
follicular cells
zona pellucida

CLEAVAGE

PRIMARY OOCYTE

follicular cells (corona radiata)

zona pellucida

LIBERATED AND FERTILIZED OOCYTE
AFTER FIRST CLEAVAGE DIVISION

MORULA

Cleavage in the human egg

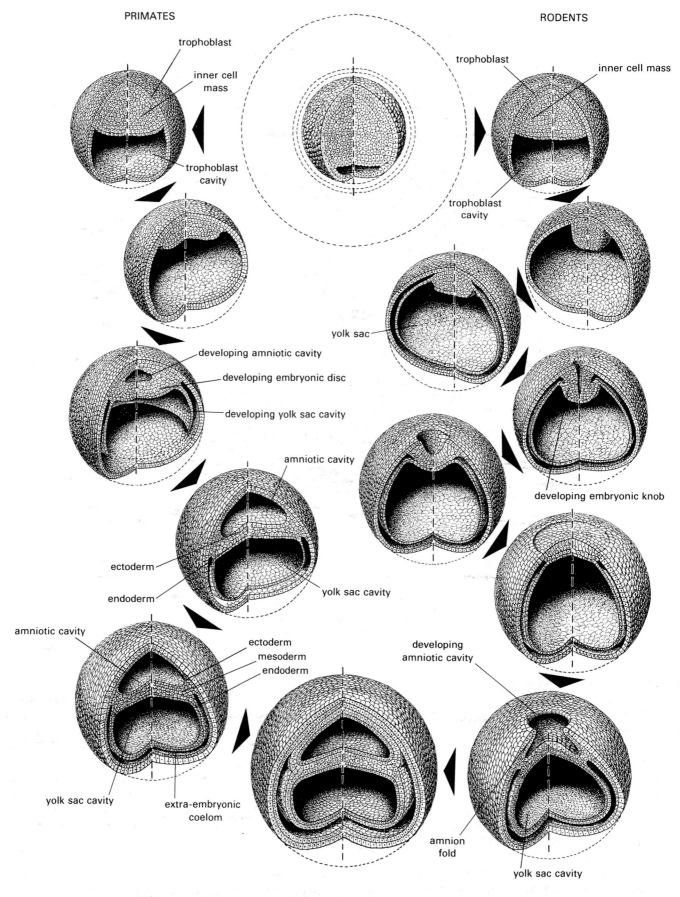

PRIMATES

RODENTS

trophoblast

inner cell mass

trophoblast cavity

trophoblast

inner cell mass

trophoblast cavity

yolk sac

developing amniotic cavity

developing embryonic disc

developing yolk sac cavity

amniotic cavity

ectoderm

endoderm

yolk sac cavity

developing embryonic knob

amniotic cavity

ectoderm
mesoderm
endoderm

yolk sac cavity

extra-embryonic coelom

developing amniotic cavity

amnion fold

yolk sac cavity

Two types of development of the extra-embryonic membranes in mammals

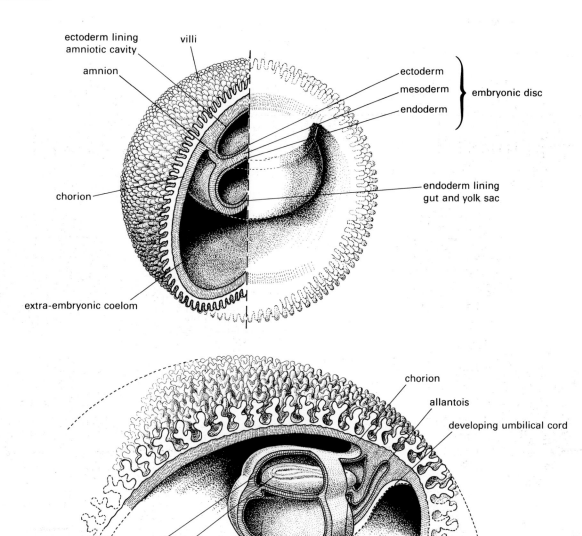

ectoderm lining
amniotic cavity

villi

amnion

chorion

extra-embryonic coelom

ectoderm ⎫
mesoderm ⎬ embryonic disc
endoderm ⎭

endoderm lining
gut and yolk sac

chorion

allantois

developing umbilical cord

amnion

embryo

yolk sac   gut

Mammalian development (based on human)
Top: extra-embryonic membranes
Bottom: primitive streak formation

The development of the placenta

The placenta is formed by a close union of the extra-embryonic membranes and the maternal tissues, and its primary embryonic component is the chorion. This develops finger-like projections, the chorionic villi, which project into the maternal tissue. However the chorion is only part of the placenta, and may in fact disappear in late embryonic development. In addition, since the chorion has no direct contact with the embryo, an additional structure is necessary for blood to circulate between the embryo and the placenta. In some mammals (marsupials) this is provided by the yolk sac, but in nearly all higher mammals, including man, the allantois develops to lie beneath the chorion in a similar manner to that already described for the birds. The placenta is thus formed by union of the chorioallantoic membrane and the uterine lining. In humans this union is particularly intimate and the chorionic villi, which contain blood capillaries, penetrate the uterine lining and break it down so that they are bathed in the maternal blood. This type of placenta, in which the chorion and maternal blood are in contact, is sometimes described as a haemochorial placenta. However no blood is exchanged between the mother and developing embryo, and gases, nutrients and waste products move by diffusion across the villi and capillary walls.

Neural tube and gut formation

The process of neural tube and gut formation is essentially the same as that already described.

Summary of chordate embryology

Egg type	Cleavage (increase in cell number)	Gastrulation (to give external ectoderm and neurectoderm, internal mesoderm and endoderm)	Early organ formation
Microlecithal e.g. amphioxus	Holoblastic. Slight size difference between cells in animal and vegetable hemispheres.	Single invagination through blastopore.	Ectoderm and neurectoderm grow together separately. Coelom develops from pouches, of mesoderm (enterocoely).
Mesolecithal e.g. the frog	Holoblastic. Considerable size difference between cells in animal and vegetable hemispheres. Blastocoele partly occupied by yolky cells.	Basically similar to amphioxus but some modification, e.g. epiboly, to allow for increased yolk.	
Macrolecithal e.g. the chick	Cleavage restricted to a disc forming the blastoderm. Flattened blastocoele.	Bird: invagination occurs through a primitive streak in the centre of the blastoderm.	Ectoderm and neuroectoderm develop as folds with sinking of neural plate. Coelom develops as split in mesoderm (schizocoely).
Eutherian mammals e.g. Man	Holoblastic cleavage giving a trophoblast. No blastocoele.	2 splits, giving amnion, blastoderm and yolk sac. Primitive streak in centre of blastoderm.	

Summary of the fate of the primary germ layers

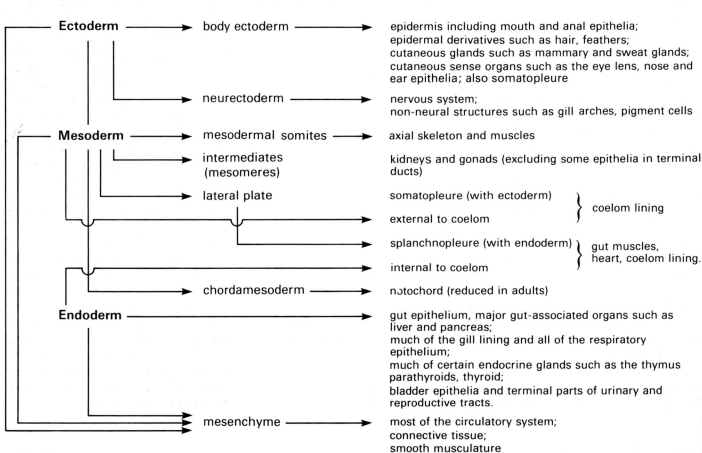

Vertebrate diversity

Introduction

Some indication of the structural diversity of vertebrates has already been given in the preceding chapter. In this section we will consider how various vertebrates have adapted their basic structure at different levels of organization to a variety of different modes of life. Vertebrate zoology is a vast subject with an extensive literature. Since vertebrates include most of the largest and most conspicuous animals, as well as man's food, and domestic animals, it is not surprising that they were the object of the earliest serious anatomical studies. The most influential of the early writings were those of the Greek philosopher and natural historian Aristotle (384–322 BC) who amongst other things showed that he was well aware of the distinction between fishes and aquatic mammals. He also described various aspects of human and mammalian anatomy although with a very superficial understanding of the structures he described; for example he believed that arteries contained air and the heart was the seat of intelligence. However the earliest anatomical writings come from Egypt in the Edwin Smith and Ebers Papyri, which were written in about 1600 BC and 1550 BC respectively but themselves represent compendia of earlier writing and traditional lore. In addition to magical formulae and folk remedies, the Ebers Papyrus includes a surprisingly accurate description of the function of the heart and the human blood system, written over 3000 years before the circulation of the blood was described by William Harvey in 1628. The Edwin Smith papyrus is believed to be a copy of earlier work dated about 3000 BC and is apparently an early textbook of surgery, revealing among other information a knowledge of the pulse as related to the heart and of the function of the stomach and intestines. In the sixth century BC a Greek philosopher, Alcmaeon of Croton is said to have dissected vertebrates, and may indeed have been the first vivisector. He described, among other structures, the optic nerve and Eustachian tubes, and inferred that the brain was the seat of intelligence. The Hippocratic writings (500–300 BC) probably formed the medical school library at Cos, where Hippocrates (regarded as the father of medicine) taught, and show that surgeons of that time had a good knowledge of the human skeleton. Vertebrate embryology was also studied by these early workers. For example an unknown Greek author of about 370 BC wrote 'On the nature of the infant' which was based on examination of various stages in the development of the hen's egg; thus the methods used by modern experimental embryologists are hardly new!

Noah's ark has been suggested as the first Natural History museum (by a seventeenth century Jesuit), and the Old Testament includes many observations on vertebrates and vertebrate behaviour. The 'fierce sorrow' of the female bear when deprived of her young was so well known that the prophet Hosea (writing from 786–746 BC), used it as an example of divine justice. Lions, leopards, rabbits, apes, mice, moles, bats, elephants, camels, oxen and sheep are among the many mammals mentioned in the Old Testament; unhappily the Leviathan (the monster of the seas) and the Behemoth (the monster of the land) cannot be related to modern fauna. Those interested will find them in the Book of Job (c. 600–500 BC).

The Chordata, the phylum to which the vertebrates belong, is only one of the animal phyla now recognized, but vertebrate animals dominate all early classifications. For example Linnaeus described in his Systemae Naturae (1735) six main classes of animals as follows:

I MAMMALIA	Lungs respire alternately; jaws: incumbent, covered; teeth usually within; teats lactiferous; organs of sense: tongue, nostrils, eyes, ears, and papillae of the skin; covering: hair, which is scanty in warm climates and hardly any in aquatics; supporters: four feet, except in aquatics, and in most a tail. Walk on the earth and speak.
II BIRDS	Lungs respire alternately; jaws: incumbent, naked, extended without teeth, covered with a calcareous shell; organs of sense: tongue, nostrils, eyes and ears without auricles; covering: incumbent, imbricate feathers; supporters: two feet, two wings and a heart-shaped rump; fly in the air and sing.
III AMPHIBIA	Jaws: incumbent; penis (frequently) double; eggs (usually membranaceous); organs of sense: tongue, nostrils, eyes, ears; covering: a naked skin; supporters: various, in some none at all; creep in warm places and hiss.
IV FISHES	Jaws incumbent; penis (usually) none; eggs without white; organs of sense: tongue, nostrils, eyes, ears; covering: imbricate scales; supporters: fins; swim in the water and smack.
V INSECTS	Spiracles: lateral pores; jaws lateral; organs of sense: tongue, eyes, antennae on the head, no brain, no ears, no nostrils; covering: a bony coat of mail; supporters: feet and in some wings; skip on dry ground and buzz.

VI WORMS Spiracles: obscure; jaws: various; frequently hermaphrodites; organs of sense: tentacles, (generally) eyes; no brain, no ears, no nostrils; covering: calcareous or none, except spines; supporters: no feet, no fins; crawl in moist places and are mute.

He placed animals within his classification using a simple key:

Heart with 2 auricles, 2 ventricles; blood warm, red.	viviparous	MAMMALIA
	oviparous	BIRDS
Heart with 1 auricle, 1 ventricle; blood cold, red.	lungs voluntary	AMPHIBIA
	external gills	FISHES
Heart with 1 auricle, ventricle 0	have antennae	INSECTS
Sanies (wound discharge): cold, white	have tentacles	WORMS

The main subdivisions of the vertebrates are broadly those recognized today, though Linnaeus did not separate amphibians (in the modern sense) from reptiles, subdivide the fishes, or recognize whales as mammals. However, the invertebrates to which over half this book has been devoted are placed in only two classes, the insects and all the rest (worms).

Formal research in both zoology and medicine was carried out at an institution founded by Ptolemy Philadelphus in 300 BC, and Aristotle may have been provided with both a zoo and museum by his former pupil, Alexander the Great. Generally museums and the formal study of vertebrates came later, in the Renaissance (1400–1600 AD). Leonardo da Vinci, working between 1482 and 1518 produced a Bestiary including observations on a great variety of vertebrates although these are mainly of an inaccurate nature. The toad is described as 'feeds on earth and always remains lean because it never satisfies itself, so great is its fear lest the supply of earth should fail!' However he gives detailed descriptions of animal locomotion, bird flight and human anatomy. Gesner, a celebrated Swiss naturalist and mountaineer, published 'Historiae Animalium Libri' between 1551–87, which summarised European knowledge and belief in animal life and was illustrated with wood cuts including probably the first figures of frogs and vipers. John Ray, the son of a village blacksmith, turned to the study of zoology at the age of 46, having first been appointed lecturer in Greek, mathematics and the humanities at Cambridge, and published extensively in Botany. He was the first to show the value of internal anatomical structures in taxonomy. By the beginning of the eighteenth century museums and collections were numerous and the curious could observe such interesting vertebrate exhibits as 'A frightful large Indian bat', 'A mermaid's hand' or 'the skeleton and stuffed skin of a woman who had eighteen husbands' in the museum of the Anatomy School at Oxford.

Tens of thousands of vertebrates have been described, many of them being fossil forms. It is hardly surprising that it is difficult for the student of vertebrates to know where to begin. Even within the various vertebrate classes: Agnatha, Elasmobranchiomorphi (Chondrichthyes), Teleostomi, Amphibia, Reptilia, Aves, and Mammalia, it is impossible to select one representative which can be regarded as in any way typical of recent vertebrates, let alone the fossil forms, and the animals selected for more detailed description have been selected as much on the basis of their availability to the student as anything else.

CLASS AGNATHA

The class Agnatha, regarded as the most primitive living vertebrates, are superficially fish-like but differ from 'true' fishes in two important respects: they have no jaws (Agnatha means 'without jaws') and no paired fins. There are rather few species of living agnathans, known collectively as cyclostomes, (meaning 'round mouthed'), representing only a remnant of a much greater fauna which flourished in the middle Palaeozoic. These fossil forms, the ostracoderms, were small (less than 36 cm length) and characterized by an extensive bony armour of plates and scales developed in the dermis. This is thought to have been defensive in function, protecting the ostracoderms from, amongst other things, the large scorpion-like eurypterids which were numerous in fresh water at that time. As many ostracoderms were freshwater dwellers the dermal armour may have had an additional advantage in reducing the loss of solutes from the body fluids by osmosis. The majority of ostracoderms were dorsoventrally flattened and were probably bottom dwellers.

The two living families of agnathans, Petromyzonidae and Myxinidae, show little resemblance to each other, either in body form or in mode of life. The Myxinidae (hagfishes), with fifteen species grouped in six genera, are slimy bottom dwelling animals which live in the sea. A hagfish can produce two gallons of slime in a few seconds as a defence mechanism, and an alternative name for the best known genus, *Myxine*, is the slime hag. They have a mouth surrounded by short tentacles and live in burrows in the mud, feeding on dead or dying fish or on slow moving invertebrates such as polychaete worms. They lay their eggs in the water and attach them to the bottom with hooks. The young develop directly from the egg.

The better known Petromyzonidae with nine genera and over thirty species are eel-like in form and are blood-sucking predators feeding on active, living fish. Most have a complicated life cycle involving a freshwater larval stage. *Lampetra fluviatilis* will be described in more detail as an example of the class Agnatha although it should be remembered that the two families which comprise the group are very different from each other.

Lampetra fluviatilis

Lampreys are best known as gastronomic delicacies, having been prized as such since Roman times. Henry I is popularly regarded as having died from eating too many of them. 'Potted Lamperns' made from lampreys caught in the River Severn were a Worcestershire speciality.

The lamprey, *Lampetra fluviatilis* lives and feeds in the sea as an adult, and in Britain swims up the rivers in spring or early summer to breed in fresh water. It may use its mouth to anchor it to stones during migration or spawning: the common name derives from the medieval Latin *lampodra*, meaning 'lick a stone'. It was formerly more common than it is now, and was caught in the rivers in large numbers.

General body form

The lamprey is easily recognized by its eel-like form and the large, circular, sucking disc which surrounds its mouth. It has a greyish colour and grows up to about a metre long with two well-developed median dorsal fins, a caudal fin, and a smaller anal fin. It has well-developed eyes and also a pineal eye, which appears as a yellow spot covered with transparent skin on the dorsal surface of the head between the eyes. There is no evidence that its function is any more than simple light detection. A single nostril lies immediately to the anterior of the pineal eye. Along the side of the head are seven pairs of gill openings.

The integumentary system

The skin is smooth and somewhat slimy and the divisions between successive myotomes can clearly be seen beneath it. Although it is entirely without scales, a study of the ostracoderms suggests that this is a derived rather than a primitive character. The epidermis is multilayered and contains numerous gland cells. The outer epidermal cells are living, and secrete a thin cuticle. The dermis is thinner than the epidermis and firmly attached to the myocommata which lie between the muscles. It contains irregularly branched chromatophores (pigment cells) which give the lamprey its characteristic colour.

The skeleto-muscular system and locomotion

Lampreys have no bone, and their skeleton consists of a notochord and various small cartilages. Although some of the cartilage is of the normal vertebrate type, some of it is rather like fibrous connective tissue. The notochord, which is comparable with that of amphioxus, is formed of vacuolated cells enclosed in a thick fibrous chordal sheath which also surrounds and protects the nerve cord.

The brain is surrounded by a simple chondrocranium, resembling that of the embryos of higher vertebrates. The floor of the chondrocranium is formed by the notochord and paired cartilages, the parachordals and trabeculae, which lie on either side of it. The single nasal capsule and the paired otic capsules are in effect cartilaginous boxes which surround the 'nose' and 'ears', while the eyes are protected by cartilaginous ridges. The chondocranium is roofed by a layer of tough, membraneous fibrocartilage.

An extensive branchial basket of continuous unjointed elastic cartilage supports the gills, and specialized cartilages

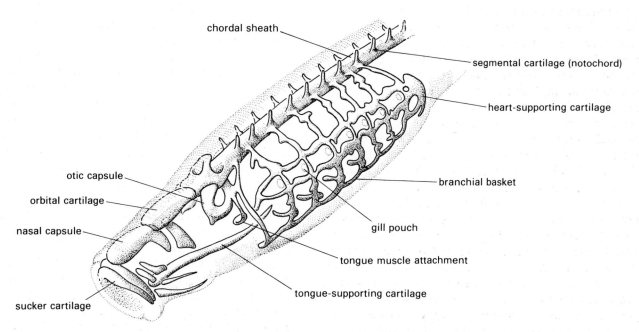

Visceral skeleton and brain case of the lamprey

support the sucker and tongue. A posterior extension of the branchial basket forms a box which supports and protects the heart. The branchial basket lies just below the skin and is not homologous with the gill arch skeleton of fishes.

Posterior to the head, small cartilages (two pairs per segment) abut onto the chordal sheath on its dorsal side while in the tail region similar cartilages also extend ventrally. The fins are supported by rod-like radial cartilages.

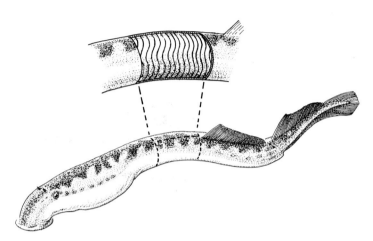

Lamprey : myotomes

Lampreys swim with eel-like lateral undulations, using alternating serial contractions of their segmentally arranged myotomes. The notochord functions in a similar way to that of amphioxus, preventing the body from shortening when the muscles contract. The myotomes are almost vertical but bend forwards at their dorsal and ventral ends. In the head they are modified to form a specialized musculature associated with the tongue and gills.

The digestive system and nutrition

The feeding apparatus of the lamprey consists of a powerful sucker and a protrusible rasping tongue, which the animal uses as a substitute for jaws. The lamprey attaches itself to living fish by means of a sucker. Very little is known of how it finds its prey or how the initial attachment is made; once made, however, it is extremely effective. The margin of the sucker is formed by a series of flexible lips which make a tight attachment to the prey and are also thought to have a sensory function. When the lamprey is attached and feeding, the margins are spread out, but they are drawn together during swimming. With the sucker held against the side of the prey the tongue is pulled backwards by the action of a large cardio-apicalis muscle, and this creates sufficient suction for attachment. Once the lamprey is attached the tongue is fixed into its withdrawn position by a ring of circular muscle fibres which contract around its base and clamp it in place. The inside of the sucker is studded with

conical keratin 'teeth' which are epidermal in origin and not homologus with the teeth of other vertebrates. The arrangement of these 'teeth' is the basis for distinguishing between the various species of lamprey.

A small round mouth, through which the tongue protrudes, lies at the base of the fleshy, funnel-like sucker. The tongue is a highly specialized structure, and, like the teeth of the sucker, is not homologous with other vertebrate tongues. It has dorsal and ventral muscle bands which, when contracted alternately, rock the tongue up and down. The tongue is studded with similar teeth to those of the sucker and the rocking action rapidly scrapes a wound in the prey's skin. An anticoagulant, known as lamphedrin, is secreted by an oral gland and once the lamprey is attached and a wound made it is able to feed continuously. Eventually the lamprey ceases to feed and detaches itself. The annular muscle, running just above the lips of the sucker, draws in the sucker margin and acts as a release mechanism. The prey is greatly weakened and left with a characteristic round wound. However, since fish have been recorded with the scars of several lamprey attacks, at least some prey do survive.

The alimentary tract is generally thin-walled and extremely distensible. The mouth opens into a short narrow oesophagus which leads directly to the intestine where digestion and absorption take place. The intestine is straight and therefore fairly short, and its internal surface area is increased by a longitudinal fold which follows an approximately spiral course. Lampreys have a small liver but there is no gall bladder, bile duct, or pancreas.

Respiratory system

The respiratory system is highly specialized and well adapted to the animal's unusual mode of life. Unlike that of most vertebrates, the lamprey pharynx does not connect the mouth with the oesophagus but is a separate diverticulum with its own opening into the oral cavity. The respiratory lamellae or gills lie within a series of spherical muscular pouches, well separated from each other. Each has a separate internal opening into the respiratory diverticulum and opens to the exterior through a small round aperture in the body wall. When the animal is not attached to its prey the respiratory current enters through the pharynx, and passes over the gills to leave through the external gill openings. However this is obviously impossible when the animal is feeding and the current is then created by muscular pumping movements of the individual pouches, with water entering and leaving through their external openings.

Circulatory system

The blood system is arranged on a similar plan to that of amphioxus. A well developed, three-chambered S-shaped heart is present with a large, thin-walled atrium in which

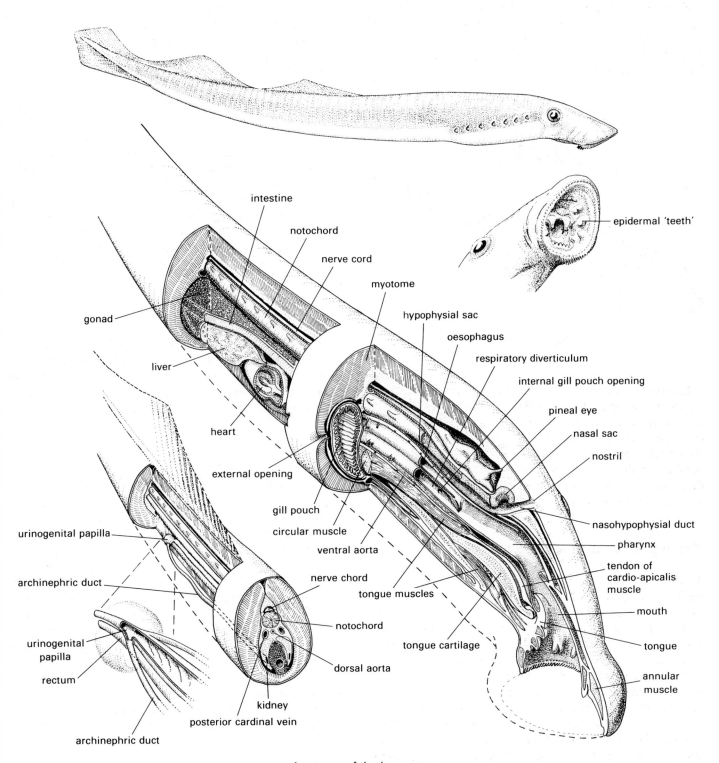

intestine

notochord

nerve cord

myotome

hypophysial sac

oesophagus

respiratory diverticulum

internal gill pouch opening

pineal eye

nasal sac

nostril

epidermal 'teeth'

gonad

liver

heart

external opening

gill pouch

circular muscle

ventral aorta

nasohypophysial duct

pharynx

tendon of cardio-apicalis muscle

mouth

tongue

annular muscle

urinogenital papilla

archinephric duct

urinogenital papilla

rectum

archinephric duct

nerve chord

notochord

tongue muscles

tongue cartilage

dorsal aorta

kidney

posterior cardinal vein

Anatomy of the lamprey

blood collects, a small muscular ventricle and a short conus. The atrium opens into the ventricle through a small aperture which is guarded by atrio-ventricular valves. Tough cords extend from the ventricle to the valves and prevent them from being pushed backwards into the atrium as the ventricle contracts. Blood is pumped from the ventricle into the conus and hence to the ventral aorta. Four pairs of afferent branchial arteries arise from the ventral aorta, before it forks at the level of the fourth gill pouch. Each branch gives rise to four further afferent branchial arteries. Together the afferent branchial arteries supply blood to the gill lamellae where gaseous exchange takes place. Blood is drained from the gills by the efferent branchial arteries which correspond in position to the afferent vessels but join a single median dorsal aorta. At its anterior end the dorsal aorta forks, the two halves joining again to form the cephalic circle from which arteries arise to supply the brain, eyes and tongue. Posteriorly, arteries from the dorsal aorta supply the body wall and viscera.

The venous system is similar to that of amphioxus in basic plan, and consists of a network of vessels and sinuses with contractile venous hearts. A large caudal vein in the tail region divides into two posterior cardinal veins on entering the abdomen, and these extend forwards collecting blood from the kidneys, gonads and body wall. Immediately posterior to the heart they unite as a single duct of Cuvier which opens into the atrium. A ventral jugular vein collects blood from the muscles of the sucker and gill pouches and paired anterior cardinal veins from the remaining anterior structures. Blood from the gut passes to the liver through a hepatic portal system and thence to the heart via the hepatic vein.

Lamprey blood, like that of other vertebrates, contains haemoglobin in corpuscles. There is no separate spleen and the red-cell-forming tissue lies below the nerve cord and in the kidneys. White blood cells are also formed in the kidneys. Lampreys have no lymphatic system.

Excretion and osmoregulation

Lamprey kidneys are elongated structures, suspended by a membrane along either side of the dorsal mid-line of the body cavity. The archinephric duct runs along the free edge of the kidney. Posteriorly a short duct from the coelom opens into it; this serves for the transfer of gonadial products which are shed directly into the coelom. The ducts from each side join posteriorly to open into a urinogenital sinus at a urinogenital papilla. The urinogenital sinus, opens into the cloaca. The adult kidney is an opisthonephros, the anterior (pronephric) portion being modified as a mass of lymphoid tissue. Lymphoid tissue and fat also form part of the opisthonephric kidney, along with a mass of excretory tubules. These are fairly rudimentary and the filtrate is expelled from the body almost unchanged.

Nervous system and sense organs

The nervous system of lampreys shows the characteristic plan present in all vertebrates. The forebrain, associated with the olfactory sense, and the midbrain, associated with the visual sense, are both moderately well developed, but the acousticolateralis and auditory systems are poorly developed and the cerebellum is small. The brain has an extensive vascular plexus overlying its upper surface and extending into the ventricles. This is thought to be related to its lack of cerebral blood vessels.

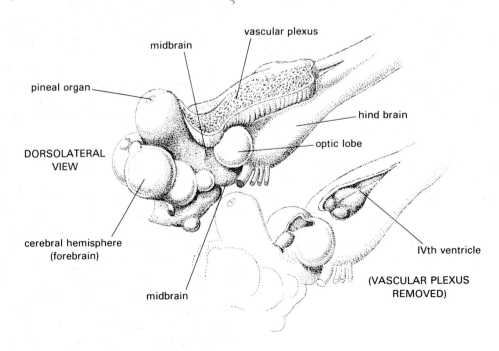

Brain (lamprey)

Unlike the great majority of vertebrates, lampreys have only two semicircular canals in each ear and since it is the horizontal canal that is missing, we might expect them to be insensitive to rotation in the horizontal plane. In fact they are sensitive to movement in all planes, because the neuromasts of the semicircular canals are not in ampullae (the more usual condition) but in the main cavity of the ear at the ends of the canals. They can therefore be displaced both by movement of endolymph through the canal and by horizontal movements of endolymph in the ear cavity. The hair cells show two different types of orientation: some are arranged so that they respond to movements along the canals, while others are arranged so that they are stimulated by movements at right angles to the canal. It is these that respond to movements in the horizontal plane. Lampreys have a single large otolith in each ear.

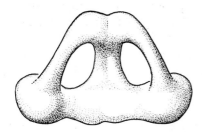

Cyclostome semicircular canals

The method of focussing the eye in the lamprey also differs from that of other vertebrates. A muscle flattens the cornea and pushes the lens towards the retina thus bringing more distant objects into focus. The reverse process occurs to focus on near objects. In addition to the eyes, and the light sensitive pineal eye, lampreys have photoreceptors in the skin. Although they appear to have a poorly developed sense of taste they are moderately sensitive to smell. The single nostril on the top of the head opens into a naso-hypophysial duct which extends to a nasal sac lying within the branchial basket; this is expanded and compressed in much the same way as the gill pouches during respiration. A sensory epithelium lies within a double diverticulum of the duct, immediately anterior to the brain. As water is sucked in and out of the duct, it passes over the olfactory epithelium to which it is directed by a flap.

The spinal cord is more or less uniform in diameter and circular in cross section. The roots of the spinal nerves do not join each other (a primitive condition), except in the gill region where they join to form the hypobranchial nerve which is composed of both sensory and motor fibres and supplies the ventral gill region. The lamprey has the typical pattern of vertebrate cranial nerves except that the terminalis nerve (0) is absent.

Reproduction

In temperate regions spawning occurs in the spring when the lampreys collect in rivers, in places where the water is shallow and swift moving. They build nests, consisting of a shallow depression in the river bottom surrounded by stones, into which they lay their eggs. Egg laying is preceded by a type of 'copulation' in which the female and male intertwine, the male attaching himself to the female with his sucker. Eggs and sperm are shed into the water, and the fertilized eggs are covered in sand for protection by movement of the female's tail fin. The adults die after spawning. The spherical eggs are about 1 mm in diameter and enclosed in delicate membranes. They are mesolecithal, and the early stages of development are not unlike those of the frog. The larvae hatch in about 10 to 15 days in temperate climates as tiny ammocoetes, about 7 mm long. The ammocoete has a rather worm-like appearance, totally unlike that of the adult, and has rudimentary eyes, a horseshoe-shaped mouth surrounded by barbels, and no teeth. The main photoreceptors are in the tail, and this ensures the ammocoete remains buried in the sand. Ammocoetes remain in the nest for about 30 days and then leave to drift downstream and search for a suitable site in which to bury themselves again in sand or mud. They remain in their burrows for three to five years, and only leave them if it becomes necessary to change their feeding site. During this period the ammocoete larva grows to about 1·7 m in length and in many respects has a similar mode of life to amphioxus. It is a filter feeder, and lives on minute organisms or organic detritus. It has seven small, round gill openings which lie within a branchial groove and are covered by flaps of skin which act as valves. Water is prevented from leaving through the mouth by a specialized flap of skin, the velum, and the water current (created by muscular action rather than by a ciliary beat as in amphioxus) is therefore normally one way. The mouth cavity and gills are surrounded by a branchial basket of cartilage, and the associated muscles compress the cavities to expel water. Expansion of the chambers is passive, by elastic recoil of the branchial basket.

Food particles are carried into the mouth by the feeding current, and the barbels act as a sieve which excludes large particles. The particles are trapped in strings of mucus secreted by the endostyle and swept into the stomach by ciliary action.

Larval life lasts for between three and five years and is followed by metamorphosis which takes about two months. The mouth contracts and a sucker, tongue and teeth develop; the branchial groove is lost. Internally, as well, the animal undergoes extensive modification; the endostyle develops as a thyroid gland and the kidneys, alimentary canal, and skeleton take on the adult form. The eyes appear, and the gills develop as sacs which open into the pharyngeal chamber. The lamprey then moves down to the sea to take up its adult mode of life as an ectoparasite of fish. The duration of the adult life is unknown, but the lamprey only returns to the rivers to spawn once. During this period in freshwater it does not feed, but lives on fat accumulated in the muscles and under the skin.

AGNATHAN DIVERSITY

The living agnatha

Modern cyclostomes are a specialized remnant of the class Agnatha which survive in competition with jawed vertebrates because they are well adapted to a highly restricted mode of life in which they have few rivals. Although many lampreys have the life style described for *Lampetra fluviatilis* spending their adult life in the sea, others are entirely freshwater, migrating only a few miles upstream to reproduce. A British example is the small brook lamprey, *Lampetra planeri*. In contrast, the hagfishes (for example *Myxine* and *Bdellostoma*) are entirely marine,

normally living at depths below 100 m, and are primarily scavengers. However they will attack disabled fishes, including those which are hooked or netted, and are therefore a nuisance to commercial fisheries.

In hagfishes, the method of feeding is totally different from that of lampreys; they cut a hole through the skin of a victim and eat out the insides, leaving little but the skeleton and skin behind. Since they have no sucker and no means of gripping their prey, their diet is restricted to either disabled and slow-moving prey or to dead animals. The mouth is surrounded by six cartilage-supported tentacles which are used to find food. Hagfish have a remarkable feeding apparatus consisting of a group of jaw-like cartilages which close the mouth sideways. The main cartilage (dental plate

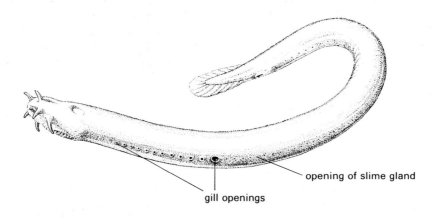

Bdellostoma

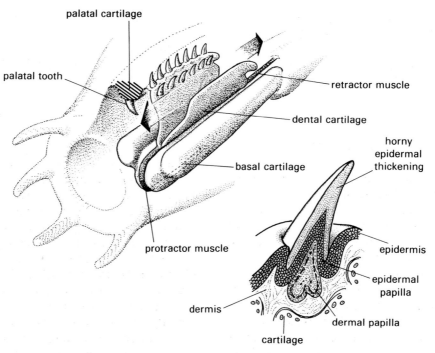

The jaw mechanism of *Myxine*

cartilage) is a flexible, trough-shaped structure with 2 rows of keratinised teeth and can be opened and closed like the leaves of a book. A palatal cartilage lies in the roof of the mouth and carries a single tooth. The dental plate is supported by a basal cartilage along which it can be pulled by the action of protractor and retractor muscles; in this way it can be flicked in and out of the mouth. As it is pulled out of the mouth it opens. Since the teeth point backwards the flicking action tears pieces off the prey while the palatal tooth holds on to the prey and prevents food from being pushed out of the mouth. Once the food enters the gut it becomes enclosed by a bag of mucus secreted by the gut wall. The food is digested within this structure, which is permeable both to enzymes and digested food, and undigested food is excreted in the bag. The significance of this method of digestion is unclear. An interesting comparison can be made between the thick folded gut wall of the hagfish, with a solid diet, and the thin-walled gut of the lamprey, which feeds on liquid.

Like the lampreys, the hagfish has a single nostril but rather than leading into a blind-ending sac it opens into the mouth cavity. There are between 5 and 15 pairs of gills which open directly into the pharynx and, in *Myxine*, to the exterior through a common duct and opening on each side. The respiratory current is created chiefly by a muscular velum at the junction of the nasal tube and the mouth cavity but also by muscular action of the gill pouch walls. At rest the velum is rolled up into a scroll and when active is alternately rolled and unrolled, pushing water over the gills. Other anatomical peculiarities of hagfish include simple sense organs (the eyes are degenerate and covered by skin, the ear has a single semicircular canal with two neuromasts) and kidneys of a primitive holonephric type.

Little is known about hagfish breeding behaviour. They lay large yolky eggs, about 20 mm long, which are roughly spindle-shaped and enclosed in tough horny capsules. At either end there are large hooks and the eggs are often linked in strings by these. Small, but completely formed hagfish hatch from the eggs.

Fossil agnathans: the ostracoderms

There are four extinct agnathan orders (Heterostraci, Osteostraci, Anaspida and Thelodonti) which are known collectively as the ostracoderms, a name derived from the Greek word meaning 'earthernware skin' and referring to their armour of bony plates or thick scales. The earliest agnathans, and also the earliest known vertebrates, are members of the Heterostraci, (different shells) but unfortunately more of the early (Ordovician) fossils are complete, and knowledge of the group is based on later specimens. Their armour consisted of large plates over the head with smaller plates over the rest of the body. Their eyes were set far apart on the sides of the head, and the pineal eye did not pierce the top of the head. There were between five and nine pairs of gills and the mouth was a

transverse slit on the under surface of the head with paired nostrils above it. The mouth slit was bounded by a series of flexible plates of bone which may have been used for nibbling or scraping or, in life, been joined by a membrane to form a scoop. Early Heterostraci have a generally rounded body shape, and in the absence of any form of lateral stabilisers, must have been rather erratic swimmers. Their rigid trunk and neck, combined with the position of the mouth suggests a sluggish bottom dwelling mode of life as scavengers or detritus feeders in the shallow seas in which they lived. Later evolutionary trends in the Heterostraci led to improved locomotion, with the development of lateral wing-like projections (cornua) from the side of the head shield which may have acted as hydrofoils, producing lift when the animal swam. Some, with dorsally directed mouths and a much reduced skeleton, were probably surface feeders, which lived on plankton. Others were clearly benthic and showed ventral flattening, and the development of teeth on their oral plates. Later heterostracans radiated into fresh water.

Pteraspis (reconstruction)

The Osteostraci (bony shells) had a trunk that was triangular in cross section and protected as far as the pectoral region by a solid carapace of bone pierced by the eyes, pineal eyes and a single nostril. The undersurface was flattened, again suggesting a bottom living habitat, and many had a pair of paddle-like flippers in the position of pectoral fins. A feature of all ostracoderms is the possession of polygonal bony plates behind the eyes and at the carapace margins; their functional importance is uncertain; though they have been interpreted as electric organs and, more recently, as areas sensitive to water vibration. The osteostracans were diverse and abundant in the Devonian, and survived the emergence of the first jawed vertebrates; possibly those unique structures contributed to their success.

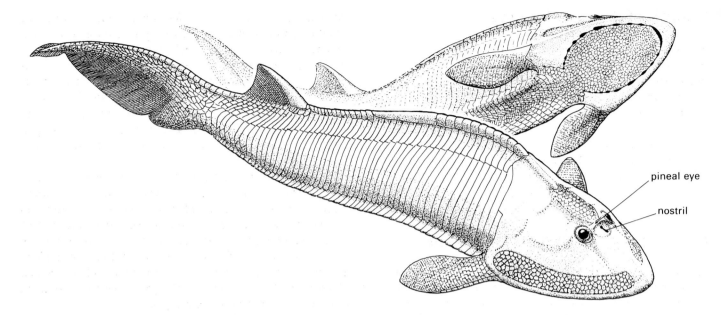

Hemicyclaspis, an osteostracon (reconstruction)

In contrast to the solid carapace over the dorsal surface of the head, the ventral surface was covered by a mosaic of small plates, probably to allow sufficient flexibility in the pharyngeal and gill region to create both respiratory and feeding currents.

The internal anatomy of the osteostracans is better known than that of other Palaeozoic agnathans. They had a cartilaginous brain case, whose inner surface was covered with bone. The bony layer extended to the walls of channels through the cartilage and we can therefore trace the shape of the brain and the passage of nerves and blood vessels in considerable detail. The anatomy of the brain and nervous system shows many similarities with that of the modern lamprey, suggesting a close evolutionary relationship. For example in both groups there are two semicircular canals in each ear, and a similar olfactory organ, and both have numerous small, round gill openings.

Osteostracans had a tail in which the dorsal lobe was both larger and stiffer than the ventral lobe. This type of tail is described as heterocercal and its significance is discussed in more detail in the following section. Generally it provides considerable lift, and would have increased the mobility of osteostracans; it contrasts with the hypocercal tail characteristic of Heterostraci, in which the lower lobe is disproportionately large and contains the main axial supporting element.

The anaspids ('shieldless ones') which occur in both Silurian and Devonian deposits were entirely freshwater animals. Like the osteostraci they had a single nasal opening anterior to the pineal foramen, but as their name suggests, they lacked the heavy bony carapace. The posterior part of their body, like that of the Osteostraci, was covered by a complex of small scales and the head was similarly covered or naked. They had flexible slender bodies and stabilising dorsal, anal and lateral fins, suggesting they were far more

Jamoytius (reconstruction)

agile than other agnathans and could escape predators more easily. They had small mouths and were probably detritus feeders.

The Thelodonti are the least well known fossil agnathans. They are known from isolated teeth from the Ordovician, but no articulated specimens are known from before the mid-Silurian and by the mid-Devonian the group was extinct. Like the heterostracans (with which they are sometimes grouped) they had lateral eyes, a pineal opening, and a hypocercal tail. They were dorsoventrally flattened anteriorly but laterally compressed posteriorly and were probably bottom-dwelling detritus feeders. They were small and lightly armoured and must have been more agile than the Heterostraci. Unlike these, they did not radiate into freshwater and are known only from shallow estuarine deposits.

Synopsis of the class Agnatha

The ostracoderms:
Order Heterostraci (*Pteraspis*)
Order Osteostraci (*Hemicyclaspis*)
Order Anaspida (*Jamoytius*)
Order Thelodonti

Order Cyclostomata, the living Agnatha
    Family Petromyzonidae (lampreys)
    Family Myxinidae (hagfishes)

Introduction to the gnathostomes

The majority of vertebrates have jaws and are sometimes discussed together as the Gnathostoma (jawed mouths). It is now generally accepted that jaws evolved as a modification of two anterior pairs of gill arches. The first pair (mandibular arch) was transformed into a pair of biting structures consisting of an upper palatoquadrate cartilage and a lower Meckel's cartilage. The second pair (hyoid arch) became modified as a supporting structure for the upper jaw, and for the tongue. A condition similar to this is found in modern cartilaginous fishes. In bony fish (as in some more advanced vertebrates) the primary upper and lower jaws have been extensively modified, and dermal bones are present in addition to the cartilage bones derived from the gill arches. Meckel's cartilage is surrounded by dermal bone but retains its function as a lower jaw. The palatoquadrate cartilage, however, ceases to function as a jaw but is transformed as the roof of the mouth, its biting function being taken over by dermal bones, the maxillae and premaxillae.

The development of jaws was an important step in vertebrate evolution enabling the animals to adapt to a wide range of diets and modes of life. The possession of movable strengthening structures, surrounding the mouth and operated by muscles, allowed vertebrates to grasp and manipulate food very effectively. With associated development of teeth, jaws made it possible for vertebrates to feed on larger food items which must first be cut up into swallowable pieces, or on harder substances which must be ground down into a more assimilable form. It is hardly surprising that throughout the vertebrates jaws show greater diversity than any other part of the body. To the systematic zoologist, and particularly to the paleontologist, jaws and teeth are of enormous value in identifying and classifying animals.

Among the early jawed forms were the placoderms, an extinct group of shark-like fishes known from the Silurian, Devonian and lower Carboniferous and characterised by an armour of bony plates. In placoderms the armour was concentrated in an anterior cephalothorax, and covered the head and thorax, with the armour of the two regions meeting at a hinge; the posterior part of the body was naked or covered by small scales. Many had peculiar, hollow, arthropod-like pectoral appendages.

Placoderms were long regarded as primitive bony fishes and were classified with that group, but present day taxonomists classify them with the Chondrichthyes as members of the class Elasmobranchiomorphii. It may seem absurd to group animals which are so conspicuously bony with others characterised by a complete absence of bone. Briefly the justification lies in the similarity of their snout anatomy, and the presence of intromittent organs (claspers) in some placoderms; these are highly unusual structures only otherwise found in cartilaginous fishes, and not found in bony fish. The placoderms were rather grotesque animals, totally unlike modern fishes, and may be regarded as an early gnathostome experiment which failed!

Paired appendages were absent in ancestral jawless vertebrates and true pectoral and pelvic appendages are only found in more advanced gnathostomes. The appendages of placoderms differ in many respects from those of other gnathostomes with variation even in number. The median fins which occur along the dorsal and ventral mid-lines arose as stabilising structures and the paired appendages probably originated as lateral stabilisers. Since

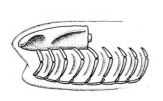

Hypothetical jawless form

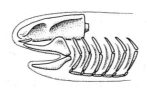

Intermediate stage showing
modified mandibular arch

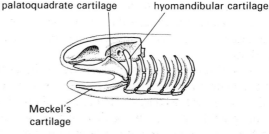

palatoquadrate cartilage   hyomandibular cartilage

Meckel's
cartilage

Primitive gnathostome with
modified mandibular
and hyoid arches

The evolution of vertebrate jaws

both median and paired fins have a similar structure with a central skeleton and a muscle layer on either side it is reasonable to assume that both originated in the same way as adaptations to improve the efficiency of swimming.

CLASS ELASMOBRANCHIOMORPHII
Subclass Chondrichthyes

From the time of their first appearance in the Devonian the cartilaginous fishes have been a marine group and remain so today, with few exceptions. Their origin is obscure; because they have a cartilaginous skeleton the fossil record is poor and the earliest traces of the group take the form of isolated spines, teeth and scales. Recent classifications divide the chondrichthyes into two groups: Elasmobranchii and Holocephali. The modern Holocephali are a remnant of a once important group, represented by only six genera of so-called chimaeras or rabbit-fishes. The Elasmobranchii fall into two broad categories: the skates and rays, which are dorsoventrally flattened and frequently bottom dwelling forms, and the free swimming, typically mid-water forms of which the sharks are the best known. The dogfish, a small shark, has been selected for more detailed study.

Scyliorhinus canicula: the dogfish

The dogfish, or lesser spotted dogfish, to give it its full common name, is known for its use in dissection, and also for appearance in the fish shop where it is sold as rock salmon. It is one of the commonest small sharks of British waters, growing to about 760 mm in length, and is familiar to sea anglers and commercial fishermen alike. Its brown spotted skin is extremely rough and gives the fish the alternative common name of 'rough hound'.

Scyliorhinus canicula can be found in all types of coastal waters at depths of between 2 and 110 metres, though it is particularly common between 18 and 110 metres over gravel or sandy bottoms. Like the vast majority of free swimming fish it has a streamlined and elongated (fusiform) body shape. This is smoothly rounded at its anterior end, expanding to a maximum a third of the way along its body and tapering to the tail. This shape is very important for movement in a dense medium since it reduces turbulence and enables the animal to move through the water with minimum expenditure of energy. The tail, as the main propulsive organ, is well developed and powerful, and the head merges imperceptibly into the trunk. The dogfish has well developed eyes, situated on the side of the head, and nostrils, associated with a well developed sense of smell, lying on the ventral surface of the head immediately anterior to the mouth to which they are connected by a pair of oronasal grooves. A number of canals ramify over the surface of the head, opening to the surface by pores. They form part of the acoustico-lateralis system, which is the means whereby the dogfish detects vibrations in water. A long canal, the lateral line canal, extends the whole length of the body, almost to the tip of the tail on either side. Additional sense organs, the ampullae, in the form of flask shaped depressions, are scattered over the surface of the snout.

Immediately behind the eye is a small modified gill cleft, the spiracle, and five functional gill clefts.

The dogfish has several fins. The median (unpaired) fins are represented by anterior and posterior dorsal fins, a ventral (anal) fin, and a bilobed tail or caudal fin. In addition the dogfish has two sets of paired fins, pectoral and pelvic, which are well-developed broad-based structures. The pectoral fins project horizontally from the body, just behind the gill region while the smaller pelvic fins lie on either side of the median cloaca on the ventral surface. In the male dogfish paired intromittent structures, the claspers, are associated with the pelvic fins.

The integumentary system

Scales are characteristically present in fish skin though their structure and origin vary considerably. Those of the dogfish are known as placoid scales or dermal denticles and as the latter name suggests, are dermal in origin and extend out through the epidermis. Placoid scales are constantly being worn out and replaced by new scales growing out from the dermis and a vertical section through dogfish skin therefore reveals scales at all stages of development. They first appear as a condensation of mesenchyme cells (dermal papilla) which later becomes overlain by a layer of epidermal cells. Each scale consists of a bone-like basal plate, embedded in the dermis, and a projecting denticle composed of an outer shiny layer of vitrodentine, which overlies a dentine tubercle. The tubercle surrounds a pulp cavity containing

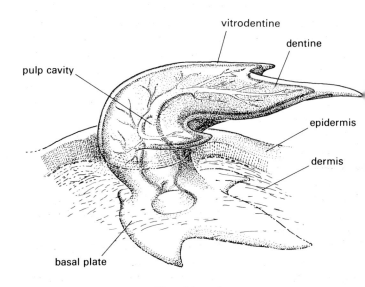

Placoid scale

blood vessels, nerves, and the cells which produce dentine. The denticle shape varies in different parts of the body; the dorsal scales have three points, the ventral scales have one, while those overlying the jaws are large and strong with five points and function as teeth.

The denticles are very numerous and are set close together, their points directed backwards; if the skin is stroked from head to tail it feels smooth, while if it is stroked in the opposite direction it is distinctly rough. The scales are so strong that dried dogfish or shark skin, known as shagreen, is used for polishing.

The structure of the integument itself is similar to that described for the cyclostomes with a thin outer epidermis composed of several layers of simple cells, and a dermis of connective tissue, with muscle fibres, blood vessels and nerves. Unicellular glands (modified single cells) are scattered among the epidermal cells, and multicellular saccular glands are also present, formed from ingrowths of the stratum germinativum into the dermis. Both types of gland secrete a protein, mucin, which forms a slimy viscid material, mucus, when mixed with water.

The skeleto-muscular system and locomotion

The dogfish skeleton, like that of the lamprey, is entirely cartilaginous. The notochord is functionally replaced by the vertebral column and is itself greatly modified by the presence of the vertebrae; it fills the regions between the centra of successive vertebra as a jelly-like substance and also extends within them as a constricted cord. If removed intact the dogfish notochord would therefore look like a string of beads. In addition a series of fibrous pads, derived from the notochordal sheath, lies between successive vertebrae. The vertebral centra are spool-shaped structures and each has a conical depression on its anterior and posterior surface, a condition termed amphicoelous. They are derived from the embryonic notochord which is invaded by cartilage secreting cells (chondroblasts). Chondroblasts also lay down a series of cartilaginous plates extending from the centra. In the trunk region of the adult dogfish each vertebra consists of a centrum bearing a pair of ventro-lateral transverse processes (basopophyses) derived from the basiventral plates. The distal portions of the transverse processes form slender ribs. The spinal cord is protected by a neural arch made up of a neural plate on each side and two median neural spines on top. The space between successive neural plates is occupied by an intervertebral neural plate. In the tail region the transverse processes are modified to form a haemal arch which surrounds the caudal artery and vein. The dogfish tail has two vertebrae per body segment (diplospondyly).

The chondrocranium consists of an incompletely roofed box surrounding the brain, with capsules (olfactory, auditory, and orbits) surrounding the major sense organs. On either side the upper jaw (palatoquadrate bar) is attached to the chondrocranium by a muscular ligament about the middle of its length, and is connected to its fellow anteriorly. The two halves of the lower jaw (Meckel's cartilages) are also joined anteriorly. The posterior ends of both jaws are slung from the back of the chondrocranium via the hyomandibular cartilages, which fit into depressions in the auditory capsules of the chondrocranium and are attached to it by ligaments. This type of jaw suspension is termed hyostyly.

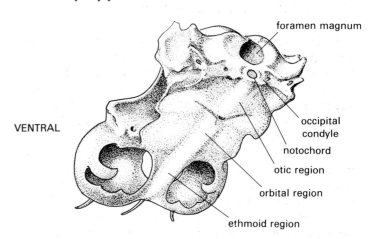

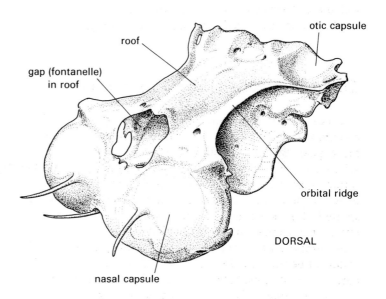

Dogfish cranial skeleton (chondrocranium)

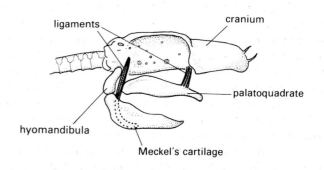

Hyostylic jaw suspension

The supporting skeleton for each pair of gill arches follows the same basic plan, which consists of a ventral median basi-branchial cartilage to which are attached on each side hypo-, cerato-, epi- and pharyngobranchial cartilages. These are angled to each other with the basibranchial cartilage directed posteriorly. However in the dogfish a considerable amount of fusion has taken place so that the basibranchials form a single cartilage, and the hypobranchials of the first branchial arch are attached to the basihyal; the pharyngobranchials of arches four and five are also fused to each other. Cartilaginous gill rays arise from the gill arches. The gill arch cartilages are associated with muscles which can act to bring them together or move them apart.

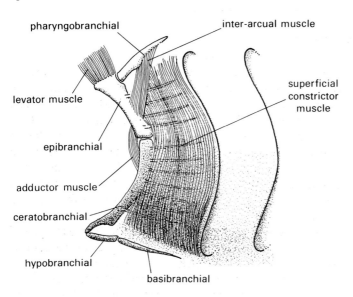

pharyngobranchial

inter-arcual muscle

levator muscle

superficial constrictor muscle

epibranchial

adductor muscle

ceratobranchial

hypobranchial

basibranchial

Gill arch skeleton and musculature

Both pectoral and pelvic girdles are present. The pectoral girdle is a hoop of cartilage, embedded in the ventrolateral body muscles. In its mid-ventral region it has a well defined pericardial depression which contains the heart. Each half hoop is sometimes referred to as a scapulo-coracoid. The fins are supported by a series of horny distal structures, called dermotrichia, and a more proximal series of plate-like radial cartilages called radialia. The radialia articulate with the girdle by three basal cartilages, the pro-, meso-, and metapterygia. The pelvic girdle is embedded in the ventral body muscles and consists of a bar of cartilage, the ischiopubic bar. The fins articulate with the pelvic girdle by a single basal cartilage on either side, the basipterygium, and are themselves supported by a uniform series of basal cartilages and dermotrichia. Associated with each basipterygium in the male there is a backwardly directed, elongated cartilage which supports each clasper, and the fins are joined to each other posteriorly.

The appendicular musculature is simple, consisting of small masses of opposed muscle running from the girdle to the base of the fin. In addition coracohyoid, coracomandibular, and coracobranchial muscles insert onto the ventral portion of the pectoral girdle. These muscles arise from the skeleton of the mandibular, hyoid and branchial arches and are responsible for enlarging the cavity of the pharynx and opening the mouth.

Locomotion

The dogfish swims by means of lateral undulations of its body and tail (axial locomotion) which form a series of waves passing from anterior to posterior. These waves are caused by serial alternating contraction and relaxation of the right and left myotomes. The vertebral column, while being flexible, prevents the body shortening, which would be the normal result of such contraction. The myotomes are composed of longitudinal muscle fibres arranged in a sharply angled shape, like a 'W' lying on its side, and each is attached to only one vertebra but overlaps a number of segments. Such an arrangement allows waves of contraction to pass along the metameric chain so that body undulation takes place in smooth waves. If the myotomes were straight, perpendicular to the longitudinal body axis, and not overlapping, the undulations would pass along the body in a series of jerks. The myotomes are separated into dorsal (epaxial) and ventral (hypoaxial) portions, separated by a lateral longitudinal connective tissue septum.

The forward thrust is provided principally by the tail and caudal fins as they are swung from side to side across the line of motion. As the caudal fin moves it acts as an inclined plane moving sideways through the water, and displacing it to the rear. The other undulations function in the same way, but to a lesser extent. Of the fins, only the caudal fin provides forward thrust but the other fins are extremely important in locomotion in providing stability and controlling the direction of movement. An animal, such as a fish, which is suspended in a fluid medium can move about three primary axes: longitudinal, horizontal and vertical. Deviation about the longitudinal axis is defined as rolling; about the horizontal axis as pitching, and about the vertical axis as yawing. The median fins, both dorsal and ventral, give stability by reducing both rolling and yawing, while the pectoral fins contribute considerably to the stability of the dogfish in the horizontal plane. The pelvic fins have little effect on the stability of the animal, but may contribute to a reduction of roll. The body shape is at its thickest towards the anterior end which itself reduces any tendency to swing from side to side as it swims.

Dogfish are more dense than water and so tend to sink slowly to the bottom. The tail fin produces an upward as well as a forward thrust as the lower (hypochordal) lobe is larger than the upper (epichordal) lobe and the tail fin bends as it is swept sideways. This type of tail is termed heterocercal and is normally associated with the generation of lift. However, lift from the tail will tend to tip the anterior part of the body downwards. The pectoral fins act as hydroplanes and give an upward thrust which serves to counteract this motion in horizontal swimming and their angle of inclination can be altered to vary the amount of lift

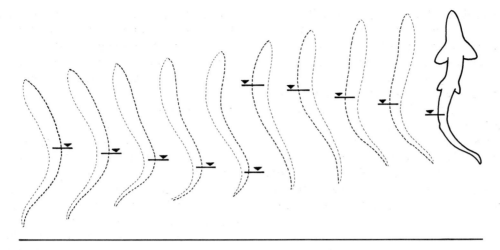

Dogfish swimming movements

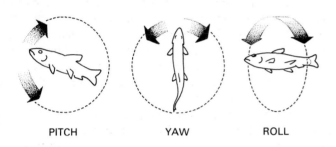

PITCH   YAW   ROLL

Deviation in a fluid medium

produced. Thus, in order to keep from sinking, the dogfish must keep swimming. The wide based pectoral fins are capable of only limited movement in the vertical plane; a dogfish cannot stop by increasing its resistance in the water along the axis of movement (braking), but must move to the side of an obstacle. U-turns are also a slow process.

The digestive system and nutrition

The dogfish is an active predator which hunts using its eyes, nostrils, and well developed acoustico-lateralis system. It is chiefly a bottom dweller, where it feeds on molluscs, crustaceans and small bottom living fish such as gobies, gurnards and dabs, but also feeds on surface living forms such as pilchards or mackerel. It has a battery of sharp teeth which serve both for biting and to prevent the escape of living prey. The teeth are enlarged placoid scales and form continuously in areas of actively growing skin inside the mouth immediately behind the upper and lower jaws. As the skin grows, the teeth are pushed over the edge of the jaw, and at the same time they increase considerably in size. During use they become worn, but by this time they have passed over the jaw surface onto its outer edge where they are resorbed. The snout region of the head is prolonged as a rostrum and the large mouth lies on the ventral surface of the head. The jaw muscles, which are simple and are all derived from the mandibular segment, show the gnathostome condition in a fairly unmodified form. The muscles which open the mouth are weak and extend ventrally between the lower jaw and the shoulder girdle. In addition a levator palatoquadrate extends between the cranium and the upper jaw which it lifts and binds to the cranium. The largest jaw muscle in the dogfish is the adductor mandibulae which extends between the cranium and the lower jaw and contracts to close the mouth.

The mouth opens into the buccal cavity. There is a pad of tissue, sometimes referred to as the tongue, in the floor of the buccal cavity which is used to press the food backwards and thus aids in swallowing. Posteriorly the pharynx opens into a short, wide oesophagus whose wall is thrown into numerous longitudinal folds, which effectively close the oesophagus except during swallowing. These folds allow considerable rapid dilation of the oesophagus to take large prey and food can be passed rapidly into the stomach. There is practically no distinction between the oseophagus and the stomach, which is basically an expanded region of the gut for food storage. It is J-shaped with a wider cardiac limb and a smaller, narrower pyloric limb. The muscular walls contract to break down and churn the food thoroughly; a pyloric sphincter muscle controls the exit of food from the stomach. The duodenum is short, and receives ducts from the liver (bile duct) and pancreas (pancreatic duct). The vertebrate intestine shows many modifications which increase the surface area without increasing the length. In the dogfish the small intestine is short, wide, and straight, but the lining has become partly separated from the wall and is twisted to form a spiral valve, a structure characteristic of (but not restricted to) elasmobranchs. The large intestine is a short passage between the intestine and the cloaca. A small brown rectal gland opens into the alimentary tract by a duct at the junction of the large and small intestine. This secretes a highly concentrated solution of sodium chloride and thus removes excess salt taken in with the food from the blood.

Respiratory system

As an aquatic animal the dogfish depends on oxygen dissolved in water for its respiratory needs and the surfaces for gaseous exchange are provided by numerous gill filaments, which are thin-walled epithelial extensions each containing a vascular network. Each gill filament (or lamella) is supported by cartilaginous gill rays and bears a series of folds, the secondary filaments (lamellae) which together greatly increase their total surface area. The fine structure of the filaments allows the blood to come into very close contact with the water and the system is so efficient that up to 80% of the oxygen dissolved in the water can be obtained by the fish. The vascular system of the gill lamellae is an extensive series of blood sinuses which are supplied by the afferent branchial vessels and drained via the efferent branchial vessels.

The side walls of the pharynx are perforated by five gill clefts and a spiracle. The partitions between the clefts (the interbranchial septae) are fairly thick and are supported by the gill arch cartilages. Each cleft consists of an internal pharyngeal opening, a pouch, and an external gill cleft which is visible on the side of the head. The interbranchial septum tapers and bends posteriorly at its distal end to form a flap or valve which can close the external gill cleft. Two groups of gill lamellae (hemibranchs) arise from each branchial arch so that each of the first four gill pouches has a hemibranch on its anterior and posterior walls while the last pouch has a hemibranch on its anterior wall only. The spiracle is a vestigial gill (pseudobranch) which has no gas exchange function.

To maintain a continuous supply of oxygenated water over the gill lamellae the pharynx functions as a force pump. The pharyngeal floor is lowered by contraction of the coracohyoid and coracobranchial muscles which extend (as their names suggest) between the hyoid and branchial arches and the shoulder girdle. The volume of the buccopharyngeal cavity is thereby increased and the internal pressure is lowered, thus drawing water into the cavity through the opened mouth and the spiracle. The valves which cover the gill clefts are closed since the pressure outside the dogfish is greater than that inside the buccopharyngeal cavity. In the second phase the mouth is closed and sealed by the lips, and the floor of the pharynx returns to its original position; the buccopharyngeal cavity is further reduced in size by contraction of a system of superficial constrictor muscles and water is therefore forced over the gill lamellae and out through the external gill clefts.

Circulatory system

The blood system of the dogfish, which is broadly similar to that of the lamprey, may be regarded as typical of aquatic gnathostomes. The heart lies within the pericardium and is protected ventrally by the coracoid region of the pectoral girdle. It is four-chambered and S-shaped, so that the sinus venosus and atrium lie ventral to the ventricle and conus arteriosus. The ducts of Cuvier, which return venous blood to the heart, are paired, and drain into the thin-walled triangular-shaped sinus venosus which is separated from the atrium by the sino-atrial valves. The atrium is rather more muscular, and the ventricle and conus arteriosus, as the main pumping chambers, are very powerfully developed. Atrio-ventricular valves separate the atrium and ventricle while the conus arteriosus has two pairs of semi-lunar valves; the combined effect of the various valves is to maintain a unidirectional anterior blood flow through the heart.

The conus arteriosus perforates the anterior wall of the pericardium and carries blood into the ventral aorta which gives off an aortic arch between each pair of gill slits. Potentially in gnathostomes there are six aortic arches, normally given the numbers I to VI from anterior to posterior. In dogfish the first gill slit is modified as a spiracle and the aortic arch in front of it is lost leaving five arches, II to VI. The aortic arches are interrupted by the gill sinuses and thus divided into afferent and efferent branchial arteries as in the lamprey. The arrangement of the rest of the arterial system is also similar to that already described for the lamprey. A cardiac artery supplies the heart muscle.

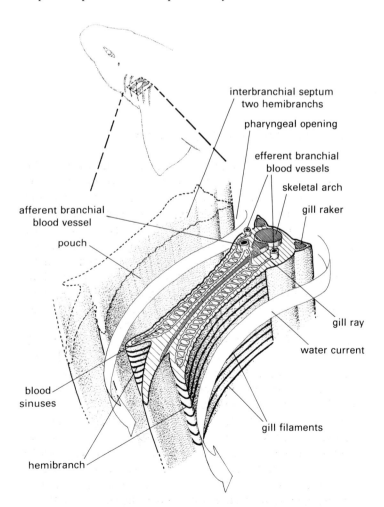

interbranchial septum
two hemibranchs

pharyngeal opening

efferent branchial
blood vessels

skeletal arch

gill raker

afferent branchial
blood vessel

pouch

gill ray

water current

blood
sinuses

gill filaments

hemibranch

Section of dogfish gill (diagrammatic)

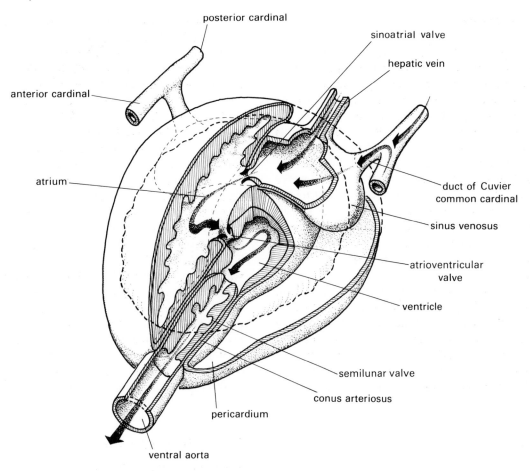

posterior cardinal

sinoatrial valve

hepatic vein

anterior cardinal

atrium

duct of Cuvier
common cardinal

sinus venosus

atrioventricular
valve

ventricle

semilunar valve

conus arteriosus

pericardium

ventral aorta

Dogfish : heart and principal vessels (diagrammatic)

In addition a hypobranchial artery, from the second or fourth efferent branchial arteries, is present in many sharks and has been described in some dogfish. A series of arteries arises from the posterior dorsal aorta to supply the body wall and the viscera. The head is supplied by a complex group of arterial branches from the paired anterior dorsal aortae and efferent branchial arteries of which the internal carotid arteries are most notable.

The venous system consists mainly of blood sinuses, but again follows the basic vertebrate plan. The caudal vein divides anteriorly to give two renal portal veins which pass through the kidneys into the posterior cardinal sinuses. The latter also drain the body wall and eventually join the anterior cardinal sinuses which drain the head, to enter the sinus venosus via the common cardinals (ducts of Cuvier). The head is also drained by the inferior jugular sinuses, which open into the sinus venosus. Veins from various parts of the alimentary canal combine to form the hepatic portal vein which transports blood to the liver. Beyond the liver this blood is collected by the hepatic sinus which empties directly into the heart, into the middle of the base of the sinus venosus.

There is also a simple lymphatic system, consisting of a series of thin-walled longitudinal sinuses which drain into the cardinal sinuses but have no arterial associations.

Excretion and osmoregulation

In the dogfish as in the majority of living gnathostomes, the excretory and reproductive systems are connected. The organization of the excretory tubules and ducts follows the basic anamniote plan, although with some modification. The dogfish has paired mesonephric kidneys, which in the male are divided into anterior genital and posterior excretory regions. In early development the nephrons have a segmental arrangement but this is obscured in the adult. The nephrons of each kidney drain into five collecting ducts which join to form paired channels opening to the outside through the urinogenital sinus and cloaca. By comparison with the lamprey these urinary ducts of the dogfish represent new structures, and the archinephric duct has an entirely genital function except in the region where it forms the urinogenital sinus. The kidneys and ducts of female dogfish are less complex, but are differentiated into anterior and posterior regions, the posterior region being the functional kidney. The archinephric ducts begin anteriorly and in the posterior (excretory) portion dilate and join to form a median urinary sinus into which drain a series of collecting ducts from the kidney. The sinus opens into a urinogenital sinus, which also receives the genital ducts at a urinary papilla.

The renal corpuscles of individual nephrons are large in the dogfish, as in all elasmobranchs, and the animal has a relatively large urine output, more typical of freshwater fishes than of marine forms. The osmotic concentration of elasmobranch body fluids is below that of sea water. A proper water balance is maintained in two ways: by the action of the salt excreting rectal gland already described, and by the retention of a high concentration of nitrogenous waste in the blood in the form of urea. The importance of this adaptation to the dogfish is emphasised if it is compared with *Pomatotrygon*, one of the few freshwater elasmobranchs. In these forms the blood contains 2 to 3 mg of urea per 100 ml as opposed to 300 to 1300 mg per 100 ml in marine elasmobranchs.

Nervous system and sense organs

The dogfish nervous system follows the characteristic vertebrate plan already described, apart from various features adapting it for its particular mode of life. Since it is a bottom living form it is adapted to low light intensities, but depends to a considerable extent on its sense of smell for hunting; it has enormous olfactory organs and correspondingly well developed olfactory lobes in the brain. The eyes are adapted for near vision in dim light, with no cones in the retina. The lens is an incompressible spherical structure of the typical fish type, light entering the eye being focussed onto the retina by moving the lens away from, or towards the cornea by the action of a protractor lentis muscle. When the muscle is relaxed, the lens is held close to the cornea so that the eye is adapted for near vision. An ordinary spherical lens would give an extremely distorted image, but in fact the fish lens does not have a uniform refractive index. Instead it shows a steady increase towards the centre and the focal length of the lens is therefore shorter than it would be if the refractive index was uniform throughout. This is an important adaptation for aquatic animals since the refractive index of the cornea is only slightly greater than that of water. Thus almost all of the focussing of light onto the retina must be done by the lens, which must therefore be nearly spherical. The optic lobes and pallium of the brain are of moderate size.

The acoustico-lateralis system (lateral line system) is of the typical fish structure and is well developed. In addition, dogfish, like other elasmobranchs, have scattered sense organs on the snout known as the ampullae of Lorenzi. These consist of tubules filled with gelatinous material and containing globular sensory cells with sensory hairs. They permit the dogfish to detect the electrical stimuli set up by the electric organs of some other fishes for detection of prey or navigation, and may also react to temperature changes. The ear, as in other fishes, is represented by a membraneous labyrinth, composed of a sacculus, and a utriculus with three semicircular canals.

The dogfish has a typical pattern of cranial nerves as outlined in the previous chapter. In addition to cranial nerves I-IX it has a terminalis nerve (0). This is a tiny nerve which runs parallel to the olfactory nerve and is apparently sensory in function but is not concerned with the sense of smell. It has been interpreted as an evolutionary remnant of the innervation in front of the first pair of gill pouches in ancestral vertebrates, and is present in all vertebrates except cyclostomes and birds.

Reproduction

Egg laying occurs throughout the year but is more frequent in the winter and spring. The eggs have large yolks, and are enclosed in horny brown capsules made of keratin. These are about 50 mm long and oblong in shape, with long elastic tendrils in the corners, which become entangled round seaweed and stones and so anchor the egg as it develops. Empty capsules, known as 'mermaid's purses', are often found on beaches along the tide mark. Dogfish development is a lengthy business, and the young do not hatch until about nine months after the egg is laid; by this time they have grown to about 100 mm in length, and have a good chance of survival.

The reproductive systems of both male and female show some specialization compared with the basic vertebrate plan. The female has a single ovary, which occupies a median position but in origin is the ovary from the right hand side. However both oviducts are present as stout tubes whose anterior openings join to form a single wide opening close to the oesophagus. Posteriorly each oviduct has a swelling about a third of its length from the anterior end marking the position of the oviducal gland. This gland is an elasmobranch specialization which secretes the egg case. Other glands in the oviduct wall surround the egg with albumen before it becomes enclosed by the egg case. The oviducts open into the cloaca by separate openings at either side of the urinary papilla.

The testes of the male are elongated structures suspended in the mesentery. Sperm are transported from the testes along vasa efferentia to the vas deferens which is the genital portion of the kidney duct. At the posterior end each vas deferens becomes swollen to form a vesicula seminalis before opening into a common urinogenital sinus, which connects with the cloaca at the urinogenital papilla. Two elongated sacs, termed the sperm sacs, open out of the urinogenital sinus.

Fertilization is internal, which is an unusual feature in aquatic animals but common in elasmobranchs. The pelvic fins of the male are adapted to form a pair of intromittent organs or claspers; each of these has a medial groove, but since the edges of the claspers are rolled together they function as a single tube. The end of the tube nearest to the cloaca is termed the apopyle and leads from a heavy muscular sac, the siphon, which lies immediately below the skin in the ventral abdominal region, and is filled with sea water. During copulation, the clasper tube is filled with spermatozoa and inserted into the female cloaca.

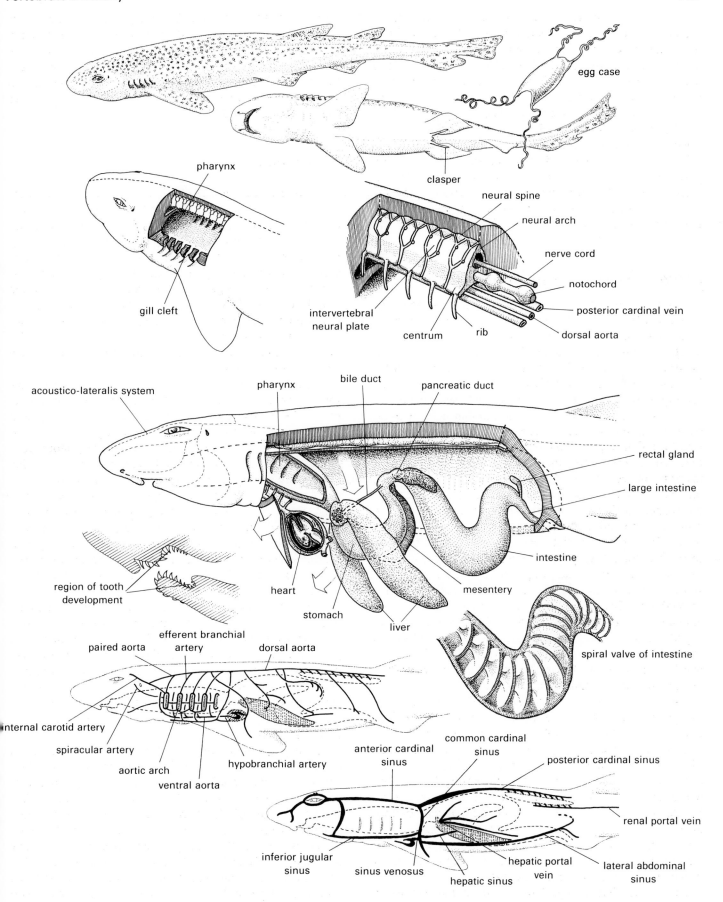

egg case

clasper

pharynx

gill cleft

neural spine

neural arch

nerve cord

notochord

posterior cardinal vein

intervertebral
neural plate

centrum

rib

dorsal aorta

acoustico-lateralis system

pharynx

bile duct

pancreatic duct

rectal gland

large intestine

intestine

mesentery

region of tooth
development

heart

stomach

liver

spiral valve of intestine

efferent branchial
artery

paired aorta

dorsal aorta

internal carotid artery

spiracular artery

aortic arch

hypobranchial artery

ventral aorta

anterior cardinal
sinus

common cardinal
sinus

posterior cardinal sinus

renal portal vein

inferior jugular
sinus

sinus venosus

hepatic sinus

hepatic portal
vein

lateral abdominal
sinus

Dissection of the dogfish, *Scyliorhynus caniculus*

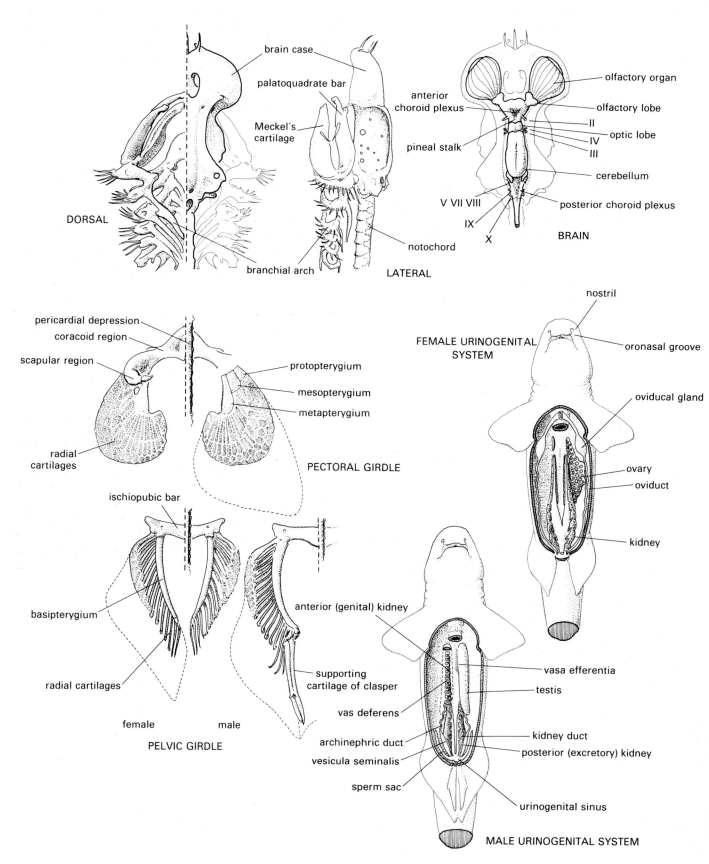

The anatomy of the dogfish, *Scyliorhinus caniculus*

Contraction of the siphon pushes sea water down the clasper tube and forces the spermatozoa into the female oviduct. Fertilization takes place in the upper part of the oviduct, above the oviducal gland.

Internal fertilization in the dogfish is an inevitable feature of a reproductive system in which the egg becomes enclosed in a tough capsule impermeable to sperm. It has already been mentioned that the dogfish adult retains high levels of urea in the blood, and that this is essential to the maintenance of its osmotic balance and hence its survival. The embryonic kidneys are at first incapable of retaining urea, and high concentrations are maintained by the egg case which is impermeable to urea. During this phase gaseous exchange takes place by diffusion through the egg case. As the kidney matures mucus plugs in the egg case dissolve and sea water is able to enter the egg case and circulate round the embryo so that oxygen can be obtained directly from the water. Thus the egg case has a complex physiological role in addition to its obvious function of protecting the embryo through its long period of development.

CHONDRICHTHYAN DIVERSITY

Living forms

The majority of cartilaginous fishes are members of the infraclass Elasmobranchii, a group which includes about 2,000 species of sharks, skates, and rays, and their extinct relatives. Within the group we can recognize three main levels of organization (two of which are only known from fossil forms) each with its own radiation. The dogfish provides examples of most of the distinguishing features of modern elasmobranchs: well developed jaws and a hyostylic jaw suspension, five pairs of gill clefts each with a separate external opening, an anterior spiracle, and double nostrils opening beneath the tip of a prolonged snout. The endoskeleton is always cartilaginous and the fins are supported by horny rays. The males have claspers. A large yolked egg is typical of many members of the group; but in others the fertilized egg develops in the oviduct and the young are born alive.

Elasmobranchs of the modern type first appeared in the Triassic, and by the Jurassic both the modern sharks (order Selachii), and the skates and rays (order Batoidea), were established and many modern genera had evolved. Modern sharks are probably the best known members of the group, as formidable predators, including some thirty species which are known to attack man, often with fatal results. Among these, the Great White Shark, or Man-eater *Carcharodon carcharias*, is one of the most dangerous, growing to a length of 6.4 m and having huge jaws with a formidable battery of sharply serrated teeth. It will feed on almost any aquatic animal; its diet includes many species of fish, other sharks among them, and also large vertebrates

such as seals, sea lions and dolphins. On the South African and Australian coasts, it often attacks swimmers and surf boarders and has been reported as having sunk small boats to get at their occupants. The stomach of one individual was found to contain a pair of trousers, three overcoats, a pair of shoes, a cow's hoof, a set of antlers, a chicken coop and twelve lobsters! The tiger shark, *Galeocerdo cuvieri*, though slightly smaller at 5.5 m, is one of the most aggressive sharks, and is responsible for many fatal attacks on humans. It normally feeds on a wide variety of marine life including fish, turtles, conch shells, crabs and porpoises, though strange items such as copper wire, leather wallets and canned salmon have been found in tiger shark stomachs. Sharks generally bite pieces out of their human prey, particularly out of the buttocks and thighs, but they also amputate limbs and sometimes the victim disappears entirely. Shark experts who advise 'do not provoke the shark in any way' are surely being ironical; however the shark expert who reduced his advice to two words 'don't bleed', was entirely serious since sharks are attracted by blood in the water which can excite them to a frenzy in which they will attack anything that moves.

Most sharks are solitary predators, feeding on whatever fish they encounter; however the thresher, *Alopias*, is a more systematic hunter which swims in packs and uses its long whip-like tail to drive the herring, mackerel, or pilchards on which it feeds into a compact shoal where they can easily be seized. Some sharks are rather sluggish bottom-dwelling forms, feeding on molluscs and crustaceans; the dogfish, *Scyliorhinus*, is an example as are the hounds (*Mustelus*) and nurse sharks (*Ginglystoma*). The frilled shark, *Chlamydoselachus* is a deep water form which lives between 200 and 1000 m and feeds on cephalopods as well as other fishes. This is one of the most primitive living sharks, and the sole surviving representative of an otherwise extinct family.

Chlamydoselachus

Despite the notorious reputation of their group, the largest living sharks are not predators but live entirely on plankton which they strain out of the water by means of specially modified gill arches. *Cetorhinus maximus*, the basking shark, grows up to 10.4 m and feeds simply by swimming around with its mouth open. *Rhincodon*, the whale shark, is related to *Carcharodon* rather than *Cetorhinus*, but has an almost exclusively planktonic diet, although it is known to take small fish. It is the largest living fish and attains lengths of up to 18 m; it produces an egg case 300 mm long.

The skates and rays, some of the commonest elasmobranchs, are specialized for life at the bottom of the ocean. They are dorsoventrally flattened, with enormous, lobe-like pectorals which are the principal means of locomotion, their caudal fins being reduced. Propulsion and steering are provided by waves of muscular contraction along the pectoral fin. The eyes are set in the top of the head and the respiratory system is modified so that water is drawn in only through the spiracles which are enlarged and occupy a dorsal position and can be closed by a special valve at expiration.

Skates and rays are primarily benthic invertebrate feeders and have a crushing, grinding dentition to cope with a diet which involves a high proportion of shelled forms such as molluscs, crustaceans and sea urchins. *Myliobatis*, the eagle ray, feeds almost exclusively on clams and oysters and has a crushing mill of flattened bar-like teeth.

Several species of rays have electric organs, which also occur in some bony fishes. In both cases they have a similar histological structure and are formed from modified muscle tissues. The electric organ consists of a stack of flattened multinucleate cells, embedded in a jelly-like material and bound together by connective tissue. Each cell is supplied by a nerve fibre and the whole forms an organic battery made effective by charge differences between the surface of the plates. In many cases the electric organs create a weak electric field around the fish which is used for navigation in poorly lit or murky water in a similar way to radar. Nearby objects distort the pattern of the electric field and the fish can detect such distortions and react accordingly. However, the electric ray, *Torpedo*, has massive electric organs and can create a discharge of up to 220 volts (as opposed to 4 volts for *Raja*) which it uses for both attack and defence.

Poison organs are found in several elasmobranchs including the spiny dogfishes (*Squalus*), sting rays (*Dasyatus*) and eagle rays. These generally consist of a stout spine associated with a poison gland in the epidermis. *Dasyatus* has a large toothed spine on its tail with which it can inflict severe wounds, and human fatalities have been recorded following attacks by the Indo-Pacific species *D. brevicaudata*. All the sting rays are fairly large and *D. brevicaudata* grows up to 4.3 m in length with a 410 mm spine.

Elasmobranch egg cases are all basically similar, however those of skates and rays have their corners produced to form horns rather than the tendrils typical of sharks. Many

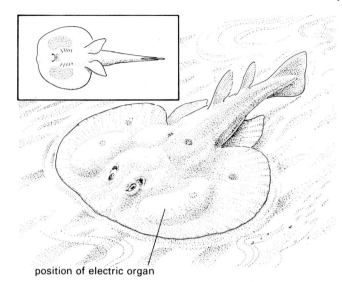

position of electric organ

Torpedo

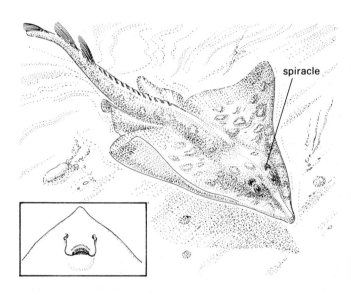

spiracle

Raja

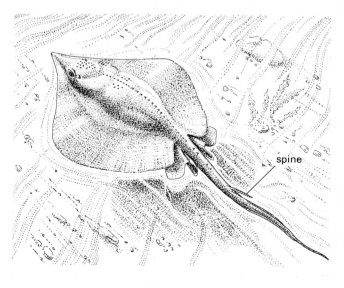

spine

Dasyatus

Raja: egg case

Chimaera

elasmobranchs are viviparous. For example the spiny dogfish (*Squalus acanthias*) has a gestation period of between 18 and 22 months during which the young are retained in the oviduct, which is modified as a uterus. This period may seem very long, but the species is long-lived and the male does not attain sexual maturity for 11, and the female for 19–20 years. The modified yolk sac lies in close contact with the uterine wall, and functions as a placenta. Similar adaptations are seen in *Sphyrna*, the hammerhead shark. In contrast, many elasmobranch embryos rely on yolk present in the original egg during their gestation period. *Lamna nasus*, the mackerel shark, has yet another method of feeding the young: eggs are released from the ovary throughout the development period and are swallowed by the embryo. The final eggs are swallowed just before birth and young mackerel sharks are released with a store of food and have pronounced pot bellies as a result. In the smooth shark, *Mustelus mustelus*, the embryo feeds on mucus ('uterine milk') which is secreted by the oviduct wall.

The second subdivision of the Chondrichthyes is the infra-class Holocephali, the rabbitfish or chimaeras, which are a curious group, showing many unusual features. These fish have a small mouth, an upper jaw which is fused to the skull, and a free hyoid arch (hyostylic jaw suspension). There is no distinct stomach and no spiral valve in the intestine. The gill openings are covered by a cartilaginous operculum, and the males have two pairs of claspers. The eggs are laid in long (60 to 100 mm) tapering cases. Rabbitfish are bottom living forms, generally found below 80 m, and feed on shrimps, gastropods, and sea urchins. They have solidly fused ripping and crushing teeth, and take in their food in small pieces, biting it with their anterior teeth which may account for their lack of stomach or intestinal valve. An alternative common name 'rat-tail' refers to their greatly elongated tails which they use, together with large pectoral fins, for propulsion. Chimaeras have various methods of defence; some have stout dorsal spines associated with a poison

gland, while the males may have mace-like structures, the so-called 'cephalic claspers', on their heads.

Fossil elasmobranchs

Two main levels of radiation are recognized among the now extinct elasmobranchs. The most primitive level is best known from *Cladoselache*, which lived in the Devonian period and may lie close to the ancestors of modern sharks. The cladodonts were probably predators in open water, feeding on other fishes such as placoderms which were abundant at that time. *Cladoselache* grew to about 2 metres and had large fins (like some modern sharks it had no anal fin), five pairs of gill openings and a terminal mouth. The jaw suspension was of a type known as amphistylic with several ligamentous attachments between the upper jaw and the chondrocranium. One was anterior, at the joint between

Cladoselache (reconstruction)

the left and right halves of the upper jaw, a second arose from a dorsally directed process to behind the eye, and a third, also from a process, extended to behind the otic region with some support from the hyoid arch. The jaws extended well to the back of the cranium, and the large gape, coupled with a battery of three, highly-characteristic pronged (cladodont) teeth must have made *Cladoselache* a formidable predator. The teeth were specialized placoid scales which were worn away and replaced in a similar way to that described for the dogfish. *Cladoselache* had a notochord rather than a vertebral column, although cartilaginous neural arches protected the spinal cord. Claspers have not been described for *Cladoselache* but were present in other cladodont species. The best known example of the second extinct elasmobranch radiation is *Hybodus*, a late genus from the Triassic and Carboniferous periods. Hybodonts shared many of the characteristics of the cladodonts, but had different feeding habits and methods of locomotion. They also show some enlargement of the snout as a rostrum. Hybodont teeth were more versatile than cladodont teeth but the relationship between the two is easy to see. *Hybodus* had two types of teeth; the anterior teeth were sharply pointed and used for biting and piercing whereas the more posterior teeth were blunter and modified for crushing. Teeth similar to those of *Hybodus* are found in a modern genus, *Heterodontus*, the Port Jackson shark, which feeds on a mixed diet of small fish, crustaceans, bivalves and sea urchins. Soft-bodied animals are seized and killed by the anterior teeth, while any shelled animals are crushed by the posterior teeth. *Hybodus* also lived in shallow seas, and probably had a similar diet available to it. Hybodonts have narrow based fins, which must have had a greater mobility than those of *Cladoselache*; they probably used their pectoral fins in a similar way to the dogfish. They also had a heterocercal tail like a dogfish, and possessed an anal fin. In spite of their greater mobility and versatility, the hybodonts were extinct by the end of the Mesozoic.

Synopsis of class Elasmobranchiomorphii

Subclass Placodermi: jawed fishes with bony armour and (often) peculiar fins; mainly from the Devonian
    Order Arthrodira (*Coccosteus*)
    Order Ptyctodontida
    Order Phyllolepida
    Order Petalichthyida
    Order Rhenanida
    Order Antiarchi (*Bothriolepis*)
Subclass Chondrichthyes: the cartilaginous fishes
  Infraclass Elasmobranchii: the sharks and related forms
    Order Cladoselachii: primitive Palaeozoic sharks
    Order Pleuracanthoidei
    Order Selachii: sharks
    Order Batoidea: skates and rays
  Infraclass Holocephali: chimaeras and related forms
    Order Chimaeriformes: chimaeras
    Various fossil orders

Coccosteus (reconstruction)

Bothriolepis (reconstruction)

Hybodus (reconstruction)

CLASS TELEOSTOMI (OSTEICHTHYES)

The remaining fishes have skeletons formed mainly of bone, and are frequently classified as Osteichthyes (literally 'bony fishes') to distinguish them from the cartilaginous fishes already discussed. However, since bone is the most commonly occurring skeletal material in vertebrates, it cannot be regarded as a distinguishing characteristic of the group. An alternative name for the group is Teleostomi, meaning 'end-mouthed', and refers to a so-called 'completion' of the jaw system: unlike those of other gnathostome fishes the upper and lower jaws are largely formed from new dermal bones (premaxilla, maxilla, and dentary) rather than from the palatoquadrate cartilages. The most distinctive feature of the group is the possession of a sac-like diverticulum of the pharynx known as the swim bladder. In most teleostomes this functions as a buoyancy organ, but in some cases it functions as a lung. It is most lung-like in certain primitive teleostomes and is therefore thought to have originated as an accessory breathing organ.

The earliest bony fishes can be traced back to the early Devonian and late Silurian by which time several lines were established. The acanthodians, which flourished in the Devonian, are characterized by a lack of true teleostome jaws and by the presence of prominent supporting spines at the anterior edge of their dorsal, anal and paired fins (the Greek *akantha* means 'spine' or 'prickle'); many acanthodians also had intermediate spines between the pectoral and pelvic fins. In addition the body was covered with heavy bony scales, which were less strongly developed in later members of the group. The majority are thought to have fed on small invertebrates which they strained out of the water, using well developed gill rakers, a method of feeding used successfully in many modern fish.

Living teleostomes belong to two main groups, one large and one very small; the Actinopterygii ('ray fins') has about 30,000 species while the Sarcopterygii ('fringe fins') has only seven, though many fossil forms are known. The actinopterygians are characterized by fins which are supported by dermal rays, and by olfactory sacs which have no connection with the mouth cavity. In contrast sarcopterygian paired fins are supported by a longitudinal axis of muscle and bone, and in many cases the olfactory organs connect both to the outside and to the mouth cavity. The sarcopterygians include two orders, Dipnoi (lungfishes) and Crossopterygii ('fringe fins') which must have diverged from each other early in the history of the group.

Actinopterygians are thought to have first appeared as the group we recognize today in the Upper Silurian. The subclass includes two infraclasses, Chondrostei and Neopterygii, which are distinguishable by their fin structure as well as other characters. The Chondrostei have dwindled to a few species of which the bichir (*Polypterus*) and the sturgeon are examples. The Neopterygii originated in the Permian and the early species, together with a few modern survivors such as the garpike, *Lepisosteus*, and bowfin *Amia*, are grouped together as the Holostei. The remainder comprise the large group Teleostei, to which most familiar fishes belong, and which in contrast has expanded continuously since it originated in the Mesozoic to include about 30,000 present day species. Several evolutionary trends, mainly connected with feeding and locomotion, have contributed to the success of this group, which will be introduced by an account of the well known example, the trout, *Salmo trutta*.

The trout: *Salmo trutta*

The trout, *Salmo trutta*, is one of the best known bony fish, both to the angler and the cook, although the related North American *S. gaidneri* (the rainbow trout) is now extensively farmed and more commonly eaten. Interest in the trout as a sporting fish is centuries old; Dame Juliana Berners described in 1496 how 'Ye may angle to hym' and Izaak Walton (1653) wrote 'The Trout is a fish highly valued both in this and foreign nations'. Much of the early literature about trout dealt with its habits and recommended methods of capture. *Salmo trutta* is essentially European in distribution and extremely variable in colour and habit with two varieties commonly occurring in English waters. The silver or sea trout is a silvery fish which grows up to 1.4 m and has a migratory life cycle in both the sea and fresh water. The smaller and darker brown trout is usually between 150 and 300 mm long and spends its entire life in lakes, rivers, and small streams. Many other varieties of *Salmo trutta*, with a more local distribution, are recognized and were once given the status of distinct species: for example the Loch Leven trout (*S. levenensis*), the brook trout (*S. fario*), and the gillaroo of Ireland (*S. stomachius*). However, modern authorities regard them all as belonging to a single, large, although very variable species, *S. trutta*.

Trout are members of the family Salmonidae which is widely distributed in the Northern Hemisphere and includes, as might be expected from its name, the salmon *Salmo salar*, as well as the char (*Salvelinus*). Since these fish are widely prized both as sporting fish and as food, some species (including *S. gaidneri* already mentioned) have been introduced to areas where they do not naturally occur.

Salmo trutta is a more slender and streamlined fish than the dogfish, and has a deeper and narrower body. The tail is shorter and is outwardly symmetrical, with dorsal and ventral lobes of the same size, although internally there are still traces of asymmetry with an upwardly turned vertebral column. It has two dorsal fins, and also a ventral anal fin. The paired pectoral and pelvic fins are small and fan shaped with a narrow base which makes them highly mobile. All fins are supported by bony fin rays with the exception of the posterior dorsal fin which is therefore soft; this is often described as an adipose fin.

The gill clefts of the trout do not open separately to the exterior, like those of the dogfish, but into a common opercular cavity covered by a bony plate, the operculum. The fish has a well developed acoustico-lateralis system, large eyes, and paired nostrils on each side.

The integumentary system

The integumentary system of the trout, which is of a typical teleost type, consists of a thin epidermis and a thicker dermis composed chiefly of connective tissue and is closely associated with the underlying myocommata. Within the epidermis there are unicellular and simple saccular mucous glands which keep the body surface covered by a thin slimy mucus. Nerves, smooth muscle and pigment cells lie within the dermis. A thin outer layer of both epidermal and dermal tissue overlies the scales, which are very different from those of the dogfish.

Two scale types occur commonly in the teleosts, with a similar basic structure. They are dermal in origin, roughly spatulate in shape and arranged with their anterior ends inserted deep into a slanting pouch in the dermis. The posterior portion overlaps the succeeding scale. The scales are termed cycloid if the free margin is smooth, and ctenoid if it is formed by a comb-like row of small spines (cteni). Trout have cycloid scales, but all gradations between the two extremes occur in fishes, and in some species both ctenoid and cycloid types are present. As the fish grows the scales also increase in size, and increments of bone are added at the margins which appear as scale rings (circuli). In many temperate fish (including trout) the growth rate is slower in winter and the circuli are closer together. By counting bands of closely and wider spaced circuli, it is possible to estimate the fish's age. The scales of the trout are extremely thin and composed of two layers, one calcified and one fibrous. This thinness is characteristic of many advanced bony fish and contributes to the great flexibility of movement they achieve.

The skeleto-muscular system and locomotion

In comparison with that of the dogfish the vertebral column of the trout is complex, although it fulfils the same function. It is extensively ossified and well adapted to resist the stresses imposed by fast swimming. The notochord is restricted to the intervertebral discs, which are pads of fibrous tissue that allow the column to retain its flexibility while resisting longitudinal compression. Each vertebra consists of a centrum, neural arch and neural spine, and there are additional haemal arches and spines in the tail region. The centra are biconcave (amphicoelous) and one is associated with each body segment. Small processes, the pre- and post-zygapophyses on the neural arches, improve articulation between adjacent vertebrae. Ribs of two kinds may be attached to the centra: slender curved ventral ribs, attached via small transverse processes (basopophyses), lie between the body wall musculature and the body cavity; intermuscular ribs are embedded within the body wall muscles. The neural and haemal spines support a series of pterygiophores which in turn support the fin rays (dermotrichia) of the median fins. Posterior epural and hypural bones, which are modified haemal arches, together with a single ossification, the urostyle, form a rigid support for the tail-fin rays. The myotomes have a plan basically similar to that of the dogfish but follow a complex zigzag

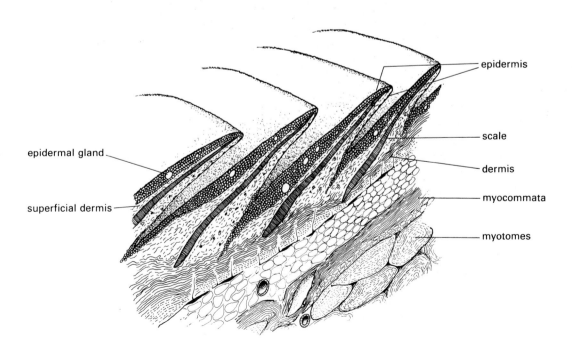

epidermal gland

superficial dermis

epidermis

scale

dermis

myocommata

myotomes

Trout integumentary system

course so that each extends over several vertebrae. As a result of this arrangement contraction of an individual myotome affects a sizeable portion of the body.

The skull of teleosts is complex and includes a lot of bones! It has as its basis a neurocranium and branchial arches comparable to the chondrocranium and arches of the dogfish. The neurocranium becomes ossified as a series of chondral bones, and the skull is completed by dermal bones. Together, the chondral and dermal bones form an effective protection for the brain and sense organs, and provide support for the gill arches and the jaws. With the exception of the anterior region, which remains almost entirely cartilaginous, the bony fish braincase is well ossified. Posteriorly a basioccipital, paired exoccipitals and a supra

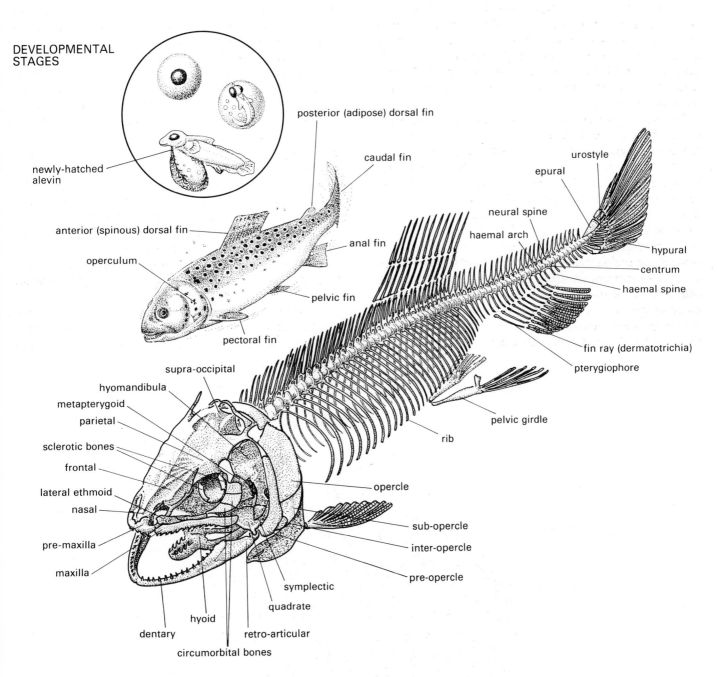

DEVELOPMENTAL STAGES

newly-hatched alevin

posterior (adipose) dorsal fin

caudal fin

urostyle

epural

neural spine

haemal arch

anterior (spinous) dorsal fin

anal fin

hypural

operculum

centrum

haemal spine

pelvic fin

pectoral fin

fin ray (dermatotrichia)

pterygiophore

supra-occipital

hyomandibula

metapterygoid

parietal

sclerotic bones

frontal

lateral ethmoid

nasal

pelvic girdle

pre-maxilla

rib

maxilla

opercle

sub-opercle

inter-opercle

pre-opercle

symplectic

quadrate

dentary

hyoid

retro-articular

circumorbital bones

The anatomy of the trout

occipital surround the foramen magnum through which the spinal cord extends, and there are three ossifications (pro-otic, epiotic, and opisthotic) surrounding the inner ear. The floor of the braincase anterior to the basioccipital is formed by a basisphenoid, with pleurosphenoids lying on either side. The orbits are separated by the orbitosphenoid and the anterior border of the neurocranium is formed by the mesethmoid and lateral ethmoids. Neither the optic nor the olfactory capsules contribute any ossifications to the neurocranium. Numerous dermal bones surround the neurocranium: it is roofed by a series of bones (nasals, frontals and parietals) and further bones (lacrimal, prefrontal and circumorbitals) surround the orbit. The side of the skull is completed posteriorly by the sphenotics and pterotics. The floor of the skull is strengthened by the parasphenoid and the palate is completed by paired palatines, and ecto- and entopterygoids.

An embryonic bony fish has both a palatoquadrate bar and a Meckel's cartilage, comparable with the upper and lower jaws of the dogfish. In the adult the palatoquadrate bar is represented by two ossifications, the quadrate and metapterygoid. These bones are articulated with the neurocranium anteriorly and suspended by the hyomandibula posteriorly, essentially the dogfish condition. The chondral elements of the skull are completed by the symplectic, a small ossification at the base of the hyomandibular which is also involved in jaw suspension.

The functional adult upper jaw is formed by dermal bones, the maxilla and premaxilla. Two ossifications also occur in Meckel's cartilage: these are the articular, which serves to articulate the lower jaw with the quadrate, and the retroarticular. In addition the lower jaw has a dermal, tooth-bearing element, the dentary. The hyoid supports the tongue and normally consists of five small bones: interhyal, epihyal, ceratohyal, hypohyals, and basihyal. Four of the remaining arches are typical gill arches. The main support of the gills is from dorsal epibranchials and ventral ceratobranchials and the gill lamellae are supported by gill rakers arising from these. The arches are completed dorsally by pharyngobranchials and ventrally by hypobranchials and median basibranchials. The fifth arch has no gills and is reduced to a ventral ceratobranchial. The operculum, which covers the gill region, is supported by preopercular, opercular, subopercular and interopercular bones. The pectoral girdle is formed of chondral ossifications of the scapula and coracoids, together with two dermal bones, the cleithrum and clavicle. Two more dermal bones, the supra-clavicle and post-temporal attach the pectoral girdle to the otic region. The pelvic girdle is small and very simple and does not include any dermal bones. A small triangular basipterygium, embedded in the body muscle, supports and provides an articulation for the pelvic fins.

Locomotion

The trout is an active swimmer capable of maintaining a speed of 8 to 10 knots and accelerating to 20 knots. The method of propulsion is essentially the same as that described for the dogfish and depends upon alternating contractions of the myotomes which throw the body into an S-shaped curve. The forward thrust comes mainly from the tail fin as it sweeps from side to side. The body muscles are well developed and make up about three fifths of the total volume of the fish.

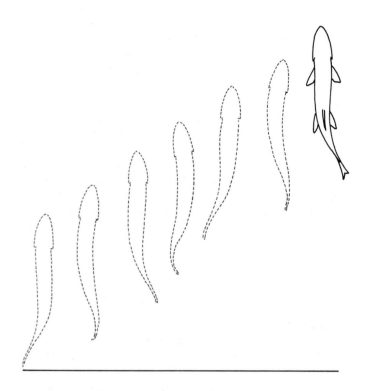

Trout : swimming movements

The median fins have a similar function to those of the dogfish, apart from the adipose fin which has no apparent function. However, the role of the paired fins, and the pectoral fins in particular, is greatly altered by the presence of a swim bladder. This develops as a ventral outgrowth from the pharynx which comes to lie dorsal to the gut as a shiny-walled sac. In most adult teleosts it loses its connection with the pharynx but in the trout the connection remains as a narrow duct to the oesophagus (the pneumatic duct). It functions as a buoyancy organ which effectively reduces the density of the fish so that it is equal to that of the water. Not surprisingly it is absent in bottom living fish and in those which live in turbulent waters such as the intertidal zone and fast flowing streams. The swim bladder has a thin elastic wall composed of two layers, and its own blood supply. The outer layer (tunica externa) has a trellis of elastic fibres while the inner layer (tunica interna) is made up of collagen and smooth muscle. Sensory nerves are present in the wall which detect volume changes in the bladder. Air can be secreted into the bladder or resorbed from it by means of special glands associated with the blood supply, and the volume is altered as a response to external pressure changes. Trout (like other fish with a pneumatic duct) can also alter the quantity of air present in the swim bladder by gulping it in through the mouth and pneumatic duct or by forcing it out by the same route. By these means the fish can achieve neutral buoyancy (so it neither rises nor sinks while stationary) at any particular depth. The significance of the swim bladder is that bony fish no longer have to depend on the pectoral fins and tail to maintain their position in the water and they can remain stationary at any level they choose. The tail can be modified to a more efficient symmetrical shape and the pectorals can be used for steering or braking or for rapid turning brought about by asymmetrical braking. Using this method a trout can turn through 180° in its own length. In most teleosts the pelvic fins are small and in the more advanced forms, for example the cod (*Gadus*), they have come to be anterior to the pectorals attached to the throat region; however in trout they occupy a posterior position. The possession of a swim bladder also frees the pectoral fins for other purposes, quite apart from their action as rudders already described. Some fish, the pike among them, when stalking their prey, creep slowly forwards by steady undulation of the pectoral and pelvic fins.

The digestive system and nutrition

Trout are carnivores and feed on a wide variety of aquatic invertebrates such as crustaceans, insects, molluscs, annelids and turbellaria, as well as small fish, tadpoles, and even adult amphibians. They hunt by sight. The teleost mouth is terminal and the jaws are smaller than those of the dogfish but form a highly efficient grasping structure. The most striking changes in the evolution of the teleost jaw are

those to the maxilla and premaxilla which form the upper jaw. The maxilla is free over its length, and forms an articulation with the skull at the anterior end about which it can pivot and swing ventrally. The premaxilla, which is also free from the skull is the tooth bearing element of the upper jaw; in the trout the vomer and palatines are also toothed. Both maxilla and premaxilla lie in the membrane which forms the side of the mouth and as the mouth opens they rotate ventrally, extending the membrane forwards and enlarging the gape. The prey is thus sucked towards the mouth (larger prey are prevented from escaping by the vomerine and palatine teeth) and swallowed whole.

The trout has a short oesophagus which opens into a highly distensible U-shaped stomach separated from the duodenum by a sphincter. About thirty or more pyloric caecae open into the duodenum and increase the effective surface area in this part of the intestine. Digestive enzymes are secreted from the rather scattered pancreatic tissue into the stomach, the pyloric caecae and the duodenum. A bilobed liver and gall bladder are also present with a bile duct which opens into the duodenum near the pyloric sphincter. The intestine is fairly long and opens at a separate anus, just behind the pelvic fins.

The respiratory system

Although the structure and function of the gills are fundamentally the same as those described for the dogfish there are differences both in the gross anatomy of the pharyngeal region and in the method by which the respiratory current is created. The trout has only four pairs of gills since there are no gills in either the fifth branchial cleft or in the spiracle though the spiracular gill clefts persist

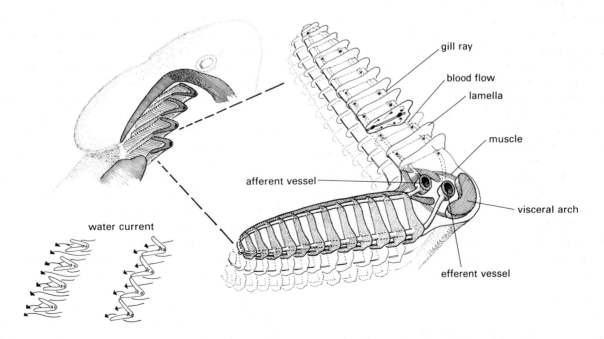

gill ray

blood flow

lamella

muscle

afferent vessel

visceral arch

water current

efferent vessel

Horizontal section of an osteichthyan gill bar

as a small glandular pseudobranch. Furthermore there is no half gill on the anterior wall of the first gill cleft. The gill clefts open into a common branchial chamber covered by the operculum. The hind and ventral margins of the operculum are free and provide a large aperture through which water leaves the branchial chamber. The gill filaments are similar in structure to those of the dogfish, but the interbranchial septae are much shorter leaving the gill lamellae as free flaps and greatly increasing the available respiratory surface.

The pharyngeal and branchial chambers function as separate pumps, the former a force pump and the latter a suction pump, and force a constant stream of water across the gills. The opercula only permit the exit of water and therefore act as valves. There are also valves inside the mouth. The valves are passive structures, which depend on the pressure gradient across them. In the initial phase of the pumping sequence, water is sucked in through the mouth and passes between the gill filaments into the branchial cavity; at this stage the mouth valves are held open and the opercula are held closed due to negative pressure in both chambers. Contraction of the pharyngeal walls then creates a positive pressure in the pharyngeal cavity, which forces more water across the gills and holds the mouth valve closed. In the third phase, both the pharyngeal and branchial chambers are constricted, and water leaves the branchial chambers through the open opercula. Finally expansion of the pharyngeal chamber draws water into the mouth, while the branchial chamber completes its contraction. The success of this double mechanism depends on the pharyngeal and branchial chambers acting out of phase; if a living trout is observed the mouth movements can be seen to be a quarter cycle ahead of those of the opercula. As active fish trout have a high oxygen requirement, and are restricted to waters such as shallow lakes or swiftly flowing rivers and streams with high oxygen concentrations.

The circulatory system

The general plan of the trout circulatory system is similar to that of the dogfish. The heart lies between the ventral ends of the two halves of the pectoral girdles. It is essentially a three-chambered structure, since the thick-walled conus arteriosus of the dogfish is absent, but a thin-walled bulbus arteriosus forms the base of the ventral aorta. The pericardium is fibrous and flexible with no communication through it to the main body cavity. The arrangement of the main blood vessels is similar to that of the dogfish and there is a well developed lymphatic system beneath the skin and in the body wall muscles and viscera.

Excretion and osmoregulation

The principal nitrogenous waste product of the trout is ammonia which is eliminated in large quantities, with smaller amounts of urea through the gills. It also has mesonephric kidneys, present as elongated dark red masses above the swim bladder. Two ducts, one from each kidney, join posteriorly to form a distensible bladder which opens at a urinogenital pore just behind the anus. Although small amounts of nitrogenous waste are eliminated through the kidney, its principal function in a freshwater fish is elimination of excess water which diffuses into the blood via the gills. The trout therefore produces copious quantities of dilute urine. Marine teleosts have the opposite problem of maintaining sufficient water in the body fluids. Unlike the dogfish they do not solve it by retaining urea in the blood, but instead take in seawater with their food, and excrete the excess sodium and chloride ions by active transport through special chloride excreting cells in the gill epithelium. Thus, in contrast, marine teleosts produce small amounts of concentrated urine. Migratory fish like the sea trout must adjust their osmoregulation according to the prevailing environmental conditions. The change cannot be made suddenly but in a normal migration gradual adjustment is possible.

The nervous system and sense organs

The nervous system is built on the same general plan as that of the dogfish though the detailed structure reflects the animal's particular mode of life. Since trout are active by day, hunting by sight and inactive in the dark, their most important sense organs are the eyes. Both the optic lobes and cerebellum are well developed. The eyes have a similar structure to those of the dogfish and are focussed in a similar way. However the trout is also sensitive to colour and has both rods and cones in the retina. The olfactory lobes are small and poorly developed. The medulla oblongata is well developed with special lobes associated with the sense of taste. Trout have taste buds inside their mouth and may have similar structures on other parts of the body.

Reproduction and life cycle

The gonads lie in the body cavity above the gut. The female has elongated ovaries which shed ripe eggs into the body cavity; these pass through paired abdominal pores at the base of the excretory duct, and thence through the urinogenital pore to the exterior. This condition is unusual, and in the majority of teleosts the ovaries are sacs whose lumens are continuous with the oviduct. In the spring the ovaries are small but they enlarge in the autumn until they almost completely fill the body cavity. The testes open into ducts which lead to the exterior through the urinogenital pore, and also change in size through the year.

Spawning occurs in the autumn, when the female 'cuts' a shallow trough (redd) about 240 mm long in the gravel with her tail. After a short courtship involving quivering movements of male and female she lays her eggs in the redd

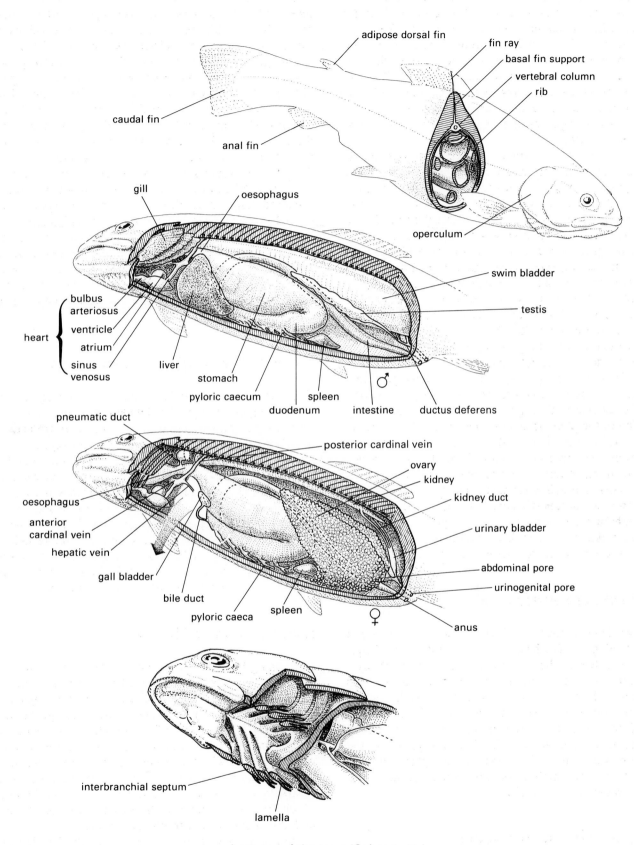

Anatomy of the trout (*Salmo trutta*)

and the male sheds sperm over them. The female then covers them with gravel. The eggs are about 5 mm in diameter with a substantial yolk and surrounded by a tough membrane. The speed of development varies according to temperature, but at 5° C the young fish (alevin) hatch about 97 days after fertilization. At hatching they are extremely small and attached to the yolk sac. At first they are inactive but after two or three days they burrow into the gravel away from the light where they continue to feed on their yolk for two to three weeks. When they are about 20 mm long, they leave the redd as parr to feed on insect larvae and other small invertebrates. By the end of the first year the survivors, for mortality in the early stages is enormous, measure between 50 and 140 mm. Sea trout remain for between one and five years in the rivers before they migrate to the sea. Here they spend from six months to five years and grow rapidly. They return to the rivers to breed and many may survive to spawn more than once.

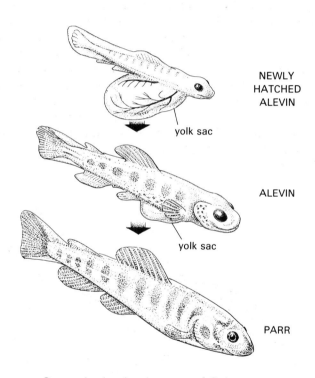

NEWLY HATCHED ALEVIN

yolk sac

ALEVIN

yolk sac

PARR

Stages in the development of *Salmo trutta*

TELEOSTOME DIVERSITY

The teleostomes must be regarded both as highly successful vertebrates, and as the most successful aquatic animals. They have effectively colonised all regions of the sea down to a depth of 9,500 metres, and are found in all types of freshwater habitat from mountain torrents to stagnant pools, in underground waters, in lakes to depths of 1600 m, and in temperatures from −18° C to 40° C. At least one species spends the greater part of its life on land and many others can leave the water for shorter periods. Teleostomes feed on every type of food and have adapted to every possible mode of life within aquatic ecosystems from commensalism to highly successful predation. They display a wide range of body form and locomotory specialization, some remarkable sensory adaptations, and a variety of complex behaviour patterns. Apart from the diversity of the group, many species are enormously abundant. For example, three thousand million herring were caught in one year in the Atlantic alone. Fish are important in a large number of food chains; and are of considerable economic importance. Many different species of fish are caught commercially and used for a wide variety of purposes. Cod, for example, is not only used as food (fresh, frozen, salted, or dried) but the waste is used for by- products (glue, fertiliser) and the liver has medicinal value (cod-liver oil).

The Actinopterygii

Actinopterygians are the main subclass of bony fishes and can be subdivided into two infraclasses: Chondrostei and Neopterygii. The Neoptergyii include the Holostei and Teleostei. Of these, the Chondrostei and Holostei both flourished for a time and then dwindled. Various evolutionary trends can be followed through the succession of fishes seen in the fossil record, relating principally to improvements in feeding mechanisms and locomotion.

The Chondrostei, which first appeared in the late Devonian, attained their maximum development and distribution in the Carboniferous. They had a swim bladder which was a functional lung rather than a hydrostatic organ, a heterocercal tail, a large non-protrusile mouth and scales of a thick bony type (ganoid scales). Today they are represented by a few species including the African genus *Polypterus* (the bichir), which retains the swim bladder as a functional lung, and *Scaphiorhyncus*, the sturgeon. *Scaphiorhyncus* lives in the Mississippi River system of

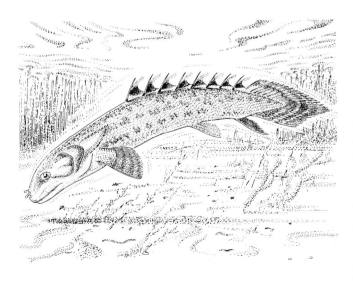

Polypterus

North America, but has declined in numbers due to extensive fishing. The flesh was eaten as food, but more important for its survival the eggs are prized as caviar! The remaining two subgroups Holostei and Teleostei are believed to be more closely related to each other than to the Chondrostei and are frequently classified together as the Neopterygii.

The Holostei evolved in the Permian and are today a relict group represented by a few species such as the garpike *Lepisosteus* and the bowfin (*Amia*). One of their most significant evolutionary developments compared with Chondrostei is the partial freeing of the upper jaw from the cheek bones. Their scales are of a thinner and lighter type known as lepisostoid. *Lepisosteus* is widely distributed in North America, where it occurs in fresh and brackish water and sometimes in the sea. Another North American genus

Amia has a swim bladder of the lung type and a rounded and superficially symmetrical caudal fin which represents a modification of the more usual heterocercal type.

Teleosts evolved in the Mesozoic and the group has expanded ever since. Apart from an extensive fossil record, they include the majority of living teleostomes, with 80,000 odd species. The group shows evolutionary trends which, like those of the Actinopterygii as a whole, are correlated with locomotion and feeding. The more primitive teleosts, of which the herring (*Clupea*) is an example, are active open water predators; they have an elongated fusiform body with a single dorsal fin, posterior pelvic fins, flexible branching fin rays, and cycloid scales. Their jaws are short but non-protrusile and there is a pneumatic duct connecting the swimbladder and the gut. In contrast, advanced teleosts, for example the perch (*Perca*), have two dorsal fins (the more anterior one being supported by spines), anteriorly positioned pelvic fins, and ctenoid scales. Their jaws are highly specialized and protusile, and they do not have a pneumatic duct. Between these extremes, many teleosts have a mixture of primitive and advanced characters, as well as their own peculiar adaptive features. As an example ostariophysians resemble the primitive teleosts in several respects but have a small chain of bones, the Weberian ossicles, extending from the swim bladder to the inner ear. These fish apparently have better hearing than other groups and the swim bladder is thought to function as an accessory hearing organ or hydrophone.

Lepisosteus

Clupea

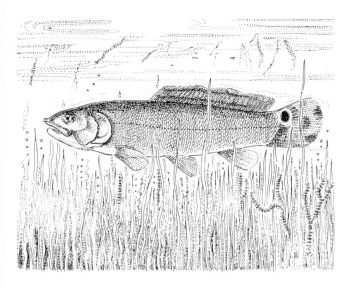

Amia

Several factors contribute to the overwhelming success of teleosts. A major factor is their size. Although teleosts range in length from 12 mm for the Philippine goby (the smallest fish and the smallest vertebrate) to 4.26 m for the blue-fin tuna, and 4.5 m for the world's largest freshwater fish, the Amazonian *Arapaima gigas*, the majority are less than 150 mm long. This means that individuals have small food requirements and can invade environments where little food is available as well as living in large numbers in richer habitats. It also enables them to live in small spaces, such as in shallow water, under stones, or in the interstices of coral reefs. Furthermore the protrusile jaw has proved to be an extremely adaptable feeding structure and has enabled the group to expand its feeding areas to include confined spaces as well as facilitating bottom and surface feeding. The development of the swim bladder as a buoyancy device made possible not only the development of a symmetrical tail fin and faster swimming, but released the paired fins (particularly the forwardly placed pelvic fins) for new roles such as slow stealthly hunting movements, mating, and aggressive displays. In addition teleosts have more complex brains than cartilaginous fishes and complex behaviour patterns: one example is the association of individuals in schools, another is the elaborate behaviour of courtship, and the building and defending of a nest as seen in the stickleback.

Modes of life

Apart from the active predatory mode of life typified by the herring, teleosts have occupied almost every aquatic ecological niche. Some have become strongly compressed laterally, a modification which allows them to live in confined spaces, and is seen in many coral reef fishes. Many fish have a tubular body form, for example pipe fish and moray eels, and though these fish may have lost speed and versatility of movement they are well adapted to life in a range of habitats such as in weed, in coral reefs, under stones and burrowing. Halibut, sole and plaice have become strongly compressed laterally to the extent that they swim and lie on one side. (In these fish the bilateral symmetry becomes distorted so that both eyes come to lie on the upper side.) This is an adaptation for bottom dwelling and in some respects parallels the adaptations seen in the skates and rays, though the animals are widely separated evolutionarily and the compression has been achieved in different planes.

Fishes show a remarkable diversity of colourings and markings. The pipe fish (*Syngnathus*) bears a strong resemblance to marine sea grass while the sargasso fish (*Histrio histrio*) has numerous seaweed-like projections which enable it to hide in the floating *Sargassum* weed in which it lives. *Monocirrhus polyacanthus*, the Amazonian leaf fish is well known for its remarkable mimicry of leaves. It has strong lateral compression, with transparent dorsal and anal fins which it can flatten tightly against its body. It habitually drifts hanging head-down from the surface or hides among dead leaves at the bottom of the river. Doubtless this mimicry has a protective function, but in addition the leaf fish is a remarkable predator. Its resemblance to a leaf allows it to approach potential prey by stealth, and it is then able to engulf suitable fish with its extremely large protrusile jaws.

Syngnathus

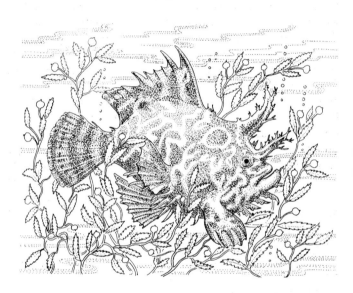

Histrio histrio

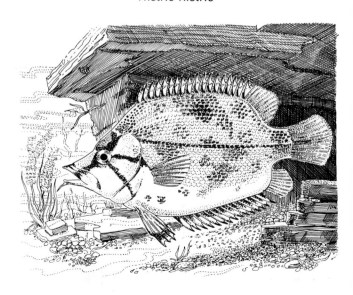

Monocirrhus

Modes of locomotion

While the normal method of progression in teleosts is by axial locomotion, as described for the trout, many fish swim solely by means of their fins. *Histrio histrio* uses its fins to scramble about in the weed. *Anabas* (the climbing perch) spends much of its life on land, and treks from one pool of water to another by a method of locomotion which is effective if rather ungainly; it pushes itself along by flexure of its tail against the ground, using its pectoral fins and opercula as props. Another adaptation to its semi-terrestrial existence are greatly reduced gills and the possession of supplementary rosette-shaped respiratory organs housed in special chambers above the gills. The common name of 'climbing perch' comes from the fact that *Anabas* was described as having been found up a tree by a Danish Naturalist, Daldorf, in 1797. Subsequently it has been demonstrated that while *Anabas* is sometimes found in trees, it is more likely to have been deposited there by a bird which may have seized it during one of its land migrations. *Periophthalmus*, the mudskipper, is also frequently seen out of water, leaping about on the mud flats of mangrove swamps using active movements of its muscular pectoral fins (which it uses as a pair of crutches) and tail. It is remarkably agile and can leap in almost any direction with the result that it can avoid most predators. It is one of the most amphibious teleostomes and can close its branchial chamber and still obtain oxygen via its gills. The pipe fishes and sea horses cruise around in an upright position, propelling themselves by rippling waves which pass along the dorsal fin. The trigger fishes (Balistidae) swim by simultaneous side to side flapping of the dorsal and anal fin. One of the most outstanding modifications of fish locomotion is seen in the flying fish. *Exocoetus* is only one example of several which have their pectoral fins greatly expanded into a wing-like form and can 'fly', or more accurately glide, for considerable distances.

Periophthalmus

Modification of spines

Fin spines can be modified in a variety of ways. Trigger fishes get their name from an interesting modification of the dorsal fin spines: the first spine is large and stout and can be raised and then locked into position by the smaller second spine. This lock must be released before the fin can be lowered. Most trigger fish live in coral reefs, and if attacked they retreat into a crevice, and wedge themselves in with their spine, snapping their teeth at the would-be predator. Another modification of the dorsal fin spines is seen in angler fish, for example *Lophius*. Here the first dorsal fin ray has a fleshy lobe which can be dangled in front of its jaws and used as a lure to bring smaller fish or crustaceans within striking distance.

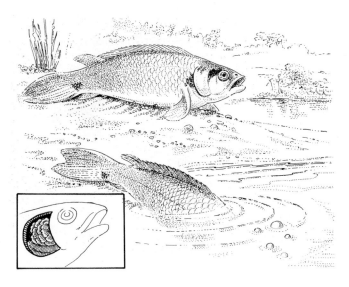

Anabas

Lophius

Feeding styles

Although the majority of fish are carnivorous, they show a wide variety of feeding adaptations which are closely reflected in their diversity of jaw shapes. The clupeids (herrings, sardines and their relatives) are specialized to feed on plankton and have small fine teeth and very long complex gill rakers which they use for filtering out microscopic organisms from the water. They are normally found in huge shoals in the plankton-rich parts of the ocean such as the Iceland coast and the north west coast of Africa. *Heterotis niloticus* (widespread in the river Nile), is another filter feeder and in this case the fourth gill arch is modified with a spiral epibranchial organ in which plankton is trapped in mucus. The diet of the plaice, *Pleuronectes*, consists largely of molluscs which it crushes using its strongly developed palatal teeth. The deep sea stomatoids have a mouth which is large and wide, and a distensible body, so that they can eat prey as large or larger than themselves. In complete contrast, the chaetodontid fishes, most of which live in coral reefs, have a small mouth at the end of a long tubular snout which can be poked into holes in the coral in search of suitable prey.

Some fish seek their food outside the water. Special mention must be made of *Anableps* ('four eyes'), a surface dwelling form which feeds on both aerial and aquatic prey, and whose eyes are divided into two, for vision above and below the water. The archer fish (*Toxotes*) also feeds on flying insects which it knocks into the water by spitting drops of water at them; this is done with great accuracy over a distance of a metre or so, but a few ranging shots may be necessary over three metres. Alternatively it may leap out of the water and seize prey from overhanging vegetation.

Some teleosts have taken to a commensal mode of life and take food that other animals have gathered. The remora (*Remora remora*) has its dorsal fin modified as a large adhesive disc on the top of its head which it uses to attach itself to sharks. It feeds on the shark's parasites or on fragments from its meals, and also 'hitches a lift' during the shark's migrations. Another member of the same genus *R. australis*, the whale sucker, has a similar mode of life, but attaches to whales. *Labroides dimidiatus*, the cleaner fish, provides a valuable service by feeding on the parasites of other larger fishes. It is extremely common in coral reefs and defends a territory where it waits for customers; its distinctive dark blue streak along its body making it easily recognisable. However there is a possible snare for the unwary because this colouration and pattern is mimicked by another fish, *Labroides mimichei*, this time a predator!

Remora

Reproductive behaviour

Teleosts show an immense range of reproductive behaviour. Most of them lay tens of thousands of eggs, each with little yolk and the young have a long larval life and little parental care. Often the minute, newly-hatched fish lives a planktonic life, though the adult form may occupy a very different habitat. However, there are numerous exceptions to this and as a general rule where there is a high level of protection or parental care, fewer and larger eggs are laid and the young fish reach a more advanced stage of development before they must fend for themselves. Some fish hide their eggs (the trout is one example) while others build more or less elaborate nests. The breeding habits of the three-spined stickleback (*Gasterosteus aculeatus*) are well known. The male builds a tubular nest out of pieces of plant material which he sticks together with a sticky secretion manufactured by the kidneys. A period of courtship follows and the female is encouraged to lay her

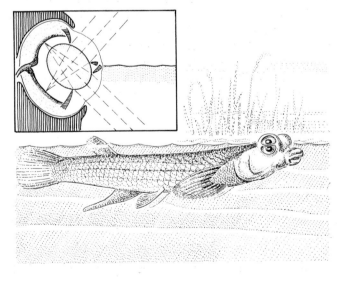

Anableps

Gasterosteus aculeatus and nest

eggs in the nest. The male then guards the eggs until they hatch, 7 or 8 days after fertilization, and takes care of the young for the first 14 days or so after hatching. Anabantids make floating nests of bubbles of mucus, and again the eggs and young are protected by the male. Male sea horses and pipefish carry their eggs in special sac-like cutaneous brood-pouches, and the mouth-brooding cichlids protect their eggs by carrying them in the mouth. In this case parental duties are undertaken by the female; even after hatching the young fry remain in her mouth. As they grow they leave her for short periods, but keep within easy reach and return to the mouth if danger threatens them. A few teleosts are viviparous. In the viviparous blenny (*Zoarces viviparus*) the ovary is hollow and the young develop in its lumen, and are born when they are about 40 mm long. *Zoarces viviparus* produces quite large numbers of young at any one time, with a range to 9 to 132 being recorded. An intermediate stage is seen in the guppies (Poeciliidae) where the early stages of development occur within the ovary and the eggs are only laid immediately before they hatch.

Electric organs, luminescence and poison glands

Electric organs, already mentioned in the elasmobranchs, are also known in actinopterygians. The structure is like that of the elasmobranchs. Among the largest such organs the electric eel (*Electrophorus*) from the Amazon basin can produce pulses in excess of 500 volts, which are sufficient to stun a would-be predator, or prey. The electric catfish, *Malapterurus*, from tropical Africa produces between 350 and 450 volts and is recorded as having knocked a fisherman unconscious! Other fishes, for example *Gymnarchus*, produce weaker pulses which they use for navigation.

Many deep-sea fish have luminescent organs (photophores); these are modified gland cells which either secrete luminous mucus or are associated with luminous bacteria. They may be relatively simple, or highly elaborate cup-shaped structures with lens and reflector. The function of bioluminescence in fish probably varies according to species, and while in some cases it enables members of the same species to recognise each other, in others its role is to attract prey. Alternatively a sudden flash of light may be emitted to confuse a would-be predator, in a similar way to the ink cloud produced by a squid.

Poison glands are found in a number of bony fishes, often associated with spines. The venom of the weaver fish (*Trachinus*) is particularly unpleasant, indeed the common name of these fish is said to be derived from the French *wivere*, a viper. In this case the poison glands are associated with long sharp spines on the gill covers. Porcupine fishes (Diodontidae) may produce toxins in their liver, gonads, intestines or skin. Their flesh is edible and the Japanese regard it as a great delicacy (fugu), though, even when it is prepared by specially trained chefs, eating fugu is a gamble, and is the cause of many fatalities each year.

Adaptation and habitat

Many teleostomes illustrate clearly the tendency for distinct organisms to converge in response to life in the same habitat. For example fish living in caves generally lose their colour, scales and eyes; the inhabitants of torrents or the turbulent intertidal zone (such as gobies) commonly have dorsoventrally flattened bodies, and suckers (often modified pelvic fins), and feed by browsing on algae. Several teleosts from diverse taxonomic groups have adapted to a bathypelagic mode of life and show similarities which can be related to a dark, cold environment, a poor food supply and high predation pressure. Generally these adaptations are designed to enable life to continue with the minimum expenditure of energy. The lateral line system is well developed for the detection of prey and predators, and may be extended by the development of a very long tail. Dwarf males, which lead a commensal mode of life attached to the female, are common. Bioluminescence is common in these forms, often associated with large eyes. These various features make the fish of deep waters appear, to our eyes, bizarre in the extreme.

Subclass Sarcopterygii

Two superorders are included in the sarcopterygii: they are the lungfish (Dipnoi) and the crossopterygians (Crossopterygii). Both were successful in the Devonian and have an extensive fossil record, but are represented today by only a few relict species. Generally they had a weak skeleton with a poorly ossified vertebral column, and specialized crushing teeth. The evolutionary trends seen in the fossil record parallel those of the Actinopterygii, with progressive reduction of the armour and dermal skull bones of the earliest forms, (living representatives have a largely cartilaginous skeleton), thinning of scales to the cycloid type, and the development of a symmetrical tail fin.

Most of the fossil lungfish have been found in freshwater deposits and the modern genera, *Neoceratodus* (Australia), *Protopterus* (Africa) and *Lepidosiren* (South America) are all fresh water animals. All three genera represent specialized derivatives of the Devonian fauna. The early lungfish had heterocercal tails and heavy, enamelled cosmoid scales. The most characteristic features are their crushing tooth-plates of an unusual structure consisting of an enameloid substance, composed largely of calcium phosphate and borne on the palatoquadrate and lower jaw. There were no teeth on the margins of either the upper or lower jaws and in many lungfish the maxillae and premaxillae were absent. Modern lungfish have similar tooth plates; *Protopterus* feeds on molluscs, and *Lepidosiren* and *Neoceratodus* on a more mixed diet of invertebrates and plant material.

Neoceratodus

Protopterus

The lungfish have paired fins quite unlike those of other bony fish, with a thick central lobe containing bone and muscle, and a thin blade supported by fin rays. Fins of this type are found in *Neoceratodus* and it can use them to row itself through the water and also propel itself over the bottom. *Protopterus* and *Lepidosiren* have vestigial paired fins consisting only of slender filaments, and none of the modern Dipnoi can walk on land. The tail is of a diphycercal type with both internal and external symmetry.

One of the most widely known facts about lungfish is that they can breath air using lungs; like the swim bladder, these are diverticula of the pharynx as are the lungs of tetrapods. *Neoceratodus*, which lives at all times in well-aerated waters and is the least highly modified, uses a full set of gills for

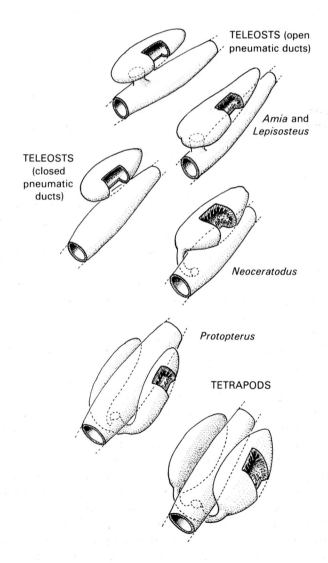

TELEOSTS (open pneumatic ducts)

Amia and *Lepisosteus*

TELEOSTS (closed pneumatic ducts)

Neoceratodus

Protopterus

TETRAPODS

Swim bladders and lungs

respiration. However at intervals (hourly) it comes to the surface and gulps air into its lungs. *Protopterus* and *Lepidosiren* often occur in less well-aerated conditions such as swamps and are therefore more dependent on their lungs for survival. The gills are reduced and *Protopterus* has no gill filaments on two of its gill bars; both genera drown if prevented from coming to the surface to breathe. *Protopterus* and *Lepidosiren* are well known for their ability to survive long periods of drought buried in the mud; indeed, *Protopterus* is recorded as having survived for four years in this way. It digs a tunnel by biting out mouthfuls of mud, and lives in it, surrounding itself in a mucous cocoon with only a small hole left for its mouth. Like the Dipnoi, the Crossopterygii appeared in the Devonian, and until fairly recently were thought to be extinct. The earliest forms (rhipidistians) were large freshwater predators, some over 4 m long, and in many ways resembled modern lungfish, with similar paired fins, scales, and tails. It seems probable that they had a similar way of life to modern lungfish; they certainly cannot have walked effectively on land. However their head structure was very different. While teleosts have two pairs of nostrils on the top of the head, water entering the anterior opening and leaving through the posterior opening, in Dipnoi the relative position of the openings has altered so that the anterior nostril is on the upper lip and the posterior nostril inside the mouth. Rhipidistians had one or two external nostrils, but in addition they had new internal openings, which would have been an aid to smelling, since water would be drawn through the nasal sacs by respiratory movements. The tooth structure of the crossopterygians is very unlike that of the Dipnoi. The rhipidistians had large teeth along the edges of the jaws with a characteristic and complicated structure in which the dentine was folded into the pulp cavity in a series of plates. This kind of tooth is described as labyrinthodont and is also found in the lower tetrapods. In addition rhipidistians had pharyngeal and palatal teeth, which in no way resemble the tooth plates of lungfish.

The rhipidistians lived in fresh water, but another crossopterygian group, the coelacanths were mainly marine. They arose in the Carboniferous and were believed to be extinct until one was caught near East London, South Africa in December 1938. That this fish was not lost must be due to the efforts of Miss Latimer, the curator of the East London Museum, who recognized its importance and preserved it; the fish is named *Latimeria* in honour of her. Since then, several specimens have been caught between Madagascar and Mozambique, and as the sole surviving crossopterygian genus it has been extensively studied. The swim bladder, unlike that of many fossils, is a peculiar fatty organ, which while giving some buoyancy, could not function either as a hydrostatic organ or a lung. There are no vertebral centra, and a long tough notochord forms the principal axial skeletal structure. The skull has the typical crossopterygian structure, in two hinged parts about an intercranial joint. *Latimeria* is a fish eater and swims, at least in aquarium conditions, by sculling with its dorsal and anal fins. It is thought to utilize urea in its blood as a means of osmoregulation in a similar way to modern sharks. The kidneys are in a ventral position unlike those of all other vertebrates and the heart is of a simple structure when compared with that of other fish. It is known to lay a few extremely large eggs (each about 160 mm long) but nothing is known of its behaviour.

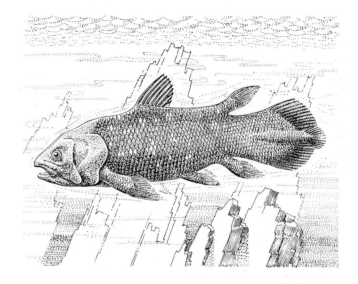

Latimeria

Synopsis of the class Teleostomi

Subclass Sarcopterygii: fleshy-finned fishes
   Superorder Dipnoi: lungfishes
   Superorder Crossopterygii: lobe-finned fishes
      Order Rhipidistia: primitive freshwater crossopterygians
      Order Coelacanthini: more specialized freshwater and marine forms such as *Latimeria*
Subclass Acanthodii
Subclass Actinopterygii: ray-finned fishes
   Infraclass Chondrostei: primitive ray-finned fishes
      Order Palaeonisciformes
      Order Polypteriformes: bichirs
      Order Acipenseriformes: sturgeons and paddle fishes
Infraclass Neopterygii
   Division Holostei: intermediate ray-finned fishes.
      Order Semiontiformes: garpikes
      Order Amiiformes: bowfin.
   Division Teleostei
   Superorder Clupeomorpha: herrings
   Superorder Elopomorpha: tarpons, eels, *Saccopharynx*, gulper eels
   Superorder Osteoglossomorpha: bony-tongued fishes

Superorder Protacanthopterygii: primitive teleosts includes salmon and trout
Superorder Ostariophysi: fishes that are primitive in most respects but have Weberian ossicles
Superorder Atherinomorpha (includes *Exocoetus*)
Superorder Paracanthopterygii (cods and angler fishes)
Superorder Acanthopterygii (advanced teleosts)

Introduction to the tetrapods: the pentadactyl limb

Locomotion brought about by bending the body (axial locomotion) is a characteristic of most aquatic vertebrates. In contrast most terrestrial animals move by appendicular locomotion, using limbs which act as levers. The evolutionary transition from axial to appendicular locomotion was a major one which necessitated modification and elaboration of the limbs and a change in the function of the axial skeleton from propulsion to support. Obviously the structural modifications were immense, and as an evolutionary process the problem could be compared to that of producing a car from a submarine without either pulling it to pieces or even putting it out of commission. The transition was a gradual one and among terrestrial vertebrates we can see a number of stages in the achievement of effective movement on dry land.

Water is a dense supporting medium whereas air is not, and in addition to providing propulsion the limbs must also take the weight of the body. On the other hand the vertebral column, rather than acting as a strut to resist compression, as in fishes, instead functions as a girder, from which the

internal organs are slung. This new function necessitates the development of a structure which is strong rather than flexible, with vertebrae which are well articulated to each other. As in the fishes, the vertebrae articulate with each other at the centra but there are also overlapping articular processes borne on the vertebral arches. In addition, there are usually vertebral spines extending dorsally from the arches and lateral transverse processes which serve for the attachment of muscles and ligaments. Associated with the independent movement of different parts of the body due to appendicular locomotion, there is regional differentiation of the vertebral column. All tetrapods have one or more neck (cervical) vertebrae, including a specialized vertebra (the atlas) which articulates with the skull and allows independent movement of the head; a series of trunk vertebrae which articulate with the pelvic girdle; and a series of caudal vertebrae.

The pectoral and pelvic girdles which connect the limbs with the vertebral column are built on basically similar plans and consist of tripartite structures. Each side of the pectoral girdle (with which the fore-limbs articulate) is made up of scapula, coracoid, and pre-coracoid bones surrounding an articulation point, the glenoid fossa. The pelvic girdle consists of an ilium, ischium and a pubis which surround and form the acetabulum. In primitive tetrapods a number of dermal bones may overlie the pectoral cartilage bones and contribute to the girdle. One of these, the clavicle, normally persists in higher vertebrates. The pectoral girdle is connected to the axial skeleton through muscles and ligaments, but does not articulate directly with it, whereas the pelvic girdle attaches to the vertebral column in the sacral region.

The skeleton of the fore and hind limbs are also based on

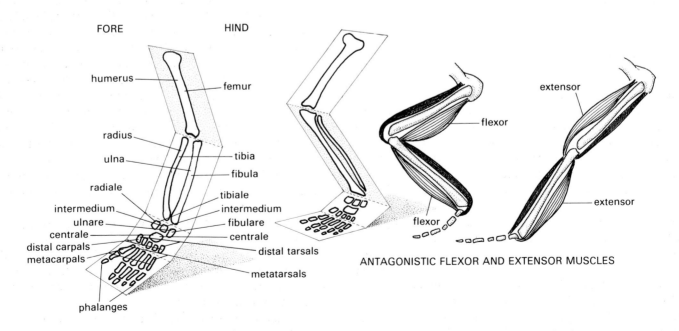

The pentadactyl limb

a common plan, normally referred to as the pentadactyl limb since the limb terminates in five digits. We can see the basic arrangement in human limbs: the fore-limb consists of four main sections, namely upper arm, forearm, wrist and hand. The hind (lower) limb has the same four sections, this time known as thigh, leg or shank, ankle and foot. In the fore-limb, the upper arm skeleton consists of a single long bone, the humerus, and that of the forearm of two bones, the radius and ulna. The wrist (carpus) has nine small carpal bones arranged in three rows. In the proximal row of three bones the radiale lies next to the radius, the ulnare at the base of the ulna, and the intermedium between the two. The middle row consists of a single centrale, and the distal row of five carpals, one at the base of each digit. The hand (manus) is composed of two series of bones: the palm is supported by five metacarpals and the fingers by a series of phalanges. The number of phalanges varies between the digits; two support the thumb (pollex), three the remainder.

The bones of the hind limb have a precisely similar arrangement. The thigh bone is the femur, the leg bones are the tibia and fibula, the ankle (tarsus) bones, the tarsals (tibiale, intermedium, fibulare, centrale and five distal tarsals), and the foot (pes) are the metatarsals and phalanges, the first digit being termed the hallux. In tetrapods there is variation in the details of the limb bone anatomy but the basic arrangement of the fore and hind limbs remains the same.

The movements of the limbs are brought about by opposing sets of muscles. The hinge joints (between thigh and leg, leg and foot in the lower limb, and the corresponding positions in the upper limb) are moved by paired antagonistic flexors and extensors. In addition a ventral, lateral muscle block draws the limb mainly forwards and towards the mid-line and a dorsomedial mass backwards and away from the body at the point between limb and girdle.

CLASS AMPHIBIA

The amphibians illustrate the beginning of the vertebrate transition from water to land. In addition to the obvious changes in locomotion and respiration this transition necessitated modifications in almost every organ of the body. Although the Amphibia were at their peak in the Carboniferous, in a quiet and unobtrusive way they remain a successful group, with a world-wide distribution though generally restricted to habitats near fresh water. The majority lay their eggs in water and their development includes an aquatic larval stage, the tadpole, which lacks limbs and breathes by means of gills. Lungs and limbs only appear when the tadpole transforms itself into an adult, at which point the animal becomes truly amphibious.

The living amphibians are normally grouped together in the subclass Lissamphibia in three orders: Urodela (the tailed amphibians, the newts and salamanders), Anura (the tail-less amphibians, the frogs and toads) and Apoda (the legless burrowing caecilians). Of these three groups the urodeles have the most generalized body form and locomotion. The anurans are highly specialized with elongated, powerful legs used for jumping and swimming. Nevertheless, since frogs and toads are very widespread and are among the most commonly dissected vertebrates a frog, *Rana temporaria*, has been selected to introduce the group.

Rana temporaria: general body form

Most terrestrial vertebrates have a body divided into three regions: head, neck and trunk. In the case of the frog, which spends much of its time in the water and has a greatly shortened body with long hind limbs as an adaptation for leaping, there is no sharp external distinction between the head and trunk. However the neck region is represented in the skeleton by an atlas vertebra which supports and permits some independent movement of the head. The trunk is short and compact and the tail reduced to a small stub.

The neck has a wide terminal mouth with a pair of external nostrils and large prominent mobile eyes positioned on the top of the head and protected by eyelids. The eyes are directed forwards giving the frog binocular vision. On each side of the head behind the eye is a circular pigmented patch of skin, the tympanic membrane, which represents the external opening to the ear and acts like a drum head, picking up sounds and transferring them to the middle and inner ears.

The integumentary system

The integumentary system of the frog is very different from that of the fishes, reflecting the partially terrestrial mode of life of the animal. It has an outer dead stratum corneum which is periodically shed, and which provides some protection against both friction and excessive water loss. Skin of this type is said to be cornified. Amphibian skin is highly unusual among vertebrates in that, for reasons which will be discussed more fully in the section on respiration, it is an important respiratory organ and therefore shows little cornification. In order to provide the necessary moisture to function in this way, the skin has numerous mucus secreting glands, which are epidermal in origin but form invaginations deep into the dermis. Each gland is flask-shaped and the expanded portion is lined with columnar cells which are continuous with the transitional epidermal layers and produce a watery secretion. This is poured onto the surface of the skin and keeps it moist and slimy. Some of the glands produce a distasteful, possibly poisonous substance. This is certainly true of the toad and accounts for its long association with witchcraft.

The dermis is composed of an outer spongy layer with abundant blood vessels, lymph spaces, and nerves, and an

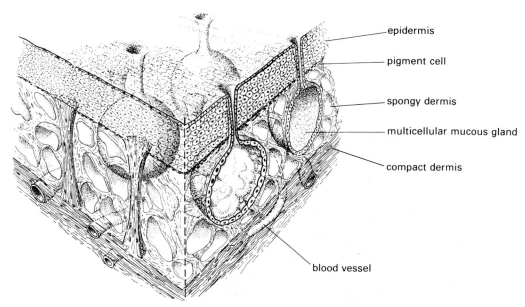

Frog integument (diagrammatic section)

inner more compact layer. Pigment cells, which are concentrated between the epidermis and dermis, can alter the colour of the animal to some extent according to whether the pigment is dispersed or concentrated about the nucleus. The skin is loosely attached by connective tissue to the underlying muscle at intervals, in contrast to the extremely tight attachment characteristic of fishes; this is associated with increased freedom of movement and the development of appendicular locomotion.

The skeleto-muscular system and locomotion

The skeleto-muscular system of the frog follows the basic tetrapod plan although with various specializations connected with locomotion. A notochord is present only in the embryo; in the adult it is replaced functionally by the vertebral column and all that remains are fibrous intervertebral discs. A ventrally placed cartilaginous structure, the sternal rod, forms a ventral articulation for the thoracic ribs and therefore a complete enclosure of the chest region. This structure is unknown in fishes and must have evolved in the lower tetrapods. Posteriorly, the vertebral column is shortened in response to the lack of a tail.

Tadpoles have a cartilaginous cranium and gill arches comparable to those of the dogfish. In contrast the adult skull, like that of bony fish, is made up of chondral and dermal bone. Like the early bony fishes, ancestral amphibians had a heavy armour of dermal bone protecting the head. Many of these bones persist in the cranium of modern Amphibia but the skull shows some reduction and

specialization in comparison with that of earlier types. Like that of bony fish it consists of a series of capsules of chondral bone which surround and protect the brain and sense organs and which are partly covered by dermal bone. The roof of the skull is reduced and largely unossified and formed of fused frontals and parietals (fronto-parietals). The olfactory capsules are reinforced dorsally by nasals and ventrally by the vomers which bear small pointed teeth. Posteriorly the ventral surface of the brain case is strengthened by a dagger-shaped parasphenoid which extends forwards between the orbits.

The upper jaw is formed of a maxilla and premaxilla, both of which bear slim, pointed teeth, and a toothless quadrato-jugal which connects to the otic capsule posteriorly by a small quadrate and the squamosal. The lower jaw is toothless and is formed of a dentary and large pre-articular with a partly ossified Meckel's cartilage (Meckelian bone). The posterior part is an articular which articulates onto the quadrate of the upper jaw. The hyomandibula is not involved in jaw suspension but is modified as a stapes (columella) in the middle ear. The remainder of the hyoid arch is represented by a broad cartilaginous hyoid plate which supports the tongue and provides for the insertion of muscles which depress the floor of the mouth. The branchial arches are modified during metamorphosis to form the cartilages which support the larynx.

Locomotion

A frog has three methods of locomotion: swimming, walking and leaping, all of which are appendicular. Both swimming and leaping depend on synchronized extension of the hind limbs. The fore-limbs are held against the side of

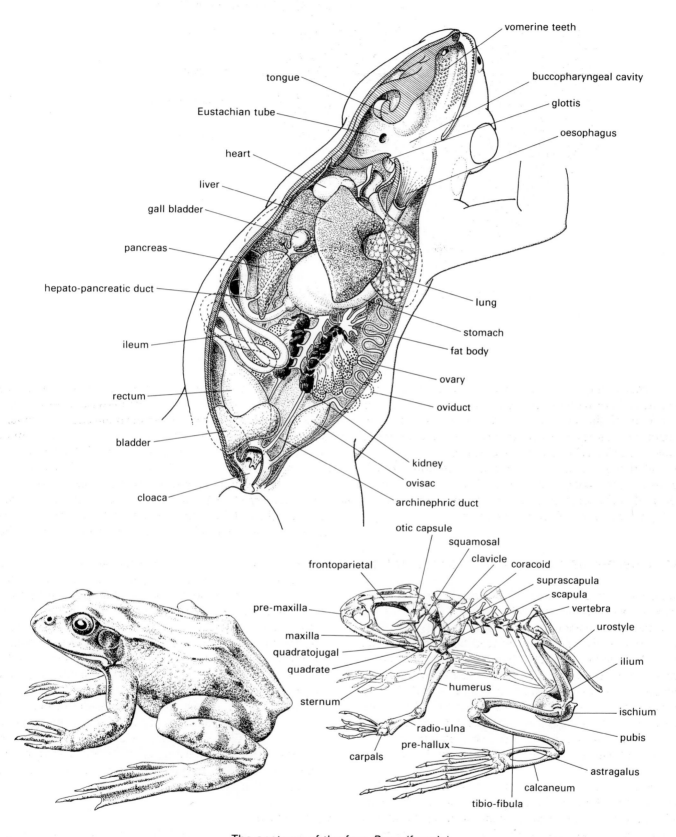

tongue

vomerine teeth

buccopharyngeal cavity

Eustachian tube

glottis

oesophagus

heart

liver

gall bladder

pancreas

hepato-pancreatic duct

lung

stomach

fat body

ileum

ovary

oviduct

rectum

bladder

kidney

ovisac

cloaca

archinephric duct

otic capsule

squamosal

frontoparietal

clavicle

coracoid

pre-maxilla

suprascapula

scapula

vertebra

maxilla

urostyle

quadratojugal

quadrate

ilium

humerus

sternum

ischium

radio-ulna

pubis

pre-hallux

carpals

astragalus

calcaneum

tibio-fibula

The anatomy of the frog, *Rana* (female)

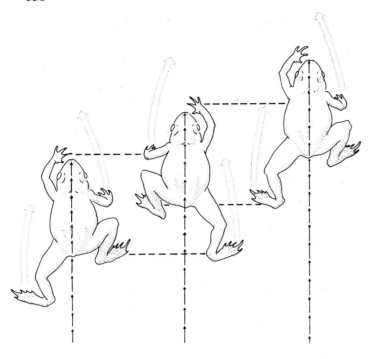

Frog : walking movements

the body during both types of movement, and are used as shock absorbers when the animal lands at the end of a leap or to support the body when it is at rest. During walking, both fore and hind limbs are used, those on one side being held in opposite positions (forward and back).

The vertebral column, which has to support most of the body weight, also has to withstand the powerful force exerted by the limbs during leaping and swimming, and the vertebrae are solid and bony with only limited movement possible between them in either a lateral or a vertical plane. A feature peculiar to the anurans is considerable shortening of the vertebral column as an adaptation to jumping. The nine vertebrae which make up the column are the atlas, seven trunk vertebrae and a sacral vertebra with well-developed posteriorly-directed transverse processes by which the ilia of the pelvic girdle are strongly attached by ligaments. In addition there are two caudal vertebrae which have fused to form a long slender urostyle to which many of the jumping muscles are attached and which lies in the space between the ilia.

The chief sources of propulsion in the frog are the considerably elongated powerful hind limbs. Their structure is that of the basic pentadactyl limb, though the tibia and fibula have fused to form a single tibio-fibula bone.

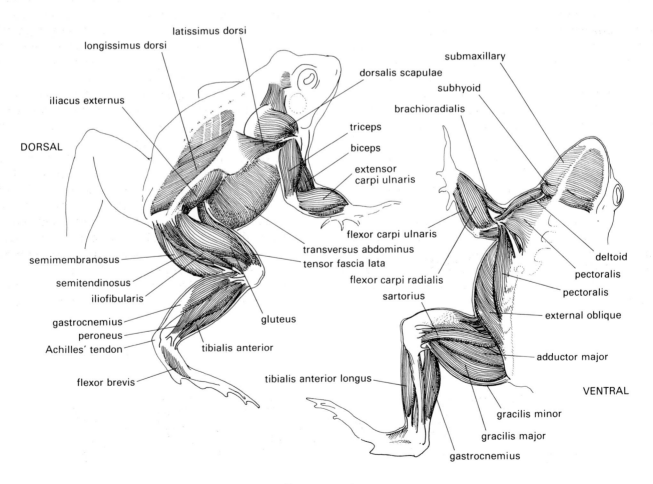

Frog : muscles

Considerable modifications are also seen in the ankle and foot. The ankle is elongated and the proximal row of tarsals is represented by only two bones, the astragalus (tibiale) and calcaneum (fibulare). Of the remaining tarsals, the centrale is absent and there are only two bones in the distal tarsal row. Both the metatarsals and the phalanges are elongated and there is an additional (sixth) digit, the prehallux. The surface area of the foot is greatly increased by a web of skin between the digits as an adaptation for swimming. In contrast to the well developed hind limbs the fore-limbs are much smaller. The humerus, and the fused radius and ulna (radio-ulna) are short, and stout. There are only six bones in the carpus because the centrale is missing and the distal row is represented by three bones. The metacarpals are short and the phalanges of the pollux are absent.

As might be expected, the pelvic girdle is well developed to transmit the thrust when the animal jumps. The ischio-pubic portion and the proximal part of the ilium on each side have fused to form a solid disc of bone set vertically in the body from which the ilia extend to the sacral vertebrae. The acetabula lie in the centre of the disc on each side. The pectoral girdle is also strongly built and forms not only a system for articulation of the fore-limbs but also for attachment of strong muscles upon which most of the strain falls when the frog lands from a leap.

The musculature is very different from that of a fish and the original metameric organization of the trunk musculature has been obscured, a trend which continues throughout the tetrapods. There is a smaller volume of axial musculature which is now less important in locomotion and the myotomes are vertically arranged rather than in the zig zag pattern of fishes. The epaxial muscles bend the vertebral column dorsally rather than laterally, and are modified anteriorly to bring about turning movements of the head.

In contrast to fishes, whose appendicular muscles are simple and composed of dorsal and ventral muscle masses, those of frogs are strongly developed. The musculature is differentiated into separate muscles which are large and have broad attachments. Furthermore, we can make a clear distinction between extrinsic muscles which attach the girdle or limb to the axial skeleton and thus move the whole appendage, and intrinsic muscles which move parts of the limb relative to each other. Details of the musculature are given in the appropriate diagrams.

The digestive system and nutrition

Although the tadpole is almost exclusively herbivorous, the adult is a carnivore which will feed on a wide range of small animals such as worms and insects. It is interesting that the vegetarian tadpole has a gut which in relative terms is considerably longer than that of the adult to allow it to feed on a diet with a high water content and a large proportion of rather indigestible material. The frog has a wide terminal mouth and the teeth, although numerous on the upper jaw and vomers (in the roof of the buccal cavity) are very small. They are used for holding the food rather than for chewing. Prey is caught by means of a highly protrusible tongue which is unusual in several respects. Instead of occupying the normal position, attached at the back of the buccopharyngeal cavity and projecting forwards, it is attached anteriorly, immediately behind the lower jaw, and lies with its free end projecting backwards. To feed, the frog waits until a suitable prey, perhaps an insect, passes close to it and then opens its mouth and flicks the tongue rapidly forwards by contraction of its muscles. The tongue is coated with a sticky secretion to which the prey adheres and so is pulled into the mouth by the return flick. The movement is extremely rapid and the tongue can be flicked out and in within about a tenth of a second. The buccopharyngeal cavity is well lubricated with saliva and any particles of food left in the mouth are entangled in the secretion and transferred by ciliary action to the short oesophagus which acts as a chute, connecting the buccopharyngeal cavity to the stomach. The oesophagus, which merges almost imperceptibly with the stomach, has a series of longitudinal folds allowing it to expand as larger food passes along it. The stomach is relatively simple and has well developed muscles and a glandular lining. Food is broken up by its churning action and mixed with hydrochloric acid and the digestive enzyme pepsin. Digestion and absorption of the food takes place principally in the intestine. This is separated from the stomach by a pyloric constriction which functions as a valve controlling the exit of food. The passage of food along the intestine is by peristaltic contractions of the gut wall. The first part of the intestine is the duodenum into which opens a common duct (the hepato-pancreatic duct) from the liver and pancreas. The liver is large and lobed and a gall bladder is present. The second part of the intestine is the greatly elongated, coiled ileum. It has a high

Frog : feeding movements

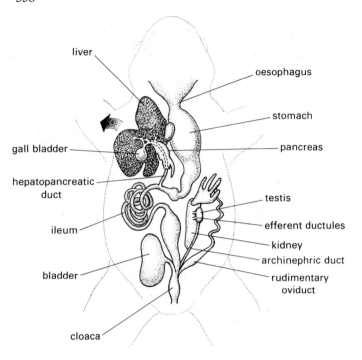

Frog : alimentary tract and left urinogenital ducts (male)

absorptive area because of its length and because its internal surface has numerous villi. The ileum leads into a short, wide, large intestine (rectum) where water is reabsorbed. The various digestive secretions add considerable amounts of fluid to the intestinal contents, and in the absence of any reabsorption the frog might soon become dehydrated. The rectum opens into a cavity known as the cloaca (from the Latin for a sewer) which also receives the ducts from the kidneys and gonads, so that at different times it acts as a route for eggs or sperm, urine and faeces.

Respiration and circulation

In many respects the problems of gaseous exchange in terrestrial vertebrates are comparable with those encountered by terrestrial arthropods and molluscs. In aquatic vertebrates, gaseous exchange involves a flow of water over the respiratory membrane which is therefore kept moist. A current of air removes moisture from the respiratory surface as it passes over it. This must be replaced, for diffusion of oxygen across the membrane to continue. In dry environments the process of breathing represents an important and unavoidable source of water loss.

The frog has three respiratory surfaces: the lining of the buccopharyngeal cavity (of least importance), the skin, and the lungs. The lungs are relatively simple structures

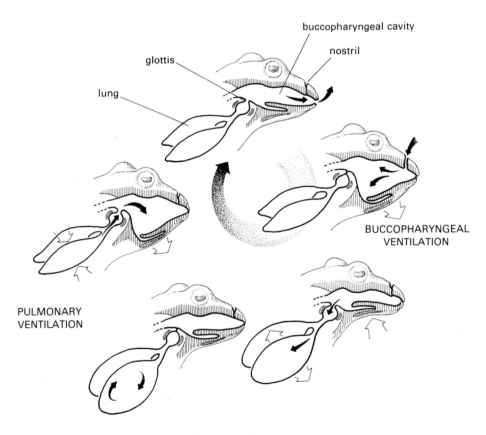

Frog : ventilation

consisting of two elastic and highly vascular sacs which lie in the abdominal cavity and open into the pharyngeal floor via a slit-like glottis. The walls are thin and slightly muscular and internally the lungs are divided into numerous simple chambers; this gives them an external appearance rather like foam rubber and increases the surface for gaseous exchange.

Air is forced into and out of the lungs by movements of the floor of the mouth which acts as a force pump. When the mouth is closed and the nostrils open, the floor of the mouth is depressed and air is drawn into the buccopharyngeal cavity where gaseous exchange takes place. For pulmonary ventilation the nostrils are closed and the floor of the mouth raised to force the air through the glottis into the lungs. The lungs are emptied by a forcible contraction of the general body wall muscles, and by the elastic recoil of the lungs themselves.

During normal activity the frog does not use its lungs, the skin and the buccal mucous membrane providing a sufficient surface for gaseous exchange. Buccopharyngeal movements which are conspicuous externally take place at between 8 and 120 per minute when the frog is at rest. These are interrupted by gulping movements accompanied by movements of the flanks which form part of pulmonary ventilation. The advantage of this type of respiration rather than pulmonary respiration is that it provides an adequate maximum oxygen supply for most circumstances with minimum water loss, the latter being a problem for the Amphibia. In addition the skin has an extensive blood supply, and is also important as a respiratory surface, though the level of gaseous exchange via this route varies. While carbon dioxide is normally eliminated through the skin, oxygen uptake is most important by this route during the winter, when cutaneous respiration accounts for about two thirds of the total oxygen uptake. In the summer, when frogs spend most of their time out of the water, only about one quarter of the total oxygen needs are obtained through the skin.

Circulation

As we saw in the previous chapter the circulatory system of the frog is rather different from that of fishes. In association with the shift from gills to lungs, many changes have occurred in the heart and aortic arches. The aortic arches are reduced in number, the first two and the fifth being lost. Those that remain are continuous tubes, uninterrupted by gill capillaries. The remaining aortic arches show further modifications: the third pair forms part

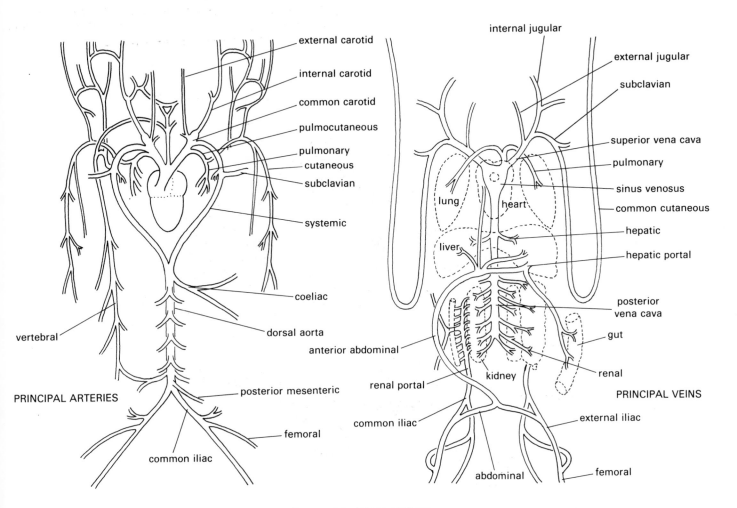

Frog : circulatory system

of the blood supply to the head (internal carotid arteries), the fourth pair (the systemic arteries) is the main channel to the body and viscera through the dorsal aorta and the sixth pair (pulmocutaneous arteries) supplies the lungs and skin. New vessels, the pulmonary veins, return aerated blood from the lungs to the heart, which therefore receives blood separately from both the body and the lungs.

The heart is modified from the single tube present in fishes. Obviously to achieve maximum efficiency of the respiratory system, blood from the body and oxygenated blood from the lungs must be kept separate as far as possible. The solution in the amphibians is not perfect, but a partial separation is achieved. The atrium is subdivided into two halves, which the pulmonary vein (which carries oxygenated blood) entering the left half, and the sinus venosus (which receives venous blood from the head via the anterior venae cavae, and the body via the posterior vena cava) opening into the right hand side. The two blood streams converge into a single ventricle. Although some mixing is inevitable, a high degree of separation is achieved by a variety of means including a spongy ventricular muscle which tends to keep the streams apart; slight differences in the times the oxygenated and deoxygenated blood enter the

ventricle; and the development of a spiral valve within the conus arteriosus with internal partitioning of the arterial trunks which lead to the vessels. Thus the pulmocutaneous arch receives blood low in oxygen which has come mainly from the right atrium, while the systemic arch receives mixed blood. The carotid arch, which supplies the head, receives oxygenated blood from the left atrium. Nevertheless, the blood recovered by the tissues contains lower concentrations of oxygen than is the case in fishes, though this is partially offset by a higher blood pressure and increased rate of circulation. Blood in the dorsal aorta of the frog has a pressure of about 30 mm Hg which is fifty per cent higher than that of the dogfish. (The major blood vessels of the frog are illustrated on page 339.)

Excretion

The kidneys of frogs are derived from the opisthonephros and contain about 2000 glomeruli. The amount of urine excreted by a frog in a 24 hour period may be as much as a third of the total body weight of the animal, and so contributes considerably to the dehydration problem. It is

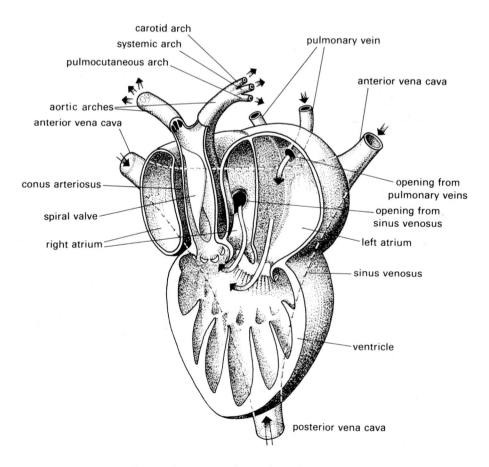

Frog : diagrammatic section of the heart

interesting that many desert amphibians lack glomeruli altogether, as an adaptation to reduce excessive water loss. The excretory product is urea, which is less toxic in higher concentrations than ammonia. In female frogs there is no connection between the kidneys and the reproductive system, but in males some of the anterior tubules are modified as vasa efferentia, which connect the testis to the kidney and the archinephric duct. Thus in males the archinephric duct serves for the transport of both sperm and urine. It is located within the lateral margin of the kidney, and leaves at the posterior end to extend to the cloaca. A bilobed urinary bladder is present in amphibians which opens out of the cloaca but forms no direct connection with the urinary ducts. Substantial quantities of urine may be stored here, particularly during periods of dormancy, and some water may be resorbed to reduce dehydration.

Nervous system

The parts of the frog brain and the arrangement of the spinal nerves are typical of those of anamniotes, though the frog, with its shortened trunk, has a shortened spinal cord.

Within the brain, the optic lobes are prominent and the olfactory lobes small, indicating a greater reliance on sight rather than smell. Although the cerebellum is small, reflecting a relatively low level of muscular activity in comparison with such active forms as the trout, the cerebral hemispheres are large, in association with a wider range of activities and more complex behavioural patterns such as different types of locomotion.

There are ten pairs of cranial nerves of which I to VI and VIII have a distribution similar to that of the dogfish. However in the adult frog those cranial nerves (VII, X, XI) which in the dogfish are associated with the gills and the acoustico-lateralis system, are modified to supply their derivatives. Cranial nerve VII, which supplies the spiracular gill cleft, of the dogfish, has visceral sensory and motor components to the taste buds and lower jaw. Cranial nerve IX (associated with the dogfish first typical gill slit) is reduced in size and supplies the tongue; cranial nerve X is a large and important visceral nerve and supplies the heart, lungs and gut.

The spinal cord is short and small, tubular, with two swellings, the branchial enlargement in the region of the fore-limbs and the sacral enlargement in the region of the

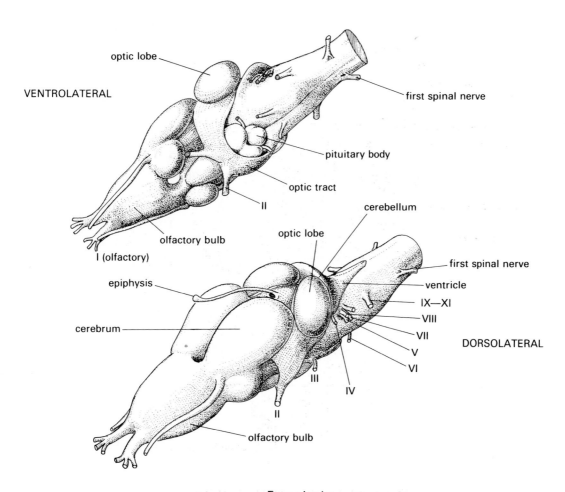

Frog : brain

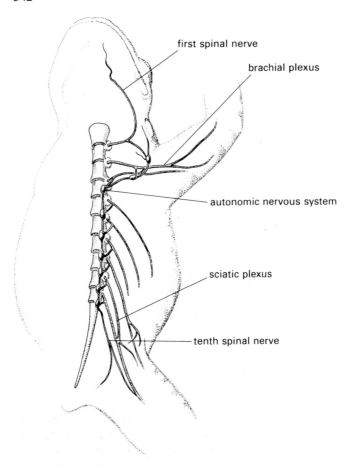

Frog : spinal nerves

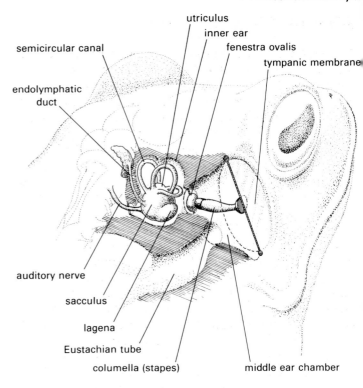

Frog : ear

closed by a small valve. Its function is to equalize pressures on either side of the tympanic membrane. Like most gnathostomes, the frog has an endolymphatic duct extending from the sacculus and utriculus to terminate as a sac within the brain case.

Reproduction

The position of the testes relative to the kidneys has already been described. In the female, the reproductive and excretory systems are completely separate. The ovaries of the frog are large and in a mature female can occupy the greater part of the body cavity, distending the abdominal wall. The oviducts have thin-walled internal openings, near to the base of the lungs, but are thickened and glandular for the greater part of their length. At their distal ends each oviduct expands into a thin-walled ovisac, narrowing again to open into the cloaca at the urinary papilla. Frogs produce large numbers of eggs which are about 1.6 mm in diameter and with a moderate amount of yolk. As the eggs pass down each oviduct they are coated with albumen and are then stored in the ovisacs. Mating always takes place in the water, with external fertilization. The male mounts the female, and clasps her behind her fore-legs, the grasp being made easier by the presence of swollen nuptial pads on the metacarpal of the first digit. As the eggs are released the male pours seminal fluid onto them and fertilization takes place. The albumen surrounding the eggs swells on contact with water and the eggs adhere together in masses forming a jelly-like frog spawn. The subsequent development to metamorphosis has already been described.

hind limbs, which indicate the presence of large numbers of nerve cells and their processes associated with the limbs. There are only ten pairs of spinal nerves in the adult frog, but the tadpole has twenty or more of which the more posterior nerves disappear when the tail is resorbed.

The eyes are well developed on the general vertebrate plan, with moveable eyelids, moistened by glandular secretions, as an adaptation to a terrestrial mode of life. Focussing is brought about by protractor lentis muscles which move the lens forwards for near vision. The ear is primarily an organ of balance, although the frog has developed some sense of hearing. In many respects it is similar to that of bony fish, with semicircular canals, a utriculus, and a sacculus bearing a lagena. In addition to the inner ear the frog has a middle ear. This is derived from the first branchial pouch and is an air-filled cavity separating the outer wall of the auditory capsule from the skin. The outer wall of the middle ear chamber is fused to the skin, forming the tympanic membrane which responds to air vibrations. A slender bone, the stapes or columella, bridges the chamber and transfers those vibrations to the inner ear. It is attached to the inner surface of the tympanic membrane at one end, and to a membrane closing a small aperture (the fenestra ovalis) in the wall of the auditory capsule. A Eustachian tube connects the middle ear with the pharynx, where it is

AMPHIBIAN DIVERSITY

The term 'amphibian' means 'double-life' (the Greek *amphi* means 'double', *bios* means 'life') and the majority of amphibians do have a double life with an aquatic larval stage and a terrestrial adult. Living amphibians are placed together in the subclass Lissamphibia with three distinct orders: Urodela (tailed amphibians), Anura (tailess amphibians) and Apoda (legless amphibians). Although in most cases the adults are adapted to life on land, they are generally restricted to moist habitats with a high relative humidity, a restriction imposed by their inadequate water conservation.

In spite of a slight cornification of the epidermis, their use of cutaneous respiration necessitates a moist, vascular skin with a thin, keratinized layer, so that considerable water loss by evaporation is inevitable in dry conditions. Although nitrogenous waste is eliminated as urea (a terrestrial adaptation) rather than as highly toxic ammonia, a considerable amount of water is also lost in the process of urine formation. Amphibians are also restricted in distribution because, with some exceptions, they must return to water to reproduce. They are ectotherms whose body temperature is always closely related to that of their environment, and at low temperatures they become inactive. As a result, in temperate climates they have to hibernate during the winter. All living adult amphibians are carnivorous, and generally the group shows little specialization for different diets. Many species will eat anything they can catch and swallow, so that one important factor restricting their diet is the size of their mouth. The tongue shows a certain amount of specialization to different conditions: in some terrestrial Amphibia, for example the frog, it is long, protrusile and highly mobile, while in aquatic forms it is broad and flat with little mobility.

Order Urodela: the salamanders and newts

The urodeles represent the least modified body form in recent Amphibia with short legs, a long trunk and well-developed tail, which allows them to walk with an undulating fish-like movement. Their clumsy gait when walking is probably similar to that employed by the early amphibians, combining lateral bending of the body with leg movements. At each step the fore and hind legs are held in opposite positions (forward and back) on the side of the animal that is convexly curved, giving the maximum length of stride. The group is largely confined to the northern hemisphere, with the greatest species diversity occurring in North and Central America.

The largest living salamanders (and the longest living amphibians) are the Chinese and Japanese giant salamanders (*Melagobatrachus*), which may reach a length of over 1½ m, but most are smaller and typically about 200 mm in length. They have secretive habits, and are generally found below stones and logs in woods. Many are

Walking movements in a urodele

very poisonous because of toxins produced by their skin glands and warn potential predators to avoid them through brilliant colouration (for example yellow and black). The European salamander, *Salamandra salamandra* (not found in Britain) is an example of a more complete adaptation to terrestrial life as it is ovoviviparous, giving birth to living larvae which undergo an aquatic larva phase. A further stage is seen in *S. atra*, the alpine salamander, which gives birth to one or two fully developed young thus eliminating the free-living aquatic larval stage. In contrast some urodeles, such as the mud puppy, *Necturus*, which is found in the lakes and streams of North America, are entirely aquatic.

Salamandra

Necturus

Necturus illustrates a frequent evolutionary trend in urodeles known as neoteny, in which metamorphosis is suppressed and sexual maturity is acquired early in development so that aquatic larval features are retained throughout life. *Necturus* retains external gills throughout its life and the lungs remain rudimentary. In some cases neoteny may be a flexible response associated with certain environmental conditions. For example in *Ambystoma tigrium*, (axolotl) a common North American species, complete metamorphosis occurs under normal conditions, but individuals living at high altitudes undergo a neotenous development to become, in effect, sexually mature larvae. Several salamanders have become specialized cave dwellers and the majority of these are also neotenous. For example *Siren* has external gills throughout life and lacks hind limbs. The european *Proteus* shows other features associated with its life in caves: it is blind, has an unpigmented skin and has a shovel-shaped snout that can be poked into small spaces between rocks in search of food.

In the British fauna, urodeles are represented by newts all belonging to the genus *Triturus*. In some respects these are the most terrestrially adapted urodeles, since their limbs are capable of supporting the body weight and the feet, unlike those of the salamander, and are held turned forwards with their soles applied to the ground. Newts are voracious feeders: when on land they feed on worms, snails, insects and slugs, while in the water their diet consist of aquatic larvae, crustaceans, molluscs, and the tadpoles of other amphibians. Newts are unusual among amphibians in that, though they mate in the water, fertilization is internal. The male deposits a spermatophore which is picked up by the female in her cloaca, and the sperms are stored in a spermatheca until the eggs are laid. To ensure that the female is receptive and will pick up the spermatophore, its deposition is preceeded by a period of courtship. Once she has been inseminated, a female will not take up another spermatophore until the eggs which were ripe at the time of insemination have been fertilized and laid. The eggs are laid singly and normally attached by the female to aquatic plants, which she selects with some care. The eggs are oval in shape and about 4 mm in length. Unlike those of the frog, newt tadpoles have a close resemblance to the adult except for their external gills, and do not show an elaborate reorganization at metamorphosis. Both adults and tadpoles swim by means of their tail, the limbs being held folded against the sides. The smooth newt (*Triturus vulgaris*) is the commonest British species and nocturnal in habit. During the breeding season, which lasts from about the end of February until May, the male smooth newt develops a large crest, running from the back of the head to the tail tip. After breeding it remains in the water until about July. The rest of the summer and the winter are spent near the pond, the animals hiding under stones or pieces of timber or in thick grass by day.

Siren

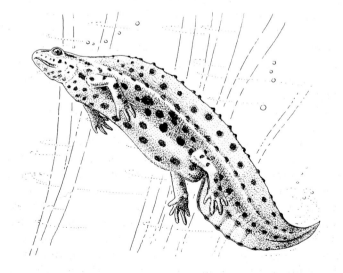

Triturus vulgaris, the smooth newt

Triturus helveticus (the palmate newt) is the smallest British newt and rarely exceeds 60 mm in length. During the breeding period the male develops a crest and webs on its hind feet. In contrast the great crested newt (*Triturus cristatus*) is relatively large, up to 14 cm in length. This species is more aquatic than other newts and has been recorded as overwintering in the water.

Order Anura: the frogs and toads

In contrast to the urodeles, anurans have highly specialized locomotion which is reflected in their body form, already described in detail for the frog and typical of the group. While many of the anurans are amphibious as adults and remain close to the water, some are more terrestrial in habit and others have adapted an arboreal mode of life (tree frogs). This diversification has taken place without great morphological adaptation and the distinction between frogs and toads is more one of common usage than anything else. Many toads (for example *Bufo*) have heavier bodies and proportionately shorter legs than frogs and move by a rather ungainly walk or by short hops. Although the skin is more cornified than that of the frogs, toads and frogs lose similar amounts of water under the same experimental conditions but toads can survive a greater water loss. Like the newts, toads are crepuscular or nocturnal in habit, sheltering during the day and coming out in the evening to feed.

Frogs and toads are the only amphibians found in deserts apart from one newt species (*Triturus vittatus*) which has been recorded in arid regions. Many are not truly desert animals, being restricted to areas of permanent water, but others survive through burrowing, and live 300 mm below the soil surface where the humidity remains high. The spadefoot toads (*Scaphiopus*), as their name suggests, have modifications for digging in the form of well developed tubercles on their hind feet. In addition, during the dry season *Scaphiopus* secretes a special layer of skin which helps to reduce water loss. *Breviceps*, another toad which remains buried for long periods, is a specialized ant-eater with a long ant-eater-like snout. The characteristic adaptations of the hylid tree frogs such as *Hyla* is the development of round expanded pads on the tips of the fingers and toes which improve the grip, and a generally long-legged, slim-waisted body form.

Most anurans, including the two most widespread genera *Rana* and *Bufo*, show the reproductive pattern described for the frog, in which large numbers of small eggs are produced with little or no attempt to protect them. However some species have evolved methods of protecting their eggs from predators or have modified their development to enable it to take place in areas other than open water. For example some frogs construct a protective nest of foam made from a mucous secretion that floats on the surface of the water, and some tree frogs place their eggs on overhanging branches, the tadpoles dropping into the water as they hatch. A higher degree of protection is produced by those species which

Hyla

brood their young in the larval stages. The female of *Pipa pipa* carries her fertilized eggs in the puffy skin on her back where development proceeds; the supply of yolk in the eggs is rather greater than in *Rana* but in other respects the larval stages are similar and the animal leaves its mother as a small toad. In the European midwife toad, *Alytes*, the eggs are carried wrapped round the male's legs. In the case of *Gastrotheca*, the marsupial frog, the young are carried by the female in a dorsal brood pouch; in one species of the genus the young are released as tadpoles, while in another they develop in the pouch until they are small frogs.

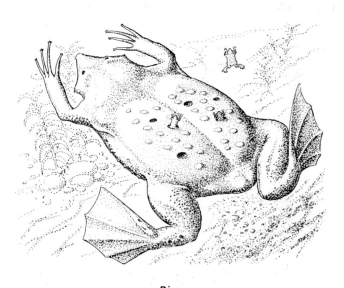

Pipa

Alytes

However the only anuran which can be regarded as truly viviparous is *Nectophrynoides occidentalis*, a West African toad which is faced with the particular problem of alternating wet and dry seasons. Fertilization takes place at the end of the wet season and the young develop in the oviduct whose walls have an increased blood supply and are expanded by villi. The developing larva receives nutrients from its mother across the oviduct walls and absorbs them through its tail.

Order Apoda

The order Apoda is the least known and most highly modified amphibian group and includes legless burrowing or aquatic forms with a world-wide tropical distribution. Apodans have very reduced eyes and a pair of protrusible tentacles lying in grooves above the maxillae which are thought to have a sensory function. Some species have small dermal scales.

The reproductive biology of the apodans is highly specialized. Internal fertilization is normal and the males have an intromittent organ which can be protruded from the cloaca. A few apodans lay eggs, but many are viviparous, with the young developing in the oviduct of the female. During the early stages of development the young feed on yolk, but in the later stages are supported by a thick secretion of the oviduct wall ('uterine milk') and by scraping the cells of the oviduct itself. The foetal gills are held closely against the highly vascular oviduct wall through which gaseous exchange also takes place. Waste products may be eliminated in the same way. The gills are absorbed before birth and subsequent foetal respiration is presumed to be cutaneous.

The fossil Amphibia: subclasses Labyrinthodontia and Lepospondyli

Many early amphibians, which must include forms similar to the ancestors of the modern Amphibia as well as of tetrapods as a whole, are known from the late Palaeozoic and have been divided into two now extinct subclasses, Labyrinthodontia and Lepospondyli. The earliest known forms occur in deposits from the late Devonian or very early Carboniferous and are members of the Labyrinthodontia group (the name comes from a somewhat fanciful similarity seen by an early palaeontologist, Sir Richard Owen, between the complex pattern of infoldings in the tooth enamel, and the Cretan labyrinth of Greek mythology). *Ichthyostega* (order Ichthyostegalia) is typical of the large newt-like forms (*Ichthyostega* was about 1 m in length) which had fish-like features including a tail fin, lateral line canals on the head, an operculum, and a covering of small scales. These features suggest a mode of life which was far more aquatic than that of present day amphibia. *Ichthyostega* has several features in common with the rhipidistians, which were predaceous members of the Crossopterygii, and are believed by many, though not all, zoologists to be the group from which the Amphibia arose. These features include a similar pattern of dermal bones distinct from that of other bony fishes, a 'tunnel' in the skull for the passage of the notochord, and similar 'labyrinthine' teeth.

The Labyrinthodontia flourished in the Permian. A good example can be seen in *Eryops* (order Temnospondyli) which was a rather dangerous looking beast, 2 m long with short sturdy limbs, a strong vertebral column, a well developed tail, and heavily armoured skin. It has a large broad head, very large mouth with extensive rows of small

Ichthyostega (reconstruction)

Eryops (reconstruction)

teeth and a small brain! In life *Eryops* must have looked rather like a modern crocodile, and probably filled a similar ecological niche, preying on fish and land vertebrates as well as on invertebrates. Most labyrinthodonts have been found in bog or pond deposits, which suggests that they probably spent much of their time in the water. The Lepospondyli which also occurred in the Permian were a group of small, more or less salamander-like forms, also from swampy deposits. They are recognized by the spool-shaped centra of their vertebrae. They are usually regarded as a side branch of amphibian evolution but may have given rise to the modern Amphibia.

The Permian was a period of extensive amphibian radiation, but their success was to be short lived. By the end of the Permian the predaceous reptiles had appeared; later labyrinthodonts developed heavy bony armour but even this did not enable them to survive in the face of increasing reptile pressure. Triassic labyrinthodonts are degenerate flat-bodied forms with tiny limbs and by the end of the period the group had died out. The third labyrinthodont order (Anthracosauria) is of particular interest since it includes forms which closely resemble the reptiles. It is from this group, although itself short-lived and with few species, that modern reptiles are thought to have evolved.

Synopsis of class Amphibia

Subclass Labyrinthodontia
   Order Ichthyostegalia
   Order Temnospondyli
   Order Anthracosauria
Subclass Lepospondyli
Subclass Lissamphibia: living amphibians
   Order Anura: frogs and toads
   Order Urodela: salamanders and newts
   Order Apoda: worm-like burrowing forms.

Introduction to the amniotes

The three higher groups of jawed vertebrates, reptiles, birds and mammals, are often considered together as the amniotes, a name derived from one of the extra-embryonic membranes in these forms and a reference to their more complicated development. Although some reptiles and most mammals bear their young alive, they share a similar general developmental pattern with the egg-laying forms, a pattern which contrasts with the simpler development of the anamniote fishes and amphibians. Even in the case of viviparous anamniotes, the distinction remains marked.

The amniote egg, already described using the chick as an example, is characteristic of the reptiles, birds, primitive (egg-laying) mammals, and, in a modified form, of advanced (placental) mammals. It is assumed to have been evolved by the reptiles and the earliest fossilized amniote eggs are found in Permian remains. Fossilized dinosaur eggs are frequently found in Mesozoic deposits. While the amniote egg is often referred to as a 'land egg', this description is misleading. Both some fishes and some amphibians have eggs which develop on land yet are of the anamniotic type; similarly many terrestrial invertebrates manage successfully with anamniotic eggs. Nonetheless, as was emphasized previously, the amniote egg is well adapted to conditions of dry land since it permits the young animal to achieve an advanced stage of development before having to face the hazards of terrestrial life.

CLASS REPTILIA

Modern reptiles (turtles, tortoises, crocodiles, lizards and snakes) are adapted far more completely for terrestrial life than the Amphibia and have successfully occupied a far wider range of terrestrial habitats as well as secondarily invading the water. Undoubtedly one important factor in this was the development of an amniote egg; this freed them from dependence on water for reproduction and is regarded as the diagnostic distinction between Amphibia and reptiles. However, it is not the only distinguishing feature and the course of evolution which produced a reptile from an amphibian was in many respects as complex as that which evolved an amphibian from a fish. The reptiles are more active and agile than amphibians, and better able to conserve water; they can also regulate their body temperature to some extent. In many respects, they can be regarded as completing the processes which partially adapted the amphibians for a terrestrial mode of life.

The reptiles are believed to have evolved from the ancient labyrinthodont amphibians. A few are known from the Carboniferous but by the early Permian they were relatively abundant. An important factor favouring the expansion of the reptile fauna, but largely unrelated to tetrapod structure, was the evolution of insects, the most successful of terrestrial invertebrates. The ancestral reptiles were carnivores and towards the end of the Carboniferous an abundance of primitive insects appeared for the first time. These, with their larvae, gave a basic terrestrial food supply for land vertebrates, which themselves in turn became food for larger flesh-eaters. The first animals are often thought to owe their existence to the abundance of plant life; in the same way the higher vertebrates owe theirs to the insects.

The Mesozoic era, which lasted for 160 million years, is often known as the Age of Reptiles. During this time a worldwide reptilian fauna diversified and radiated into both aquatic and terrestrial habitats. For a long period dinosaurs, ichthyosaurs, plesiosaurs and pterosaurs, including among them some of the largest animals that have ever lived, dominated the earth. In the case of most extinct animals it is easy to see how and why they were replaced by forms which were better adapted to prevailing conditions. However, in the case of the reptiles it is less so, and a wide variety of explanations have been offered from climatic change to mass dinosaur constipation, to explain their abrupt decline. The present day reptile fauna is limited to about 6000 species, mainly lizards, snakes and turtles, which are abundant in the tropics but unimportant in temperate climates. Although the class name is derived from the Latin *repere* meaning 'to creep', the lizards, with 3,000 species, are the largest single group. While generally remaining closest to the basic tetrapod plan. Three lizard species, *Lacerta agilis* (sand lizard) and *L. vivipara* (viviparous lizard), with the highly modified slow-worm (*Anguis*) are found in Britain, and a general description of the anatomy of *Lacerta* will be used to introduce the group.

Lacerta

Both *Lacerta vivipara* and *L. agilis* are widely distributed in Europe and the range of *L. vivipara* extends to the Arctic Circle. In the British Isles the common lizard (*L. vivipara*) is the most likely to be seen, and occurs in all types of country including gardens, commons, open woods and sandy and marshy areas. The sand lizard on the other hand is scarce, and as its name suggests, it is restricted to dry open country such as sand dunes or sandy heathland. In Europe however, it is less restricted in its habitats. Both are very variable in their colouration. While the male sand lizard is typically brown or greenish, it is sometimes a brilliant green. In common lizards a brown, yellow or dull red or, more rarely, a greenish hue may be the predominant colour. Both species often have either black or dark brown markings.

Lacerta has a typical tetrapod body form, with a long tail and two pairs of well developed legs, recalling that of the urodele amphibians. However, there are some important points of contrast. The head is elongated, deep and narrow, and is separated from the body by a well defined neck which gives it greater independence of movement. Urodeles have a rather clumsy, sprawling movement, with side to side bending of the body during locomotion. In lizards nearly all

trace of this is lost and the animal moves by rotating the humerus and femur backwards and forwards about their articulations. Well developed claws on the digits provide a firmer contact between the foot and the ground, and permit faster and more efficient movement. Other important but less conspicuous points of contrast between *Lacerta* and the urodeles can be found in the skin, and in the respiratory, excretory and reproductive systems.

The integumentary system

The skin of *Lacerta*, like that of all amniotes, is characterized by substantial thickening of the keratinized layer which overlies the epidermis. Since keratin is insoluble, and therefore water-proofing, this has the effect of minimizing water loss through the skin. Highly developed overlapping epidermal scales are characteristic of reptiles. They first appear during development as symmetrical thickened elevated areas in the integument, the papillae, each consisting of a core of dermis covered by epidermis. These grow and eventually tilt backwards. The stratum germinativum of the epidermis and the dermis gradually retract leaving a hollow keratin structure which collapses to form the characteristic thin, flat overlapping reptile scales. Unlike fish scales, reptile scales are continuous with each other at their bases. In microscopic sections, the scale can be seen to be composed of two regions, the outermost being thicker and harder. Periodically, the lizard forms a new set of scales to replace those damaged by wear and tear; these develop under the old layer which is shed as a whole or in large irregular pieces when the new development is completed, in a process known as ecdysis or moulting. Snakes show a similar moult but generally shed their skin entire. The frequency of moulting varies between species and individuals.

Reptile skin also differs from amphibian skin in an almost total absence of integumentary glands, which in the frog keep the skin permanently moist. However, both the sand and common lizard have a row of 7 to 13 femoral glands on

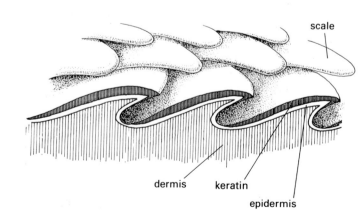

Lacertid skin

the under surface of the thigh. Each is a small blind sac, with an epidermal lining, whose cells are shed into the sac as they die, with the cellular debris accumulating as a little waxy core. The glandular contents are therefore derived from cellular breakdown rather than secretion. The debris is scentless and rough to touch and the glands are present in both sexes. The function of these organs is not known, although as they increase in size in the breeding season it is assumed to be sexual; one suggestion is that they prevent partners from sliding apart during copulation. In some lizards they are absent or vestigial in females.

The skeleto-muscular system and locomotion

Many of the adaptations of the amphibian skeleto-muscular system to the demands of terrestrial life are carried further in the reptiles, especially those concerned with the vertebral column, walking and breathing. The vertebrae articulate with each other by a system of interlocking processes, which are more elaborate than those of the Amphibia and increase the weight bearing capacity of the vertebral column. In *Lacerta* the centra are procoelous, a condition in which each has an anterior concave facet which fits into the convex socket of the preceding vertebra; this is the most usual reptile condition although a variety of types of centra may be found. The vertebrae are also united, as in all amniotes, by anterior and posterior (pre- and post-) zygapophyses. The notochord is represented by intervertebral discs.

The lizard shows considerable regional variation in the structure of the vertebrae and cervical, dorsal (combining the thoracic and lumbar regions of higher vertebrates), sacral and caudal vertebrae are clearly distinguishable.

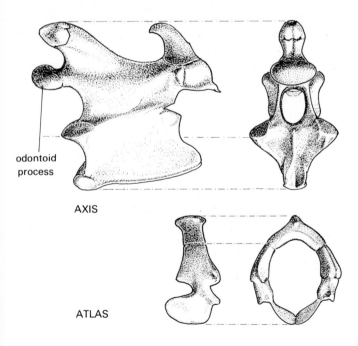

odontoid process

AXIS

ATLAS

Mammalian atlas and axis vertebrae

Among reptiles, and also in birds and mammals, the first two cervical vertebrae (atlas and axis vertebrae) are different from the rest. The centrum of the axis vertebra (the second vertebra) bears an anterior projection, the odontoid process, which represents the centrum of the atlas vertebra (the first vertebra); this process forms a pivot which allows considerable freedom of movement for the head. The atlas has an anterior concavity with which it articulates with a single occipital condyle of the skull. The remaining six neck vertebrae are freely movable, and bear small cervical ribs. The dorsal (thoraco-lumbar) vertebrae have wide neural spines and distinct pre- and post-zygapophyses. Broad headed ribs articulate with the vertebrae in this region and curve round to articulate ventrally at a sternum or breast bone. Since all those vertebrae bear ribs, there is no distinct lumbar series. The two sacral vertebrae, which support the pelvic girdle, have two short, broad sacral ribs articulating with the ilia. The caudal vertebrae become progressively reduced towards the tip of the tail; long transverse processes disappear about half way along the tail and neural spines disappear shortly afterwards, but neural arches and pre- and post-zygapophyses continue nearly to its tip. A peculiarity of many lizards is that the caudal vertebrae have a vertical unossified region, the fracture plane, across the middle of the centrum. In this region the vertebral column is easily broken and the tail can be cast off (autotomy) by sudden violent contraction of the caudal muscles. This is a defensive response, because the tail continues to wriggle for some time after it has been shed, thus distracting the enemy while the animal escapes. A shorter tail subsequently regenerates but the vertebral column is never fully replaced. It is represented by a tube of cartilage; the spinal cord is much reduced.

The basic plan of the reptile skull is similar to that of primitive amphibians though there is a greater degree of ossification and an increased density of bone. The ethmoidal and interorbital parts of the chondrocranium are almost entirely unossified. Reptiles also show a development of holes (fossae) in the temporal region of the skull to provide space and attachment for more strongly developed jaw muscles, and to allow them to extend to the outer surface of the skull.

The skull roof is formed of a series of dermal bones (nasals, prefrontals, frontals, supraorbitals and parietals) and the skull floor strengthened by an elongated parasphenoid which is fused to the basisphenoid. The sides of the skull are formed by the tooth-bearing elements, maxilla and premaxilla, together with the lacrymal, jugal, postorbital, squamosal, supratemporal and quadrate. Flanges from the premaxillae and maxillae, with a large toothless vomer, pterygoid, ectopterygoid and quadrate bones form the palate. Large holes (fenestrae) are present in the palate and the internal nostrils are typically, and in *Lacerta*, positioned anteriorly between the maxillae, vomers and palatines. Posteriorly, the skull floor is formed of the pterygoid bones and the parasphenoid. In *Lacerta* a

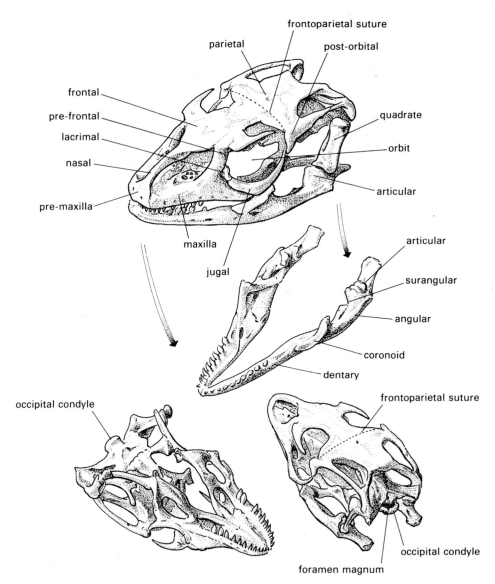

A representative lizard skull : *Varanus* (short-snouted form)

small epipterygoid extends from the pterygoid behind the orbits, the eye is surrounded by a ring of sclerotic plates and a thin interorbital septum lies between the eyes. The foramen magnum is surrounded by the occipital complex of bones and the skull articulates with the atlas at a single occipital condyle.

The lower jaw consists of several bones of which the dentary is the tooth bearing element and the articular and quadrate form the jaw joint.

The hyoid arch is represented by a basal plate (copula), often with the remains of three visceral arches attached to it. As in the frog a bone (stapes) is present in the middle ear. In the lizard the upper jaw and front of the skull can move relative to the occipital region of the skull about a straight transverse suture between the frontals and parietals. Skulls of this type are said to be kinetic; it is an adaptation which both widens the gape, and also perhaps helps to reduce the shock when the jaws are snapped together. When the lizard opens its mouth the snout swings up as the lower jaw is

moved down, and the reverse occurs as the jaw is closed. It is possible that this double action enables the animal to close its mouth more quickly, and allows the lizard to bite more than once to kill its prey without its escaping.

The limbs and girdles show the same general structural features as those of amphibians though without the extreme modifications for jumping found in the frog. The humerus and femur are normally held bent at an angle to the body, and the foot and hand are held bent forwards at right angles to rest on the ground. The principal muscles are therefore those that pull the humerus and femur backwards and forwards as well as downwards and the ventral regions of the girdles are highly developed for their attachment. The pectoral girdle, which is closely connected with the sternum, is formed of paired scapulae, coracoids and clavicles and a median interclavicle. The usual ilium, ischium and pubic bones form each half of the pelvic girdle.

Although the organization of the appendicular musculature is similar to that of the amphibians, the trunk

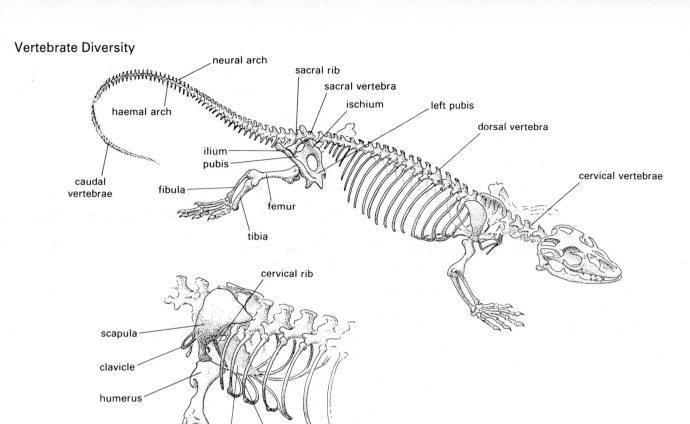

The skeleton of a representative lizard (*Varanus*)

muscles are highly modified; the differences reflect changes in their function as a result of the generally increased size and importance of the limbs and girdles, and the increased importance of the ribs. The longitudinal septum separating the epaxial and hypaxial muscles is generally absent in amniotes except in certain regions, such as the lizard's tail. The epaxial muscles are differentiated into several groups of muscles: they are retained in a largely unmodified form (longissimus dorsi) along the vertebral column but are split in their deeper regions into numerous small muscles between adjacent vertebrae or vertebrae and ribs. The longissimus dorsi control dorsal flexure of the vertebral column and are opposed by a subvertebralis muscle derived from part of the hypaxial muscle. The hypaxial muscles show additional and greater changes. Internal and external intercostal muscles connect adjacent ribs and are used to expand and contract the rib cage in breathing, while the body wall muscles (rectus abdominus, transverse and oblique muscles) act as a sling suspending the viscera. The main muscle involved in closing the mouth is derived from the adductor mandibulae of fish and is complex. A superficial portion extends from outside the lower jaw to insert on the maxilla, jugal and quadratojugal. A much larger part runs from the inside of the lower jaw to the brain case: it is in response to the development of this muscle that the skull has become fenestrated to allow the muscle to bulge as it contracts.

Locomotion

Locomotion in lizards is essentially similar to that of urodele amphibians, although side to side movement of the body is less pronounced. When the animal is hurrying, some bending of the body occurs which both lengthens the stride and increases the leverage. The legs are used as levers and are alternately advanced, placed on the ground, and pulled backwards. In this manner lizards can achieve short bursts of extremely rapid movement.

The digestive system and nutrition

Both sand and common lizards are carnivores and feed chiefly on insects and insect larvae which they hunt mainly by sight. Their prey are seized in the mouth and either crushed, or if larger, shaken violently until they are stunned. We have seen how lizards have greatly strengthened jaw muscles compared with Amphibia; they also have numerous small conical teeth along the edges of the jaws which not only grip the prey but are also used for chewing. There is a well developed protrusible tongue which in addition to manipulating the food has a sensory function as an accessory to the sense of smell. It can pick up particles and wipe them off onto the roof of the mouth where they can be 'tasted' by a pair of specialized and isolated structures below the nose known as the organs of Jacobson.

The oesophagus is elongated because of the development of the neck, and capable of distension to take large pieces of food. Digestion begins in the long, spindle-shaped stomach. The alimentary canal is completed on the typical vertebrate plan with an elongated coiled intestine opening into a cloaca. An ileocolic valve separates the ileum and colon and here a small pocket, the colic caecum, arises in the majority of reptiles. The function of this structure is not known but since it is largest in herbivores it is thought to act as a site of bacterial digestion. A lobed liver with a gall bladder, and a pancreas are present and both open into the duodenum by separate ducts. The cloaca is subdivided into three by folds of mucous membrane. One section, the coprodaeum, serves as a passage for faeces and is also a region of water reabsorption, the other, urodaeum, is for the passage of urine and eggs or sperm. These open into a common proctodaeum.

Respiratory system

In the lizard, as in the majority of amniotes, virtually all gaseous exchange takes place via the lungs. In contrast with the Amphibia there is no significant respiration through the mouth epithelium or the skin and the presence of internal nostrils means air is taken in through the nose and mouth. Thus the nose is adapted not only for smelling but as part of the respiratory system. The simplest reptile lungs are simple sacs with the respiratory epithelium in pockets in the walls; in *Lacerta* these pockets are deepened as alveoli, thus greatly extending the surface available for gaseous exchange.

The amphibian method of ventilation by movement of the pharyngeal floor is abandoned to be replaced by a more powerful suction pump mechanism: air is drawn in and out of the lungs by expansion and contraction of the thorax caused by backward movement of the ribs. The lungs lie in a continuous body cavity which also contains the viscera, and thus expansion of the thoracic region would tend to pull the abdominal wall inwards, reducing the effect of thoracic expansion. This does not occur because the body wall muscles are held rigid during inspiration.

Circulatory system

In association with their development of lungs as the only respiratory organs, reptiles have developed an efficient pulmonary circulation involving modification of the heart. The heart is three-chambered with two atria and a ventricle, but the ventricle is partially subdivided by an incomplete inter-ventricular septum providing a more efficient separation of oxygenated and deoxygenated blood than in Amphibia. The conus anteriosus is modified and no longer exists as a separate structure: its distal end is split into three

arterial trunks, the right and left aortae, and the pulmonary aorta which subsequently divides to form two pulmonary arteries carrying blood to the lungs. The opening of the pulmonary aorta lies opposite the left hand side of the ventricle and receives mainly deoxygenated blood. The left aorta also leads from the right side of the ventricle, almost opposite the septum, but crosses to the left side of the body; it receives mixed blood. The right aorta opens from the left hand side of the ventricle and receives oxygenated blood. Like amphibians therefore, the lizards retain aortic arches III (internal carotids), IV (right and left aortae) and VI pulmonary aorta). The venous system is based on a similar plan to that of the frog.

Excretion and osmoregulation

The urinary and genital systems of reptiles are completely separated in the adult. As is typical in amniotes, the posterior or metanephric portion of the kidney develops as the functional adult excretory organ. The kidneys are bulky and lobed structures which lie in the posterior part of the body cavity, close to the dorsal body wall. Lizards have between 3,000 and 30,000 nephrons in their kidneys; ultimately they discharge into a single ureter on each side which opens into the urodaeum. A urinary bladder arises from the ventral surface of the urodaeum and water is reabsorbed both in the urodaeum and the bladder. Many of the modifications of the excretory system are concerned with water conservation. The glomerulus is rather small, and, in some reptiles is lost altogether, and uric acid, rather than urea, is the final excretory product. Uric acid is sufficiently insoluble to precipitate as a white crystalline mass at concentrations attained in the kidneys, and its production is a valuable adaptation not only to the adult, but to the developing embryo. It would be impossible to store the nitrogenous waste of an embryo as urea within the confines of an egg.

Thermoregulation

Reptiles, like fish and amphibians and all invertebrates, cannot maintain their body temperature at a constant level by physiological methods as do birds and mammals, but instead cool down or heat up according to environmental conditions. (They are therefore said, rather misleadingly, to be cold blooded, or poikilothermic). This means that at low temperatures, for example in winter or at night, they are sluggish or totally inactive. Hence lizards and snakes can often be seen in the early morning especially in spring, basking in the sun to warm up sufficiently to move around and hunt. Despite this major physiological limitation lizards exercise considerable behavioural control by moving into and out of the sun, and orientating their body to absorb more or less of the sun's rays. In addition they show a limited physiological control by dilating or contracting the

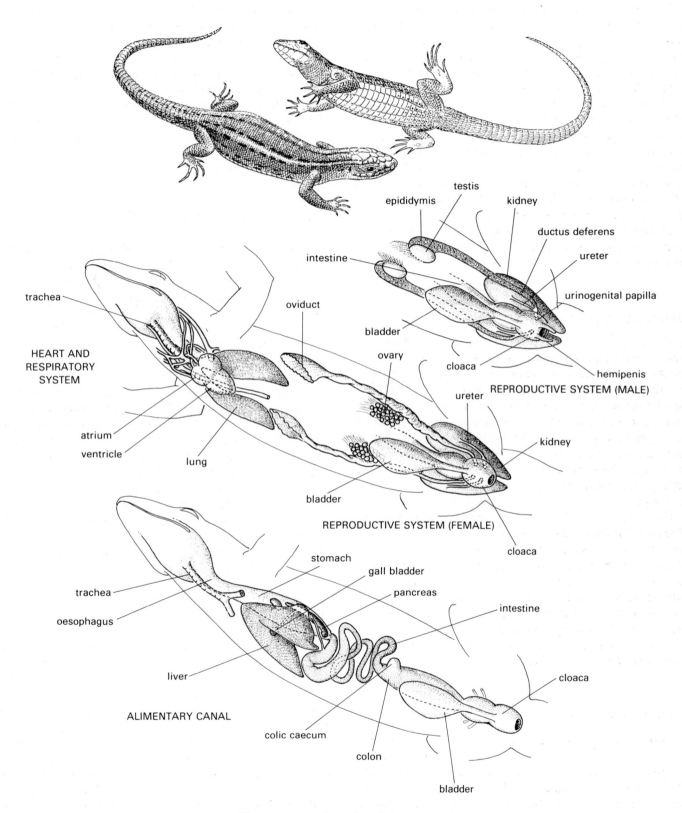

HEART AND
RESPIRATORY
SYSTEM

trachea

atrium

ventricle

lung

testis

epididymis

kidney

ductus deferens

ureter

intestine

urinogenital papilla

bladder

cloaca

hemipenis

REPRODUCTIVE SYSTEM (MALE)

oviduct

ovary

ureter

kidney

bladder

cloaca

REPRODUCTIVE SYSTEM (FEMALE)

trachea

oesophagus

liver

ALIMENTARY CANAL

stomach

gall bladder

pancreas

intestine

cloaca

colic caecum

colon

bladder

The anatomy of the lizard, *Lacerta*

blood capillaries in the skin and in some cases by colour change. A combination of these methods enables the lizard to remain active for the greater part of the day.

Nervous system and sense organs

The reptile brain is broadly similar to that of the amphibians but shows some interesting developments. The cerebral hemispheres are relatively larger, due to considerable expansion of the corpus striatum. The thalamus is also well developed, and receives connections from the optic tracts, which no longer run to the midbrain. Generally, nervous functions appear to be transferred from other regions of the nervous system to the forebrain. The olfactory region is well developed and has, in addition to the olfactory bulbs, a small swelling behind each main olfactory bulb which receives fibres from a separate vomero-nasal organ, the organ of Jacobson. These are paired structures, with a lumen lined by a sensory epithelium, which lie above the palate and open into the mouth cavity through paired ducts. Their function is olfactory and they smell particles of food which are transferred to the ducts on the tips of the tongue.

The eyes are the principal sense organs and are of the basic vertebrate pattern. In *Lacerta*, as in most lizards, they are protected by eyelids (including a third or nictitating membrane) and moistened by secretions from the lachrymal glands, which drain into the mouth through tear ducts; as these open near Jacobson's organs, the lachrymal secretion also acts as a solvent for scent particles. The eyeball is strengthened and supported by scleral bones. The lens is supple and focused by contraction or relaxation of the ciliary muscles; contraction makes the lens more spherical and therefore adapted to close vision. *Lacerta* has some colour vision and both rods and cones are present in the retina, with cones predominating. The eyes are positioned so that the field of vision overlaps by about 25°, giving the animal limited stereoscopic vision.

In many lizards, though not *Lacerta*, a functional pineal eye is present in the middle of the head, beneath a pineal foramen in the skull. The eye has a retina and a lens-like structure, and while it must be a light detector of some kind its function is uncertain. It may detect direct solar radiation and therefore influence the animal's thermoregulation behaviour.

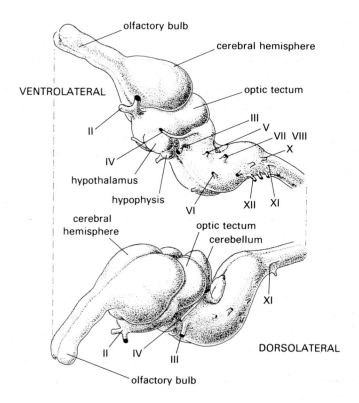

Lizard : diagram of brain showing principal regions and cranial nerves

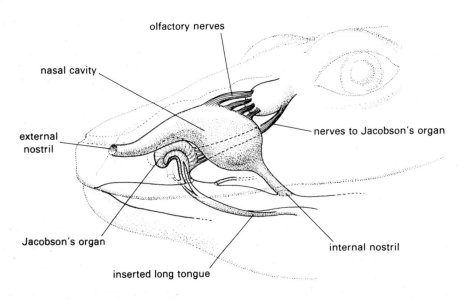

Jacobson's organ

The ears show little modifications. The sacculus, utriculus and semicircular canals are not essentially different from those of anamniotes although the lagena is more developed; a single ossicle (stapes) is present as in Amphibia. The tympanum is a thin membrane, devoid of scales, which lies at the back of the jaws and is sunk a little below the surface. (It is absent in some lizards and in all snakes.)

Reproduction

Lacerta vivipara produces living young but is the only member of the genus to do so. The gestation period is about 3 months and the litter of 4 to 8 young are normally born in July. The young are usually born in their embryonic membranes from which they soon escape. The sand lizard (*Lacerta agilis*) lays eggs in groups of 6 to 13, normally in June or July, and development again takes about 3 months. The ovaries are similar to those of amphibians and produce large yolky eggs. Fertilization takes place in the upper part of the oviduct and the various membranes are deposited around the egg as it passes down the oviduct. Neither the eggs of lizards nor snakes have albumen, and albumen glands are absent from the oviduct.

The testes of the male are compact and vary in size, becoming larger towards the breeding season. A vestigial Mullerian duct (oviduct) is present. The vas deferens from the testis joins the ureter of its side and the two open at a urogenital papilla. The intromittent organs are the hemipenes, paired hollow tubular cloacal sacs, which when not in use extend posteriorly within the tail base. These can be turned inside out and everted, one at a time, through the cloacal aperture into the cloaca of the female; the surface of the hemipenis is grooved to facilitate sperm transfer.

Copulation is normally preceded by a courtship which, in the sand lizard, takes place in May or June, when the males chase off possible rivals. The male seizes the female by her flank and curves his body around her, bringing his cloaca into position next to hers. In the viviparous lizard the copulation position is similar to that described, but rivalry between males and courtship behaviour is less marked.

REPTILE DIVERSITY

The emergence of reptiles in the late Carboniferous coincided with the evolution and radiation of terrestrial insects; the earliest reptiles were small, superficially lizard-like animals which, like the terrestrial amphibians of the time, diversified to exploit this abundant source of food. Having evolved at a time when there was little terrestrial competition the reptiles multiplied rapidly and spread into a wide variety of ecological niches; larger carnivores (2 to 3 m in length) developed and preyed upon the smaller insectivorous reptiles and amphibians. From these early beginnings, two main lineages arose: one of large

herbivores and one of large carnivores. Much of the adaptive radiation centred on changes in locomotion, and particularly in feeding: adaptations to new food sources involved changes in the morphology of the head and particularly in the jaw muscles. Because of this, and since the majority of reptiles are in any case known only from their fossils, it has become established practice to trace the various lines of reptile evolution primarily on the basis of their skull bones.

The dermal bones of amphibians not only roof the brain case, but also cover the jaw muscles which occupy the region posterior to the orbits and at the side of the brain case. In early reptiles, as in the labyrinthodont Amphibia, this region was covered dorsally and laterally by a solid layer of dermal bone and the only skull openings were those for the sense organs. This type of skull is described as anapsid (literally 'without an opening') and is characteristic of the stem reptiles (cotylosaurs) as well as modern turtles and tortoises which are believed to be their direct descendants (where the anapsid condition is often in a modified form). The jaw muscles originate on the medial side of the temporal bones, occupying a position between them and the brain case, and pass down through holes in the palate to insert on the medial surface of the lower jaw. Although this type of skull is strong and rigid it has certain practical disadvantages: in particular it is heavy and the total mass of jaw muscle is limited by the space available between the temporal bones and the braincase. In the other reptile subclasses, Euryapsida, Lepidosauria, Archosauria and Synapsida, these problems are alleviated by the evolution of openings (fossae) in the temporal region of the skull, surrounded by bony arches. These openings lighten the skull (although without reducing its strength since the arches provided rigidity), and both increase the area of

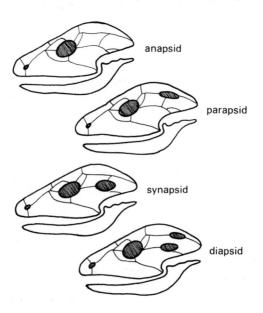

Reptilian skulls showing different arrangements of temporal openings

muscle attachment and provide space for the muscles to bulge when contracted. The number and position of the fossae vary (one or two may be present in either a high or low position) and they provide a basis for dividing the reptiles into subclasses. The Euryapsida, Ichthyopterygia, and Synapsida all have a single temporal opening: in euryapsids and ichthyopterygians this is located high in the skull, a condition sometimes known as parapsid. The synapsids (mammal-like reptiles) have a rather larger fossa, positioned low in the temporal region; the group derives its name from the original interpretation of this opening as the merging of two separate fossae. A diapsid skull, with two temporal openings, is found in two reptile subclasses: the more primitive Lepidosauria, to which the order Squamata (the lizards and snakes) belongs, and the Archosauria represented in recent fauna by the order Crocodilia (crocodiles).

Living reptiles

Although modern reptiles represent a limited remnant of the previously extensive Mesozoic groups, they do exhibit considerable morphological and ecological diversity. The turtles and tortoises, easily recognizable by their unique outer armour or shell (order Testudinata or Chelonia) have changed little since the Triassic. Within the group there is relatively little variation in form: terrestrial herbivorous tortoises, with a high domed shell and sturdy, somewhat elephant-like feet represent one adaptive extreme, while aquatic turtles, with a flattened shell and flipper-like feet represent the other.

Testudinata: turtles, terrapins and tortoises

The evolution of the Testudinata in the Triassic necessitated drastic reorganization of the reptile body form. The body was shortened and broadened, and the animal developed a box-like armour into which the head and limbs could be withdrawn. The shell consists of a dorsal carapace and a ventral plastron, each made up of bony plates, overlain by modified horny scales. The bony plates of the carapace are attached to, and incorporate the vertebral column and ribs, while those of the plastron are expanded dermal bones of the pectoral girdle and comparable to the gastralia (abdominal ribs) of crocodiles. The limb girdles have taken up an unusual position within the ribs and are much modified; the limbs themselves are short and stout especially in terrestrial forms but otherwise retain the basic reptilian structure. The skull has no true temporal openings and for this reason testudinates are placed in the Anapsida. The turtle *Chelone* shows the anapsid condition in its least modified form; in other forms the dermal bones are reduced by deepening the margins to form a long deep notch on either side, giving improved attachment for the jaw and neck muscles. Tortoises and turtles have no teeth, but these are effectively replaced by a horny beak. The soft anatomy shows several primitive features suggesting that the group

has changed little since the Permian; for example the brain and nasal regions are more comparable with those of the Amphibia than of other reptiles. However the evolution of a shell necessitated adaptations in breathing which is brought about by the action of the abdominal muscles and by pumping of the pharynx. There are about 250 species of living testudinates, popularly divided into tortoises (terrestrial forms), terrapins (freshwater forms) and turtles (marine forms). Systematists subdivide them into rather different groups on the method used to retract their head into the shell. The majority are cryptodires (suborder Cryptodira) which bend the neck into a vertical S-shape; pleurodires (suborder Pleurodira), restricted to two freshwater families, retract their head by bending the neck horizontally. Ancestral testudinates, about which relatively little is known, could not retract their heads and had teeth, unlike the modern forms which use a horny beak.

Tortoises, for example *Testudo*, are mainly herbivorous, and live in deserts, grasslands and woods in many regions of the world. Their proverbial slowness is a necessary consequence of their heavy armour. Their metabolism is also slow but some of them are very long-lived and though the records are unreliable, some are known to have reached an age of 152 years or more. Among the longest lived are the giant tortoises of the Galapagos Islands, where slightly different subspecies are present on the various islands of the archipelago. Even larger fossil forms are known (*Testudo atlas* was about 2.4 m long) while at the other extreme the smallest species may be less than 150 mm long. The African pancake tortoise has a flattened flexible shell with reduced ossification; however in compensation it can scramble over the rocks in which it lives with considerable agility and protects itself from predators by hiding in rocky cracks, wedging itself in position with its legs. Snapping turtles (*Chelydra, Macroclemys*) are aquatic forms which prowl about in ponds and rivers. They are effective predators, stalking and capturing such prey as fish and water birds.

Macroclemys

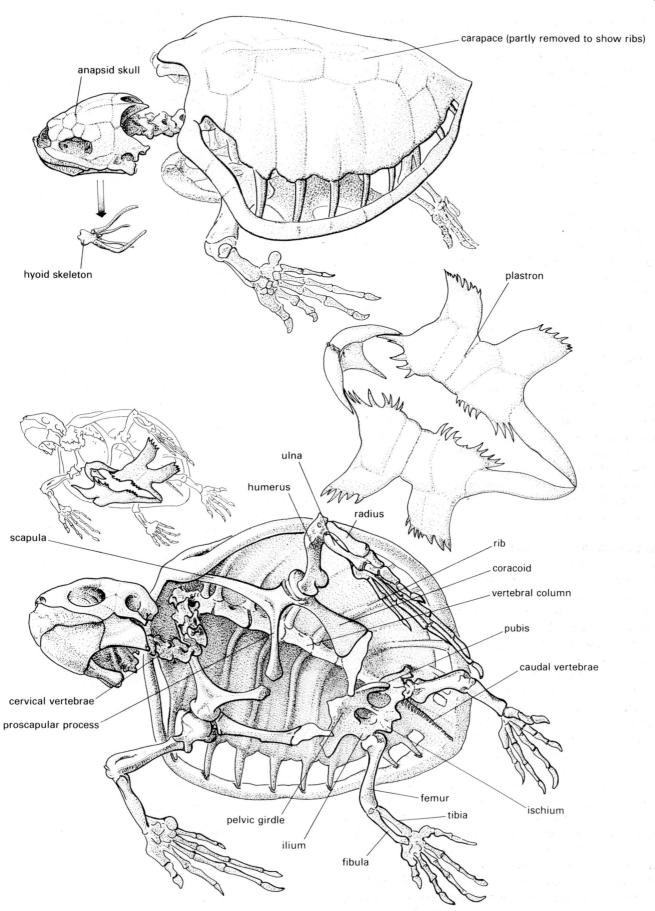

carapace (partly removed to show ribs)

anapsid skull

hyoid skeleton

plastron

scapula

ulna

humerus

radius

rib

coracoid

vertebral column

pubis

caudal vertebrae

cervical vertebrae

proscapular process

pelvic girdle

femur

ischium

ilium

tibia

fibula

The testudinate skeleton

Soft-shell turtles (Trionychidae) are specialized as fast swimmers with much less bony shells, a streamlined shape and large webbed feet.

The Squamata

The great majority of living reptiles (more than 95%) are members of the Squamata (subclass Lepidosauria), the scaled reptiles, a group which includes four living suborders, though two of them (snakes and lizards) include 98% of the species in roughly equal numbers. The suborder Rhynchocephalia (the tuataras or sphenodons) contains only a single living species (*Sphenodon*) and is restricted in distribution to a few islands of New Zealand. It is regarded as the most primitive and generalized living reptile group, the nearest living approach to a stem reptile. It is superficially lizard-like, and its skull is of the diapsid type in its most basic form. It has a long life span (one specimen is recorded as having lived for fifty years in captivity) and its eggs require a year to develop. It feeds largely on invertebrates and is nocturnal in habit, but raises its body temperature by day by basking in the sun. The amphisbaenians (suborder Amphisbaenia) are a small group of highly modified, usually legless reptiles with highly specialized skulls and teeth and much reduced eyes. They are burrowers and at a quick glance bear a strong resemblance to large earthworms.

Of the two main suborders of Squamata, the lizards (Lacertilia which includes the genus *Lacerta* already described) are the older and more generalized. The lizards range in size from the smallest geckos, a mere 20 mm long, to monitor lizards 3 m in length, and have undergone extensive adaptive radiation into a variety of habitats, including terrestrial, arboreal, burrowing and aquatic forms. Many are highly adaptable and the group occurs in conditions ranging from swamp to desert. While the majority are carnivores, most of the larger species are partly or wholly herbivores. Iguanids are mainly arboreal inhabitants of Central and South America, but the great Iguanas of the Galapagos (archipelago) and South Caribbean are terrestrial, probably because natural predators are absent on these islands. The marine Iguana of the Galapagos (*Amblyrhynchus*) is highly unusual; it lives on the beach and feeds on seaweed and will enter the water to feed, diving up to 10 m to reach suitable food. Chameleons are highly specialized arboreal forms with peculiar feet in which the toes are fused into two opposing groups (zygodactylous) and a long prehensile tail. They are very slow moving but can cling in any position on twigs and foliage and their skin colour changes within a few minutes to blend with their surroundings. They have large, independently movable eyes and feed on insects which they catch by means of a long sticky tongue which is protracted at high speed. In contrast another arboreal group, the geckos, have evolved expanded digital pads furnished with fine ridges of bristles; these are so effective at clinging to smooth surfaces that geckos, which are ubiquitous in homes in the tropics, can run up window panes and even over the ceiling.

Chamaeleo, an arboreal lizard

Gekko

Many lizards are adapted for life in dense herbage or in rock crevices, and some species burrow. In such habitats normal locomotion is impossible and many of these species have evolved an elongated snake-like body form with the loss of their appendages. The slow worm *Anguis*, like *Amphisbaena* shown here, is a familiar example of a reptile which has lost all external signs of limbs, but the various species of burrowing skinks show various levels of limb loss.

The large monitor lizards (Varanidae) are a group of active predators, feeding on a large variety of vertebrates and invertebrates. The Komodo monitor lizard (sometimes called the Komodo dragon) grows to about 3 m long and is a formidable predator which waits in ambush for such prey as deer, goats and even water buffalo. This systematic hunting is a contrast to the more opportunistic behaviour of most reptiles and almost comparable with that seen in some mammalian carnivores such as cats.

Amphisbaena

The snakes (suborder Ophidia) are extremely specialized reptiles, limbless and with greatly elongated bodies. They have an increased number of vertebrae (up to 400), with additional articulations on the vertebral centra, the ventral zygospheres, and move by lateral undulation. A comparative study of the anatomy of lizard and snake eyes suggests that snakes evolved from lizards via a burrowing (fossorial) phase in which the eyes were reduced. In modern snakes the eye is covered by a transparent spectacle instead of movable eyelids. Their highly developed Jacobson's organ, working in conjunction with the protrusible forked tongue also suggests a phase in which taste and smell become the primary senses. Snakes have far more flexible bodies than legless lizards and taking advantage of small irregularities, can crawl easily even over fairly smooth surfaces. Most have enlarged transverse ventral scales which help to prevent slipping and a few (some boas and vipers) can progress in a straight line by muscular movements using these scales as fixed points (rectilinear locomotion).

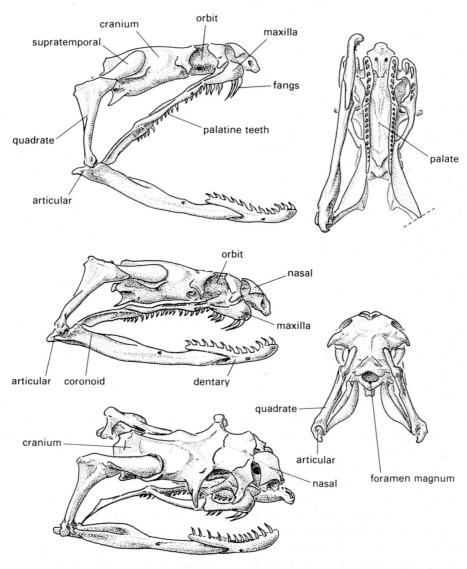

Skull of a representative snake, *Bungarus* (the krait)

Snakes are notable, even notorious, for their ability to swallow prey which is several times larger than themselves. Having taken a large meal, for example a small pig, a python may go without food for many months. Swallowing those outsize meals is made possible by a highly flexible jaw mechanism, combining a kinetic skull with other joints, including one between the anterior ends of the lower jar which are not united, and a joint between the quadrate and squamosal. The brain case is strengthened ventrally to avoid damage to the brain during swallowing. In addition they have very flexible and elastic skin over the chin and throat. In association with this ability to take large prey, snakes have evolved various methods of immobilising their victims; obviously swallowing a very large and angry animal is not a practical proposition! Some snakes (boa constrictors, pythons) coil themselves round their prey and suffocate it by contracting the powerful muscles of the body wall. Others have developed mechanisms for injecting poison into their prey with specialized oral glands associated with hollow or grooved fangs (cobras, mambas, vipers). Not all snakes are big; the smallest grow to 100 mm and feed on termites, the biggest large constrictors to 10 m.

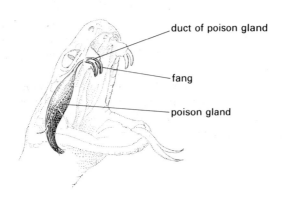

duct of poison gland

fang

poison gland

Venom apparatus

Living archosaurs: the crocodiles

The crocodiles (order Crocodilia) are the sole surviving representatives of the large and impressive order Archosauria. They are in many respects the living reptiles most like the Mesozoic forms. They are specialized for a partially aquatic mode of life, having external nostrils at the tip of the snout, with a secondary palate carrying air to internal nostrils which open at the back of the mouth cavity. A crocodile can therefore breathe with only its nostrils exposed. The largest crocodiles are the Indo-Pacific saltwater fish-eating crocodiles which live in estuaries and mangrove swamps and grow up to 7 m long or more. Crocodiles show relatively little structural diversity, though a distinction can be made between broad-and narrow-jawed forms. Narrow-jawed crocodiles are generally fish-eaters, the commoner broad-nosed forms having a more generalized carnivorous diet of fish, turtles, birds and mammals.

Fossil reptiles

The earliest reptiles, cotylosaurs, were small, lightly built animals which occupied a similar ecological niche to modern lizards, but exhibited many primitive features. They radiated into both carnivorous and herbivorous modes of life and produced a number of specialized lineages which died out without leaving any descendants, as well as giving rise to the remaining reptile subclasses and, through them, to the birds and mammals.

The parapsid (euryapsid) skull type, with a single temporal opening in a high position, occurred in two reptile subclasses, neither of which have any living representatives. The Euryapsida (plesiosaurs) and Ichthopterygia (ichthyosaurs) were prominent groups in the Mesozoic which gave up life on land in favour of the sea. The plesiosaur has been described as 'a snake strung through the body of a turtle', a description which is apt, although at first reading improbable. It had a short, broad, inflexible body, with a short tail, and either a very long neck or an elongated snout or both. With this ungainly shape, fish-like locomotion was impossible and the plesiosaurs rowed themselves about, using powerful oar-like limbs. They preyed upon bony fish, using their flexible neck and/or elongated head as an effective means of chasing their agile prey. In contrast the ichthyosaurs were almost ideally fitted for a marine mode of life, and parallel the fishes and whales in the possession of a streamlined body with no neck, and a dorsal fin. The limbs were reduced to steering structures and the tail provided forward thrust. Fossil ichthyosaurs have been found with squid pens in their body cavities and squid may have been their chief source of food. Plesiosaurs may well have come ashore to lay their eggs, in the same way as modern marine turtles; ichthyosaurs however, could not have done this and are thought (from the discovery of a preserved mother with young) to have been viviparous.

The lepidosaurs (subclass Lepidosauria) and archosaurs (subclass Archosauria) are groups which include both fossil

Plesiosaurus (reconstruction)

Stenopterygius, an ichthyosaur (reconstruction)

Tyrannosaurus, an archosaur (reconstruction)

and living representatives. The earliest lepidosaurs were more or less lizard-like, and their descendants include the Rhynchocephalia which was never a major group but survives today represented by a single form *Sphenodon*, as well as modern lizards and snakes. Archosaurs have a diapsid skull similar to that of *Sphenodon* but with additional fossae in front of the eye (antorbital fossa) and in the lower jaw at the junction of the dentary, angular, and surangular (mandibular fossa). Generally the legs were held beneath the body and later forms showed increased bipedalism, an adaptation that is still seen in a modern crocodile when on land and in a hurry. The tail, which was generally well developed in these animals, probably evolved as a counter weight to facilitate bipedal locomotion. The archosaurs were the ruling reptiles of the Mesozoic, including the well known dinosaurs (literally 'terrible lizards'), as well as giving rise not only to the crocodiles but to modern birds. If the tortoise is the most familiar living reptile, *Tyrannosaurus rex* must be, in the popular imagination, the most formidable. The largest flesh eater ever known, though thought to have been mainly a scavenger, it was 15 m long and stood over 6 m high, with a skull length of about 1.75 m and fearsome shearing teeth; it must have been an impressive sight! The ancestral archosaurs were a Triassic group known as the thecodonts and may have early showed a tendency (not always preserved in their descendants) to bipedalism. Generally their hind legs were considerably longer than the fore legs. Their evolutionary importance is considerable since they gave rise to several lines of adaptive radiation. These include aquatic forms with a similar mode of life to the crocodiles, two lines of flying forms, pterosaurs and birds (Aves), and the saurischians and ornithischians which together are the animals popularly known as dinosaurs. The pterosaurs show a high degree of convergent evolution with the birds in response to the same problems posed by flight and life in the air, but they developed wings of an entirely

different design. The fourth finger of the fore-limbs was greatly elongated and supported a membrane of skin attached to the body and hind legs. Primitive forms had a long tail expanded as a vane, which they used for steering, but later forms had no tail. The wing structure must have been extremely vulnerable to damage, and it is hard to see how they could have healed effectively in a short space of time. Nevertheless the pterosaurs, judging from the diversity of fossils were quite successful in the Jurassic and Cretaceous, ranging in size from the sparrow-sized *Pterodactylus* to *Quetzalcoatlus* which had an almost unbelievable 15 m wing span.

Rhamphorhynchus, a pterosaur (reconstruction)

The ornithischians were a group of herbivores which radiated into a wide range of body forms. The group includes large bipedal animals such as *Iguanodon*, with a length of 10 m, as well as the bizarre, quadrupedal

Iguanodon, an ornithischian (reconstruction)

Dimetrodon, a pelycosaur (reconstruction)

ankylosaurs covered with a heavy armour of plates and spines; in some of these the tail was modified as an enormous spiked club. By contrast the ceratopsians (horned dinosaurs) had large defensive horns on the head as well as bony frills over the neck region. Generally these last two were specialized for a peaceful, sedentary mode of life though it has been suggested from studies of their skeleton, that they could achieve speeds of 30 miles an hour. We might gain the impression, from the size to which herbivores grew, their defensive armour and adaptations for running, that the age of reptiles must have been a dangerous time to be alive!

The saurischians include two groups, one of bipedal carnivores and one of quadrupedal herbivores. Among those are the large dinosaurs which attract attention in museum displays: *Tyrannosaurus, Gorgosaurus,* and the giant amphibious *Diplodocus. Diplodocus* is common in museum collections for a most unscientific reason; it was given the specific name *D. carnegiei* in the honour of millionaire Andrew Carnegie, who was highly delighted and distributed casts of it to a number of museums. It was a spectacular animal and a strong contender for the length record, with a total measurement of 29 m; however the head is small and the teeth feeble, and it probably lived in swamps feeding on soft vegetation. Certainly it seems unlikely that an animal of such bulk could have moved effectively on dry land. The Synapsida (mammal-like reptiles) are characterized by a single temporal opening on the side of the skull. Pelycosaurs were the early members of the group, and apart from their fossae, differed little from the stem reptiles. Their advanced mammal-like descendants (Therapsida) flourished from the mid-Permian to the mid-Triassic, and like them radiated successfully into both carnivorous and herbivorous forms. *Dimetrodon* (a pelycosaur) was unusual in having a large 'sail' along its back, in the form of a web of skin, supported by greatly elongated neural spines. Many

ideas have been put forward as an explanation for this structure ranging (*sans* web) as camouflage in reed beds, to a sail by which the animal moved over the water. Somewhat prosaically, it is now generally accepted as a heat-regulating device.

The therapsids get their common name from the various mammal-like features found in the group. One of these is the possession of distinct, anterior piercing-teeth (canines) and posterior chewing-teeth which is seen in *Cynognathus* (translated as 'dog-jaw') with its superficially dog-like skull and heterodont teeth. In therapsids the limbs are held in a different position with the knees and elbows pressed into the sides of the body to give a typically mammalian posture

Cynognathus, a therapsid (reconstruction)

and gait associated with another mammalian tendency to equalize the length of the digits. Perhaps more significantly, advanced therapsids have developed a secondary palate, a structure normally associated with endothermy. Probably later theriodont therapsids were cursorial predators, capable of sustained running. The therapsids are a tantalizing group in that they leave so many questions unanswered; did they, for instance, have hair, or nurse their young? There is no evidence for either suggestion, but the widespread therapsid fauna suggests that they were successful animals in their time. However, these animals evolved in the Permian and early Triassic and then apparently vanished, to be superseded by reptiles of archosaurian lineage, especially the 'dinosaur' fauna which, by any criteria of success: size, species number, or diversity, dominated the world until the end of the Cretaceous. Nevertheless we assume that they left descendants, which however insignificant in their own time, were the ancestors of the mammals.

Synopsis of class Reptilia

Subclass Anapsida: characterized by a solid roof in the temporal region of the skull
   Order Cotylosauria: primitive reptiles
   Order Testudinata (Chelonia): tortoises and turtles
Subclass Euryapsida
   Order Sauropterygia: the plesiosaurs
Subclass Ichthyopterygia
   Order Ichthyosauria: the ichthyosaurs
Subclass Lepidosauria
   Order Eosuchia: ancient lepidosaurs
   Order Rhynchocephalia: *Sphenodon*
   Order Squamata: lizards and snakes
Subclass Archosauria
   Order Thecodontia
   Order Saurischia
   Order Ornithischia
   Order Pterosauria
   Order Crocodilia: crocodiles
Subclass Synapsida
   Order Pelycosauria
   Order Therapsida

CLASS AVES

The two most advanced groups of vertebrates, the birds (class Aves) and the mammals (class Mammalia) are descendants of reptiles which underwent further adaptations to the terrestrial environment. The birds are believed to have evolved from the archosaurs in the Jurassic, and were the second archosaurian group to develop flight. In many respects birds are merely specialized reptiles since most of their peculiarities are associated with flight, which requires a high level of metabolism and a body of minimum weight. Undoubtedly birds are the most successful flying vertebrates; both the pterosaurs and the mammals (bats) have evolved flight, but neither group achieved the same diversity or domination of the air as the birds. They are represented today by about 8,700 living species (with tens of thousands of fossil forms) occurring abundantly in every climate from the polar regions to the tropics, and are found on the continents, over the sea and on small islands. Hardly surprisingly, birds are not adapted to life totally underground or underwater (although some are remarkable swimmers and divers), but apart from this they have invaded most vertebrate habitats. Since flight necessitates a large surface to weight ratio, birds do not attain the size of larger mammals (apart from some flightless forms). Nevertheless, many species occur in large numbers (estimates include 100 million individuals in Britain, with 100,000 million world wide) and are a highly important component of many ecosystems.

Feathers

As a class, birds are remarkably uniform in their gross structure, though diverse in their behaviour and ecology. The main differences from reptiles are various adaptations for endothermy and for flight. Both of these depend on, among other things, the possession of feathers, which is the main feature which distinguishes birds from all other animals. Feathers are interpreted as modified reptilian scales, and probably evolved as insulation, the adaptation for flight being secondary. They are composed mainly of β-keratin, similar to that found in reptile scales and mammal hair. This homology is emphasized by a study of their development. In addition to their feathers, birds have epidermal scales of the reptilian type on their legs and feet. The beak is an expanded and enlarged keratinous structure, in a sense comparable with an epidermal scale.

Feathers are the most conspicuous features of the bird integumentary system, providing a covering which gives the bird its characteristic shape (birds look very different when plucked) as well as providing for insulation, flight, camouflage and sexual display. Four types of feathers: contour, flight, down and filoplumes are present. Contour feathers and flight feathers have essentially the same structure, but contour feathers cover and shape the bird while flight feathers provide the necessary light yet strong membrane, impermeable to air, which is essential for flight. These feathers have a central stem or rachis carrying a broad flat vane, made up of numerous barbs, held together by barbules which carry hooklets. The barbules of one barb attach by their hooklets into grooves of the next, to give a flat, wide, resilient and impermeable surface. An advantage of this structure over a membrane of skin is that if the vane surface is damaged the bird can easily relock the barbules by running the feather through its beak. (Birds spend a considerable amount of time preening their plumage.)

Flight and contour feathers arise from distinct regions of skin known as pterylae or feather tracts. The other two kinds of feather are down feathers and filoplumes. The insulating properties of down are well known, and exploited by man in the eiderdown which, in the past at least, was filled with down feathers of eider ducks. Nowadays the down from several species of domestic birds is used for the same purpose. Each down feather (plumule) consists of a short hollow quill bearing a number of barbs. They form the first fluffy covering of young birds but are also abundant beneath the contour feathers of the adult. Filoplumes are fine and hair-like in appearance and of uncertain function; they are particularly conspicuous in the peacock where they are of unusual length and stick out above the contour feathers.

Feathers have muscles at their base and their positions can be altered in response to the demands of heat regulation, flight, or sexual display. They also function as sense organs, and nerve fibres are associated with the papillae (which are the living dermal bases of feathers).

The plumage of all birds has characteristic colours and patterns, some of which play an essential part in recognition, behaviour and camouflage and may be remarkably beautiful. The pigments are present in the keratin and are principally melanins and carotenoids: melanins range from black to light tan, and are manufactured by the bird; carotenoids provide the bright yellows, oranges, and reds and are taken in with food and deposited in the feathers or skin. Other colours, particularly the transient iridescence of plumage, are caused by variations in the microscopic surface structure of the feathers which reflect various portions of the light spectrum.

The early growth of feathers is comparable to that of reptilian scales, each one beginning as a dermal papilla with an outer covering of epidermis. The papilla is conical

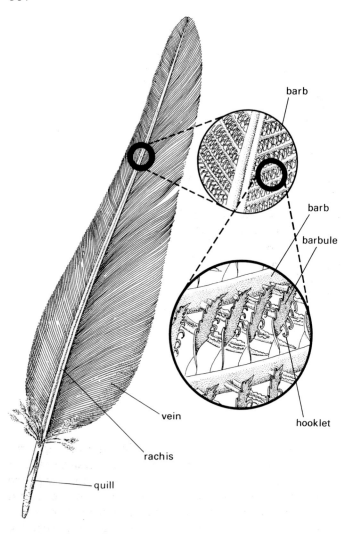

Flight feather

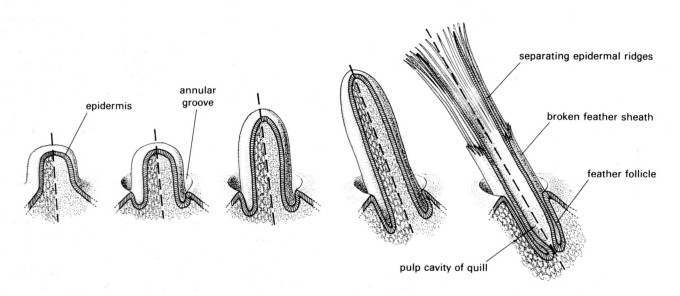

The development of a down feather

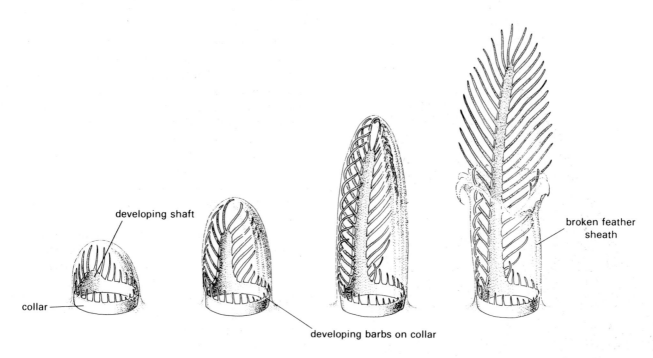

developing shaft

collar

developing barbs on collar

broken feather sheath

Later stages in the development of a contour feather

rather than flattened but otherwise the structure is the same as a developing scale. As growth continues an annular groove appears at the base of the papilla which deepens, forming a pit into which the base of the papilla sinks, and ultimately forming the feather follicle. To form a down feather the inner keratinised epidermal layer (periderm) forms a series of longitudinal ridges and the feather grows rapidly from its basal collar to project above the skin. The dermis and stratum germinativum retract, the outer layer is shed, and the epidermal ridges dry, crack, and separate to form a feather. Contour and flight feathers have a basically similar development but the basal collar puts forth an especially heavy growth which forms the shaft and pulls with it the barbs, which form continually on the collar, as it grows. When the feather is mature the periderm splits and falls away or is removed by preening and the vane unrolls. Feathers are shed at moulting, which occurs periodically often before or after the breeding season or at both times. It is a gradual process, unlike moulting in reptiles, and the feathers are not shed all at once, even from the same area of skin. The new feathers are produced from the old follicles. If a feather is plucked, the replacement starts to develop immediately.

Flight

The skeleto-muscular system of birds is highly modified for the demands of flight and shows many features unique to the class. To understand the problems of flight, and hence the significance of these anatomical modifications of birds, we must consider briefly some aspects of aerodynamics. A bird's wing (like that of an aircraft), is an aerofoil, shaped so that it is thick in front, and tapered behind; moreover it is usually cambered so that it is convex on the upper surface and slightly concave or flat on the lower surface. When the wing is moving forwards relative to the air the flow of air across the wing will be greater over the longer, upper surface than over the shorter, lower surface. Bernoulli's principle states that in a fluid stream the pressure is least where the velocity is greatest; thus in the case of an aerofoil the pressure beneath the wing is greater than that above it and the difference provides the lift necessary for flight. For a bird to fly the lift force must exceed the force of gravity on the bird. The amount of lift can depend on several factors. The angle made between an aerofoil and the horizontal is the angle of attack. If this is increased up to about 15° the lifting force also increases: however above 15° the air ceases to flow smoothly over the wing surface and becomes turbulent and the aerofoil may fall. An aerofoil can be modified to function at higher angles of attack if a subsidiary aerofoil is fitted ahead of the wing, which deflects the airstream onto the wings and restores a smooth flow.

To commence flight a bird requires some kind of propulsive force to create airflow over the wing's surface which provides the lift. This may be achieved in several ways. The simplest type of flight is gliding in which the forward motion comes from falling in a controlled manner through the air. Altitude is inevitably lost, albeit slowly, but can be regained if the bird soars. In soaring the bird

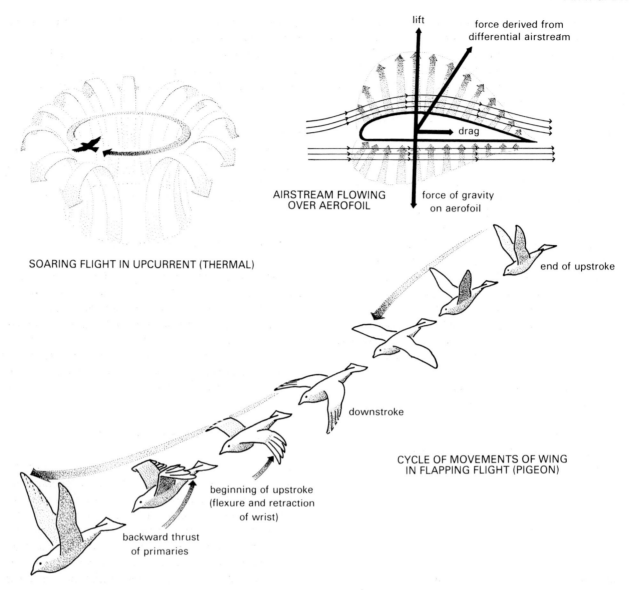

lift

force derived from
differential airstream

drag

AIRSTREAM FLOWING
OVER AEROFOIL

force of gravity
on aerofoil

SOARING FLIGHT IN UPCURRENT (THERMAL)

end of upstroke

downstroke

CYCLE OF MOVEMENTS OF WING
IN FLAPPING FLIGHT (PIGEON)

beginning of upstroke
(flexure and retraction
of wrist)

backward thrust
of primaries

Flight

increases or maintains its height by exploiting winds in which the direction is not horizontal (up-currents) or the velocity not uniform. Large birds such as sea gulls can often soar by circling and manœuvring within an upwardly directed air stream, caused either by an obstacle such as a cliff or by rising warm air from heating over sun-warmed ground (thermals). This type of soaring is static soaring and is particularly important for very large birds such as eagles or vultures. On their migration flights across stretches of sea where such up-currents do not occur, for example the Straits of Gibraltar, eagles must wait over the land and take advantage of up-currents to give them sufficient altitude to be able to glide across the open water. Dynamic soaring makes use of changes in air speed with increased height. Over the ocean surface, in conditions of a steady wind, the lower layers of air are slowed down by friction with the ocean surface. Such differences can be quite large and extend up to about 30 m above the sea. Static soarers such as the golden eagle have short, broad wings that enable them to manœuvre easily in a confined space, dynamic soarers have long narrow wings with a maximum lift. They include birds such as albatrosses and petrels which may spend months out over the oceans always flying low over the water where they can make the best use of the gradients of wind speed.

In flapping flight the wings provide both propulsion and lift. The wing movements involved are complex and variable to achieve flight, ranging from the simple hovering movement of humming birds to a complex figure of eight movement used by a pigeon. In summary, the wings move forwards on the downstroke and upwards on the backstroke. In addition the posterior margin of the distal part of the wing is twisted up on the downstroke and down on the upstroke. Forward movement results from pushing the wing against the air (providing thrust), the direction of this thrust being controlled by the wing's pitch. In all types of flight the tail provides support and balance and functions as a rudder.

Birds are endotherms with a high, usually constant, temperature of between 40 and 41° C maintained independently of their surroundings and derived from internal heat production. This high body temperature is achieved partly by high rates of resting metabolism (birds average five to ten times the metabolic rate of a reptile per unit of body weight at the same body temperature) and partly by the insulation provided by the feathers which reduces heat loss. The advantages of a high body temperature to a bird are obvious: the output of power from a muscle is approximately double for every 10° C rise in temperature; consequently a bird at 40° C has potentially about eight times as much locomotory power as a reptile with comparable muscles working at 10° C. In addition to the need for insulation the maintenance of a high body temperature requires anatomical adaptations related to a high and constant supply of oxygen and a regular supply of calorific food.

The various anatomical specializations of birds will be discussed with reference to a well known and widely distributed bird, the pigeon (*Columba*).

Columba: the pigeon

The pigeon, *Columba*, is a widely spread genus both in the wild and as a domesticated or feral animal. The majority of species are at least partly arboral, but a few are terrestrial or cliff dwelling, including the rock pigeon, *Columba livia*, which is the ancestor of the domestic pigeon and is a native of Europe, Western Asia, India, and North Africa. In the wild it is found associated with cliffs and rocks, and may come into towns, particularly ports, where it nests on the ledges of buildings. The wood pigeon, *Columba palumbus*, is larger (410 mm rather than 300 mm long) and is found in wooded country or farmland, as well as in town parks and gardens. The colour of the plumage is variable but based on black, white or blue-grey with iridescent particles on the neck. As a domestic animal *Columba livia* has a very long history of use in 'homing', either as a carrier of messages or for the sport of pigeon racing. It was widely used as a messenger by Ancient Greeks and Egyptians. In the second world war 17,000 pigeons were parachuted to the Resistance forces in occupied Europe and 2,000 returned safely carrying their messages; many more were used to bring SOS messages from aircraft which had crashed into the sea. Pigeon racing as a sport developed with the expansion of the railways and was established by 1875,

normal distances are about 500 miles and the course is always completed in less than a day.

Pigeons have long been reared for the table. Until the seventeenth century pigeon-cotes were found on most farms and great estates. These cotes housed hundreds of birds which formed a valuable source of meat in the winter as well as manure for crops. Numerous recipes exist for stewed pigeons (often in blood!), or pigeon pie and many different pigeon varieties were recognized. They were eaten from the squab (nesting stage) onwards. On Sark, because of their damaging effects on corn crops, only the seigneur of Sark has the 'droit de colombier'; the right to rear pigeons.

The integumentary system

Bird skin is thin, loosely fitting, and loosely attached to the underlying fascia. It is dry and almost totally glandless. The only cutaneous gland is the uropygial or preen gland, which is a large compound gland, lying at the base of the tail on its dorsal side and secreting an oily waterproof substance used by the bird when it preens its plumage. The uropygial gland is particularly well developed in aquatic birds which must pay particular attention to the condition of their plumage.

The skeleto-muscular system

The skeleto-muscular system has many unique features adapting birds for two distinct types of locomotion: flight and terrestrial locomotion on two legs. Although birds are truly bipedal it should be noted that the body is carried in the horizontal position rather than the upright position characteristic of bipedal mammals. A particular feature of all the bony structures is that they are very light and slender, as if pared down to the absolute minimum, yet remarkably strong. Many birds have air spaces in their long bones (pneumatic bone).

The axial skeleton

The notochord is restricted to the intervertebral discs and the vertebral column shows extensive modification. The longitudinal axis of birds, and therefore the vertebral column (apart from the neck) is shorter in relative terms than that of other vertebrates, except frogs and tortoises, and is generally rigid in response to the demands of flight. However the neck is long and mobile, compensating for a lack of mobility in the rest of the axial skeleton. Bird cervical vertebrae are more numerous and varied than those of most vertebrates: the pigeon has a total of fourteen including an atlas, which articulates with the single occipital condyle of the skull allowing considerable freedom of movement between the head and neck, and an axis with an odontoid peg.

The centra have special saddle-shaped articulations (heterocoelous), and are arranged so that the neck region forms an S-shaped curve which allows greater flexibility.

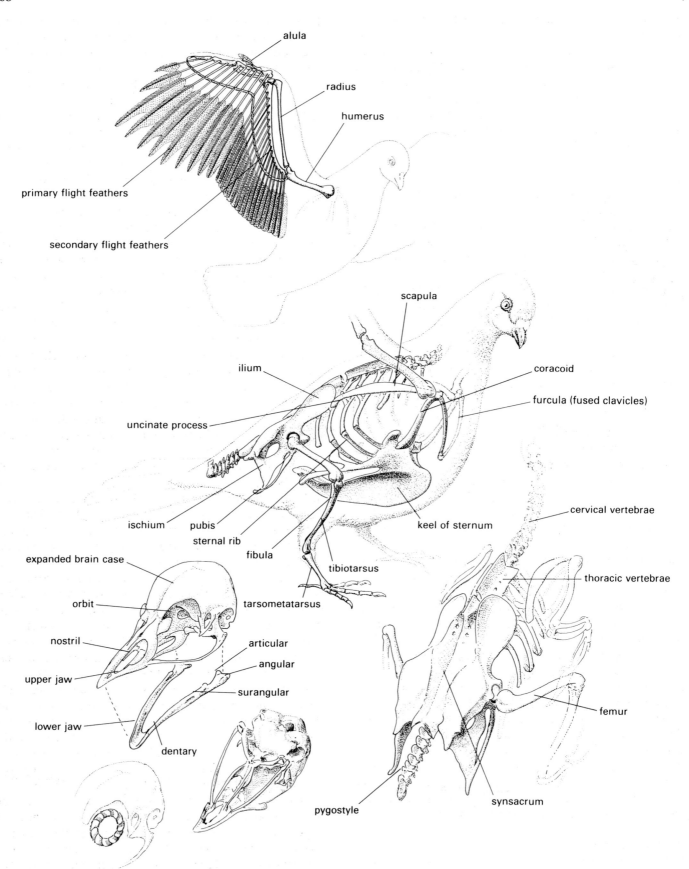

alula

radius

humerus

primary flight feathers

secondary flight feathers

scapula

ilium

coracoid

furcula (fused clavicles)

uncinate process

ischium

pubis

sternal rib

fibula

tibiotarsus

keel of sternum

cervical vertebrae

thoracic vertebrae

expanded brain case

orbit

nostril

upper jaw

articular

angular

surangular

lower jaw

dentary

tarsometatarsus

femur

pygostyle

synsacrum

The skeleton of the pigeon, *Columba*

Apart from the usual spines and projections characteristic of most vertebrae: a neural arch and spine, pre- and post zygapophyses; the cervical vertebrae have additional mid-ventral articular facets, the hypopophyses. The pigeon has a series of cervical ribs which have two heads, the tuberculum and capitulum, which articulate with a dorsal transverse process and a ventral parapophysis respectively, forming a vertebrarterial canal which protects the vertebral artery. Cervical ribs 3 to 12 are fused to their vertebrae.

The remaining vertebrae show reduction and fusion. Of the five thoracic vertebrae, the first four are immovably fused together, and the last is fused to the first lumbar vertebra, forming part of the synsacrum. These vertebrae bear flattened thoracic ribs which are jointed and consist of a dorsal vertebral portion and a ventral sternal portion. Backwardly directed uncinate processes on the dorsal ribs overlap with the adjacent ribs and both strengthen the rib cage and provide attachments for the muscles binding the scapula to the thorax. The sternum is greatly expanded, and has a large mid-ventral keel (carina) which increases the area for the attachment of the powerful pectoral muscles used in flight. A large section of the posterior part of the vertebral column is fused together to form the synsacrum, composed of one thoracic vertebra and six lumbar, two sacral and the first five caudal vertebrae; the pelvic girdle is also fused to this complex. Six free caudal vertebrae follow (permitting movement of the tail) and the column is completed by the pygostyle, representing four fused vertebrae, which supports the tail feathers.

The skull is remarkable for the large size of the brain case and the large orbits, reflecting the highly developed nervous system essential for complex activities such as flight. It is strongly ossified to the extent that the component bones are fused together and the sutures disappear. They are also relatively thin and the combination of these features with toothless jaws gives a structure which is both light and strong.

In spite of their diapsid origin the birds have a single temporal arch, since the post temporal fossa has been eliminated by the lateral expansion of the brain case. The orbits are separated only by a thin inter-orbital septum and the brain case does not extend into this region. Some bones are lost (notably the lacrymal, post-frontal, post-orbital, supra-temporal, ecto-pterygoid and epi-pterygoid) and the squamosal forms part of the brain case. The skull is kinetic with flexion at the hinge joint between the nasals and frontals and is functionally similar to that of lizards. The hyoid arch is represented by the stapes of the middle ear and a hyoid which supports the tongue. The jaws are greatly elongated, and the lower jaw is formed of five bones: angular, articular, surangular, dentary, splenial.

The limbs and girdles

Although the limbs and girdles are extensively modified, especially the wrist and hand regions of the forelimb, their homology with other vertebrates is clear and the appendicular skeleton can be related to the basic pentadactyl plan. However, there is a general reduction in the number of bones thus reducing the weight of the skeleton, and fusion to give increased strength. The pectoral girdle is basically reptilian and with each side formed of a long, slender, blade-like scapula, a short, stout coracoid which attaches to the sternum and a clavicle; the latter fuses with its fellow in the mid-line to form the wishbone (furcula) which acts as a strut holding the two halves of the girdle apart against the strong downward pull of the pectoral muscles during flapping flight. The keel of the sternum may represent the reptilian interclavicles in a highly modified form.

The humerus, which is short and stout, and the radius and ulna are fairly unmodified but there are only two free proximal wrist bones, the radiale and ulnare. The metacarpals (1 to 3) are represented by a composite carpometacarpus which is shaped remarkably like a safety pin and there are four bones representing the remains of three digits buried in the flesh of the wing. The primary flight feathers (of which there are ten) are attached to the hand, and secondary flight feathers (about thirteen of them) to the ulna; there are also a few tertiaries attached to the humerus. The anterior digit bears several small flight feathers which form the alula (bastard wing), an anterior subsidiary aerofoil. Webs of skin (patagia) extend between the shoulder and wrist (prepatagium) and body and elbow (postpatagium). The dorsal surface of the wing is covered by contour feathers which give it a streamlined curve backwards from its anterior margin and contributes to its shape as an aerofoil.

The pelvic girdle is formed of the usual three bones (ilium ischium and pubis) but is extensively modified in comparison with the reptilian pattern, and joined immovably with the synsacrum. The greatly elongated ilia extend from the thoracic region to the base of the tail and are fused along their entire length with the synsacrum. The ischia are fused with the ilia (apart from a large sciatic foramen) and the pubis is reduced to a splint-like bone directed along the border of the ischium to which it is partially fused. The whole complex forms a relatively massive and rigid structure adapted to the strains of bipedal locomotion as well as those of bending during flight.

The hind limbs are less highly modified than the forelimbs but show interesting specializations. The femur is short and thick, and the fibula reduced and partly fused to the tibia. Some of the promixal tarsal bones are also fused onto the fibia, forming a tibiotarsus, and the remainder of the tarsals are fused with metatarsals 2, 3, and 4 to form a single tarsometatarsus. The first metatarsal is small and free and attached to the tarsometatarsus by ligaments. The fifth toe is absent, and the first is directed backwards so that it is opposable to the other digits; this forms the perching mechanism which enables the bird to stand securely on a narrow twig with three toes in front and one behind. The toes are bent by tendons which run beneath the foot and behind the tarsometatarsus. If the tarsal joint is bent, the tendons are stretched and the opposing toes pulled more

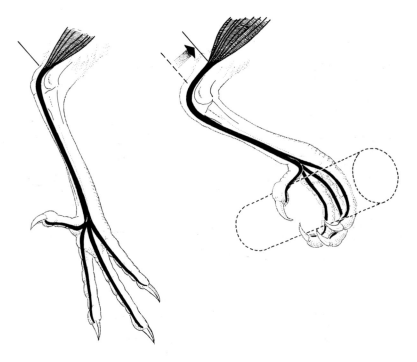

Perching mechanism

tightly together. The bird can therefore remain securely on its perch in a squatting position, even when asleep.

The muscular system

The muscular system of birds is the most extensively modified of the tetrapods, with the trunk musculature reduced and the extrinsic muscles of the wings greatly developed. The ventral pectoralis muscles which produce the powerful downstroke of flapping flight, may contribute as much as a fifth of the body weight. They take origin from the sternum, keel, and furcula and insert along the ventral surface of the humerus. The supra-coracoideus, another ventral muscle, which lifts the wing, also arises on the sternum ventral to the pectoralis and sends a tendon through a foramen formed by the scapula, coracoid, and clavicle to insert on the upper surface of the humerus; thus the direction of the muscle's action is altered by a kind of pulley mechanism. In birds like the pigeon which use flapping flight a great deal these major flight muscles contain a large quantity of myoglobin and are bright red in colour; those like the domestic fowl which use their muscles infrequently have pale pectoral muscles. In addition the bird has numerous small muscles within the wing which are responsible for folding and unfolding it, and which alter its position during flight.

The digestive system and nutrition

Pigeons feed principally on plant material with a high energy content such as seeds, fruits, berries and buds, and may also take snails and other invertebrates. The bird digestive system is compact, but highly effective in meeting the demands of a high metabolic rate and adapted to compensate for the lack of teeth. The pigeon beak is rather small and stout and does not show marked specialization for a particular kind of food, unlike that of many birds. Once the food is in the mouth it is manipulated by the small pointed tongue and moistened with saliva. The tongue is covered with a layer of keratin, and like that of many birds has no intrinsic muscles, but is moved by extrinsic muscles acting on the underlying hyoid apparatus. The oesophagus swells out at its lower end to form a thin-walled crop which provides a temporary food store. This is particularly common in grain-eating birds and has several advantages: in the crop hard food can be softened with water, which the pigeon obtains by immersing its beak and sucking. In addition, a crop enables the bird to secure a lot of food in a limited time, allowing it to compete successfully for limited resources and reducing the time which it spends on the ground where it is most vulnerable to predators. The stomach is divided into two regions, the proventriculus and the gizzard, of which the proventriculus has a glandular lining and secretes peptic enzymes. The gizzard is highly modified for grinding; it has thick muscular walls and glandular cells which secrete a tough, horny layer, protecting the gizzard from damage and aiding the grinding process. In addition pigeons swallow small stones which are retained in the gizzard and help it to reduce food to a pulp. The intestine is elongated compared with that of the reptiles and has its surface area increased by villi allowing faster and more efficient digestion and absorption. Two bile and three pancreatic ducts open into the duodenum. The pigeon has a

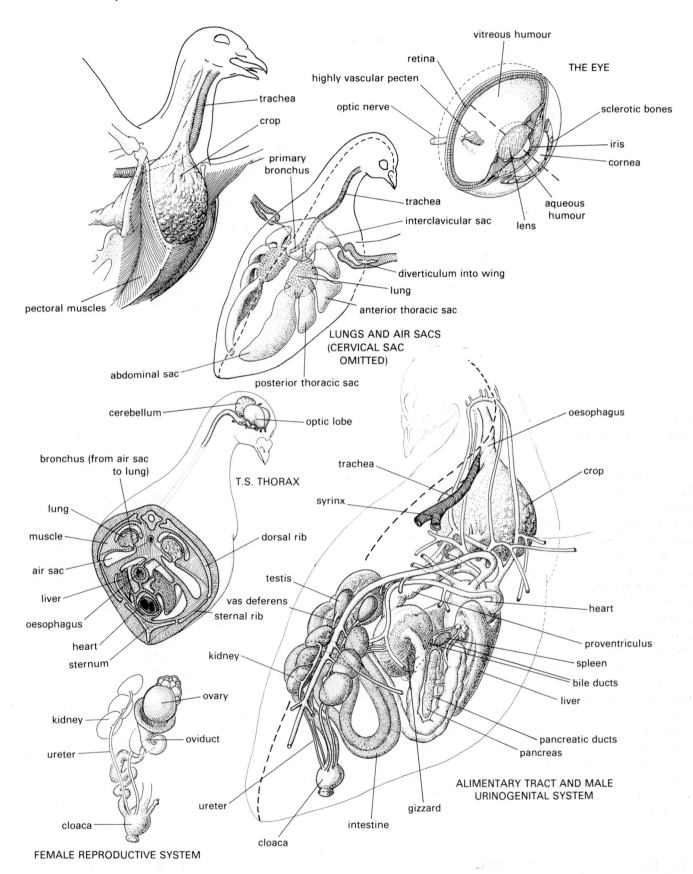

THE EYE

vitreous humour

retina

highly vascular pecten

optic nerve

sclerotic bones

iris

cornea

aqueous humour

lens

trachea

crop

primary bronchus

trachea

interclavicular sac

diverticulum into wing

lung

anterior thoracic sac

pectoral muscles

LUNGS AND AIR SACS
(CERVICAL SAC
OMITTED)

abdominal sac

posterior thoracic sac

cerebellum

optic lobe

bronchus (from air sac to lung)

T.S. THORAX

oesophagus

trachea

crop

syrinx

lung

muscle

air sac

liver

oesophagus

heart

sternum

dorsal rib

testis

vas deferens

sternal rib

kidney

heart

proventriculus

spleen

bile ducts

liver

pancreatic ducts

pancreas

kidney

ovary

oviduct

kidney

ureter

ureter

cloaca

cloaca

intestine

gizzard

ALIMENTARY TRACT AND MALE
URINOGENITAL SYSTEM

FEMALE REPRODUCTIVE SYSTEM

The anatomy of the pigeon, *Columba*

pair of colic caeca, which may function for water resorption. A short rectum leads to the cloaca, which is similar to that of the reptiles with three sections (coprodaeum, urodaeum and proctodaeum) and where further water is reabsorbed. A sac-like organ, the duct of Fabricius, opens into the proctodaeum and since it contains a lot of lymphoid tissue may serve to combat local infections.

Respiratory system

The oxygen requirements of a bird in flight are very high and birds have evolved a unique respiratory system to meet this demand. The lungs are small compact organs situated dorsally in the thoracic region and closely adhering to the overlying ribs and thoracic vertebrae. The lungs give off a number of air sacs which extend into many parts of the body. The lungs themselves are non-distensible and self-supporting so that unlike those of mammals they do not collapse when removed from the body. Since lung tissue is heavier than air sac tissue the advantages of partially replacing lungs by air sacs is obvious where lightness is crucial. In all there are nine air sacs: a pair of cervical air sacs at the base of the neck, two pairs of thoracic sacs, a pair of abdominal sacs and an interclavicular sac which lies in the region of the furcula and gives off diverticula into the bones of the wings and pectoral girdle. Each connects with a large mesobronchus (an air tube) which extends through the lung from one of the paired primary bronchi which arise at the trachea. The primary bronchi and trachea are supported by cartilaginous rings derived from the 6th and 7th visceral arches as in reptiles. In addition recurrent bronchi return air from the air sacs to the lungs. In the lungs the larger bronchi connect with small unsupported tubes, the secondary bronchi, from which in turn arise the smaller parabronchi; these give rise to an anastomosing network of air capillaries with highly vascular walls where gaseous exchange occurs. The air sacs have thin, non-vascular walls and are not regions of gaseous exchange. The advantage of this system is that residual gas volumes in the lungs, the organs concerned with actual gaseous exchange, can be kept very low, and the connecting series of tubes and air sacs allow air to flow through the system in one direction making the process highly efficient. The flow of air is complex, and depends basically on a suction pump comparable to that of the reptiles. During inspiration, the thoracic cage is expanded by action of the intercostal muscles which depress the sternum and move the ribs forwards and outwards. The air sacs expand, but the lungs are compressed slightly by muscles which contract and pull a thin, overlying membrane against their ventral surface. Stale air from the lungs moves into the anterior air sacs, and fresh air is drawn into the trachea, lungs and posterior air sacs. During expiration (which results mainly from relaxation of the intercostal muscles) the sternum returns to its original position, and the air sacs are compressed, while the muscles associated with the lung membrane relax and the lungs expand. Air moves

from the posterior air sacs to the lungs, and from the anterior sacs into the mesobronchi and out of the body. As the direction of air flow through the lungs is counter to that of the blood flow in the capillaries birds have a counter-current exchange mechanism, comparable to that of fish gills. During flight, when the oxygen demands are at their greatest, ventilation is assisted by movements of the sternum brought about by contraction of the flight muscles. There is considerable separation of the anterior part of the respiratory and feeding canals, which is provided by shelf-like outgrowths of the palatal bones comparable to those of crocodiles.

The region of sound production in birds is known as the syrinx and is associated with the trachea. The sound comes from a membrane lying at its posterior end which can be caused to vibrate by the passage of air past it. The pitch of the note can be altered by contraction and relaxation of muscles associated with the membranes, permitting, in many birds, astonishing versatility of song.

Finally the respiratory system of birds also plays a role in thermoregulation since endothermy requires not only a method of conserving heat, but also of losing it. This is achieved in the birds (as in the mammals) by evaporation of water, which has a high specific heat and therefore provides an efficient cooling mechanism. The air sacs provide a large internal area for evaporation, and when necessary heat loss can be increased by panting.

Circulatory system

Birds have evolved a double circulatory system with complete separation of venous and arterial blood, and a double pump. The heart shows some similarities with that of the lizard, but the ventricle is completely divided by the interventricular septum. The left aortic arch, which in the reptile arises from the right side of the ventricle and receives mainly venous blood, is absent. Birds have only the pulmonary arch, which carries blood to the lungs from the right ventricle, and the right aortic (systemic) arch which arises from the left ventricle to carry oxygenated blood to the head, front limbs, and the rest of the body. Because of the slenderness of the bird's neck the carotid arteries run extremely close together and in some birds (although not the pigeon) one is reduced or absent. There is no sinus venosus and all the venous blood returns via the superior and inferior vena cavae directly into the right atrium. Pulmonary veins return oxygenated blood from the lungs to the left atrium. The bird's heart has a well-developed blood supply (coronary system) with coronary arteries arising from the systemic aorta and a venous coronary sinus entering the right auricle. The combination of a double circulation, rapid heart beat (192 times per minute) and an increased blood pressure make the bird's circulation very efficient. This is important for a very active endothermic animal whose tissues need large supplies of food and oxygen, and the rapid removal of waste products. Birds

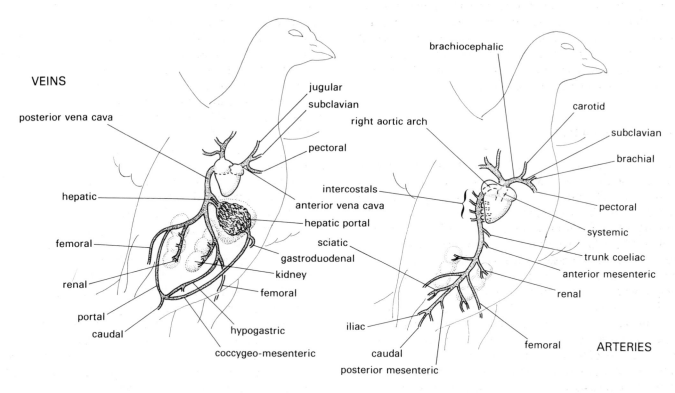

VEINS

posterior vena cava

jugular
subclavian
right aortic arch
pectoral

hepatic

intercostals
anterior vena cava
hepatic portal
sciatic

femoral

gastroduodenal
kidney
femoral

renal

portal
caudal

hypogastric

coccygeo-mesenteric

iliac

caudal

posterior mesenteric

brachiocephalic

carotid
subclavian
brachial

pectoral

systemic

trunk coeliac
anterior mesenteric
renal

femoral

ARTERIES

Pigeon : principal arteries and veins

have a well developed lymph system, the lymphatic vessels ultimately entering two thoracic ducts which join the anterior vena cava.

Excretion and osmoregulation

Birds have a pair of metanephric kidneys which are essentially the same as those of the reptiles, though with many more nephrons. They are three-lobed structures, lying in the roof of the body cavity and connected to the urodaeum by ureters. As in the reptiles, water is resorbed, and most of the nitrogenous waste is in the form of uric acid; there is no bladder which is possibly another adaptation to reduce weight.

The small volume of urine produced by birds tends to encourage retention of salts in the body. For sea birds which take in large quantities of salt with their food, and may also have to drink sea water to compensate for water loss by evaporation, the excess salt poses a considerable problem. Their intake of salt far exceeds that which can be disposed of by the kidneys, and sea birds have special salt-excreting glands. In the herring gull, these are bean-shaped structures, composed of a mass of secretory tubules radiating from a central canal and lie on the inner side of each orbit. They discharge a concentrated salt solution into the nasal cavities which leaves through the external nares.

Nervous system and sense organs

Adaptations of the bird's nervous system associated with flight are reflected in the structure of the brain, which is well developed, particularly in those parts associated with sight, muscle co-ordination, and complex behaviour patterns. Smell is less important to animals which spend much of their time in the air, and it is hardly surprising that the olfactory organs and olfactory portions of the brain are reduced. On the other hand, sight is very important and both the eyes and optic regions of the brain are well developed; in proportion to their over-all size, birds' eyes are larger than those of other vertebrates and together normally out-weigh their entire brain. The visual activity of birds such as falcons and owls, which search for their prey 'on the wing' and often at night, is probably unequalled in the animal kingdom. Focussing is brought about, as in most reptiles, by changing the shape of the lens, which in its resting position is adjusted for looking at distant objects. Birds must be able to focus (accommodate) their eyes extremely rapidly, to enable them to change from distant to near vision as they swoop to the ground or weave their way through the trees. They have both rods and cones in their retina and thus are capable of colour vision. In species which are active during the day the cones are important, giving good visual activity and colour vision, whereas in nocturnal species rods predominate giving increased sensitivity to low light intensities. In

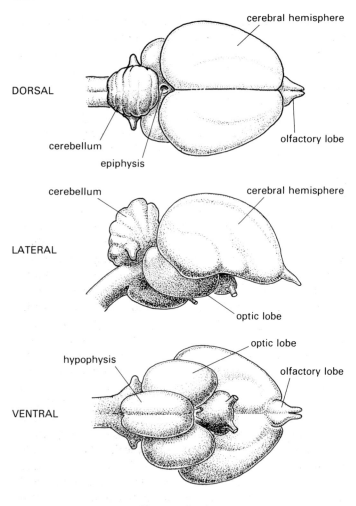

DORSAL

cerebral hemisphere

cerebellum

epiphysis

olfactory lobe

cerebellum

cerebral hemisphere

LATERAL

optic lobe

optic lobe

hypophysis

olfactory lobe

VENTRAL

Pigeon : brain

Reproduction

The male has a pair of compact whitish testes lying in a ventral position anterior to the kidneys, and a convoluted vas deferens extends from each to the urodaeum where it opens at a papilla. There are no Mullerian ducts and no intromittent organs. Only the left ovary and oviduct develop in the mature female, though the right ovary and duct are present in an early embryonic stage. This adaptation can be related both to the need to minimise weight and to the fact that there is only space for one large egg to pass through the pelvis at a time. Fertilization is internal, and takes place high in the oviduct. As the fertilized egg passes down the oviduct, albumen is deposited around it, with shell membranes and shell being added in the lower section. Since a clutch of pigeon eggs requires a third as much calcium as the skeleton, reserve deposits are accumulated in the bones prior to breeding. During coitus, which is preceded by a period of courtship the posterior parts of both the male and female cloacae are turned outwards and held together for transfer of sperm.

The female pigeon builds a nest of interwoven twigs and generally lays two pale unmarked eggs. The job of incubation is shared by both male and female as is the care of the young. On hatching the young are at a relatively advanced stage of development and are covered by pale yellow down. For the first few days after hatching the young are fed on 'pigeon's milk', which is a curd-like substance, similar in chemical constitution to mammalian milk, with a high proportion of protein and fat. It is produced by both male and female by proliferation and sloughing off of the cells lining the crop. The low clutch size of pigeons compared with many birds may be associated with this feeding method which improves the chances of survival but represents a physiological burden on the parents. After this, the young are fed on food regurgitated by the parents. In both cases the young pigeon or squab sticks its head into the parent's throat to feed. Brooding of the young is continuous until about the tenth day after hatching when the parent's attentions become progressively less marked. Pigeons grow rapidly, and may begin to fly as early as two weeks after hatching.

BIRD DIVERSITY

Birds have been abundant and diverse since the late Cretaceous, expanding to fill a wide variety of ecological niches and as a result are found almost everywhere in the world. The main radiation took place in the early Tertiary, with the majority of the thirty or more modern bird orders being established by the Eocene. Over half the living bird species are members of one highly successful order, the Passeriformes or perching birds. Nevertheless, the adaptive possibilities of birds are not without their limitations. Apart from the obvious restrictions laid on most birds by

addition each retina has one or more areas of densely packed rods (foveae) which provide for high resolution of images. A highly vascular fold, the pecten, projects from the retina. Its functions are uncertain; probably its main function is to supply oxygen and metabolites to the retina. The eyeball is protected with a ring of sclerotic bones and the front of the eye has eyelids and a nicitating membrane.

The inner ear has the typical semicircular canals of advanced vertebrates. A highly developed sense of hearing is reflected in the structure of the cochlea, which has developed partial spiralling. The ear drum is large, and a columella (stapes) is present. Pigeons have a song with a narrow range of pitch, and their ability to discriminate between notes of different pitch is correspondingly small.

Muscular co-ordination is of vital importance to birds, and the cerebellum is large and well developed. The cerebral hemispheres are also large, but in this case it is the deeper, grey matter, the corpus striatum, that is enlarged, the opposite to the condition found in the mammalian brain. This can be related to bird behaviour patterns which are largely governed by inherited and instinctive actions, with rather little flexibility to learn new responses.

Birds have the basic vertebrate plan of cranial nerves.

their adaptation to flight, all are oviparous and, as endotherms, obligate brooders of their eggs. This precludes them from a sub-aquatic mode of life. Again, bird skeletons and feathers are poor adaptations for burrowing, and there are no truly fossorial birds though many, e.g. sand martins, nest in burrows, and puffins may use those of other animals, e.g. rabbits. Furthermore most birds fall within a fairly small size range. Their maximum size is determined by their surface volume ratio, and by the demands of flight, a larger bird having to fly faster than a smaller one to remain airborne. Equally a minimum size is determined by a balance between food requirements and heat loss; a warm blooded animal can only become extremely small if it is able to feed continuously, or to allow its body temperature to fall during periods of inactivity. Avian adaptive radiation falls into two broad but related categories: locomotion and feeding. Birds have available to them two separate locomotory systems, each of which shows a wide range of modifications, and have also adopted an extremely wide range of different diets.

Locomotion

Flight

The most common adaptations of the pectoral girdle and fore limbs are those associated with flight. As we have already seen, three main kinds of flight, flapping (forward or hovering), static soaring and dynamic soaring, can be recognised, and though an individual bird can employ more than one, most birds are specialised for particular kinds of flight. A glance at any bird book will show how variable are the shapes of wings, adapted for a particular environment and way of life. The best known and most expert hovering birds are the humming birds which feed on nectar while hovering close to flowers in tropical regions. These hold their body axis vertical as they fly, and the fully extended wing is moved backwards and forwards in a horizontal plane, but rotated so that the dorsal and ventral surfaces face downwards alternately. The wing beats are powerful and exceptionally fast. The shoulder joints are modified to give greater freedom of rotation, and since the maximum propulsive force comes from the primaries, the hand bones are longer than the fore arm and upper arm combined. Birds adapted to forest habitats have wings adapted for slow flight and high manœuvrability; examples include pigeons, some passerine birds, crows, magpies and woodpeckers. In these the wing has a low length to width ratio, (aspect ratio) and is highly cambered; it usually has a greatly emarginated posterior edge with gaps between the spread primaries forming subsidiary aerofoils, which prevent the wing from stalling at low speed. This type of wing is called elliptical. In contrast, many birds, particularly those which are heavy, or attack prey in flight, or make long migrations, have wings adapted for fast flight. Examples are found in the falcons, swifts, and geese. Wings of this type are characterised by a moderately high aspect ratio (rather long and narrow shape)

with little camber, pointed tips, and a smooth posterior edge. Dynamic soarers, such as the albatross, have wings which are flat and narrow, and very long, with an exceptionally high aspect ratio (25:1). Static soarers such as eagles have a wing type intermediate between that of the elliptical and that of the dynamic soaring type, with a moderate aspect ratio, a deep camber and pronounced slotting of the trailing edges; in this way they combine a small turning circle with a low forward speed.

Land birds

As a group, birds have fully exploited the possibilities of bipedal locomotion and show a variety of methods of adaptation for walking or running, hopping, climbing, wading, walking over soft ground and swimming. Adaptation for a cursorial (running or walking) mode of life is restricted to some extent by bipedalism since for stability the centre of gravity of the bird must always lie over the feet. While reduction in toe length and in the area of the foot in contact with the ground is obviously a desirable adaptation for a cursorial bird, this can only be achieved at the expense of stability. The constant need to retain equilibrium is seen when a bird such as a domestic fowl walks and moves its head backwards and forwards in sequence with its steps as a balancing motion. The ostrich, one of the most completely adapted cursorial forms with a running speed of 60 to 70 km/hr, has greatly elongated legs to enable it to take long strides and cover larger distances; its toes, of which there are only two on each foot, are short and heavy. To enable the bird to squat or stand up without losing its balance the tibiotarsi and tarsometatarsi are almost the same length and it rises from the ground in two stages, resting on the tibiometatarsi in the process. Ostriches have fairly heavy leg bones, arranged in vertical columns to support their considerable body weight, and even more extreme adaptations are seen in larger, but now extinct, cursorial birds.

The ostrich, a ratite (order Struthioniformes)

Hopping is primarily an adaptation for moving about on trees or bushes and is found principally in perching arboreal birds (passerines). Birds do not show the extreme adaptations to jumping seen in some mammals such as the kangaroo, perhaps because of the inadequacies of the pygostyle as a posterior balancing organ. However some climbing birds (woodpeckers, tree-creepers) show some enlargement of the pygostyle as a prop, which bears especially strong tail feathers. They habitually start at the bottom of a tree and climb upwards gripping the bark with enlarged claws. On the other hand nuthatches, which use only their feet during climbing, since they have no supporting tail, can climb up and down. They have an exceptionally large claw on their hallus (hind toe). Parrots additionally use their beaks as a hand to help them climb.

A heron (order Ciconiiformes)

A woodpecker (order Piciformes)

The wide diversity of wading birds to be seen in estuaries and marshes are variously adapted for walking in different depths of water (the longer the leg, the greater the depth) and over different substrates. The deep water forms (for example, herons and flamingos) have modifications of the legs similar to those of cursorial birds, and have correspondingly elongated necks and beaks. These function as a counterbalance to the legs and feet during flight and for feeding in deep water. The majority of long-legged birds fly with their legs and necks outstretched; the heron, which is an exception, has very broad wings. Many wading birds search for their food in soft substrates, and to avoid sinking the surface area of their feet is increased by webbing, or by elongation of the toes. Similar adaptations are seen in birds which live on sand or snow; both the sand grouse and the ptarmigan have especially broad toes and claws, with a dense covering of feathers. It is interesting that many of the birds which are most versatile in the air are particularly ill-adapted for terrestrial locomotion; thus the feet of swallows

and swifts are of little use for anything but perching. At the opposite extreme penguins, which do not fly at all, have such short legs that they cannot walk at all fast. Nonetheless some species have to make long migrations from their breeding grounds to the sea, which they accomplish whenever possible by tobogganing over the snow, and sometimes using their flippers (wings) to gain extra speed.

Male emperor penguin incubating an egg
(order Sphenisciformes)

Aquatic birds

Although no bird is entirely independent of the land, there are a large number of more or less aquatic birds which can be conveniently grouped as swimmers and divers, though many birds can do both. Swimmers have a wide body to increase their stability in the water, and a dense plumage and large uropygial gland, to give buoyancy and insulation; their feet are essentially webbed or lobed. For maximum advantage when paddling the legs need to be near the rear of the body so that the leg muscles do not interfere with streamlining and the bird can steer easily. Some birds, especially those which also dive (grebes and guillemots), have their legs positioned so far back that they must hold themselves nearly upright to walk on land. Others such as seagulls, which swim, but also walk and run and spend much of their time far from water, do not show such an extreme adaptation to swimming.

A grebe (order Podiapediformes)

Diving requires even greater specializations. The buoyancy of the bird is reduced with reduced air sacs and pneumatic bones, and the bird forces air out from between its feathers as it dives. Penguins even swallow small stones which act as ballast! The plumage is smooth to improve streamlining. Diving birds show a variety of adaptations concerned with maintaining an adequate oxygen supply. They normally have a large blood volume with an exceptionally high oxygen carrying capacity, a large amount of myoglobin in the muscle and a high tolerance of carbon dioxide. Furthermore the metabolic demand for oxygen is reduced while the bird is underwater by a reduction in heart beat and constriction of peripheral blood vessels. Even so the diving time is usually between one and three minutes, with a maximum recorded survival time of fifteen minutes. Many divers are propelled by their feet when underwater, and show special adaptations of the legs. The grebes, for example, have a long narrow pelvis and a short femur with a hinge-like double articulation at the acetabulum; they have a long tibiotarsus and laterally compressed tarso-metatarsus, and their leg musculature is incorporated into the main body mass nearly to the ankle joint. In contrast the penguins propel themselves when they dive using their wings, which have become modified as flippers.

Feeding

Early birds probably had a diet of insects and other small animals, similar to that of early reptiles, but their diets have since diversified to include almost every imaginable source of food. This variety is clearly reflected in the diversity of beak structure, but adaptations to diet go far further than this, and affect in particular the digestive tract and sense organs. In many birds, hind limbs have taken on grasping and holding functions. The shape and size of the beaks and tongues of birds which feed on fish show various features which fit them for catching and holding their slippery prey. The majority of piscivores (such as gannets and king-fishers) use their beaks as forceps to grip the fish, and have their edges modified to prevent it from slipping: piscivorous geese have keratin hooks on their beaks, while puffins have serrated edges. The western grebe on the other hand uses its beak as a spear, a comparatively rare adaptation. Cormorants and pelicans have highly distensible throat pouches, in which they first hold their prey, rather as in a dip net. Some birds fish with their feet; the osprey or fish hawk (*Pandion haliaetus*) is one such bird, and has spiny scales on the soles of its feet with which it holds securely a fish it has snatched from the water while flying over. The fish eagles (*Haliaetus*) have less specialised feet but also manage to take fish in a similar way. Many large gulls, skuas and the giant fulmar feed on injured or dead animals which float at the surface and will often mob and harry other birds such as the fish eagles, forcing them to drop their prey.

The vast amounts of minute planktonic invertebrates in lakes and seas represents an almost limitless source of food, but to take advantage of it requires some type of filtering device. The spoonbill (*Platalea leucorodia*) is a large wading bird with a flattened spatulate bill which it moves sideways through the water to capture invertebrates. Flamingoes have a series of lamellae in the beak. They take a mouthful of water and invertebrates and strain out the food, forcing the water out through the lamellae by arching their powerful muscular tongue against the palate.

Flesh-eating birds

Raptorial birds (hawks, owls, falcons and kestrels for example) are hunters, and commonly capture and kill prey which is too large to swallow whole. Most raptors have keen eyesight and the nocturnal forms, notably the owls, also have extremely acute hearing which enables them to locate prey such as small mammals by the noise they make as they scurry through the undergrowth. Hawks and falcons hunt in various ways for prey while soaring high in the air, while kestrels usually hover closer to the ground and many owls await their victims on a stationary perch. When a suitable prey is spotted the bird may either pounce upon it, or fold its wings to its side and dive steeply. The power dive or 'stoop' of a peregrine falcon is spectacularly fast and precise, as it plunges downward to catch some other bird on the wing, killing it by biting and severing the back of the neck. In contrast the merlin uses fast flapping flight as it pursues

An osprey (order Falconiformes)

small birds, though it may also hover. The prey is caught and killed by the talons which are long and sharp, or by the beak which in raptorial birds is hooked and modified for tearing. Normally the prey is grasped and pierced by the long dagger-like talons, but less specialised raptors may jab it with their beaks or beat it against the ground. Web-footed raptors such as skuas both catch and kill their prey with their beaks. Once dead, the prey is torn into bite-sized pieces with the beak, or swallowed intact.

Carrion eaters, of which the vultures are the best known examples, generally have beaks and feet which are adapted for tearing flesh and are similar to those of raptors. They search for their food by soaring often very high in the air from where they can survey a wide area, and watching not only for carcasses but also for other vultures which may have found a source of food. Many forms have reduced or no feathers around the head and neck, thus reducing the chance of fouling their plumage with half decayed flesh but giving them a singularly unlovely appearance. In areas where vultures are numerous large numbers rapidly gather around a new carcass in a short time. The order in which the species feed depends on the stoutness of the bill since only the larger vultures can rip open a tough hide.

Insectivores
About 60% of birds are insectivorous to some degree, obtaining their food in a vast variety of ways. Some are gleaners, hunting by sight and with short broad beaks. Others, such as swallows and martins, are highly specialised to catch insects in flight and have a broad bill with a wide gape surrounded by bristles. They feed by flying rapidly through insect swarms. Some birds probe for their prey in crevices of bark and wood and, like the tree-creepers, have long, thin, decurved beaks. Woodpeckers take more active measures to remove invertebrates from wood by pounding

holes or at least enlarging them with a powerful chisel-shaped beak; the skull is strengthened to withstand the stresses of hammering wood, and they have a long protrusible barbed tongue to extract their prey. Starlings may forage by diving the beak into the earth and opening it to create a hole which can be searched for suitable food. The jaw muscles are extremely strong and the bird's eyes are positioned well forwards so that it can peer into its hole. Rock nuthatches hunt in rock crevices; the white breast of these birds is thought to act as a mirror, to reflect light into the crevice.

Vegetarian birds
Birds are adapted to a variety of vegetarian diets, including pollen and nectar, fruit, seeds, leaves and roots. The majority of birds which feed on pollen and nectar (for example humming birds, honey creepers) are probers with long beaks and tongues. However the beak structure is highly variable and is correlated to the particular type of flower. In many cases birds with this habit pollinate the flowers in a similar way to pollen-feeding insects and many tropical plants are clearly adapted to this method of pollination with tough, often bright red or yellow flowers held well away from the foliage. Fruit-eating species are widely spread, and generally have short, wide beaks like those of many finches. Otherwise the main specialisations for fruit-eating are found in the digestive tract. The stomach is normally poorly developed, and the remainder of the gut is short and tubular, so that food passes rapidly through the system, with only the pulp being digested. It is interesting that many birds can eat fruit such as the berries of deadly nightshade which are poisonous to man. Plant food is generally of relatively low calorific nutritional value which means that herbivores must consume more food than carnivores, and fruit eating birds commonly feed insects to their young.

Seeds are among the most nutritious of plant food, and usually have a high energy content, with the additional advantage of being available outside the growing season. The diversity of seeds is enormous, both in terms of size and in the hardness of their coatings, consequently seed eaters show a wide range of adaptations. Some swallow their seeds whole (for example gallinaceous birds: such as turkeys, domestic hens, pigeons, sand grouse) and grind them in their gizzards which are well developed and muscular; larger birds such as turkeys and pheasants can swallow whole acorns and dispose of them in this manner. The finches are a large group of birds, many of which are predominantly seed eaters. They crack seeds with their short, heavy, arched beaks which vary in size according to the type of seed eaten. The cutting edge of the beak may be sharpened, and the jaw muscles are particularly well developed. The small-beaked goldfinch feeds on birch or dandelion seeds, while the hawfinch has a massive beak and can feed on the seeds of cherries and similar stoned fruit. Some forms hammer seeds open with blows of the beak (for example nuthatches), a method reminiscent of that used by a thrush to obtain molluscs by holding it in its beak and bashing it against a stone (thrushes' anvil).

Only a few birds graze on grass, leaves, twigs or roots as their main source of food, probably because of the low digestibility and low nutritional value of such food. The holoarctic grouse has large storage caecae associated with its small intestine in which its low calorie diet is digested by symbiotic bacteria, and in this respect resembles many mammalian herbivores. Geese may feed by cropping grass, ducks on aquatic plants, and some pheasants scratch for roots with their feet.

In all animal groups there are a few species which have adapted to a highly bizarre mode of life, often in response to an environment in which food is not freely available; the birds are no exception. A Galapagos finch (*Geospiza difficilis*) bites the bases of growing feathers of other birds and then eats the oozing blood: in the Arctic the ivory gull feeds on dung deposited on the ice while puffins and petrels will take dung floating on the water.

Migration

Undoubtedly much of the world-wide distribution of birds comes from their abilities to make long migrations, during which they display an almost uncanny ability to navigate. Many birds migrate seasonally. Although there are obvious hazards in this, many migrants fall to storms or predators on the way, it has advantages and indeed may be essential for animals which must maintain a high metabolic rate at all times. The migration of birds from winter quarters in the tropics or temperate regions to the north for breeding enables them to breed in an area where the days are long, food is plentiful and territories easy to establish and maintain. In addition migration helps birds to avoid climatic extremes and to spread their distribution beyond areas where they could survive throughout the year. It also prevents predatory populations increasing, since migratory birds do not provide a constant food supply for the predators in any particular region. The migratory record is held by the Arctic tern which nests in the Arctic and winters in the southern hemisphere, travelling 22,000 miles. In the Antarctic most species of penguins move northwards in the winter, travelling as much as 400 miles by swimming and floating on ice rafts to remain in open water. Penguins breed in the winter so that the young are able to feed and grow sufficiently to survive the next winter. Emperor penguins have a complicated breeding programme involving a trek of as much as 100 miles over the snow. The female lays a single egg (in May, at the beginning of the Antarctic winter) and gives it to the male who holds it on his feet, covering it with a special fold of abdominal skin. He then incubates it for 60 days, during which he remains hunched over the egg eating only snow. Meanwhile the female crosses the ice to the sea and feeds, returning with a crop full of squid and fish to feed the chick. The male makes the same journey, a process which is repeated until December when the ice breaks up.

By January the chick is able to feed itself. Without these long journeys, and unusual distribution of labour, neither parents nor chick would survive.

Adaptability

Throughout this section on modern birds we have emphasised the specializations of birds which adapt them to particular habitats or ways of life. However in a final word on the subject it is worth mentioning that the success of many birds has been their relative lack of specialization and thus their adaptability. In the recent past man has drastically altered the environment by creating totally new habitats such as farmland and towns. Most of the common birds we see in towns, such as the starling or black-headed gull, are abundant because they have been able to adapt so readily in a behavioural sense to the new conditions, to exploit new habitats and modify their diets to take advantage of what is available. In contrast those birds which are more closely specialized to a particular way of life, for example corncrakes and bitterns, have become increasingly rare as their habitats disappear.

Fossil birds

The oldest known birds are recorded from the upper Jurassic and are more obviously reptile-like than modern birds to the extent that they are placed in a separate subclass Archaeornithes. The first entire specimen *Archaeopteryx lithographica* (the generic name means 'ancient wing') was found in 1861 in Bavaria by a Dr. Haberlain, who sold it to the British Museum for £700, a then princely sum, which he used for his daughter's dowry. The first description was made by Sir Richard Owen and published by the Royal Society, though his work was incomplete; because of the twisted position of the specimen he failed to find its head and jaws. Fortunately this significant omission was speedily rectified by John Evans, who wrote to the Royal Society saying he had examined the specimen and found not only the skull, but the brain case and jaws as well. Since John Evans was regarded by the scientific establishment as a nobody the first reaction to his discovery was withering scorn. Part of the secretary's letter is worthy of quotation: 'Hail Prince of Audacious Palaeontologists! Tell me all about it. I hear you have today discovered the teeth and jaws of *Archaeopteryx*. Tomorrow I expect to hear of you having discovered the liver and lights. And who knows but that, in the long run, you may get hold of the fossil song of the same creature impressed by harmonic vibration in the matrix . . .'

In fact, had it not been for the association of feathers with *Archaeopteryx* it would have undoubtedly been classified as an archosaur since it had a reptilian skull, brain, vertebrae, and teeth; there was no keel on its sternum and it had an archosaurian pelvis. However it did have well-developed

Archaeopteryx

clavicles forming a furcula. It is thought to have had limited flying powers and the specimens found were probably land birds that got blown away in a gale and drowned in a Jurassic lake; unusual circumstances which may account for the poor fossil record of only five specimens and a single feather.

Since the discovery of *Archaeopteryx*, many attempts have been made to discover the reptilian ancestry of the bird. Three suggestions have been made: that they were cursorial, running along and flapping or spreading their wings as they ran, and perhaps gliding (so that the forelimb developed as a wing); that they were arboreal and took to gliding as they leapt from branch to branch rather like a flying squirrel; or that they were ground living forms which evolved wings as a means of catching insect prey using a fly swat method. It seems likely that feathers originated as heat insulators and were later elaborated for gliding, but there is no direct evidence for this either. The structure of *Archaeopteryx*, with a long trailing tail, suggests that it is unlikely ever to have been cursorial, but there is no reason to suppose that it had to lie in the main line of bird evolution.

All other birds, both fossil and recent, form the subclass Neornithes. The next fossil birds are known from the lower Cretaceous, about 10 million years later than *Archaeopteryx*. Apart from the presence of teeth in a few specimens (*Ichthyornis, Hesperornis*) they are almost fully modern in every respect and no intermediate form has been found, which supports the view that *Archaeopteryx* was an evolutionary side branch. It may even have been antedated by the first 'birds' by several million years; the fossil record is not good enough for us to know. Both *Ichthyornis* and *Hesperornis* were aquatic, *Ichthyornis* being gull-like while *Hesperornis* was a flightless diving form.

The toothed birds are grouped together in super-order Odontognathae. The remaining birds fall into two groups, the Palaeognathae or ratites which is a large group of flightless cursorial forms, and the Neognathae, which includes all the toothless flying forms, and the great majority of modern birds. The palaeognathae retain a somewhat reptilian (palaeognathous) palate, and have well developed and powerful legs, but vestigial wings, a keel-less sternum and fluffy, down-like feathers. They can be traced back to the early Cainozoic, a time when there were also large numbers of ground-dwelling neognathous birds. It seems likely that at this time there was increasing competition between the birds and the mammals for the terrestrial habitats, which the mammals won. Only a few ground-dwelling birds survived to the present line: ostriches in Africa, rheas in South America, cassowaries in Australia and New Guinea, and kiwis in New Zealand.

Synopsis of class Aves

Subclass Archaeornithes: ancestral birds with many reptilian features.
Subclass Neornithes
  Super-order Odontognathae: Cretaceous birds.
  Super-order Palaeognathae: modern toothless birds with a primitive palate, mainly flightless.
Super-order Neognathae: modern birds.

CLASS MAMMALIA

Mammals, the most familiar of all vertebrates, are a distinctive group of hair or fur-covered creatures including most domestic animals, and some of the most familiar pests, as well as human beings. They are also the most cosmopolitan of all animals with the greatest variety of habitats. There are flying, terrestrial, arboreal and totally aquatic mammals. Mammals are found in every corner of the globe, from Arctic to Antarctic, from the oceans to mountain tops. Many of their distinguishing features are related to two evolutionary adaptations: endothermy and the care of the young. In these they resemble the birds, but both adaptations developed independently of the birds and in rather different ways. Like the birds, mammals appear to be derived from the reptiles, but the first mammal fossils, from the Triassic, pre-date the earliest birds. By the early Jurassic five distinctive mammal orders were established. The group survived the Age of reptiles apparently because they were small, unobstrusive, nocturnal insectivores and thus not direct competitors. However at the end of the Mesozoic they radiated rapidly to become the dominant animals of terrestrial environments, which is why the period from the beginning of the Cainozoic to the present day is often known as the Age of Mammals. The reason for dinosaur extinction and mammalian success is not easy to find; a current theory is that the former contributed to the latter. Climatic changes at the end of the Cretaceous meant that the seasons became more extreme, and this must have adversely affected the ruling reptiles. If these changes caused increased competition for food then smaller mammals had a better chance of survival: a three ton dinosaur would have required as much food as 10,000 rats, and had less chance of getting it. As the age of the reptiles came to an end the mammals radiated rapidly to occupy, among others, the ecological roles vacated by the dinosaurs. Today there are about 9,000 species of mammals, the majority of which belong to the infraclass Eutheria, the advanced placental mammals.

Mammary glands

Mammals are defined as animals which suckle their young, and their name is derived from the Latin *mammae*, meaning breasts (mammary glands). Mammals have abundant sebaceous and sweat glands in their skin, and the mammary glands are modified sweat glands. The primary function of sweat glands is to cool the animal by means of evaporation, a method of heat regulation that is best developed in man. Sweat glands are long tubular structures with coiled ends and produce a water secretion. Cooling can also be brought about by increased blood flow to the cutaneous vessels, also highly developed in man.

Mammary glands are modified sweat glands which become active immediately after the young are born, continuing to function as long as suckling continues. Both their development and functioning are under hormonal control. The actively secreting gland is made up of small masses (lobules) of alveoli which are composed of secretory cells. Ducts lead from the alveoli and unite to form a larger duct which themselves join to form a common duct or ducts opening to the outside. In the simplest condition, as seen in the egg-laying mammals, the ducts open into a depressed region of skin and the young feed by lapping up the milky secretion as it spreads onto the surface. In other mammals the ducts open at a raised region, the breast; many mammals have a single duct opening at a teat, while others, eg *Homo sapiens*, have several ducts opening at a nipple. The breasts are completed by adipose tissue which surrounds the alveoli and ducts. The nipples or teats are normally paired and vary in number roughly in proportion to the number of young born in a single litter, ranging from one pair in *Homo* to eleven pairs in some insectivores. The position of the mammary glands varies according to the position normally adopted by the mother when feeding her young. Animals which nurse their young while standing (for example sheep, cows, horses, deer) have their mammary glands in a protected position between the hind legs, whereas arboreal animals which nurse in a sitting position have them on the chest (primates). Cats, dogs, and pigs, which lie on their side to nurse, have the teats arranged in two ventrolateral rows.

Hair

Hair is one of the diagnostic features of mammals and has the same function as feathers, trapping an insulating layer of air. The sebaceous glands are in many ways equivalent in function to the uropygial gland of birds but are normally found associated with individual hairs. They secrete a waxy substance, sebum, formed by disintegration of the lining cells, which prevents the skin and hair from drying and may also act as a water-proofing layer. Grease recovered from sheep's wool is purified to form lanolin which is the basis of most ointments and cosmetics. Hair is an epidermal structure, and is thought to have arisen as a new structure in the mammals. Probably the individual hair evolved to occupy a position between those parts of reptile skin that have scales; thus in mice and rats, with scales and hairs on their tails, the hairs develop in trios behind the scales (a triplet arrangement is quite common in mammalian hair). Although the primary function of hair in modern mammals is as an insulating layer, it may have first evolved as a sense organ. Hair grows from follicles which develop as a deep down-growth into the epidermis. A dome of specialised dermal cells in the base of the follicle forms the dermal papilla, and the hair is produced by the multiplication of epidermal cells to form a tubular structure. Hair is made principally of the protein keratin and has a thin outer layer, the cuticle, an inner cortex of hard keratin and in many cases a central medulla of soft keratin. (Both scales and feathers are composed of hard keratin). The growing part of the hair is nourished from blood vessels in the dermal papilla and its diameter is associated with the size and shape

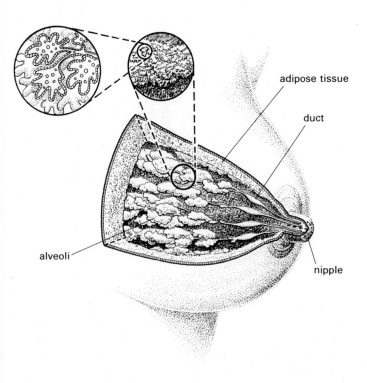

adipose tissue

duct

alveoli

nipple

Mammary gland (human)

of the papilla. Mammals get additional insulation from a layer of subcutaneous fat deep to the dermis. While the majority of mammals have both hair and subcutaneous fat, in some one or other may be emphasised. For example in the whales the hair layer is greatly reduced since the insulation provided by hair depends upon trapped air which cannot persist if the animal is permanently immersed in water.

The skull

The skull of mammals is instantly distinguishable from that of all other vertebrates by several characteristics: the brain case is enlarged; the teeth are usually strongly heterodont (of different types); and the lower jaw is formed from a single bone, the dentary, giving it a stronger structure than that of other vertebrates.

To accommodate the larger brain the brain case is enlarged to the extent that both the otic capsule and the squamosal form part of the cranial wall. Articulation of the lower jaw is between the squamosal and the dentary (craniostyly); both the quadrate and the articular becoming part of the middle ear as ear ossicles, along with the stapes (modified hyomandibular) of lower vertebrates. Furthermore a bony shelf, the secondary palate, has developed giving a total separation between the nasal passages and the mouth cavity. Air entering the nasal passages is warmed and moistened by mucous membranes supported by a complex set of thin bony plates, the turbinal bones.

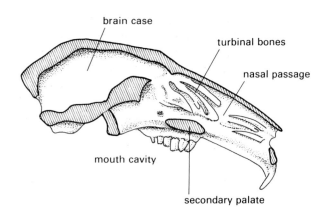

Longitudinal section of a mammalian skull showing secondary palate

Teeth

Mammals have a series of teeth of different form and function which are a source of considerable variation according to diet. The anterior incisors are used for biting food and have a sharp chisel edge. The canines are long, pointed teeth, usually conical in shape, and derive their name from their condition in dogs (*canis* is Latin for 'dog') in which they are well developed and used for killing prey and fighting. The cheek teeth, normally molars and premolars, are broad teeth adapted for chewing and grinding. Generally mammals have fewer, larger teeth than other vertebrates, and these are not continuously replaced.

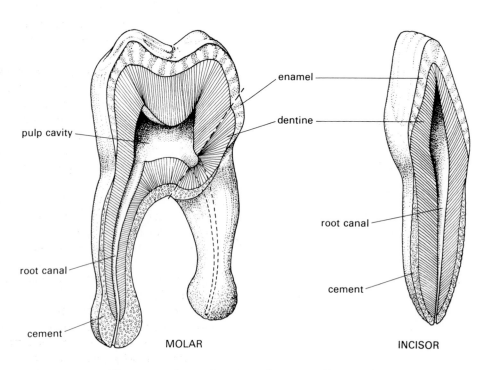

Diagrammatic section through mammalian teeth

In most mammals no further growth takes place once the teeth are fully formed although in some cases, notably rodent incisors, they continue to grow throughout life. In most other vertebrates the teeth are continually replaced, but in a sequence so that the animal retains a functional set of teeth at all times. Mammals develop a first or milk dentition, generally after birth (although Napoleon is a famous example of a mammal with a tooth present at birth!), which erupts from anterior to posterior. Before the milk dentition molars have erupted, the incisors are replaced by the second and final set. The molars of the milk dentition are functional throughout life. Mammalian teeth have a similar structure to those of most vertebrates, and can be used as an example for tooth structure in the subphylum. They are derived from both the epidermis and the dermis, and are similar in basic structure to dogfish scales. Three regions can be distinguished: the crown is the visible projecting part of the tooth and is covered by a layer of enamel; the neck is surrounded by the gum; and the root or roots embedded in sockets in the jaw, by a layer of cement. The centre of the tooth is a pulp cavity, containing nerves and blood vessels to nourish the developing tooth. In teeth which continue to grow throughout life this remains open but in the majority of mammals the root canals (extensions of the pulp cavity in the roots) close almost totally when the tooth is fully formed. The greater part of the tooth is made of dentine, which is similar in composition to bone. Enamel, which only covers the crown, is made up almost entirely of inorganic matter, chiefly calcium phosphates and is the hardest substance in the body; calcium fluoride and phosphorus are both essential for enamel formation.

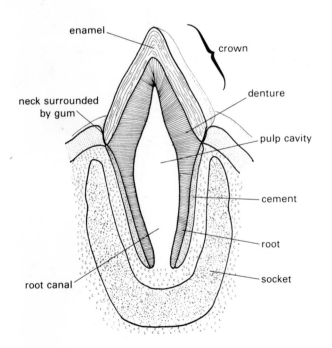

Diagrammatic section through a mammalian tooth

enamel
crown
neck surrounded by gum
denture
pulp cavity
cement
root
socket
root canal

Distinguishing features

Another character which distinguishes mammals from all other vertebrates is that their body cavity is divided by a sheet of muscle, the diaphragm, into the thoracic and abdominal cavities. The function of the structure is in improving the efficiency of breathing. Contraction of the diaphragm muscle flattens and helps to enlarge the volume of the thorax and so draw air into the lungs.

The majority of the distinguishing features of mammals can be related to their development of endothermy. However, a large part of their success must be related to the fact that they have a long period of infancy during which there is close parental care. With their well developed brain mammals show increased ability to learn from experience and much of this acquired knowledge can be passed onto the young. This gives them an adaptive plasticity not found in any other group. Fishes, amphibians, and reptiles show no trend to increase their brain size, but even the archaic mammals had a brain to body weight ratio of four or five times that of their reptilian ancestors. As in the birds the mammalian cerebellum is enlarged, but the principal expansion is concentrated in the surface grey matter (cerebral or neo-cortex) of the cerebral hemispheres, which functions not only as a major co-ordinating centre but as the seat of learning, intelligence, and consciousness.

Living mammals

Living mammals separate into three main taxa: Prototheria (monotremes), Metatheria (marsupials) and Eutheria (advanced placental mammals), of which the Prototheria are egg-laying mammals and the other groups are viviparous. Marsupials are born alive, but at an early stage of development since the placenta is not strongly developed, and they complete a large part of their development in a pouch (marsupium) on the mother's abdomen. The eutherians are the most successful mammals with about 8,800 species of recent mammals (compared with 242 marsupials and a mere six monotremes). Reproduction in the placental mammals involves a refinement of viviparity, to a stage when birth occurs at a relatively advanced stage of development, to be followed by a long period of parental care. Mammals are more structurally complex than other vertebrates and their development can take a long time (an African elephant has a gestation period of 660 days). Consequently a ready source of energy, provided by the placenta during the foetal stages and by lactation after birth, is essential. At birth the mammal is often poorly fitted for an independent life, although the degree of maturity varies. Horses, which have to run with the herd to escape predators, can walk within minutes of birth, whereas a human child lives for a year or so before taking his first steps. Neither could survive in the absence of parental care. Mammals give birth to relatively few young at a time, with many species producing only one or two offspring. Of all the placental mammals, the rodents, with about 1700 species (40 per cent of all mammal species), are the most successful in terms of diversity and total numbers, flourishing in almost all parts of the world and in most habitats. Although generally small, they are of enormous ecological importance, both because of the food they eat, and because they provide important links in the food chains of carnivorous reptiles, birds and mammals. The name of the order, Rodentia, is derived from the Latin *rodere*, to gnaw, referring to their unique incisors. They can, in general anatomy and body form, be regarded as representative mammals, with a long body, short legs, four feet with five digits and a long tail. For this reason, and because it is readily available *Rattus*, the rat, will be used as an introduction to the mammals.

Rattus

The rat has a world wide distribution and a long association with man, making at various times significant contributions to world history. The genus *Rattus* includes about 120 species. The black rat *Rattus rattus* was the major reservoir of bubonic plague, (although the disease was actually transmitted by the rat flea) and as man unintentionally carried the rat with him on his voyages and travels its spread can be followed by historical references to this disease from AD 160. It was first recorded in the British Isles in 1187 when 'large mice, popularly called rats, have been expelled from

the districts of Fers in Leinsters by the curse of the Bishop Ivov whose books they had gnawed.' Chaucer mentions rat poison being sold in the Pardoner's tale, and an Elizabethan 'Acte for preservation of Grayne' authorised church warders to levy tax to provide a reward for vermin catchers: one penny for every three rat heads. Plague first reached England in 1348, travelling from Weymouth to Oxford and London. Today the most common and important pest in England is the brown rat, *Rattus norvegicus*, which commonly lives in cellars and sewers (the black rat prefers attics), and is highly aggressive, attacking and killing the black rat. *Rattus norvegicus* is estimated to infest one per cent of urban property, 50 per cent of farms, and 60 per cent of sewers, and millions of pounds are spent each year in trying to control it.

The integumentary system

Unlike birds, mammals have thick skin with a thin multilayered epidermis and a thick dermis attached to the underlying muscles by loose connective tissue or fascia. The skin of a rat is fairly typical, with glands and subcutaneous fat. Two types of hair are present, outer, long, silky, guard hairs which cause water to run off the fur easily, and a shorter layer of insulating fur. The hair follicles are associated with slips of smooth muscle, the erector pili muscles, which when contracted bring the follicle and projecting hair shaft to occupy a position almost perpendicular to the surface. This is a response to cold,

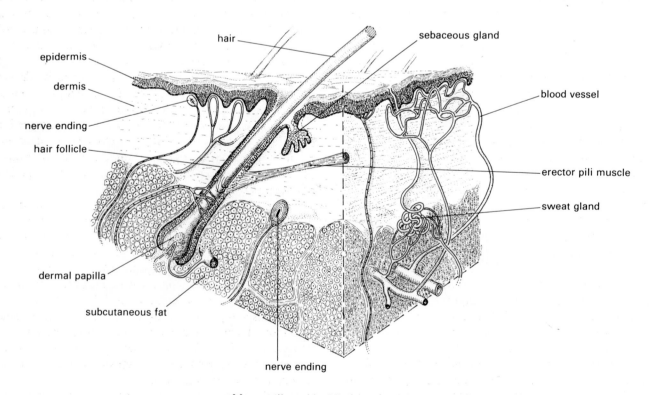

Mammalian skin (diagrammatic)

increasing the thickness of the insulating layer, but may also be invoked by other stimuli such as fear; as a consequence the animal appears larger and perhaps more formidable to its adversary. Both sebaceous and sweat glands are present, the former associated with the individual hairs and the latter distributed fairly evenly over the body surface and producing a typical watery secretion. Although both types of gland are epidermal structures, they are sunk into the dermis. The dermis also has nerves, some of which extend to receptor organs in the epidermis, and blood vessels.

Some regions of the skin show special modifications. The hairs round the mouth are long and strong forming sensitive structures, the vibrissae or whiskers, whose follicles are associated with a sensory nerve ring. The under-surfaces of the feet are hairless but have greatly thickened regions of stratum corneum forming the pads, and the tail, which is an appendage rather than a continuation of the body, is covered by both scales and hairs. The feet terminate in short sharp keratin claws which cap the terminal phalanx of each digit.

The skeleto-muscular system and locomotion

The skull
The rat, like other rodents, is a specialized herbivore with a skull in which the basic mammalian form is adapted for gnawing and grinding. The principal features which distinguish mammalian skulls from those of other amniotes have already been outlined, but a little more detail is appropriate here. The bones of mammal skulls are fewer in number than those of lower tetrapods (there are generally about 28 to 35 including ear ossicles, as opposed to as many as 180 in some bony fishes) due to loss or fusion. They are generally strong, well ossified and firmly united, (with the exception of the ear ossicles, dentary and hyoid) and there is close association between bones of dermal and cartilage origin. A single occipital ring (formed from the fusion of four separate ossifications found in lower vertebrates) surrounds the foramen magnum, with paired occipital condyles derived mainly from the exoccipitals, articulating the skull with the atlas. The skull is elongated anteriorly and is roofed by paired long, narrow nasals, frontals, and parietals. The maxillae and premaxillae are greatly enlarged and form both the sides and part of the roof of the nasal region of the skull, and an interparietal lies posteriorly between the two parietals. Both the maxillae and premaxillae are tooth-bearing. The temporal fossae are greatly enlarged and extend almost to the dorsal mid-line, leaving a narrow band of bone, the zygomatic arch, formed by the squamosal and jugal. The anterior part of this arch forms the lower margin of the orbit, and the jaw muscles lie between its posterior part and the brain case.

The anterior floor of the skull is formed by the secondary palate which develops as ventral shelf-like folds of the premaxillae, maxillae and palatines. Posteriorly and medially the secondary palate is completed by a vomer, a single bone representing a fusion of the paired structures of lower forms. The respiratory canal is divided medially by an internasal septum, and the greater part of the nasal cavity is filled by the complex of turbinal bones. The nasal cavities are large, reflecting the importance of olfaction in mammals. An auditory bulla surrounds the middle ear cavity and the chain of three auditory ossicles, incus (quadrate), malleus (articular), and stapes. A tympanic bone, thought to be a modification of the angular bone of lower tetrapods, supports the ear drum and forms part of the bulla. The inner ear is surrounded by the petrosal which represents the ossified otic capsule. The lower jaw is formed of the dentary which articulates with the mandibular fossa in the squamosal part of the temporal bone. A coronoid process is received by the fossa and articulation is completed by an angular process. The hyoid is represented by a plate from which arise a series of slender bones (ceratohyals, epihyals, stylohyals, and tympanohyals) which form the anterior cornua. The posterior cornua is represented by a thyrohyal on each side.

Teeth and jaws
The skull and teeth of the rat are specialized for gnawing and grinding. A single pair of very large incisors are present in both the upper and lower jaws; these grow throughout the animal's life, balancing the wear to which they are subjected. The anterior surface is coated with a thick layer of enamel but the posterior surface is not. The upper and lower incisors grind together and in this way they are kept sharpened to a chisel-like edge of enamel. Rodents can cut through all kinds of material from grass to wood, and rats have even gnawed through lead pipes or out of cages of iron wire. There are no canine teeth, and the anterior premolars are also absent, leaving a long space, the diastema, between the gnawing and grinding teeth. The advantage of this arrangement is that the sides of the lips can be drawn in while the animal gnaws, preventing shavings of material from flying into the mouth. The remaining cheek teeth are high crowned and grow continuously, again an adaptation to grinding in which the teeth are subjected to continuous wear. The grinding surface is formed of a series of enamel ridges, and these tend to get worn away and run together as the teeth are used.

Rats have two main jaw actions: biting and gnawing with the incisors, and grinding and chewing with the cheek teeth. When biting and gnawing the lower jaw is moved forwards so that the incisors are brought edge to edge, the lower ones passing posteriorly; in this position the upper and lower cheek teeth are out of contact with each other as the jaws close. For chewing, the lower jaw is slid backwards, to bring the molars and premolars together, and grinding is brought about by a combination of forward and back and side to side movements. *Rattus* is unusual among mammals in that the upper and lower jaws are of equal width and so both sides can be used at the same time.

In addition to the temporal muscles which close the jaw in

other tetrapods, the mammals have an additional series of muscles, the masseters, which pass from the zygomatic arch to the outer surface of the lower jaw. The masseters in rodents are especially well developed and divided into two parts: the lateral portion produces a simple up and down scissor-like action of the jaw while the medial portion pulls the lower jaw forwards. In the rats these muscles have acquired extra insertions and the lateral part extends forwards onto the face. The infra-orbital canal, which is present in all mammals, is a passage-way extending from the

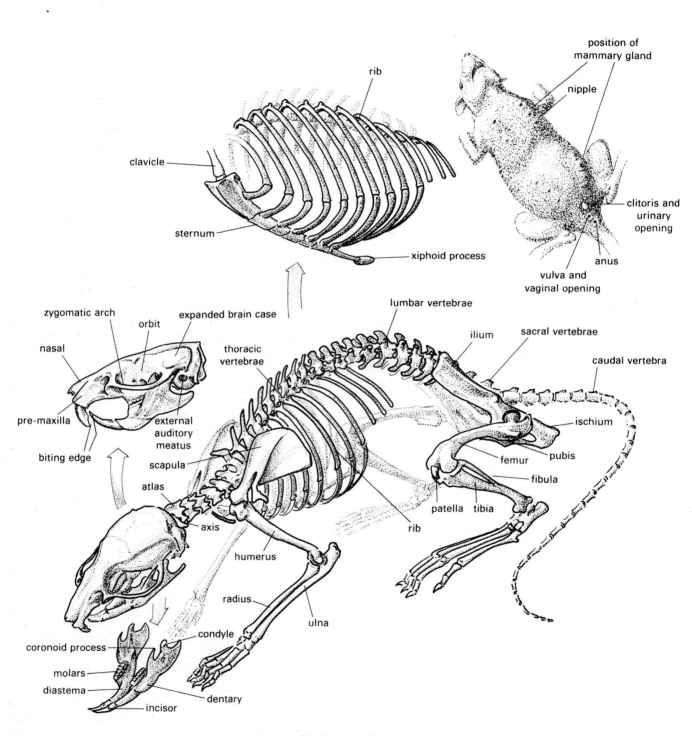

Rattus: skeleton

orbit to carry nerves and blood vessels from the orbit to the snout. In rats it is enlarged so that it also provides a passage for the medial portion of the masseter muscles. The angular process of the lower jaw is greatly enlarged to provide extra insertions for the jaw muscles and its articulation with the skull is long and low, allowing the jaw considerable freedom of movement.

The post-cranial axial skeleton

The post-cranial skeleton shows comparatively little specialization, and the vertebral column can be differentiated into the usual five regions: cervical, thoracic, lumbar, sacral and caudal. The main part of the column, the body region, forms a strong arch from which the viscera are suspended, with flexibility in the vertical plane allowing the bending required by mammalian locomotion. Rather little lateral flexion of the column is possible except in the neck region which is very flexible. The notochord is partly represented by intervertebral discs and the articular surfaces of the centra are flattened (acoelous).

Like the majority of mammals the rat has seven cervical vertebrae which can be distinguished from other vertebrae by their lack of ribs, and by the possession of paired vertebral foramina which together form the transverse canal for the vertebral artery and vein. The atlas and axis vertebrae show typical amniote modifications and the atlas has two large concave articular facets for the paired occipital condyles of the skull. Variations in neck length in mammals (and in body length in general) are generally the results of differences in individual vertebrae rather than an increase or decrease in vertebral number. As mammals grow, the vertebral centra, which have an epiphysis at each end, increase in length. In the adult the epiphyses become fused to the rest of the centrum.

There are twelve thoracic vertebrae in the rat (the number ranges between nine and twenty-five in mammals). They each have special articular facets which articulate with two parts of each rib (the tubercle and the capitulum): the tubercle articulates with a tubercular facet on the ventral side of the transverse process (diapophysis) and the capitulum fits into a depression formed between the costal demifacets (parapophyses) of succeeding vertebrae. This is the more common mammalian condition in which each capitulum articulates with two adjacent centra. The thoracic vertebrae also have large neural spines. The anterior ribs articulate with the sternum which is formed of a longitudinal fused series of sternebrae; the most posterior ribs, which do not reach the sternum, are said to be floating ribs.

The lumbar vertebrae are typically large and strong in mammals and there are usually between four and seven of them, the rat has seven. They each have a long neural spine which is directed anteriorly and well developed pre- and post zygaphophyses; the transverse processes are drawn down and forwards and there is a separate posterior process. The sacrum consists of a group of between three and five highly modified vertebrae (four in the rat) which are closely united and form a firm articulation for the pelvic girdle. The anterior sacral vertebra is particularly important in this, and its transverse process is greatly thickened and spread out to articulate with the ilium. The transverse process of the second sacral vertebra is also thickened and contributes to the support of the ilium. A secondary articulation, the costal element, develops below the transverse process and is also involved in pelvic girdle articulation. The caudal vertebrae run out into the long slender tail, and show a gradual diminution of size so that the terminal vertebrae consist of little more than centra. Tail length in mammals is highly variable and the number of vertebrae ranges from 3 to around 50 with rats having approximately 28. The anterior haemal arches are directed forwards and serve for muscle attachment.

The limb girdles

The appendicular skeleton of rats is fairly close to the basic pentadactyl plan, with the hind limbs, which provide much of the power in mammalian locomotion, showing the greater deviation. Mammalian limbs are generally longer and thinner than those of reptiles, supporting the body well away from the ground, and this is reflected in the skeleton.

The pectoral girdle is composed of a scapula, coracoid process and clavicle and shows modifications associated with the position of the limbs which are held beneath the body and support its weight during locomotion (this is also seen in some reptiles). The structure of the pectoral girdle also allows greater flexibility of the neck, and permits expansion and contraction of the thorax during breathing. The scapula is a triangular bone with a well developed spine for muscle attachment and is free from the axial skeleton, apart from its ligaments and muscles. A ventral extension of the ridge forms the acromial process. The coracoid is reduced to a process on the ventral edge of the scapula which completes the glenoid fossa. The clavicle articulates between the acromial fossa and the sternum. The pelvic girdle consists of three basic elements, the ilium, ischium and pubis, which show total fusion to form the innominate bones on either side. The ilium is elongated anteroposteriorly and the acetabulum, which is a deep cavity to accommodate the head of the femur, lies well posterior to the junction of the ilium and the sacral vertebrae.

The limbs

The legs of the rat are relatively unmodified, and are held in the typical mammalian position with elbows and knees turned in towards the body. Progression involves swinging the legs forwards and backwards in the vertical longitudinal plane and straightening and bending the elbow and knee joints. The fore-limb has a long straight humerus, with proximal articulation (with the glenoid fossa) which is nearly hemispherical and at right angles to the rest of the bone. Two projections (the greater and lesser tuberosities) provide for the attachment of the pectoralis muscle (which provides upwards and forwards movement) and the latissimus dorsi (which provides backwards and downwards movement). The forearm elements are the radius and ulna;

the proximal end of the ulna is extended posteriorly as the olecranon process to which the tendon of the triceps, which opens the elbow joint, is attached. As a result of the position of the elbow below the body the radius and ulna are twisted over each other by almost 180°. The wrist bones show reduction of the centralia to a single bone, and the digits are elongated but otherwise unmodified. The hind limb shows similar modifications but is rather more strongly developed. The head of the femur is at right angles to its shaft and has projections for muscle attachment, the greater and lesser trochanters which correspond to the tuberosities of the humerus. The tibia and fibula are parallel to each other and a sesamoid bone, the patella (knee cap) is formed in the tendon of the rectus femoris muscle which straightens the leg at the knee joint. The ankle joint, like the wrist, also shows reduction of bones, to a single centrale and two bones (astragalus and calcaneum) in the proximal row. The calcaneum is prolonged backwards as the heel bone to provide for the insertion of the tendon of the gastrocnemius muscle which straightens the ankle joint.

Locomotion

Mammals, like birds, are generally much more agile and active than the lower tetrapods. This is partly correlated with their maintenance of a high body temperature and with greater muscle power, but also with the position of the limbs. Since the legs extend almost directly below the body they not only provide better support but also permit a longer swing of the leg, and thus increased stride length and greater speed. In fast movement this can be further emphasised by alternately bending and straightening the vertebral column in the vertical plane. As a result of their change in posture, they are generally less stable transversely than the reptiles. This is more than offset by their increased level of dynamic stability, provided by more highly developed muscle responses to changes in balance.

Rats show simple quadrupedal locomotion with most of the thrust for fast movement coming from the hind limbs. The fore-limbs are somewhat shorter than the hind limbs, and are also used for handling food, for scratching, and for burrowing: the tail serves for balancing and steering. The feet and hands are held flat on the ground (plantigrade). Rats are versatile in their locomotion and can either walk, climb, jump or run and when necessary they can move remarkably fast. In walking the order of lifting the legs conforms to the normal quadrupedal rule, right fore, left hind, left fore, right hind; this is the only sequence which allows the centre of gravity to remain over a triangle provided by three legs, and so allows the animal to be sufficiently stable to stop at any instant without falling over. Rapid walking sacrifices this stability to some extent as each foot is raised before the one ahead of it reaches the ground, and there are short periods when only two feet are on the ground. Faster movement is provided by the trot in which the period of unilateral support (right fore, right hind) is shortened, and that of diagonal support (right fore, left hind) is slightly lengthened. When moving very rapidly, the

animal proceeds in a series of bends in which the original foot order is lost and the sequence becomes: left fore, right fore, left hind, right hind, with periods when no feet touch the ground at all. At this stage the stride is lengthened by alternate lengthening and straightening of the vertebral column.

Rats are remarkably agile and depend on their speed and agility to escape from predators. They normally run using an accelerated walk, their limbs can move so quickly that they seem to scurry about like mechanical toys. When moving very quickly they bound and when running along rope or twine, they swing their tails from side to side to help their balance. They are good climbers and use their claws, digits and pads to grip when on vertical surfaces. The black rat, *Rattus rattus* is predominantly a climber which, coupled with the fact that it is exceptionally unafraid of man, makes it well suited for life on ships and accounts for its wide distribution. It invades ships by running along the moving ropes, an easy feat for an animal that can run along a telephone wire less than 2 mm thick. Very often ships carry baffle plates on their mooring ropes to prevent the rats from reaching the vessel.

The digestive system and nutrition

Apart from the adaptations for gnawing, the alimentary tract of the rat can be regarded as fairly representative of mammals. It is a highly efficient system, capable of digesting large quantities of food of various kinds. Rats are typically herbivores but feed on almost anything, including a wide variety of food made available by man. As in other gnathostomes the boundaries of the mouth are surrounded by folds of skin, but in the rat (as in the majority of mammals) these are modified as movable, sensitive lips. The skin on the outside is highly sensitive, while on the inside there is a mucous membrane. The spaces between the lips and jaws are the vestibules, and rodents, as foragers, have well developed cheek pouches which serve as temporary storage spaces for food. In particular they damage stored foods and are a menace in granaries and barns. Where they have been introduced to isolated regions such as islands they have often had a devastating effect on indigenous wild life, especially on animals such as tortoises and birds, which lay eggs on the ground. In 1918 rats swam ashore to Lord Howe Island in the Pacific and eliminated four species of birds. *Rattus norvegicus* is particularly aggressive and will attack and kill *Rattus rattus*. Captive rats frequently are cannibalistic and may dispose of their victim completely, starting with the brain and leaving only a skin, neatly turned inside out. The incisors are well adapted for attack and defence as well as feeding. The mouth cavity is separated from the nasal cavity by the secondary palate and therefore functions primarily for the passage of food, which can be chewed without interfering with respiration. Mammals do not swallow their food whole, but break it up with their teeth. It is mixed with mucus from many small

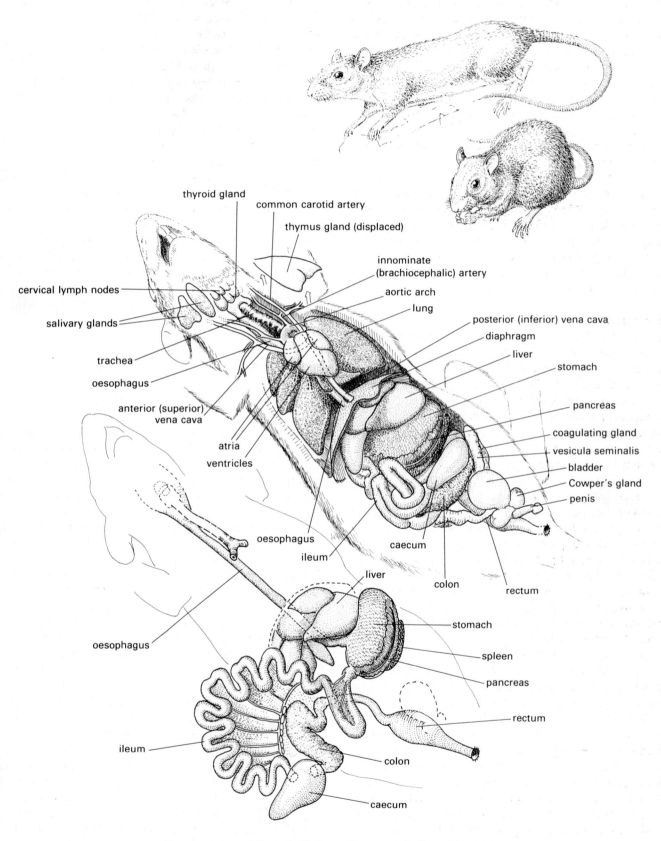

thyroid gland

common carotid artery

thymus gland (displaced)

innominate
(brachiocephalic) artery

aortic arch

lung

cervical lymph nodes

salivary glands

posterior (inferior) vena cava

diaphragm

liver

stomach

trachea

oesophagus

pancreas

anterior (superior)
vena cava

coagulating gland

vesicula seminalis

bladder

atria

Cowper's gland

ventricles

penis

oesophagus

ileum

caecum

liver

colon

rectum

oesophagus

stomach

spleen

pancreas

rectum

ileum

colon

caecum

The anatomy of the rat, *Rattus :* thorax and alimentary tract

mucous glands on the tongue and palate, and with saliva produced by the conspicuous salivary glands, which open into the mouth by ducts. Mammals have three pairs of salivary glands: parotid, submandibular, and sublingual. In addition to acting as a lubricating agent the saliva contains the digestive enzyme amylase which begins the process of starch digestion. The tongue is highly mobile, with muscles which can move it in all planes, and is used to manipulate food within the mouth, as well as to prevent regurgitation.

The swallowing process is fairly complex and as food passes into the posterior part of the pharynx on its way to the oesophagus a number of events occur. The muscular soft plate (which completes the secondary palate) rises, and prevents food from entering the nasal cavities. Breathing stops momentarily and the glottis is closed by the epiglottis. Muscular contractions of the posterior part of the pharynx move the food into the oesophagus and thence to the stomach. In mammals the stomach is quite distinct from the oesophagus and lies beneath the diaphragm, through which the oesophagus passes. It is a capacious, J-shaped bag, with a constriction separating the anterior (cardiac) and posterior (pyloric) regions. Mammals have a large, lobed liver, and the gall bladder usually lies between the lobes on the right-hand side, though it is absent in rodents as in most other herbivores. The bile and pancreatic ducts join to enter the duodenum as a single open duct. Here the stomach and intestine are supported by mesentery (the omentum) which is studded with masses of fat. At the junction between the ileum and colon (ileocolic valve) there is a well developed caecum which functions, together with the capacious colon and ileum, as a site for bacterial digestion. The alimentary tract is completed by a rectum which opens at a separate anus; there is no cloaca in eutherian mammals.

As a further adaptation to their herbivorous diet, rodents pass food through the alimentary canal twice by reingesting the faecal pellets (coprophagy). After the first passage through the gut the faeces are soft, moist, and have a high quantity of Vitamin B_1. Their consumption enables the animal to absorb valuable water and vitamins and to feed rapidly, digesting the food in the safety of its shelter.

Respiratory system

Rats have a typically mammalian respiratory system, which is highly efficient and adapted to give the maximum possible surface for gaseous exchange (estimated at 55 square metres for humans). The lungs are large bilateral, bilobed structures which lie in pleural cavities, separated from the heart by the pericardial sac and from the abdominal cavity by the diaphragm. They are formed largely of small, thin-walled sacs, the alveoli, which have a dense capillary network and are the site of gaseous exchange. A thin film of lipoprotein is secreted by the alveoli, which reduces the surface tension of the respiratory surface, aiding the flow of air over it and reducing water loss. The alveoli are the terminal divisions of a series of branching tubes.

The trachea opens into the pharynx by the glottis, and can be closed momentarily in swallowing by the epiglottis, which forms part of the larynx. This is a complex region of the trachea, supported by cartilages and containing the vocal chords. The trachea branches to give two bronchi, one to each lung, which are also supported by rings of cartilage. Repeated branchings produce the bronchi and bronchioles which end ultimately at the alveoli. As the tubes become smaller the cartilaginous rings are lost and the walls become progressively thinner.

The lungs are ventilated by changing the volume of the thoracic cavity, and thus the pressure within the lungs. Contraction of the radial muscles of the diaphragm, which is dome-shaped in a relaxed position, flatten and lower it and increase the volume of the cavity: at the same time the external intercostal muscles contract to expand the rib cage and draw air into the lungs. Expiration is brought about by relaxation of the diaphragm and the external intercostals, allowing elastic recoil of the lungs and chest wall, and also by contraction of the internal intercostal muscles. Quadrupeds rely heavily on the diaphragm in respiration as the thoracic cage is partly involved in the suspension of the fore-limbs. In bipeds this is not a factor and movements of both diaphragm and thorax are important. Unlike the birds, mammals have a tidal flow in and out of the lungs. There is therefore always a volume of residual air since even the strongest respiratory movements cannot empty the alveoli completely.

Air is drawn into the nasal cavities through the external nares, and may also be drawn in through the mouth. The surface area of the nasal cavities is expanded by the turbinal bones or conchae which support the nasal mucosa; this has olfactory receptors and also contains numerous mucous glands. Here the air is warmed and moistened, and foreign particles are trapped in the mucus. The mucosa is ciliated, and particles and mucus are moved into the pharynx and swallowed or expectorated.

The process of making sounds is highly developed in some mammals. The vocal cords are a pair of folds of the lateral pharyngeal wall in the larynx which vibrate to make sounds. Their vibration sets up oscillations in the air column in the upper respiratory tract, and these are modified into recognisable patterns of sound by the combined effects of the mouth, tongue and lips.

Circulatory system

The mammals have a double circulatory system and four chambered heart that is essentially the same as that of the birds. However the right aorta is eliminated and only the left aorta persists to supply blood to the body (in the birds the right aorta persists rather than the left). Otherwise the main features of the arterial system are strikingly similar in the two groups. The main venous canals in mammals show a definite shift to the right-hand side of the body. There is no trace of a renal portal system, and all blood from the

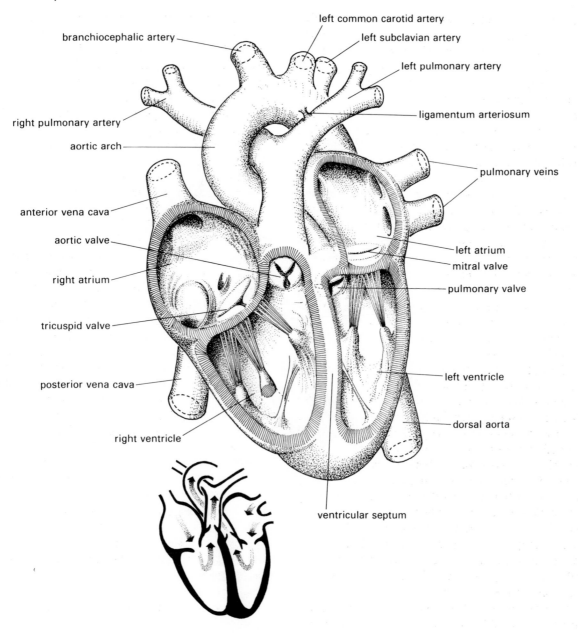

Mammalian heart (human)

Excretion and osmoregulation

The typically mammalian metanephric kidney is a bean-shaped organ, with a medial depression, the hilus, at the point where the ureter and renal vein leave the kidney and the renal artery enters it. It is a highly sophisticated structure, capable, like that of birds, of producing urine which is more concentrated than the blood. In both birds

posterior part of the body is collected by the posterior vena cava. The hepatic portal system is similar to that of lower forms. The lymphatic system is well developed with numerous lymph nodes.

and mammals the nephron is very long, with an extra region, the loop of Henlé, in which the tube folds back along itself producing a counter flow in each section and thus enabling the animal to produce a more concentrated urine than would otherwise be possible (counter-current multiplier). This loop lies between the proximal and distal convoluted tubules and is shaped like a straight hair pin; as a general rule the longer the loop of Henlé, the higher the concentration of urine that can be produced. *Rattus norvegicus* can produce urine of a concentration of 2900 mOsm litre as opposed to 600 mOsm litre in a frog, and 2000 mOsm litre in a bird. As in other vertebrates, an ultrafiltrate, with the same osmotic concentration as the blood, enters the nephron at the Bowman's capsule. In the

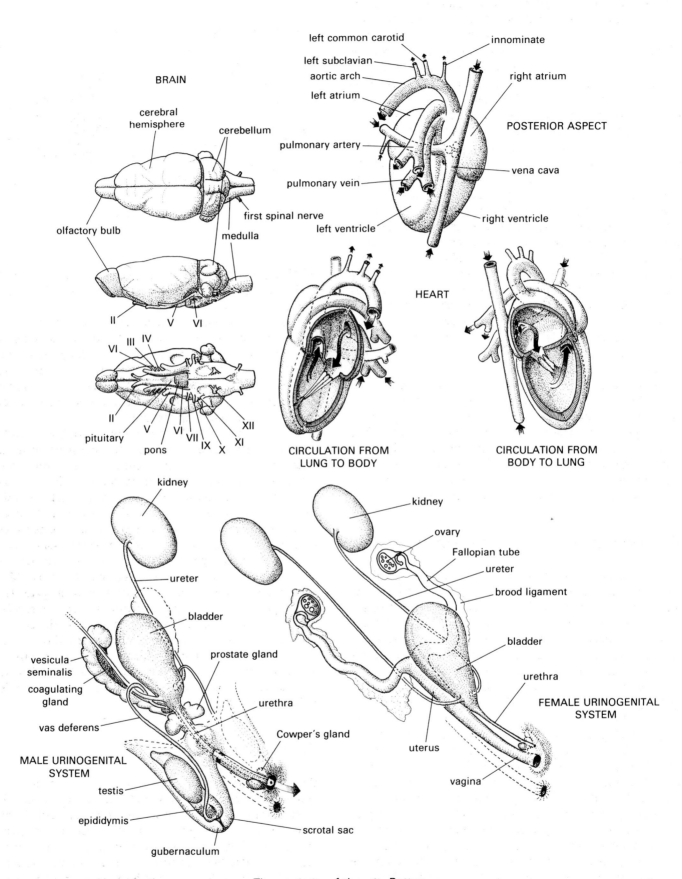

BRAIN

cerebral
hemisphere
cerebellum

olfactory bulb

first spinal nerve

medulla

II   V   VI

VI   III   IV

II

pituitary

pons   VI   VII
            IX   X   XI

V

XII

left common carotid   innominate
left subclavian
aortic arch   right atrium
left atrium   POSTERIOR ASPECT

pulmonary artery

vena cava

pulmonary vein

right ventricle
left ventricle

HEART

CIRCULATION FROM
LUNG TO BODY

CIRCULATION FROM
BODY TO LUNG

kidney

kidney

ovary

Fallopian tube

ureter

ureter

brood ligament

bladder

bladder

prostate gland

vesicula
seminalis

urethra

urethra

coagulating
gland

FEMALE URINOGENITAL
SYSTEM

vas deferens

Cowper's gland

MALE URINOGENITAL
SYSTEM

uterus

testis

vagina

epididymis

scrotal sac

gubernaculum

The anatomy of the rat, *Rattus*

proximal convoluted tubule glucose and other essential metabolites are reabsorbed by the general circulation and some water is taken up, reducing the volume of filtrate. However, at the time when it enters the loop of Henlé it is still more or less isosmotic with the blood plasma; as it passes down the descending arm of the loop it becomes increasingly concentrated. The thin walled cells of this region allow free diffusion of sodium and water, and, since the arm passes through tissues with a steadily increasing osmotic concentration, the fluid is concentrated. The cells of the ascending arm actively pump sodium from the filtrate, but are impermeable to water. It is these ions, and the chloride ions which follow them, that pass into the tissue fluids surrounding the ascending arm and the distal convoluted tubule, and are responsible for the high osmotic concentrations of these tissues. In the final region, the distal convoluted tubules, the cells are permeable to water, which therefore diffuses outwards to bring about equilibrium with the adjacent cells, further concentrating the urine. The urine drains into the collecting ducts, where some water may be resorbed if the animal is dehydrated, and passes down the ureter to the bladder. This is a muscular sac capable of great distension which opens to the exterior through a single tube, the urethra. In males the urethra opens at the tip of the intromittent organ, the penis.

The kidney is enclosed in a capsule of connective tissue, and in longitudinal section can be seen to be composed of two regions, the outer cortex and inner medulla. The medulla is composed of the glomeruli, and the proximal and distal convoluted tubules. Capillaries enter and leave through the cortex in which the loope of Henlé lie, and thus the blood vessels also have a counter-current flow.

The nervous system and sense organs

The mammalian brain and nervous system is more highly developed and is proportionately larger than that of any other kind of vertebrate. Indeed, the enlargement of the brain is so great that its resemblance to the basic vertebrate pattern is almost completely obscured. The cerebellum is enlarged like that of the birds, and mammals also show enormous enlargement of the cerebral hemispheres which are the principal coordinating centres of the nervous system. However the expansion of the cerebral hemispheres is different from that of the birds. The inner corpus striatum is only moderately developed and the greatest expansion is of the surface grey matter, the cerebral cortex or neocortex. This disproportionate development of the surface tends to cause the brain to appear as a convoluted mass with surface folds and wrinkles. The expansion of the neocortex in mammals gives them their learning ability and the ability to alter behaviour to meet the demands of different situations. The cerebellum, which is also well developed in mammals, has strong connections in both directions between the cerebral cortex. Like other amniotes mammals have 12 pairs of cranial nerves. The spinal nerves show intermixing

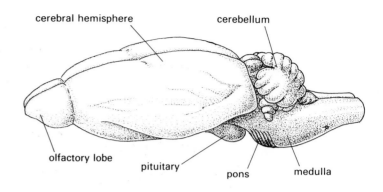

Rattus : brain

of fibres forming complicated plexuses named after the regions of the spine in which they occur: cervical, brachial, lumbar and sacral.

The sense organs of mammals have already been described and will be reviewed only briefly. Although the olfactory lobes of the brain are not as large as in some vertebrates, the sense of smell is highly important to most mammals, including the rat. Its efficiency is greatly increased by the large surface area of the nasal mucosa and hence of the olfactory epithelium.

The eye is basically like that of other vertebrates, and its adaptations are those associated with a partly nocturnal life style; it has more rods than cones in the retina and a reflecting layer, the tapetum, on the choroid coat. The mammalian ear shows advances over that of other vertebrates in several respects: the cochlea is long and coiled, and the middle ear has a chain of three bony ossicles (malleus, incus, stapes). There is an external auditory canal and well developed pinna which can be moved and directed to funnel the sound into the external auditory canal. Many small nocturnal rodents use echo-location to navigate. They emit high frequency sounds and from the echoes received learn about the size and position of nearby objects. The principles involved are simple: since the speed of sound is constant, the time between emission of a short burst of sound and reception of its echo depends upon its distance; the quality of the echo depends on the nature of the object reflecting the sound.

Reproduction

By mammalian standards, rodents are fast breeders with enormous reproductive potential; the rat has a breeding (oestrous) cycle of 4 to 5 days, a gestation period of 22 days, and the young complete their development to sexual maturity in about 5 to 6 weeks. Polyoestrous cycles in rats can be related to their almost continuous food supply, and in general mammals breed to produce young at a time when food is most abundant.

The male reproductive system

The male has paired testes which develop in the abdominal cavity and migrate downwards into special folds of skin, the scrotal sacs, at or soon after birth. The most widely accepted explanation for this is that the temperature in the thin-walled scrotal sacs is several degrees lower than in the body cavity, and spermatogenesis will not take place at the higher temperature. However in birds and a few mammals (elephants) the testes are retained in the body cavity and the process remains something of a mystery. As the testis descends it is guided by a long cord, the gubernaculum, which anchors it into position in the scrotal sac. Efferent ductules connect the seminiferous tubules of the testes to a compactly coiled epididymis where sperm mature and are stored. The epididymis joins the vas deferens which enters the body cavity, loops over the ureter, and joins the urethra. To creat the fluid medium necessary for sperm transfer a glandular structure, the seminal vesicle, joins the vas deferens. Various other accessory sex glands, including the prostate gland and Cowper's glands, are associated with the urethra. The secretions from these glands activate and nourish the sperm, as well as neutralizing the acidity of the female vagina. 400 million sperm may be transferred to the female at one ejaculation.

The intromittent organ of mammals is highly developed as a tubular penis, which evolved from the walls of the cloaca in ancestral forms. It contains the terminal part of the urethra, which is surrounded by a mass of erectile tissue, the tip of which contains highly sensitive skin and which contains large blood vessels. When these are filled with blood they cause the penis to become rigid, and, like some other mammals, some rodents also have an os penis, a supporting bone.

The female reproductive system

The female reproductive tract follows the basic vertebrate plan, with a pair of solid ovaries each with Fallopian tubes (oviducts) leading to the exterior. The lower part of the Fallopian tubes fuse to form a uterus and this opens as a single vagina which receives the penis during copulation. The vagina is lined with a squamous epithelium and its opening is associated with Bartholin's glands which produce a lubricating secretion.

The female reproductive system undergoes regular cyclical changes in its readiness to receive sperm. The period of oestrus or heat is the time at which eggs are shed from the ovary and the female is receptive to the male (every 4 to 5 days in the rat). The word is derived from the Greek for the gadfly, whose sting was said to drive cattle crazy. In Greek mythology Hera, having discovered that her husband Zeus had been unfaithful to her, sent a gadfly to sting her rival.

The oestrous cycle is under hormonal control and can be subdivided into four periods: pro-oestrus, oestrus, metoestrus and dioestrus, each defined by events occurring in the ovary. In the pro-oestrus phase an ovarian follicle begins to grow to produce a mature ovum, and as it does so it secretes the hormone oestrogen. At oestrus the ovum is shed and the ovarian follicle becomes modified as a corpus luteum which secretes a different hormone, progesterone. In metoestrus the corpus luteum regresses, and the final phase of the cycle, dioestrus, is a resting period. This sequence of events is regulated by four hormones: two of these, follicle stimulating hormone (FSH) and luteinising hormone (LH), are produced by the pituitary, while oestrogen and progesterone are produced in the ovary. Ovulation is triggered by luteinising hormone (LH) which is secreted by the pituitary under the control of the hypothalamus. This process is triggered by increased oestrogen, which also suppresses the production of FSH. Growth of the ovarian follicles is controlled by FSH.

The events in the uterus that follow fertilization are also under hormonal control. Progesterone stimulates the uterine lining to thicken and thus prepares it to receive the fertilized egg. After implantation the corpus luteum continues to develop, and its hormone production causes further growth of the uterine lining and also effects glands, muscles and blood vessels of the uterus.

The early development of the mammal and placenta formation have already been described. Young rats are born 22 days after fertilization and at birth they are helpless and blind. During pregnancy the mammary glands have been prepared by the action of both oestrogen and progesterone; oestrogen causes a growth of the ducts, and progesterone stimulates the glands themselves. However milk is not produced, as the presence of oestrogen in the bloodstream inhibits the production of the hormone prolactin which stimulates milk production. At birth, following expulsion of the placenta, the levels of both oestrogen and progesterone fall and prolactin is secreted by the pituitary under the control of the hypothalamus; the glandular cells of the mammary glands then begin to secrete milk. Yet another hormone, oxytocin, also produced by the pituitary under the control of the hypothalamus, controls the release of milk from the breast in response to the stimulus of the young suckling the nipple. Milk is a water-based and highly nutritious solution of proteins, lactose (a sugar) and fats; whose exact proportions vary between species. Generally animals with a high growth rate, like rats, have a higher proportion of protein.

MAMMALIAN DIVERSITY

The main diversification of mammals has occurred in the last 64 million years, which in terms of animal evolution is comparatively recent. Much of their radiation can be traced, like that of the birds, to specializations for feeding and locomotion. Unfortunately their early fossil record is poor and of the three main groups of mammals the Prototheria are almost unknown as fossils. Since the living forms are highly specialized it is difficult to determine their ancestors. True placental mammals probably evolved at the end of the Triassic, but very few Mesozoic fossils have been found. They are fragmentary and include no complete skeletons and hardly a complete skull. Many early mammals are known only from their teeth. To judge from this record they were small carnivores which fed on insects, grubs and worms: they were obviously inconspicuous, and their poor fossil record suggests that they may have been rather scarce.

The major mammal groupings are based on their reproductive biology. The class is divided into three subclasses as follows: Prototheria (a restricted group of egg-laying mammals); Metatheria (the marsupials) and Eutheria (the great majority of modern mammals).

Subclass Prototheria (the monotremes)

The three genera of monotremes, the duck-billed platypus (*Ornithorhynchus*) and the echidnas or spiny anteaters (*Tachyglossus* and *Zaglossus*) are the most primitive modern mammals, surviving only in Australia, Tasmania and New Guinea. They share various characteristics with reptiles and birds which are lacking in other mammals, and in particular the reproductive system is strongly reminiscent of reptiles. Both males and females have a cloaca: in the male the testes do not descend to occupy an external position, and the penis is normally held inside the body cavity. However the most striking difference from other mammals is that monotremes lay eggs which they incubate. There are also similarities with reptiles in other respects, such as in the structure of the pectoral girdle and the organisation of the brain. Although they are endothermic, monotremes have a low basal metabolic rate and maintain a lower body temperature (30 to 33° C) than most animals.

In spite of these resemblances to reptiles, the monotremes do possess the characters by which we distinguish mammals from other vertebrates. Thus they are hairy endotherms which suckle their young and have a single bony element in the lower jaw.

Monotreme eggs are about 20 mm long with a whitish shell. The newly hatched young are necessarily very small and weigh about 0.4 g. The young of the duck-billed platypus are nursed within a burrow, while those of echidnas are carried by the mother within a temporary pouch. The mammary glands of monotremes have no nipples, and those of echidnas open within the pouch. The duck-billed platypus lives a semi-aquatic life and is found in rivers and lakes in Australia, feeding on invertebrates which it finds by grubbing about at the bottom with its bill. It has a broad bill which is indeed like a duck's in shape but is covered with leathery skin and has dorsal nostrils. Although teeth appear during development, they are absent in the adult which has horny plates instead. The platypus is a good swimmer and well adapted for life in the water with its webbed feet, thick tail, short fur and lack of external ears. Echidnas follow a totally different life style and are specialised for feeding on ants. They have a greatly elongated snout with a long sticky tongue but no teeth. They use their strong, clawed feet for digging into anthills and for

Tachyglossus, a spiny ant-eater

Ornithorhynchus, the duck-billed platypus

burrowing. They have a formidable battery of spines rather like those of a porcupine, which are modified hairs. Echidnas may be close relations of an extinct group of early mammals, the multituberculates, which lie on a separate evolutionary line from the advanced placental and marsupial mammals.

Subclass Theria
Infraclass Metatheria: the marsupials

Like the monotremes, the marsupials are also very restricted in their distribution, occurring in Australia, and in South, Central, and Southern North America, though they are known from the fossil record to have had a much wider distribution in the Eocene. However the marsupials are a more diverse group, and they parallel to a great extent the radiation of eutherian mammals. There are four main groups, each specialized for a different kind of diet. The carnivores of the marsupial world include the American opossums and the Australasian 'cats' and 'wolves' which bear a remarkable resemblance to eutherian mammals with the same common names. Among marsupial insectivores are the South American rat opossums, and the Australasian 'mice and moles'. The herbivores include kangaroos, wallabies, wombats, and koalas. The bandicoots (Australasia) are omnivores which feed on insects, small vertebrates, and roots.

As a group, the marsupials are characterised by a short gestation period so that they give birth to their young at an early stage of development. The placenta is of a simple yolk sac type. The female of most marsupials has a permanent pouch (marsupium) in which the young develop, this covers the mammary glands and is supported by epipubic bones. The kangaroo (*Macropus*) provides one of the best known examples of marsupial development. The young are born about one month after fertilization, when they weigh about 0.7 g and are about 25 mm long. At birth the young kangaroo has strongly developed forelimbs with claws, and using these it crawls its way up to the pouch where it finds and attaches itself to one of the nipples. This swells so that attachment is secure. (In forms which have no marsupium, such as *Marmosa*, the American mouse opossum, the young hang from the nipples). The kangaroo lives for about 7 months in the pouch and weighs about 2 kg when it first leaves. For a short period it returns to the pouch at intervals, but later it lives outside, feeding partly on milk and increasingly on grass.

Marsupials are thought to be more closely related to the advanced placental mammals (Eutheria) than to the monotremes, but the two groups have separate evolutionary histories since the Cretaceous. Their common ancestors are probably to be found in a Jurassic group, the pantotheres. The primitive marsupials of the Americas were small and opossum-like; they flourished and were widely distributed in the early Cainozoic but were unable to compete with

Didelphis, an opossum

eutherian mammals and by the early Tertiary were restricted to regions isolated from the impact of eutherians. There they radiated as a group of successful marsupial carnivores which ranged in size from the opossum to a kind of marsupial 'bear' and included marsupial 'cats' and 'wolves'. However by the end of the Tertiary they had largely disappeared in the face of unequal competition with placental carnivores. Living opossums, such as *Didelphis*, are principally carnivorous although they have a tendency towards an omnivorous diet.

The earliest Australian marsupials were also opossum-like. Their subsequent adaptive radiation gave rise to a wide range of forms fulfilling very different ecological roles. The list of Australian marsupials includes carnivores, a marsupial anteater, omnivorous bandicoots, flying forms (phalangers) a marsupial 'mole' and herbivores such as koalas and wombats. Until recent introductions by man the only placental mammals to enter this region were bats and rodents which explains why most present day marsupials are concentrated there. However the introduction of mammals such as rodents and the dog to Australia has had a disastrous effect upon the populations of smaller native marsupials.

Subclass Theria
Infraclass Eutheria

The eutherians or placental mammals include about 94% of living mammals, with 3,938 living species. The subclass has been the dominant group of mammals amongst the Cainozoic species. Of the 17 orders into which modern mammals are divided, some contain only a few species (the aardvark, for example, is the only member of the order Tubulidentata), and most mammalian species belong to about half a dozen orders. The largest is the order Rodentia, with 1,729 species which include the rats (*Rattus* sp.) Other major groups are the Chiroptera (bats) 981 species; the Insectivora (moles, shrews and hedgehogs) with 374 species; the Carnivora (dogs, cats and bears) 252 species; the Artiodactyla (camels, deer, sheep, goats and cattle) with 194 species; and the primates (man, apes and monkeys) with 161 species. All of these have young which are born at an advanced stage of development, and are generally weaned at an earlier age than marsupial young. The diversity of the placental mammals is enormous, but some appear to have changed little since the group evolved; thus the insectivores are regarded as survivors of the primitive placental mammals from which the rest later evolved.

Order Insectivora

The insectivores are mainly small and inconspicuous animals with retiring habits and are often nocturnal. The majority hibernate in winter. The various species are classified together because they retain several primitive features which include a primitive skull pattern, and molars and limbs rather like those of the marsupial insectivore, *Didelphis*. The brain is relatively simple, with large olfactory bulbs and rather small cerebral hemispheres. The testicles, unlike those of most placental mammals, remain in an internal position and the females give birth to large litters of up to 32 young.

The most numerous and least modified insectivores are the shrews, a group which includes the smallest mammals. Fossils of *Sorex* (the shrew) are known from the Miocene and modern animals show little change. Many, including *Sorex*, are superficially mouse-like and spend much of their time in tunnels in the soil or in runways in dense vegetation. Others are aquatic, and may, like *Myogale*, have webbed feet for swimming. In spite of the name of the order,

insectivores are not restricted to an insect diet but take a variety of small prey. *Sorex* feeds on a variety of insects but also earthworms. It hunts intermittently both by day and night, finding its prey principally by means of a long probing snout with well developed vibrissae.

The mole *Talpa* feeds mainly on earthworms and is highly specialized for a fossorial mode of life. The fore-limbs are adapted for burrowing, and have broad, spade-like hands and digging claws; the cervical vertebrae are fused. It has small eyes and no external ears and hunts using its nose and vibrissae. Although moles spend most of their time underground they appear on the surface sufficiently often to be preyed on extensively by nocturnal birds such as tawny owls. Underground, the mole makes an elaborate system of tunnels in which it wanders about, finding and killing any soil animals which enter the system.

Talpa, the mole (order Insectivora)

Another familiar insectivore, the hedgehog (*Erinaceus*) is regarded as the most primitive living member of the group. Undoubtedly its protective spines have contributed to its success but it also appears to have a remarkable physiological immunity to bacteria and toxins generally, including those from snake bites.

Order Chiroptera

The bats are the only mammals capable of true flight as opposed to the gliding seen in forms such as flying squirrels. They probably evolved from arboreal insectivores, and are regarded as being closely related to that order. The majority, including all the British species, feed on insects which they catch in the air, but a large number of tropical bats, including the large flying foxes, *Pteropus*, feed on fruit. Some bats feed on nectar or pollen, and many tropical flowers are clearly adapted to facilitate bat pollination. The vampire bat (*Desmodus*) drinks the blood of sleeping

Sorex, a shrew (order Insectivora)

mammals and has thus acquired a notorious reputation in fiction. Perhaps most remarkable of all, fish-eating bats (for example *Phizonyz*) catch fish in their heavily clawed feet as they swoop over the surface of the water.

Bats have well developed wings (Chiroptera means 'hand-wing') which are formed by a membrane (patagium) stretched between the limbs and body. All the digits of the hand except the first, the legs (but not the feet) and usually the tail, are involved in supporting the wing. The hands are extensively modified for this role, with greatly elongated metacarpals in all but the first digit. The second digit, which forms the loading edge of the wing, has no phalanges and those of the remaining digits are elongated and reduced in number. The pectoral girdle has a stout clavicle which helps to anchor the shoulder joint, and the sternum is well developed with ridges and a keel for the attachment of the powerful pectoralis muscles which pull the wing downwards providing lift during flight. These muscles total 10 to 12% of the total body weight, much less than the 20% which is typical for birds. The thoracic vertebrae of bats are fused, and the ribs are flattened and fused, thus providing a firm, broad surface for muscle attachment. As a result of this fusion, breathing is almost entirely by means of the diaphragm. The legs and pelvic girdle are also modified in association with flight. The leg is rotated at the hip through 180° so that the knee joint flexion is downwards; the pelvis is rotated so that the acetabulum is dorsal and the pelvic girdle does not join ventrally at a symphysis. Bats have claws on the first digit of the fore-limbs, and on their feet. (In one major subdivision, the Megachiroptera, the second digit of the fore limb bears a claw.) Normally they hang upside-down gripping the rock or branch with the claws of their feet, but for excretion they rotate and hang by the wing claws to avoid soiling the patagium.

During flight, bats use their wings in a similar way to birds, responding to the same aerodynamic forces. They are not particularly fast fliers, but have amazing

A long-eared bat, order Chiroptera

manoeuvrability. Some bats have large eyes and acute vision, but many others depend on echo-location to navigate and find their food. The larynx is modified for emission of ultrasonic sounds. This requires more energy than the emission of lower-pitched sound, and the larynx is large and in some cases the cartilage is wholly or partly replaced by bone. The ears are well developed, often with large mobile external ears (pinnae) which can be used to focus the sound.

The carnivorous mammals

Carnivores evolved from the basic insectivore stock early in the Cainozoic. Since the insectivores had developed teeth suitable for a predatory mode of life and were already carnivores of a kind, the changes needed to feed on larger prey were simple, with only one major specialization: this was the modification of the last premolar of the upper jaw and first molar of the lower row to give a pair of shearing carnassial teeth used to cut across tendons or to strip meat off bone as the animal fed. Carnivores fall into three main groups: the extinct creodonts, the fissipedes (dogs, cats, bears, hyaenas, weasels), and the pinnipedes which include the majority of aquatic carnivores such as walruses, seals, and sea lions. The creodonts had an archaic build and were probably slow and rather stupid; they could not survive with the evolution of fast running herbivores. The fissipedes and pinnipedes represent a separate development from the insectivores and together form the order Carnivora.

Order Carnivora

The fissipedes include a number of predators dangerous to man as well as some more harmless forms which feed largely on invertebrates or even plants. They fall into two main groups (felids and canids) of which the cat and dog are convenient, though not primitive representatives of each. Both are adapted for fast movement with a digitigrade stance ('standing on the digits'), in which the metatarsals and metacarpals are held almost vertically and the animals stand on their phalanges. The cats (Felidae), which are the most specialized and successful carnivores, are partly arboreal and often occur in heavily wooded areas. The viverrids (civets and mongooses) show the characters of early felids. Felids are extremely agile and generally stalk their prey, making a sudden final dash when they may attain immense speeds over short distances. (Cheetah have been recorded to move at speeds of 90 km/hr). They are the most strictly carnivorous members of the order, feeding largely on birds and other mammals which they usually drag down and bite at the throat. Felids include some of the most formidable predators: lions (*Panthera leo*) can attack and kill prey of twice their own body weight. The extremely sharp claws of most cats can be withdrawn into bony sheaths when not in use, thus protecting them from wear and tear

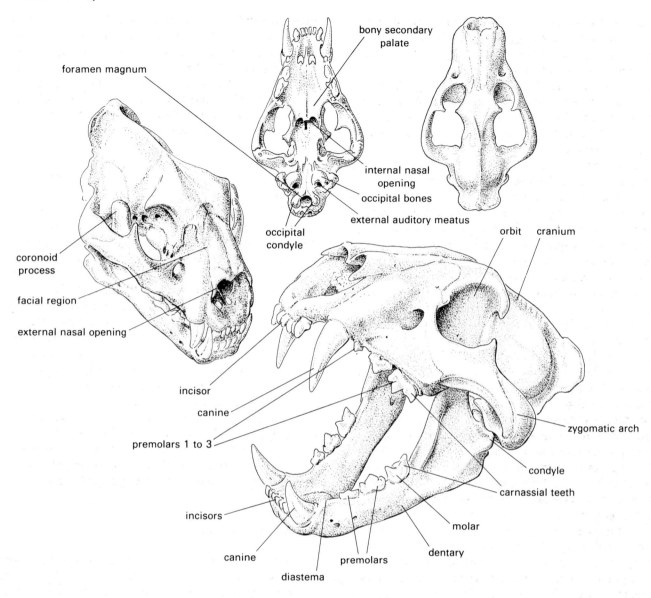

Skull of a representative carnivore, *Panthera leo* (the lion)

during locomotion and from inflicting unintentional damage during play. They have forwardly positioned eyes with an overlap of about 130°, giving them good stereoscopic vision; this is an adaptation also found in carnivorous birds and is probably important in judging distances for pouncing on prey.

By contrast the dogs (Canidae) are less agile and have a different approach to catching their prey. They are adapted to chasing their victim until it becomes weary, and are therefore primarily plains dwellers. They are fast moving, and can run up to about 50 km/hour, but do not attain the speeds of the fastest cats. Greyhounds, which have been specially selected and bred by man for speed, can manage 65 km/hour. Dogs tend to hunt in packs and tear open their prey at the belly, eating it with no further killing. They have blunt claws which cannot be retracted, and tend to select young, old, or sick animals to pursue. Although primarily

carnivores, dogs show a tendency towards a more mixed diet than the felids; and have more generalised teeth; for example the golden jackal, *Canis aureus*, will include insects and fruit in its diet.

Bears (Ursidae) show further movement away from a strictly carnivorous mode of life. They have a crushing dentition and will eat mammals, insects, honey, fruit and roots, and the polar bear is also a fish eater. They are slower moving, with plantigrade feet, (standing on the sole of the foot), in contrast with the digitigrade condition of both dogs and cats. They probably arose in the late Tertiary as an off-shoot of the canids. Another group of dog relatives, the procyonids, have departed still further from the meat eating habit and include the arboreal and entirely herbivorous pandas and giant pandas. In fact the giant panda (*Ailuropoda melanoleuca*) is extremely specialized in its diet and feeds only upon the foliage of certain bamboos.

The stoats and badgers belong to the Mustelidae, which are primitive members of the dog group. The stoat (*Mustela criminae*) is an active and aggressive hunter which feeds on mammals and birds, but the badger has a more mixed diet of beetles, small mammals, bulbs and acorns. The Mustelids also include an aquatic piscivore, the otter (*Lutra*).

The aquatic carnivores, the pinnipedes, are adapted for marine life and are believed to have evolved from the fissipedes although their fossil record is poor. Seals (Phocidae) have a streamlined body with dense fur, and an insulating layer of subcutaneous fat which in *Phoca* is 40 mm thick and makes up a quarter of the animal's total body weight. Although they leave the water to breed, they are highly adapted for an aquatic mode of life and are rather clumsy when on land. The limbs are reduced to paddles, or flippers, and the tail is also reduced; the vertebral column is simplified and highly flexible posteriorly so that the animal can swim using fish-like movements. In this way it can reach speeds of up to 14 knots. Seals feed on fish and cephalopods and have teeth with similar, laterally compressed points, adapted for holding slippery prey. The canines are long and sharp. Seals have no external ears, although hearing is thought to be acute, and have large, upwardly directed eyes. They are adapted for diving and have large lungs with strong cartilage supporting the bronchi and bronchioles. The blood volume is also large and some veins are enlarged as sinuses to give a high oxygen carrying capacity. Other adaptations to periods of submersion are a high myoglobin content in the muscles and tolerance of high levels of carbon dioxide. As a result they can survive under water for over 45 minutes, and can dive to depths of 600 m.

Phoca, the common seal (order Carnivora)

Sea lions (Otariidae) can still turn their legs forwards to move on land, and are therefore more mobile than the seals. However in the water they are less accomplished swimmers, relying on their fore-limbs a paddles. They are generally more primitive and retain external ears. Walruses (*Odobenus*) are related to the sea lions and are specialized for a diet of bivalve molluscs. The upper canines are modified as tusks which they use for digging, and the cheek teeth are modified for crushing.

Odobenus, the walrus (order Carnivora)

Order Cetacea: the whales

The 92 species of the order Cetacea (the whales, dolphins and porpoises) are highly specialized marine mammals which, unlike the pinnipedes, are entirely aquatic and never leave the water. In fact some whales die after stranding since they are unable to move and air provides insufficient support for their bulk. The whales include the largest living animals and indeed the largest animal known, the blue whale (*Balaenoptera musculus*), which grows to a length of over 30 m and weighs more than 135 tonnes. Their origin is uncertain, but may lie in primitive terrestrial carnivores. They have a fish-shaped body, in which the hind legs are entirely embedded in the body wall, and the fore-limbs are adapted as flippers. The neck is greatly reduced. Whales propel themselves by means of up and down movements of the posterior part of the body, and can swim at speeds of up to 30 km/hour. They have a dorsal fin and a pair of posterior horizontal tail flukes which are not modified limbs, but entirely new structures. The young are born and suckled in the water. Cetaceans have little or no body hair as adults and its insulating function is replaced by a layer of blubber, which also serves to reduce the specific gravity of the animal. Whales have similar adaptations to diving as seals, with capacious veins, a high concentration of myoglobin, and bronchi and bronchioles which are supported by cartilage. The oxygen capacity of the blood, however, is little higher than that of terrestrial vertebrates. Whales are

Balaenoptera musculus, the blue whale

specialized for two totally different kinds of diet. The largest whales, including the blue whale, have teeth only in the foetus, and in the adult these are replaced by fringed horny plates (baleen) composed of keratin. These are suspended transversely from the palate and used to strain zooplankton including krill (*Euphausia*) from the water. Baleen whales (Mysticeti) simply swim along with their mouths open, but in rorquals, which have shorter baleen, the muscular tongue is used to force water through the baleen, and the organisms are scraped off and swallowed. The toothed whales or Odontoceti (dolphins and porpoises) feed mainly on fish and have long rows of simple conical teeth, not differentiated into incisors, canines, and cheek teeth. Dolphins feed on small fish and cephalopods, but the killer whale (*Orcinus*), which grows to 9 m, is a fierce predator with fewer and larger teeth. It feeds on seals and smaller whales. The porpoise *Phocaena* feeds on fish, squid, and crustaceans, and the sperm and bottle-nosed whales feed largely on squids.

Whales appear to have a remarkable 'language' of sounds whose significance is poorly understood; however some species including dolphins are known to use echo-location for finding their food in a similar way to that described for bats.

The small herbivorous mammals

Order Rodentia

A member of the Rodentia (*Rattus*) has already been used to introduce the mammals, and can be regarded as representative of this highly successful order. Rodents are a diverse group and include more species, and probably more individuals than all the other mammals combined. Their body form is generally suited for scrambling, with feet capable of limited digging, and they have evolved specializations suitable for a wide variety of ecological niches including ground dwelling, climbing or arboreal, and semi-aquatic forms. As a group they are hardy animals, and species occur adapted to almost all climatic conditions from the Arctic to the Tropics, swamps to deserts. Neither are they restricted to an entirely herbivorous diet, but can live on a wide range of food. For example some deer mice (*Peromyscus*) include insects in their diet, while ground squirrels such as *Citellus* feed on mice and caterpillars; perhaps surprisingly the aquatic beaver is predominantly herbivorous. The majority of rodents are small but the South American capybara (*Hydrochoerus*) can weigh 50 kg.

The key characters which distinguish the rodents are the possession of specialized gnawing incisors and complex jaw muscles. Within the group three habit-groupings can be recognized: Sciuromorpha, represented by the squirrels and including beavers and marmots; Caviomorpha, represented by the familiar guinea pig or cavy and also including the porcupines; and Myomorpha, the mice and rats. These groups are distinguished on the basis of their jaw muscle attachment: the simplest, sciuromorphs, have a jaw muscle

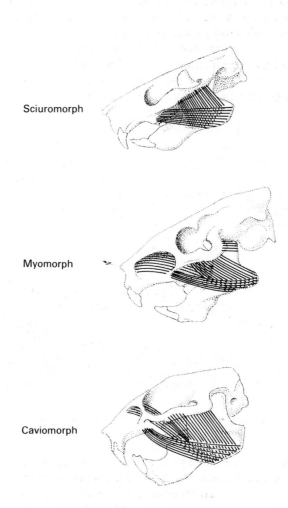

Sciuromorph

Myomorph

Caviomorph

Rodent jaw muscle attachments

attached to the zygomatic arch (cheekbone) as in other mammals, but divided into lateral and medial portions; myomorphs have extra insertions on the side of the snout; caviomorphs have massive cheek bones and a large medial insertion on the face.

Many rodents, such as squirrels and marmots, excavate elaborate burrows in which they live. Burrowing is particularly characteristic of the various rodent species which live in deserts where there is little vegetation cover. Another adaptation to desert life is seen in gerbils and jerboas whose limbs are specialized for bipedal hopping which is extremely effective over soft ground. These animals are also highly adapted for life at low humidities, producing highly concentrated urine and with reduced water loss.

Although no rodent is entirely aquatic, several have adapted to a semi-aquatic existence and almost all can swim. for example the brown rat, with no special swimming adaptations, has successfully adapted to life in sewers and on the water front, where it lives as a scavenger. The majority of more specialized semi-aquatic rodents have thick, fine fur, broad and often webbed feet, and small eyes and ears. The water voles show limited adaptation to an aquatic life, and are highly successful. The South American fish-eating rats (*Ichthyomys*), are more strongly adapted and in addition to the features listed above have a fringe of stiff hairs to expand the surface of the tail for swimming and upper incisors expanded at the corners to sharp points used to catch and kill fish.

The beavers are successful aquatic rodents with conspicuous adaptations for swimming in the form of webbed hind feet and a large flat tail. Other less visible adaptations include a lining of dense fluffy hair in the ears, which forms a seal to exclude water and special muscles for closing both the ears and the nostrils in a dive. Beavers have a well deserved reputation for industry, since they build massive lodges and dens of debris, sticks and logs. These structures are built over several generations and can be more than 30 metres long and 4 metres high. They use their gnawing teeth to fell trees and cut branches and can also use their fore-pads with great dexterity. Beavers hide large food stores against the winter, and in the winter the lodges both keep them warm and protect them against predators. Like many rodents, beavers are highly territorial and mark their territory by smell, using the oil from large anal glands. In the past, beavers were important economically, the American Indians trapped them for their flesh and fur, and regarded them with great respect. For example Grey Owl, a half-breed trapper writing in the 1930s said they '... are so much like indians'.

Many rodents climb or live in trees. Some are capable of some form of gliding with a gliding membrane on either side of the body, extending from the wrists to the ankles, for example *Petaurista*. At least some species can use ascending air currents and are capable of limited flapping flight though this is far less developed than in bats. The specializations found in most arboreal rodents however are rather less

Petaurista (order Rodentia)

outstanding. The squirrel's dexterity is aided by long limbs, flexible digits, sharp claws and a tail which can be used as a balancer as the animal leaps along branches, and as a rudder and parachute enabling the animal to make extended leaps.

Order Lagomorpha: the rabbits and hares

The lagomorphs are a small group of mammalian herbivores (63 species), which show superficial similarities to the rodents. Like them they have enlarged and continuously growing incisors (for this reason they were formerly classified with the rodents); however while rodents have a single pair of upper incisors, lagomorphs have a smaller pair of accessory incisors behind the main pair. The cheek teeth also show little similarity since they are modified for cutting rather than grinding and lagomorphs have two or three premolars. Their principal radiation has involved specializations for hopping, with long, strong hind legs and a vestigial tail. Lagomorphs typically hop or leap and in this way move remarkably fast; they never run and rarely walk. The front legs of rabbits (*Oryctolagus*) are adapted for burrowing and are used to dig extensive warrens in which they live in groups. In contrast, hares (*Lepus*) nest in depressions in the vegetation known as forms. The lagomorph diet is more exclusively herbivorous than that of rodents but they show similar adaptations to feeding on plants; like the rodents they have an elongated caecum inhabited by bacteria and they practice coprophagy. Lagomorphs are relatively defenceless animals and depend for their survival on early warning of predators and a quick get-a-way. Their eyes lie on the sides of the head, which gives them good all round vision, and they have elongated ear pinnae which function as effective sound gathering devices. Rabbits will warn each other of the approach of enemies by banging their feet on the ground, and the white underside of the tail also serves as a warning 'flag'. Like rodents they are territorial and are supplied with scent glands which are used for marking territories, as well as in courtship.

The large herbivores

The mammals include three groups of large herbivores, artiodactyls (camels, pigs, cattle, deer, antelopes), the perissodactyls (horses, rhinoceroses) and the proboscoids (elephants); of these the artiodactyls and perissodactyls combine a herbivorous mode of life with fast running which allows them to migrate in search of food, and also escape from predators. Like the fast running carnivores they stand on their toes, but they have taken the process even further, walking on their toe-tips with the phalanges as well as the metacarpals and metatarsals held vertically. Associated with this type of posture, known as unguligrade, the number of digits has become reduced and the claws of the remaining digits have become modified as broad flattened hooves which support the animal's weight. This is a modification which would be impossible in carnivorous animals, which use their claws to wound their prey and then to hold and tear it.

With the adoption of an unguligrade position the proximal segment of the limb (the humerus or the femur) is relatively short in comparison with the overall leg length. The retractor muscles are closer to the fulcrum and slight muscular contraction can produce a considerable movement at the end of the leg, which comes to function as a jointed pendulum with a rapid vertical swinging action. As a result some members of the group can move very fast indeed; gazelles can travel at 70 to 80 km/hr, and an antelope (*Antilocapra americana*) is recorded as galloping at 98 km/hour, faster than the carnivore record of 90 km/hr for a cheetah.

The two orders (known collectively as ungulates) are separated on the basis of the type of toe reduction. In the order Perissodactyla the axis of the foot passes through the third toe, which is always the largest and sometimes, as in modern horses, the only toe. In contrast, the members of the order Artiodactyla have the foot axis passing between the third and fourth toes which are always equal in size and importance. Naturally the best method of defence in these animals is running away, but many ungulates can use their strong hind legs to kick; in addition the many artiodactyls have integumentary structures in the form of antlers or horns which can be used for defence, and a few have canine teeth. The ungulates probably evolved from heavy bodied Eocene forms, similar in shape to the carnivorous creodonts which radiated to the swifter and more modern ungulates in the Miocene.

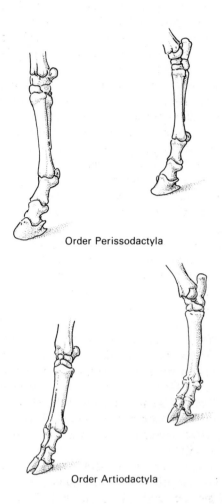

Order Perissodactyla

Order Artiodactyla

Ungulate toe reduction

Order Perissodactyla

The Perissodactyla, or even-toed ungulates, are exclusively herbivorous, and have an elongated skull which accommodates a long, broad row of high-crowned teeth. Although they were once a large and widespread order, living perissodactyls comprise only three families, tapirs, rhinoceroses and horses, and include several species which are rare and on the verge of extinction. Modern tapirs all belong to a single genus, *Tapirus*. They are shy, unobtrusive forest browsers, once widely distributed but now restricted to parts of Asia and South America. Like many large herbivores of forest they have dappled markings which help to camouflage them in their natural surroundings. They are stout bodied, moderately large and have short legs. The hind legs end in three digits, the fore-legs in four, but in both cases they are shod with small hooves supported by a single sole. Tapirs have rounded ears, small eyes, and the nose and upper lip are prolonged slightly as a proboscis. They feed on

Tapirus, a tapir (order Perissodactyla)

a wide range of softer plants, roots, and fruit, and have a dentition of 44 simple teeth.

Modern rhinoceroses are also a very restricted group, both in species number and distribution, and show the large heavy body form characteristic of several of the extinct perissodactyls. Their feet are similar to those of tapirs, though the fore-legs can have either three or four digits. They have an extremely thick skin which is sparsely haired and horns (one or two) which are dermal growths set on bony thickenings of the skull (thought by some cultures to have an aphrodisiac quality!). Although the animals look extremely ponderous, they can move with considerable speed and if startled will charge, causing considerable damage to anything that gets in the way. The largest species, the white rhinoceros weighs over 3 tons and is second only to the elephant as the largest land animal.

The horses and zebras (family Equidae) are the swiftest moving and most successful perissodactyls, although their survival is in large part due to domestication; with the exception of the zebras, wild equids are rare. They have a remarkably complete fossil record which allows us to trace their evolution from small tapir-like Eocene forms to modern equids, highly specialized for swift movement and a grazing mode of life. Speed is attained by reduction to a single functional digit on each limb (lightening the limb) with increased limb length. The teeth are reduced in number to three incisors on the upper and lower jaws, for cropping, and a battery of large high-crowned grinding, cheek teeth. Males have a small canine, and a vestigial first premolar (wolf tooth). The cheek teeth are tall, square in cross section, and have elongated cusps which when worn form long convoluted ridges of enamel (hyposodont teeth). The teeth erupt slowly and the root does not form at once (after about 5 years in a horse); the masseter muscles are well developed for a lateral grinding action and the angle of the jaw is deep to provide for their insertion. Horses have a

well developed large intestine and caecum, which together account for 60 per cent of the gut volume, and permit considerable bacterial decomposition of plant material. Like many plains dwelling herbivores they have well developed senses of sight, smell and hearing, with well developed touch receptors in the skin. Horses have an elaborate communication system within the herd.

Order Artiodactyla

The artiodactyls, with about 171 species including such diverse forms as pigs, hippos, camels, deer, antelopes, giraffes, sheep, goats and cows, are a far more successful group of large herbivores than the perissodactyls. The earliest Eocene groups showed little evolutionary change from their condylarth ancestors; they lost the pollux and hallux but they remained at a four-toed stage for a considerable time. However a distinctive feature is present in the ankle joint of even the most primitive members of the group. Whereas the astragalus of mammals is normally flattened below and pulley-shaped above, in artiodactyls both ends are curved, giving great flexibility and efficiency in both running and leaping. The pigs, like the tapirs, are relatively primitive hoofed mammals retaining essentially the Eocene condition. They have adapted an unguligrade stance, but are otherwise fairly unmodified, with four digits on each foot, though even in pigs digits 3 and 4 are both stronger and larger than 2 and 5. The dentition is nearly complete and the cheek teeth, which have low crowns and rounded cusps, are of a type associated with a mixed diet. Modern wild pigs have canine tusks which grow throughout the animal's life and are used for digging and defence. They have an interesting modification of the stomach, which forms a small pouch near the oesophagus giving it a two chambered structure. Wild pigs live in marshy or forest conditions, searching for plant or animal food by smell. The snout ends as a muscular and highly mobile disc which is used both as a tactile organ and for rooting in the soil for roots and fungi.

The hippopotamus is a large amphibious form which arose as a side branch from the early artiodactyls. The modern hippopotamus can weigh up to three tons; it is instantly recognizable from its heavy head with small ears, protruding bulbous eyes, and a huge mouth, and its enormous barrel-like body supported by relatively thin, short legs. During the day hippopotamuses live in the water with their heads just above the surface (their sense organs are on top of their heads) or bask on the shore. They have large lungs and can submerge themselves for up to five minutes at a time, closing their nostrils. At night they go inland to feed on foliage. They have a complete dentition with low-crowned teeth. The stomach is enormous, is divided into three chambers including two huge diverticula and a central chamber which extends as far as the pelvis; it contains large numbers of ciliated protozoa which break down cellulose in a fermentation process.

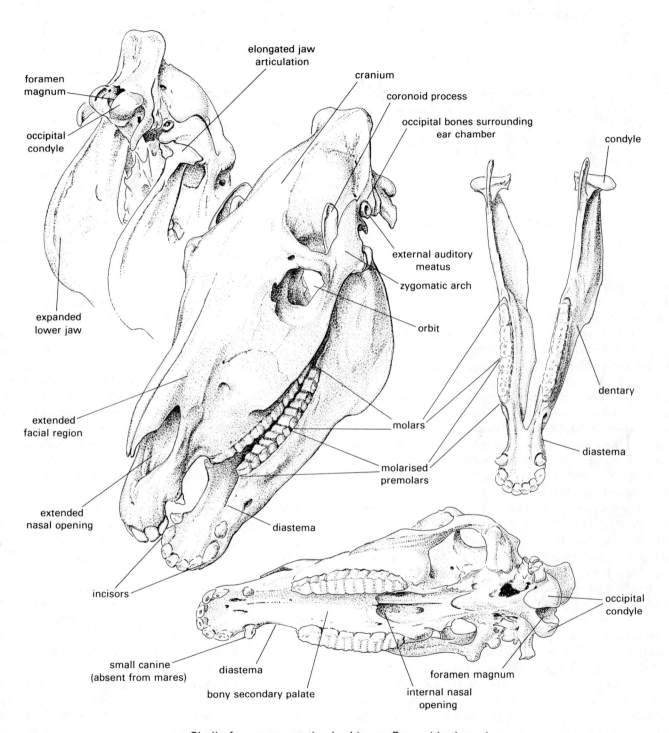

foramen
magnum

occipital
condyle

elongated jaw
articulation

cranium

coronoid process

occipital bones surrounding
ear chamber

external auditory
meatus

zygomatic arch

orbit

condyle

expanded
lower jaw

extended
facial region

extended
nasal opening

incisors

molars

molarised
premolars

diastema

dentary

diastema

small canine
(absent from mares)

diastema

bony secondary palate

internal nasal
opening

foramen magnum

occipital
condyle

Skull of a representative herbivore, *Equus* (the horse)

The Ruminantia

The remaining artiodactyls evolved in a different direction, developing high-crowned, ridged teeth for dealing with tough plant materials, and greatly elongated limbs with two digits. In these respects and in the shape of the face and jaw, they parallel the horses; however they have also developed an elaborate digestive sytem which involves chewing the cud. The main digestive process is known as rumination and is a highly efficient method of dealing with a diet of plant material. The cow, which can be taken as an example, has a large 'stomach' consisting of four connecting chambers: in order from the oesophageal end these are the reticulum, rumen (80 per cent of the total volume), omasum and obomasum. The cow spends only part of its time in feeding and the remainder in rechewing herbage from the rumen (chewing the cud). At this stage grass is compacted into balls in the reticulum to be returned to the mouth, rechewed, and then swallowed again. The rumen has a large population of ciliated protozoa and bacteria which break down plant material to yield volatile fatty acids, which are absorbed directly by the rumen lining. Undigested food passes through the omasum, which functions as a strainer, to the obomasum. This is the true stomach where peptic enzymes are secreted, the other chambers being sacculations of the oesophagus.

The advanced artiodactyls, grouped together as the Ruminantia, have only two digits and their metacarpals and metatarsals are fused forming a single cannon bone. Another ruminant character is the absence of the upper front teeth, which are replaced by a horny pad. Two groups, the deer (Cervidae) and giraffes (Giraffidae) have retained a forest browsing habit, and have low crowned teeth. The others, including cows and sheep, are highly successful grazers showing modifications of the jaws and cheek teeth similar to those of horses. Horns and antlers are characteristic of the artiodactyls, but these are very different in structure from the horns of rhinoceroses. Antlers are normally present in male deer, and are made of bone. Whilst growing they are covered in highly vascular skin bearing short fine hairs (velvet) but when they mature in late summer or autumn their blood supply is cut off, the velvet is shed and the bone exposed. Their significance is chiefly in social behaviour and they are used by the males as they fight over the females. After about one and a half years they are shed and a new pair grows, in general the number of points (tines) increases with age, and while young bucks have 1 to 3 points the older ones may have up to 12. Giraffes and okapi have short horns which develop from cartilages on the top of the head which ossify and fuse to the skull. They are permanently covered by living skin and hair. Bovids (cattle, sheep, goats) have true horns consisting of an inner bony core which develops as an outgrowth from the frontal bone and is covered by an outer epidermal layer of keratin. Again these are permanent structures which grow throughout life. The Antilocapridae, of which the prongbuck antelope is the only living member, have similar horns but shed the epidermal layer annually.

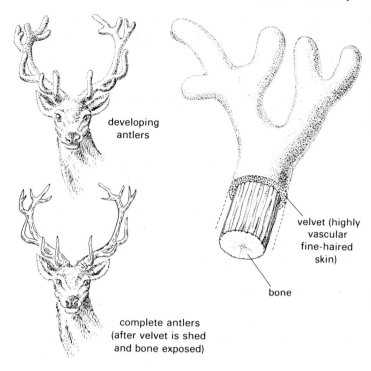

developing antlers

velvet (highly vascular fine-haired skin)

bone

complete antlers (after velvet is shed and bone exposed)

Growth of antlers

The Tylopoda

The members of the suborder Tylopoda, the camels and llamas, differ from true ruminants in being digitigrade rather than unguligrade. Like the ruminants they have no teeth at the front of the upper jaw but they retain the third upper incisor as a caniniform tooth. They were established as a distinct group from the Eocene and probably developed the ruminant habit separately. Llamas are mountain dwellers from South America; one species, the vicuna, occurs up to an altitude of 5000 m and has blood with an especially high affinity for oxygen. Llamas are long-haired and include domesticated forms used for their valuable silky wool (the alpaca) as well as for hides, meat and as beasts of burden. They have the reputation of being ill tempered, and have an effective method of defence which involves spitting the entire stomach contents at an attacker. Camels are particularly well adapted for life in arid conditions, and can survive a water loss of 25 per cent of their total weight; they conserve water by reduced evaporation from sweating and by production of exceptionally concentrated urine. They can tolerate a wide range of temperature and allow their bodies to heat up during the day, dropping their body temperatures below 'normal' at night. The famous hump of the camel is a fat store, but since it is concentrated in one place the insulating effects are minimised while retaining it as a source of energy and metabolic water; thus it is an adaptation to high temperatures and deserts where the camel may have to go for days without water. Camels have a variety of adaptations to sandy conditions which include broad feet, long eye lashes, and nostrils which can be closed.

They have a long history as domesticated animals: as beasts of burden, for pulling the plough, and as a source of meat, milk and wool.

The subungulates (orders Proboscoidea, Hyracoidea, Sirenia)

The subungulates are a highly diverse group of three orders of herbivores, which nevertheless have some basic features in common and may form a natural group. In some respects they are comparable with the ungulates. Normally they have molar-like premolars which contribute to a battery of grinding cheek teeth, and an enlarged pair of incisors or canines on either jaw. They normally retain all their digits and have flattened nail-like claws.

Order Hyracoidea

The hyraxes are the most generalized subungulates and are not unlike rabbits in size, habit, and appearance. There are six species, which are found living in rocks and mountain regions in Africa. They have plantigrade feet with four anterior and three posterior digits, tipped by nails. They are herbivores, with a battery of grinding teeth, and a pair of continuously growing incisors on upper and lower jaws. They have a well developed masseter muscle, and intestinal caecae.

Procavia, the hyrax (order Hyracoidea)

Order Proboscoidea

The elephants (two species) are the largest land animals; an African elephant can weigh up to 6 tonnes. They show many adaptations typical of very heavy animals such as thick, straight, column-like limbs. They walk to some extent on their toes, but support most of their weight on a large pad of elastic tissue posterior to the digits. Their size necessitates vast quantities of food and they are highly efficient food gatherers, with their major modifications seen in the head. Their front teeth are reduced in number to a pair of upper incisors, which are modified as enormous tusks. These are almost solid dentine (ivory) and may be as much as 4 m in length, and can weight up to 159 kg. They may function partly as counterweights to the body. The cheek teeth are high crowned with fused cross ridges which makes them very effective grinding organs. Elephant jaws are short and there is room for only one large tooth at a time on each side and the cheek teeth are therefore replaced sequentially; as one is worn down it is replaced from behind. This is an unusual but effective way of prolonging the total tooth life for an animal which relies heavily upon efficient grinding of food. The trunk of elephants represents a greatly drawn out nose and upper lip and is a highly efficient food gathering device for pulling down branches, but even so elephants spend up to eighteen hours a day eating. The stomach and intestine are relatively unspecialized, but they have a long, sacculated caecum.

Elephants have a long gestation period (660 days), and a long period of post-natal development. They are the only domesticated animals which are regularly caught in the wild to be tamed as they reach physical maturity at 8–10 years, and sexual maturity between 12 and 14 years.

Order Sirenia

The sirenians or sea cows are the only aquatic mammals that are entirely herbivorous. They are extremely aquatic, live on coasts and in estuaries and rivers, and are said to have given rise to the legend of the mermaids. Their name refers to the beautiful sirens of mythology, who were said to sing so persuasively that they lured sailors to their death. They

Dugong dugon, the sea cow (order Sirenia)

have a fusiform body, with no visible neck, and flipper-like fore-limbs, but no hind limbs; like the whales their tail is expanded into horizontal flukes. The skin is sparsely haired and they have a thick layer of blubber. The thorax is barrel-like and they probably rely largely on their intercostal muscles for breathing. They have small eyes, reduced external ears and highly modified teeth and jaws. The dentition is reduced, but supplemented by horny plates and fleshy lips which are used for cropping herbage.

Some minor mammalian orders

Three orders: Edentata, Pholidota and Tubulidentata are each represented by very few species (31, 8 and one respectively) some of which show convergences in relation to a diet consisting largely of ants and termites.

Order Edentata

The edentates (ant-eaters, tree sloths and armadillos) are not all toothless as their name suggests, but they do have a reduced and degenerate dentition. There are no teeth in the front of the mouth and the cheek teeth consist of blocks of dentine without any covering of enamel. Generally they have well developed claws. They are all South American forms.

Armadillos are characterized by an armour of dermal plates, covered by horny epidermis, which form shields protecting the head, shoulders, and back. The plates are separated by soft skin which allows considerable flexibility, and some species can even roll up into a ball, rather like a woodlouse. They are harmless and defenceless animals whose only protection is their armour and their secretive habits. They have short muscular legs, with which they dig for food and burrow. Most species are omnivores, feeding on anything that they can dig out, and a few species feed on

ants which they catch with a long, narrow sticky tongue. They have between 3 and 25 teeth on each side which are small and peg-like in shape.

The tree sloths, quite different in appearance and habits from armadillos, are arboreal, and live hanging upside-down on branches by the two or three hooked claws on their feet. Both the toes and fingers are fused, and their fur is reversed from the more normal position so that it hangs from the under surface as an adaptation for their upside-down mode of life. Tree sloths are leaf eaters but have few teeth. They have rather weakly developed muscles and are slow moving, with slow digestion and generally slow metabolism; for example it takes about a week for a meal to pass through the gut of a sloth.

Choleopus, a two-toed sloth (order Edentata)

Dasypus, an armadillo (order Edentata)

The ant-eaters are toothless, but like other ant-eating mammals (such as the spiny ant-eaters) have an elongated probing snout and a long tongue which is liberally coated with a sticky saliva, secreted by large salivary glands. They are insectivorous, feeding chiefly on termites and ants, which they dig out with powerfully developed claws on their front legs. These claws are also used for defence. They have long tails and thick shaggy fur which may protect them to some extent against ant bites.

Order Pholidota

The pangolins or scaly ant-eaters are found in tropical Africa and Asia, and show several parallels with the edentates. Like the edentate ant-eaters they have a long tail, well developed fore-claws and a long, sticky tongue associated with their insectivorous habits, but they resemble the armadillos in their dermal armour and rolled up defence posture. However, these similarities are regarded as convergences and pangolins are placed in a separate order.

Manis, a pangolin (order Pholidota)

Order Tubilidentata

The tubilidentates certainly have no relationship with either of the two preceeding orders and are placed here as a matter of convenience. The only living species, the aardvark (*Orycteropus afer*) comes from Africa and feeds on ants and termites which it catches with a long, extensible sticky tongue. It has highly unusual teeth, with no incisors or canines, and cheek teeth which grow throughout life and have no enamel. The dentine is arranged in numerous hexagonal prisms. Aardvarks are burrowing animals with well developed claws on their fore-limbs. They make tunnels up to 4.5 m long in which they hide, coming out at night to dig up the nests of ants and termites. They are unusual among mammals in lacking subcutaneous fat.

Orycteropus, the aardvark (order Tubulidentata)

Order Primates

Primates comprises 166 species of lemurs, monkeys and apes. It is also the order to which man belongs which probably accounts for its name which implies animals of the first rank (*primus* means 'first')! It diverged from the insectivores, probably in the late Cretaceous, and in a properly taxonomic sequence would be placed near that group. The most primitive primates, tree shrews (such as *Tupaia*), are difficult to distinguish from the insectivores and indeed are included with them by some mammalogists. Their most striking primate character is a bar of bone behind the eye, the post-orbital bar, formed from a post-orbital process which meets a corresponding upward extension of the jugal. While this is not an exclusively primate character (it is found in ruminants as well) it is found in all primates and is a strong clue to the affinities of the tree shrews.

Tupaia, a tree shrew (order Primates)

As a group the primates are adapted to an arboreal mode of life and those that have gone back to a terrestrial life show many traces of an arboreal ancestry. They have grasping hands and feet, often with nails rather than claws, and flexible limbs which make them unusually agile. The dentition is of the crushing type associated with a mixed diet of insects, leaves, shoots and fruit. The sense of smell is reduced, but that of sight increased, with the development of stereoscopic vision. The cerebellum is usually well developed and the brain generally is larger in proportion to body size than in other mammals. All these are adaptations for life off the ground, and many parallel features have already been seen in the birds. Primates generally have fewer offspring, and sometimes only a single infant, and there is a long period of post-natal care. This close care of the young is a characteristic frequently seen among mammals living in a precarious environment and may be related to an arboreal mode of life. Part of the primate definition is negative: a lack of specializations found in many other mammal groups.

Primates are usually divided into two suborders, the prosimians (Prosimii) and the Anthropoidea. The prosimians, which include the lorises, bush babies and lemurs, are thought to show the beginnings of primate adaptations. Lemurs (Lemuroidea) are small, long-tailed forms with long snouts related to a good sense of smell, large eyes and primitive molars. They have thick fur (the lower incisors are modified for grooming) and primitive molars. The lorises (Lorisiformes) represent a parallel evolution to the lemurs. The lemurs are restricted to Madagascar, and the lorises occur on mainland Africa.

Tarsius, the tarsier (order Primates)

Lemur cattus, the ring-tailed lemur (order Primates)

Tarsiers (*Tarsius*) represent a stage in evolution above that of the lemurs though well below that of the anthropoids. They are nocturnal and have large forwardly facing eyes, and a reduced snout. Combined with a larger braincase these features give *Tarsius* a 'face'. *Tarsius* has elongated feet (due to long tarsal bones) and legs for hopping about in trees; it feeds on insects which it picks out of the bank.

The anthropoids are the man-like primates: monkeys, apes and man. They are believed to have evolved from the prosimians, though the precise group is not known, and they appear in the fossil record in the early Tertiary. All have a flat face that is at least partly lacking in fur, well developed stereoscopic vision, and a large brain within a globular brain case. They have a degree of manual dexterity, not seen in other groups, and have the ability to sit on their haunches.

Monkeys show the fulfilment of some of the evolutionary trends initiated in *Tarsius*. The sense of smell is reduced and the snout is short. Their eyes are turned forwards and the retina is subdivided to give detailed central vision with less detailed peripheral vision. New World monkeys have three

premolars whilst Old World monkeys (such as apes and man) have two. Although monkeys are generally quadrupedal, they show a tendency towards upright sitting. They are highly agile animals, and many have a prehensile tail which acts as a fifth limb for climbing and swinging in trees. The pollux and hallux are opposable which allows them to grip branches and catch objects with their hands and feet. They are largely plant eaters, and their teeth are modified for grinding; many have cheek pouches for temporary food storage.

The apes have skeletons, organs and muscles that are closer to the human type than those of monkeys, although the skeletal proportions differ. They are represented by four living types: the gibbons of Malaysia (*Hylobates*), the orang-utan of Borneo and Sumatra (*Pongo*), the chimpanzee (*Pan*) and gorilla (*Gorilla*) of tropical Africa.

Hylobates, the gibbon

Pongo, the orang-utan

Pan, the chimpanzee

Among the various species certain important evolutionary trends can be detected. These include an increased body size (the gorilla grows to 270 kg), and particularly an increased brain size and development of the cranium. Another trend is the development of brachiation which is a type of movement involving swinging from branch to branch by the arms. This necessitates the development of elongated arms, hands which are used as hooks, a broadening of the chest and the loss of the external tail. The gibbons are the best brachiators, and some can move three metres or more at a swing. Some of the larger apes, the larger chimps and gorillas, spend more time on the ground, and only climb into the trees to sleep. When on the ground they can walk bipedally, but normally use a quarupedal bounding form of locomotion in which they support the front half of their body on their knuckles.

Apes are herbivores with robust teeth, powerful jaws and well developed jaw muscles. The skull has strengthening ridges over the orbits and the gorilla skull has sagittal and nuchal crests which increase the area available for jaw muscle attachment. They have well developed canines which they use for defence. Apes are generally highly intelligent, using simple 'tools' and with an elaborate social structure. The well developed face devoid of hair means that facial expressions can play an important part in communication between the animals.

The family Hominidae: the man-like apes

Hominids differ from contemporary apes in having adopted an exclusively bipedal walk associated with loss of the opposable hallux of the foot. With the hands freed from any direct role in locomotion they have become specialized for elaborate manipulation. Hominids have an unspecialized omnivorous diet.

Homo

Many of the anatomical adaptations of *Homo* are related to an upright stance. The back is sinuously curved, swinging the centre of gravity over the pelvis and hind legs, and raising the head; the ilium is broad and flared to provide areas for extra muscle attachment. The legs are elongated and strengthened, and the calcaneum of the foot expanded to form a prop at the back to improve balance. The tarsals and metatarsals form strong supporting arches, and the toes are short, parallel, and help to give a strong forward thrust. The distal ends of the femurs lie closer to the mid-line to permit balance on only one leg. The position of the head on the vertebral column is altered, with the foramen magnum lying at the base of the skull. Humans have reduced body hair and increased sweat glands, thought to be associated with the increased muscular activity required for fast running.

The diet of *Homo* is varied, and the jaws and teeth are reduced in size. The teeth are rounded, the canines small. However, with their grasping hands, hominids can use tools rather than teeth for attack and defence. The shorter tooth row and reduced jaws leave the nasal region of the face as an isolated projecting structure.

Finally hominids show growth and reorganization of the brain. The regions of the brain important in sensory and motor integration are enlarged, as are those involved in association, memory, and speech. Brain size and complexity evolved rapidly, from 700 cm^3 cranial capacity in the early species to the 1400 cm^3 average of modern man. From the anatomical point of view the factors of prime importance in human evolution can be summarized as the perfection of an erect posture, an extensive post-natal learning period and the growth and elaboration of the brain; associated with these developments growth in human social organization and communication has allowed man to benefit from the accumulated knowledge and experience of the population.

Man is a highly successful animal and an extreme example of a type of adaptation seen in several groups of mammals. Man, like many other mammals including dogs, cats, pigs, goats, rats and mice, is essentially a non-specialized opportunist able to exploit a wide variety of situations successfully.

Synopsis of the class Mammalia

Various early fossil forms of uncertain classification.
Subclass Prototheria
   Order Monotremata: the egg laying mammals
Subclass Allotheria: the multituberculate mammals, a
                      group of gnawing Mesozoic and early
                      Tertiary forms.
Subclass Theria
  Infraclass Trituberculata: the ancestors to modern
                      mammals
  Infraclass Metatheria
   Order Marsupialia: the pouched mammals
  Infraclass Eutheria: the placental mammals
   Order Insectivora: the insectivores, including moles,
                  shrews and hedgehog.
Order Dermoptera: the flying lemurs
Order Chiroptera: the bats
Order Primates: the primates
   Suborder Prosimii
   Suborder Anthropoidea
  Order Carnivora: the flesh-eating mammals
   Suborder Creodontia: the ancestral carnivores
   Suborder Fissipedia: modern terrestrial carnivores
   Suborder Pinnipedia: the marine carnivores
  Order Condylartha: the ancestral ungulates
  Order Proboscoidea: the elephants
  Order Sirenia: the sea cows
  Order Hyracoidea: the hyraxes
  Order Perissodactyla: odd-toed ungulates
   Suborder Ceratomorpha: tapirs and rhinoceroses
   Suborder Hippomorpha: horses
Order Artiodactyla
   Suborder Suiformes: the pigs, peccaries and the
                  hippopotamus
   Suborder Ruminantia
   Suborder Tylopoda
Order Edentata: the sloths, anteaters and armadillos
Order Tubulidentata: the aardvark
Order Cetacea: the whales
   Suborder Odontoceti: the toothed whales
   Suborder Mysticeti: the whalebone whales
Order Rodentia: the rodents
Order Lagomorpha: the rabbits and hares

Index

References in *italics* indicate illustrations.

413

NOTES

NOTES